Intracellular Thermometry with Fluorescent Molecular Thermometers

Intracellular Thermometry with Fluorescent Molecular Thermometers

Seiichi Uchiyama
Graduate School of Pharmaceutical Sciences
The University of Tokyo
Tokyo, Japan

Author

Dr. Seiichi Uchiyama
Graduate School of Pharmaceutical
Sciences
The University of Tokyo
Tokyo, Japan

All books published by **WILEY-VCH** are carefully produced. Nevertheless, authors, editors, and publisher do not warrant the information contained in these books, including this book, to be free of errors. Readers are advised to keep in mind that statements, data, illustrations, procedural details or other items may inadvertently be inaccurate.

Library of Congress Card No.: applied for

British Library Cataloguing-in-Publication Data
A catalogue record for this book is available from the British Library.

Bibliographic information published by the Deutsche Nationalbibliothek
The Deutsche Nationalbibliothek lists this publication in the Deutsche Nationalbibliografie; detailed bibliographic data are available on the Internet at <http://dnb.d-nb.de>.

© 2024 Wiley-VCH GmbH, Boschstraße 12, 69469 Weinheim, Germany

Hardback ISBN: 9783527350322
ePub ISBN: 9783527836857
ePDF ISBN: 9783527836864
oBook ISBN: 9783527836840

Cover Design: Wiley
Cover Image: © Kohki Okabe

Typesetting 10/12pt Joanna MT Std by Integra Software Services Pvt. Ltd, Pondicherry, India
Printing and Binding CPI Group (UK) Ltd, Croydon, CR0 4YY

C9783527350322_200324

Contents

Preface

I feel great honor to write this book on intracellular temperature measurements with designed fluorescent molecules. Since I received the exciting offer from Wiley-VCH, it took me over two and a half years to collect the related papers (more than 900!), read them repeatedly, and construct the text of the book with devoted support from the editorial team. During the writing process, I recognized that the field of intracellular thermometry quickly expanded more than my initial expectation. It involves organic chemistry, polymer chemistry, photophysical chemistry, analytical chemistry, fluorescence imaging, cell biology, cell physiology, medicine, and even ecology. Theoretical concepts based on thermodynamics and thermo-chemistry have strengthened the field. Then, a new field of thermal signaling is being proposed, further advancing the philosophy of applied research areas, thermal biology, and thermal medicine.

Let me introduce myself briefly. As some readers may hear in my presentations at international conferences and symposiums, I am a pure chemist, working primarily in the field of analytical chemistry, and have been fascinated by fantastic illumination from blight fluorescent molecules with different colors since I belonged to a laboratory in the final year of the bachelor course at The University of Tokyo. Luckily, I have been able to continue to make new molecules in a flask and evaluate their complex photophysical properties using lovely spectrophotometers, which has been the biggest motivation in my career. My interest has been extremely biased toward chemistry. Sometimes, as a student I achieved a perfect score in chemistry but failed in biology. This tendency remains, and thus, I prefer to put many chemical structures of fluorescent molecular thermometers in the present book with my policy that the chemical structure is the most essential information of a synthesized molecule. Similarly, I provide many fluorescence spectra of different fluorescent molecular thermometers as the most basic characteristics of fluorescent sensors. Overall, it is a curious coincidence for me to write a whole book involving profound biology. Many biologists have patiently explained the fun and mystery of biology to me, which removed the aversion to biology from my mind. In this book, the chemical parts are described in relative detail with my viewpoints based on personal experiences, and the biological and medical aspects are written in a more general style because of my background. Also, I have tried to clearly indicate my position regarding subjective opinions in the book since the content of this medium is completely constructed by my own creativity. In this sense, the style of the book is different from peer-reviewed scientific papers reporting some achievements obtained in my laboratory.

I must note that three years of indispensable postdoctoral time in Nara Women's University and Queen's University of Belfast (QUB). The first step of intracellular thermometry, i.e., the development of highly sensitive fluorescent polymeric thermometers functioning in aqueous solution, was inspired by the serial works of Professor Kaoru Iwai at Nara Women's University. A QUB professor A. Prasanna de Silva encouraged me to learn my poor subjects, biology and English communication, by showing me his unique and exciting established concept of molecular logic operation. So, this book is also my personal reply to them. Again, I feel great happiness to have contributed to intracellular thermometry with my chemical skills at the beginning and, finally, with this book. Now, I am thrilled to know attractive phenomena that will be reported from laboratories utilizing intracellular thermometry in the next decade. Hopefully, my research group will be able to join the discoverers. Above all, enjoy your science as you like, and share it with us for the fantastic future of intracellular thermometry!

Seiichi Uchiyama
in scorching summer in Tokyo, 2023

1

Temperature for Living Things

Temperature is one of the most influential physical parameters in our daily lives. For example, in Tokyo, where the author lives, there are four seasons throughout the year. When it is cold in the winter, people travel to tropical countries to search for warmth. The arrival of spring can be recognized by the warmth of the sunshine filtering through the leaves of trees and the gentle breeze. On the other hand, recent summers have been too hot to relax without an air conditioner. A French chef always pays attention to the temperature of a frying pan and an oven, and a Japanese chef checks the oil temperature in a pot to deep-fry crispy tempura.

In the same way, temperature is intensely involved in the activity of living things from a biological viewpoint (Cossins and Bowler, 1987). Humans sweat in summer and shiver in winter to maintain their body temperature. When we feel ill, we measure the body temperature first. When I was a child, we used an mercury-filled thermometer to measure the body temperature, but now a thermistor has replaced it. Accordingly, it is only natural that there is a long history of comprehending temperature and measuring it. Some intriguing literature is available for the historical background on temperature measurements from the 16th century (Middleton, 1966; Chang, 2004). Here, as the first chapter of this textbook, the relationships between temperature and organisms are summarized. As indicated in Figure 1.1, temperature-related biological phenomena can be categorized into "spontaneous thermogenesis" and "response to environmental temperature" at both an individual body level and a single cell level. Each category will be introduced with relevant examples. Of them all, the spontaneous thermogenesis in single living cells will be highlighted in Chapter 3 as a new outcome brought about by intracellular thermometry.

Intracellular Thermometry with Fluorescent Molecular Thermometers, First Edition. Seiichi Uchiyama.
© 2024 Wiley-VCH GmbH. Published 2024 by Wiley-VCH GmbH.

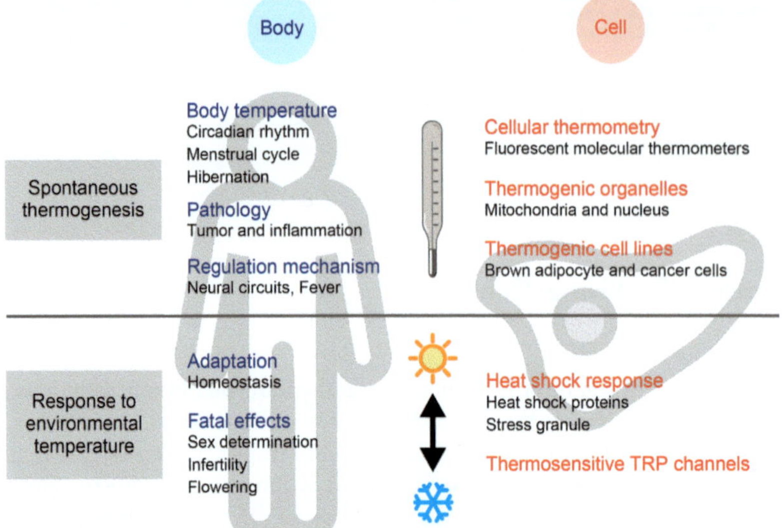

Figure 1.1 Keywords in biological studies that correlate temperature with physiological events. The original figure is from Okabe and Uchiyama (2021) *Commun. Biol.*, **4**, 1377 and is updated here.

1.1
Temperature of Individuals

1.1.1
Spontaneous Heat Generation

1.1.1.1 Human

A human with a body temperature of 36–37 °C is a representative of an endotherm; the body temperature of endotherms is kept higher than that of the environment by spontaneous heat generation (called thermogenesis) (Geneva et al., 2019). Although we were told that humans are homeotherms in our childhood, the actual body temperature of humans considerably varies within a few degrees in relation to circadian rhythms, including ultradian and infradian rhythms (Figure 1.2a) (Zulley et al., 1981). Misconceptions by the use of the terms "endotherm" and "warm-blooded" for mammals have been pointed out for correct comprehension of organismal thermoregulation (Brack Jr. et al., 2022). Women have another circadian rhythm of basal body temperature due to the ovulation cycle (Figure 1.2b) (Lee, 1988; Baker et al., 2020). The menstrual cycle-dependent variability in basal body temperature is affected by aging. However, seasonal variation is slight, and thus, basal body temperature is being proposed as a non-invasive diagnostic indicator of ovulation (Tatsumi et al., 2020). The body temperature of humans is also influenced by race, age, voluntary exercise, and disease (Refinetti and Menaker, 1992).

Body temperature is one of the most basic vital signs in clinical diagnosis and routine human health care. To measure human body temperature with high

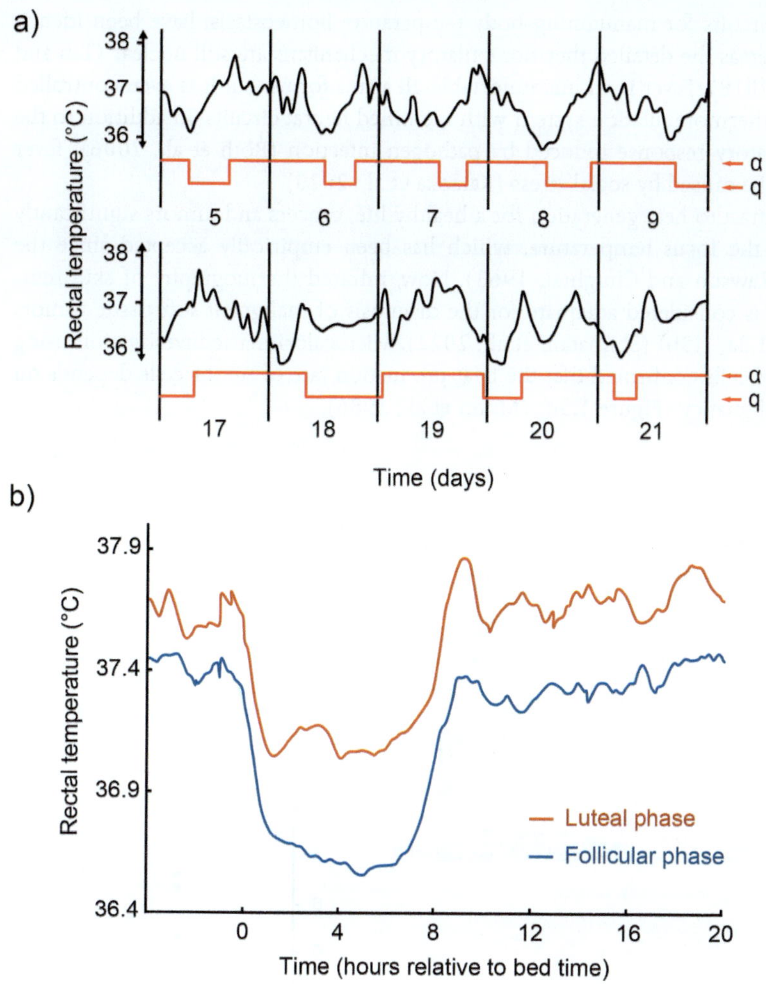

Figure 1.2 Variation in human body temperature. (a) Rectal temperature (black) and sleep–wake cycle (red) recorded in an isolated subject during internal synchronization (upper) and internal desynchronization (lower). α: wakefulness; q: bedrest. Adapted from Zulley et al. (1981) *Pflügers Arch.*, **391**, 314–318. (b) Averaged rectal temperatures of 15 young women (age: 22 ± 4 years) at the mid-follicular and mid-luteal phases in their menstrual cycles. Subjects followed their usual daytime schedules and spent the nights in a sleep laboratory. Adapted from Baker et al. (2020) *Temperature*, **7**, 226–262.

accuracy, actual and predictive measurements using instruments or mathematics, either invasive or non-invasive, are used (Childs, 2018; Chen, 2019). The methods for measuring the human body temperature involve wearable temperature sensors, and information technology-based real-time and long-term recording and expand to the application of life-critical decision-making and mass screening of diseases.

The regulation of human body temperature has been the central subject of physiology (Houdas and Ring, 1982; Jessen, 2001). It functions with the nervous system. Some molecules involved in temperature sensing in the periphery and the neural circuits that transmit temperature information to the brain, as well as the central circuits for maintaining body temperature homeostasis, have been identified, whereas the detailed thermoregulatory mechanisms are still unclear (Tan and Knight, 2018). Fever is an uncomfortable, ill status for us, but it is even controlled under a thermoregulation system with identified neural circuits. In addition to the inflammatory response induced by pathogen infection (Roth et al., 2006), fever can also be caused by social stress (Kataoka et al., 2020).

In contrast to heat generation for a healthy life, cancers and tumors significantly increase the focus temperature, which has been empirically accepted since the 1960s (Lawson and Chughtai, 1963). Now, infrared thermography of skin temperature is considered adequate for the diagnosis of malignant soft-tissue tumors (Figure 1.3a, 1.3b) (Shimatani et al., 2022). Microcalorimetric investigation using a mass of cells confirmed that the heat production rate of tumor cells depends on their malignancy (Figure 1.3c) (Monti et al., 1986).

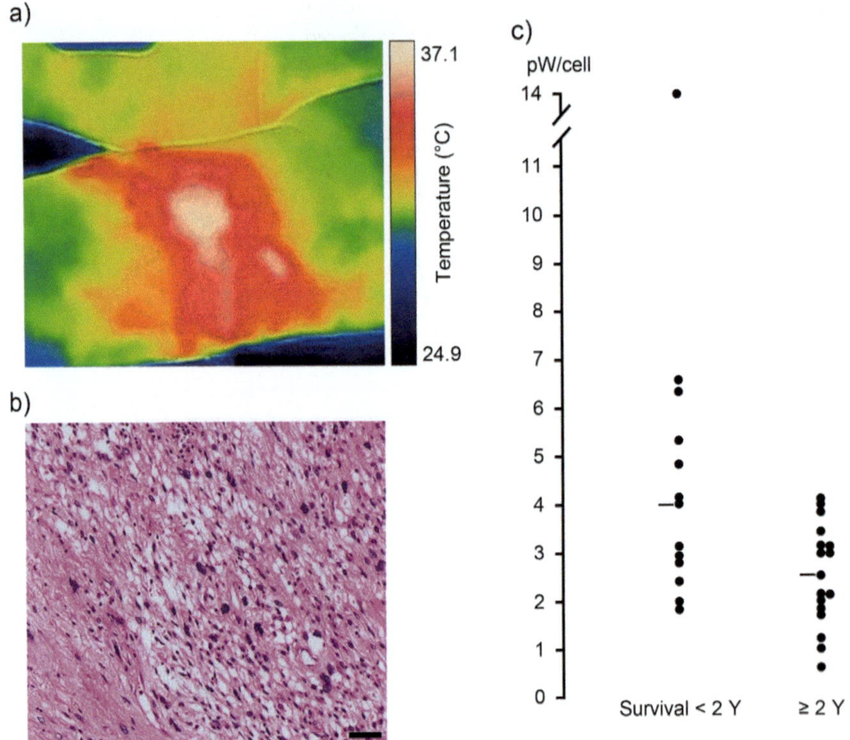

Figure 1.3 Heat production in tumors. (a,b) Representative case of myxofibrosarcoma. (a) Skin temperature map by infrared thermography. A malignant soft-tissue tumor in the right thigh was diagnosed as myxofibrosarcoma. The skin temperature on the affected part is higher than the surrounding areas. (b) Sections of the myxofibrosarcoma stained with hematoxylin and eosin (magnification: ×200). Panels (a,b) are adapted from Shimatani et al. (2022) *Int. J. Clin. Oncol.*, **27**, 234–243. (c) Heat production rate per lymphoma cell from non-Hodgkin lymphoma patients. The left group of 13 patients died within two years, while the right group of 17 patients survived two years or longer. Horizontal bars indicate median values. Panel (c) Monti et al. (1986) *Scand. J. Haematol.*, **36**, 353–357 / John Wiley & Sons.

1.1.1.2 Bear

An extreme case of animal spontaneous thermogenesis is found in hibernation (Kosara, 2011), especially in bears (Harlow et al., 2004). The body temperature of a hibernating black bear (*Ursus americanus*) is displayed in Figure 1.4 (Tøien et al., 2011). During hibernation in winter, even in cold environments (−40–0 °C) where metabolic activity is markedly suppressed, *Ursus americanus* maintains a body temperature of 32–38 °C by performing regular muscular exercises without waking up from sleep.

1.1.1.3 Oceanic Lives

Due to the high thermal conductivity of water, it is harder for marine creatures to keep their body temperature higher than their surroundings compared to animals living on land. Nevertheless, various fish maintain their body temperature higher than the surrounding sea. Figure 1.5 shows the relationship between the ambient temperature and the body temperature of bigeye tuna (*Thunnus obesus*) (Holland et al., 1992). The ambient temperature fluctuations were due to vertical excursions of *Thunnus obesus* of more than 100 meters. Analyzing these temperature data suggested that the whole-body thermal conductivity of *Thunnus obesus* was altered by two orders of magnitude between the warming and cooling phases. A mesopelagic fish, opah (*Lampris guttatus*),

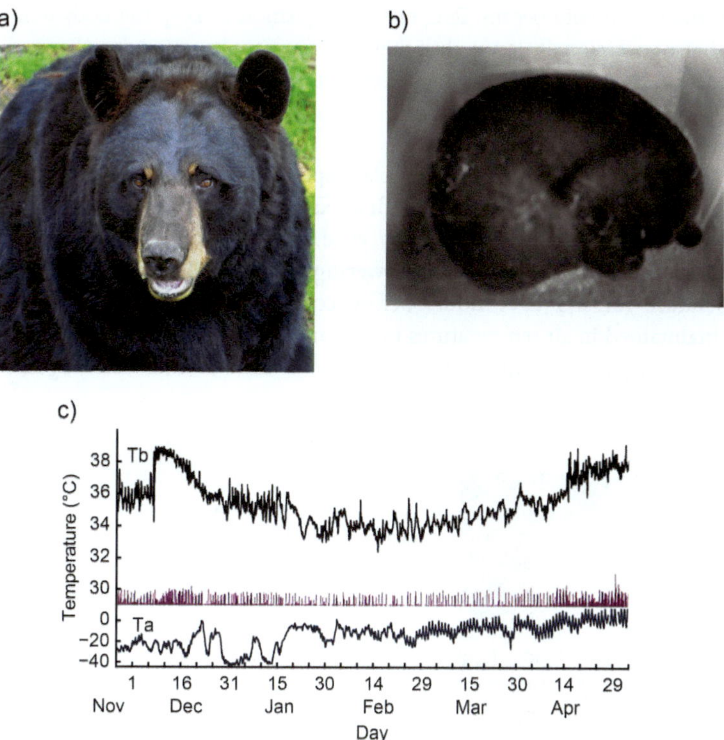

Figure 1.4 Thermogenesis of a hibernating bear. (a) A black bear, *Ursus americanus* (© Meunierd). (b) Hibernation of *Ursus americanus* in an artificial den. (c) Temperature patterns of the core body temperature of a hibernating *Ursus americanus* (Tb, black) and the outside of a cave (Ta, dark blue). Purple lines indicate the movements of the black bear. Panels (b,c) Tøien et al. (2011) *Science*, **331**, 906–909 / American Association for the Advancement of Science.

a)

b)

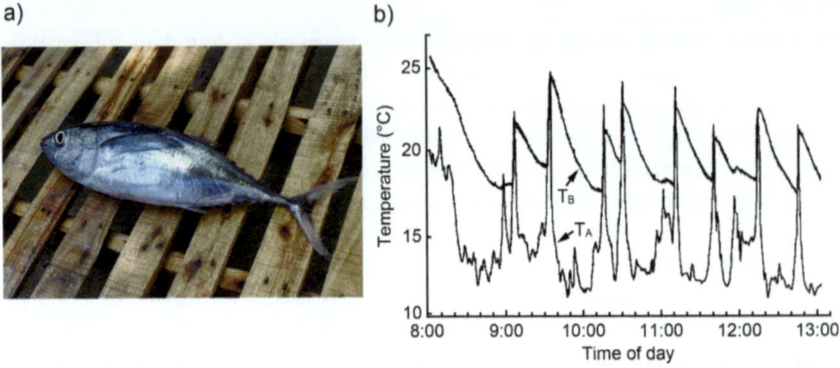

Figure 1.5 Thermoregulation of a tuna. (a) A bigeye tuna, *Thunnus obesus* (© MikeCloud). (b) The body temperature of a swimming *Thunnus obesus* (T_B) and ambient temperature (T_A). Panel (b) is adapted from Holland et al. (1992) *Nature*, **358**, 410–412.

also shows temperature elevations above ambient by 3.2–6.0 °C in its heart, viscera, cranial region, and pectoral muscle (Wegner et al., 2015). It has been assumed that the endothermic properties of these living things are beneficial for enhancing physiological performance in cold oceans. Temperature regulation to keep the body temperatures higher, or occasionally lower, than their environment has also been observed in king penguins (Handrich et al., 1997) and sea turtles (Sato, 2014).

1.1.1.4 **Plants**

It might be surprising to readers that some plant species also produce remarkable heat (Knutson, 1979; Seymour, 2001). The first example is the garden philodendron, *Philodendron selloum* (Figure 1.6a) (Nagy et al., 1972). The inflorescence of *Philodendron selloum* produces heat during the flowering sequence and remains at maximum temperature for 0.3 to 4 hours before cooling. A core temperature of 38–46 °C is maintained in air temperatures in the range 4–39 °C through a variable metabolic (i.e., oxygen consumption) rate (Figure 1.6b). Another thermogenic

a)

b)

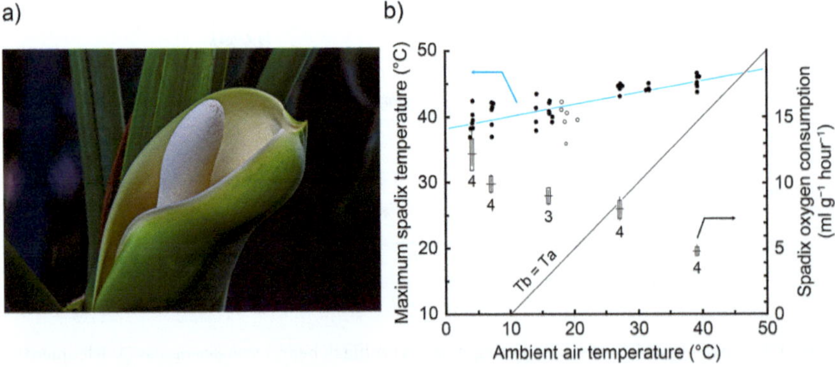

Figure 1.6 Thermogenesis of the inflorescence of philodendrons. (a) A garden philodendron, *Philodendron selloum* (© Kobus Peché). (b) Core temperatures (Tb) of philodendron inflorescences outdoors (open circle) and in incubators (closed circle) at different ambient temperatures (Ta). Solid horizontal lines indicate the mean spadix oxygen consumption rates. Vertical lines show the ranges. Rectangles are ± standard deviations. The numbers below the rectangles are sample sizes. Panel (b) is adapted from Nagy et al. (1972) *Science*, **178**, 1195–1197.

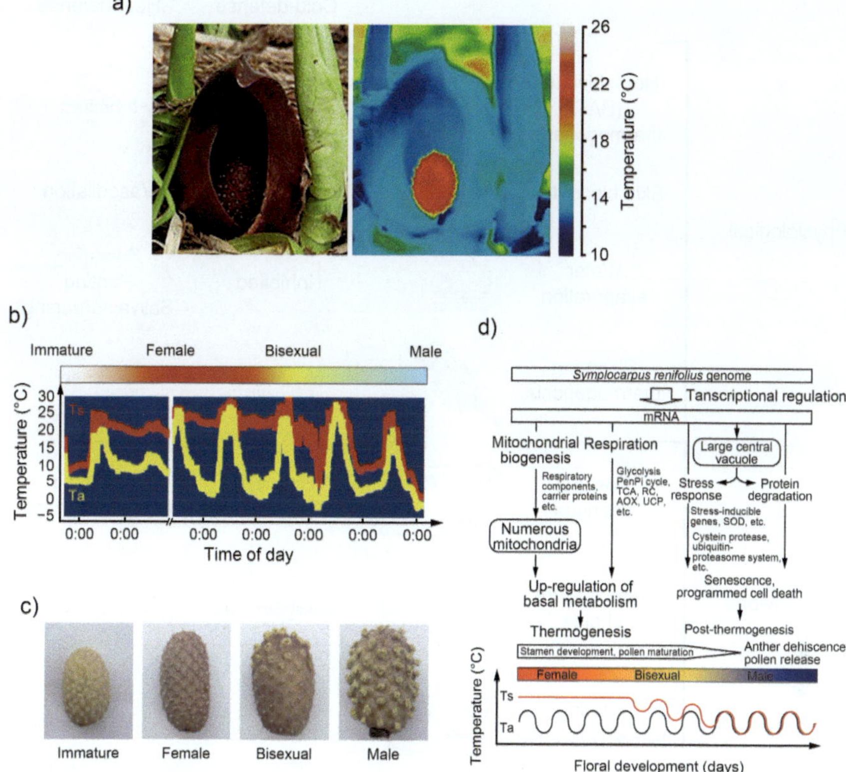

Figure 1.7 Thermogenesis of the spadix of skunk cabbages. (a) Visible image (left) and thermographic image (right) of a skunk cabbage, *Symplocarpus renifolius*. Adapted from Ito-Inaba et al. (2009) *Planta*, **231**, 121–130 / Springer Nature. (b) Variations in ambient air temperature (Ta, yellow) and the spadix temperature (Ts, red) of *Symplocarpus renifolius* during floral development. (c) Visible images of the spadices of *Symplocarpus renifolius* during floral development (immature, female, bisexual, and male). Panels (b,c) are adapted from Ito-Inaba et al. (2009) *J. Exp. Bot.*, **60**, 3909–3922 / Oxford University Press. (d) Hypothetical model of floral thermoregulations. PenPi: pentose phosphate, TCA: tricarboxylic acid, RC: respiratory chain, AOC: alternative oxidase, UCP: uncoupling protein, SOD: superoxide dismutase. Adapted from Ito-Inaba et al. (2012) *Plant Cell Environ.*, **35**, 554–566.

plant is the skunk cabbage (*Symplocarpus renifolius*) (Figure 1.7a) (Ito-Inaba et al., 2009a). The development of an inflorescence of skunk cabbage is divided into four stages: immature, female, bisexual, and male. Within these four stages, only at the female stage, the spadix generates massive heat and can maintain its internal temperature at 20–25 °C despite the environmental temperature at night falling to almost 0 °C (Figure 1.7b, 1.7c) (Ito-Inaba et al., 2009b). Studies on mitochondrial activity and protein and gene expression hypothesized thermoregulation mechanisms of *Symplocarpus renifolius*, as indicated in Figure 1.7d. However, the exact mechanisms by which the plant senses changes in ambient temperature, produces heat, and stops thermogenesis have not yet been elucidated (Ito-Inaba et al., 2012).

1.1.2

Responses to Environmental Temperature Change

The adaptation of body temperature in response to environmental temperature variation is an essential function of life for endotherms and is remarkably diverse

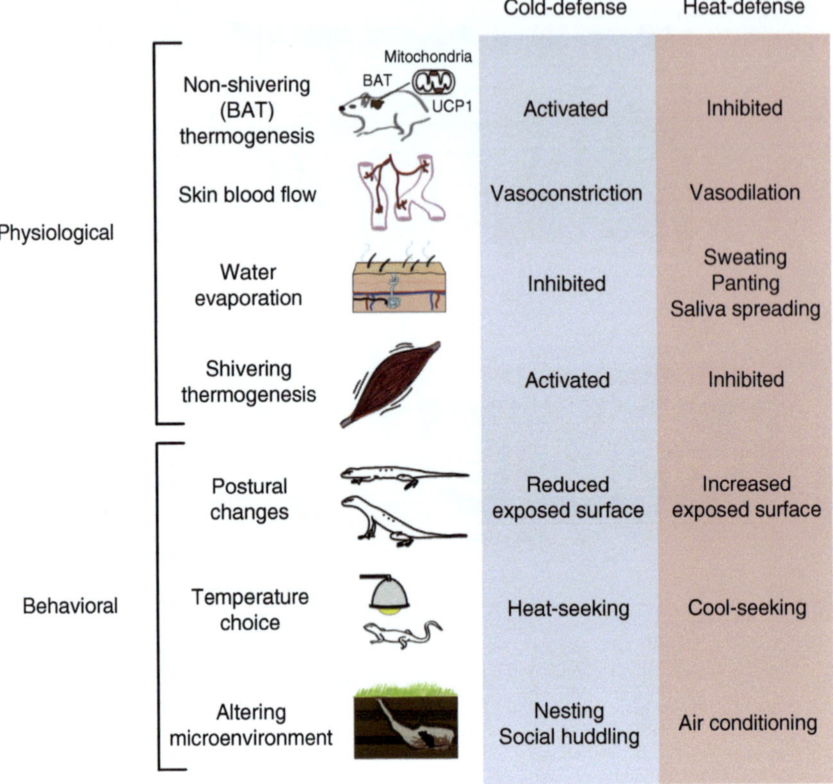

Figure 1.8 Physiological and behavioral strategies for controlling body temperature under cold and hot environments. BAT: brown adipose tissue; UCP1: uncoupling protein-1. Adapted from Tan and Knight (2018) *Neuron*, **98**, 31–48.

(Figure 1.8) (Tan and Knight, 2018). Mechanisms of adaptive non-shivering thermogenesis have been discussed in the molecular level (Lowell and Spiegelman, 2000). Humans can maintain homeostasis even in a cold, aqueous environment such as a swimming pool by generating heat (Fox et al., 1979). The body temperature is not vastly changeable even in a hot aqueous environment like a bath (Ohnaka et al., 1995). The disruption of homeostasis can lead to serious threats to life.

There are some biological phenomena in which environmental temperature is utilized to determine fate. Temperature-dependent sex determination in reptiles is an example (Bull and Vogt, 1979; Crain and Guillette Jr., 1998). As displayed in Figure 1.9, the temperature at which eggs are incubated determines gender in three reptiles: alligator snapping turtle (*Macroclemys temminckii*) (Ewert et al., 1994), a false map turtle (*Graptemys pseudogeographica*) (Bull et al., 1982), and the American alligator (*Alligator mississippiensis*) (Lang and Andrews, 1994). A slight shift in incubation temperature can cause a dramatic change in their sex ratio. Due to temperature-dependent sex determination and environmental warming, serious feminization has been observed in turtle populations in the Great Barrier Reef in Australia (Jensen et al., 2018). In a red-eared slider turtle (*Trachemys scripta elegans*),

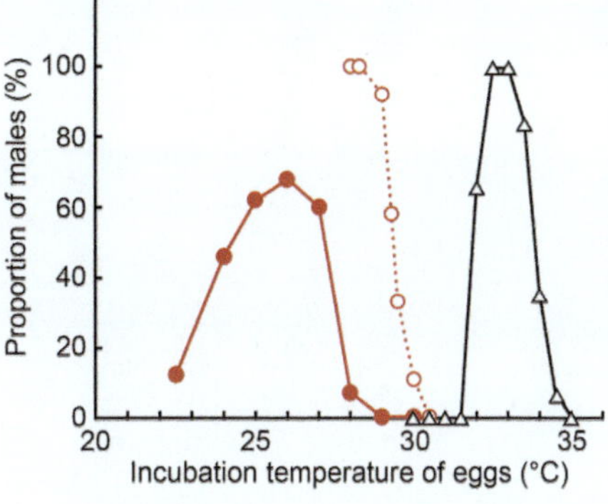

Δ *Alligator mississippiensis*

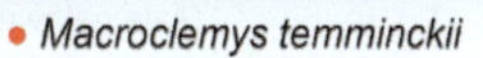
● *Macroclemys temminckii*

○ *Graptemys pseudogeographica*

Figure 1.9 Temperature-dependent sex determination in three reptiles, an alligator snapping turtle, *Macroclemys temminckii* (© Matthijs Kuijpers) (closed red circle), a false map turtle, *Graptemys pseudogeographica* (© pisces2386) (open red circle), and an American alligator, *Alligator mississippiensis* (© Richard Mcmillin) (open black triangle). Adapted from Ewert et al. (1994) *J. Exp. Zool.*, **270**, 3–15 (alligator snapping turtle), Bull et al. (1982) *Evolution*, **36**, 326–332 (false map turtle), and Lang and Andrews (1994) *J. Exp. Zool.*, **270**, 28–44 (American alligator).

the corresponding gene responsible for the temperature-dependent sex determination has been identified: under warmer environments, signal transducer and activator of transcription 3 (STAT3) is more phosphorylated through a Ca^{2+} ion influx, then represses *Kdm6b* transcription to block the male pathway (Figure 1.10) (Weber et al., 2020). In a short-lived lizard jacky dragon (*Amphibolurus muricatus*), the effect of incubation temperature on the reproductive success of males differs from the effect on females (Warner and Shine, 2008). The temperature-dependent sex determination is also observed in certain fish (Shen and Wang, 2014).

Infertility is another reproduction matter that is sensitive to environmental temperatures (Mieusset and Bujan, 1995). In mammalian species, spermatogenesis is a heat-vulnerable process, and the scrotal temperature is maintained at 2–8 °C lower than that of the core body (Moore and Quick, 1924). If the testes go through a temperature elevation beyond a threshold, spermatogenesis is interrupted, resulting in male infertility. Recently, it was revealed that heat-induced meiotic failure is one of the reasons for heat vulnerability in mouse spermatogenesis (Hirano et al., 2022).

a)

b)

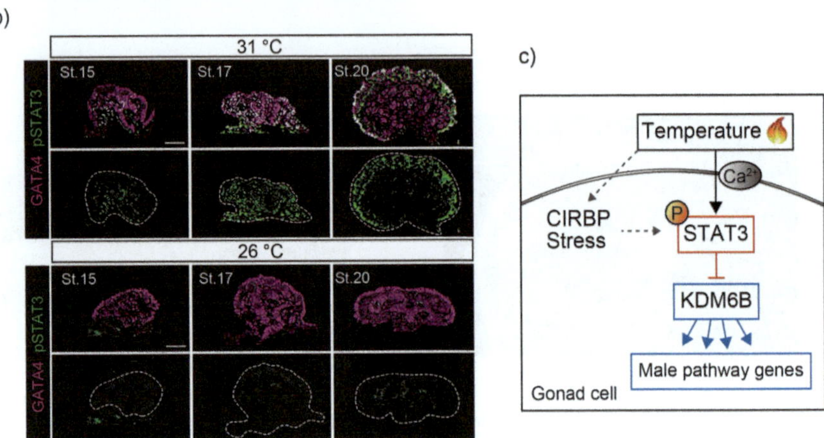

Figure 1.10 Molecular mechanisms of temperature-dependent sex determination in red-eared slider turtles. (a) A red-eared slider turtle, *Trachemys scripta elegans* (© Slowmotiongli). (b) Immunofluorescence images of phosphorylated STAT3 (pSTAT3) (green) and GATA4 (magenta, somatic gonad marker) in gonadal cross sections from embryos at 26 and 31 °C throughout the temperature-sensing window (from stage (St.) 14 to 20). pSTAT3 expression is nuclear and restricted to the sex cords and cortical domain. Dotted lines are outlines of embryonic gonads. Scale bars: 50 μm. (c) A molecular model for temperature-dependent sex determination in *T. scripta*. High temperatures initiate a rise in intracellular calcium ions that promotes STAT3 phosphorylation. pSTAT3 binds *Kdm6b* to repress activation of *Dmrt1* and the male pathway. Low temperature may activate cold-inducible RNA-binding protein (CIRBP) and stress response pathways, activating STAT3. Panels (b,c) are adapted from Weber et al. (2020) *Science*, **368**, 303–306.

Cherry blossoms (Figure 1.11) are symbolic flowers in Japan. People look forward to their short bloom as a sign of warm spring. Interestingly, the flowering day of cherry blossoms (e.g., *Prunus serrulata*) can be accurately estimated by temperature sums during February and March, implying that the flowering of cherry blossoms is a temperature-dependent phenomenon (Lindsey, 1963). In general, plants have evolved excellent plasticity to adapt to their surrounding temperature environment (Patel and Franklin, 2009; Wigge, 2013). The modulation of plant architecture by environmental temperature is an essential issue concerning crop productivity and global climate change. Gene expression is induced in plants' low-temperature perception for chilling and freezing tolerance (Knight and Knight, 2012). Nevertheless, the mechanisms through which plants sense ambient temperature variation remain elusive.

Figure 1.11   Yoshino cherry, *Prunus × Yedoensis*. © Monika Baumbach.

1.2
Responses to Temperature Variation at the Cellular Level

Disclosed responses to environmental temperature variations are not exclusive to individuals, which are described in Section 1.1.2, but also inclusive of live cells in diverse ways. For example, circadian gene expression in peripheral cells is synchronized with rhythms of host body temperature (Brown et al., 2002). Nucleic acids, proteins, and membranes are representative biomacromolecules that sense temperature variations within cells (Sengupta and Garrity, 2013).

When living organisms are subjected to an environmental temperature change, temperature-sensitive proteins initially respond to it within cells. One of the typical thermal responses in living cells is the heat shock response (Richter et al., 2010). When an organism is exposed to an excessive temperature of 5–20 °C above the normal growth temperature, the expression of heat shock proteins is induced in the cell. The heat shock proteins protect cells from heat damage by preventing the aggregation of functional proteins through modulation of the transcription of related genes.

The cellular response to thermal stimuli also includes fast translational regulation by changes in the state of translating mRNAs, which are mediated by stress granule formation (Kedersha et al., 1999). RNA-binding proteins, TIAR and TIA-1, concentrated in the nuclei of DU145 (human prostate carcinoma) cells rapidly accumulate into stress granules in the cytoplasm in response to a mild heat shock at 44 °C for 20 minutes (Figure 1.12).

Thermosensitive transient receptor potential (TRP) channels also play important roles in temperature sensing at cell membranes exposed to large temperature fluctuations (Figure 1.13a) (Clapham, 2003; Castillo et al., 2018). Three-dimensional structures of the TRP channel family have been clarified one after another to yield mechanistic insights (Figure 1.13b). Noticeably, the Nobel Prize in Physiology or Medicine 2021 was given for discoveries of a TRP channel member, TRPV1, as well as mechanically activated ion channels (Ernfors et al., 2021). A variety of TRP family proteins are activated either by molecules in stimulating foodstuffs such as capsaicin and menthol or by temperature variations in specific ranges. The latter activations are considered implicated in temperature-related physiological functions.

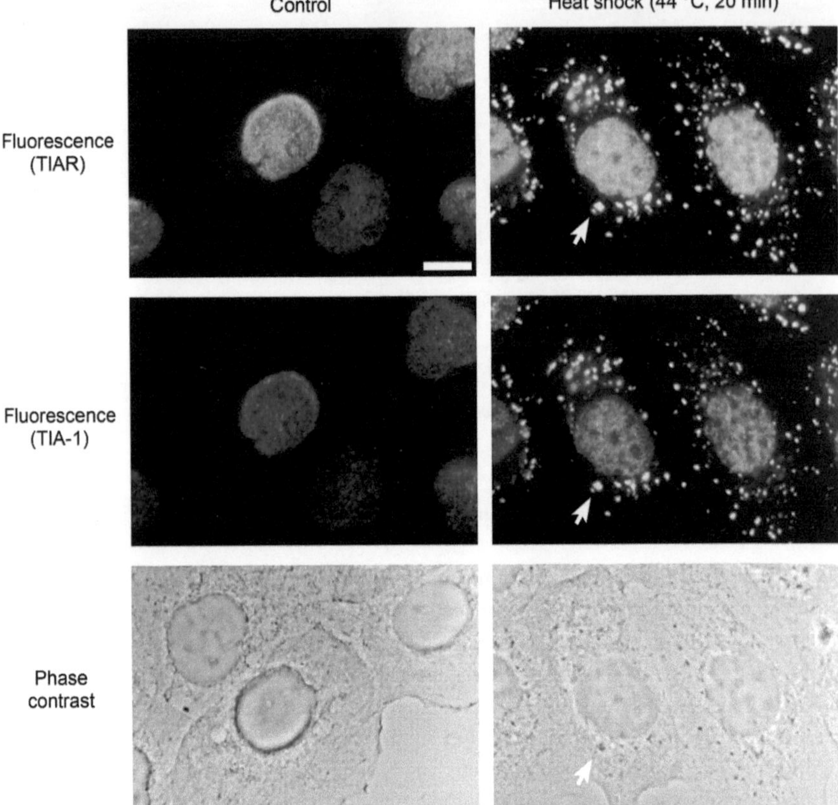

Figure 1.12 Coaggregation of TIAR and TIA-1 at stress granules. Fluorescence images of TIAR-specific antibody 6E3 (upper) and TIA-1-specific antibody ML29 (middle) and phase-contrast images (lower) of DU145 cells without (left) and with (right) mild heat shock (44 °C for 20 min). Cells were immediately fixed after the heat shock and processed for immunostainings. TIAR and TIA-1 colocalized at stress granules (arrows). Scale bar: 10 μm. Adapted from Kedersha et al. (1999) *J. Cell. Biol.*, **147**, 1431–1441.

Another interesting temperature-responsive protein discovered is the splicing factor (Preußner et al., 2017). This protein can detect a temperature change of 1 °C for use as an input factor in regulating gene expression. In cellular biotechnology, spatiotemporally controlled heating within a cell has been applied for gene induction based on a heat shock promoter (Kamei et al., 2009) and manipulation of a temperature-sensitive mutant protein that functions depending on temperature (Hirsch et al., 2018).

Neural developments are also temperature-dependent; the significant effects of environmental temperatures were reported on axonal outgrowth of cortical neurons isolated from rat embryos (Black et al., 2016) and proliferation and differentiation of neural stem/progenitor cells isolated from forebrain cortices of rat fetuses (Hossain et al., 2017). In addition, temperature-dependent migration was observed in neutrophil-differentiated HL-60 cells (derived from human peripheral blood leukocytes) (Khachaturyan et al., 2022).

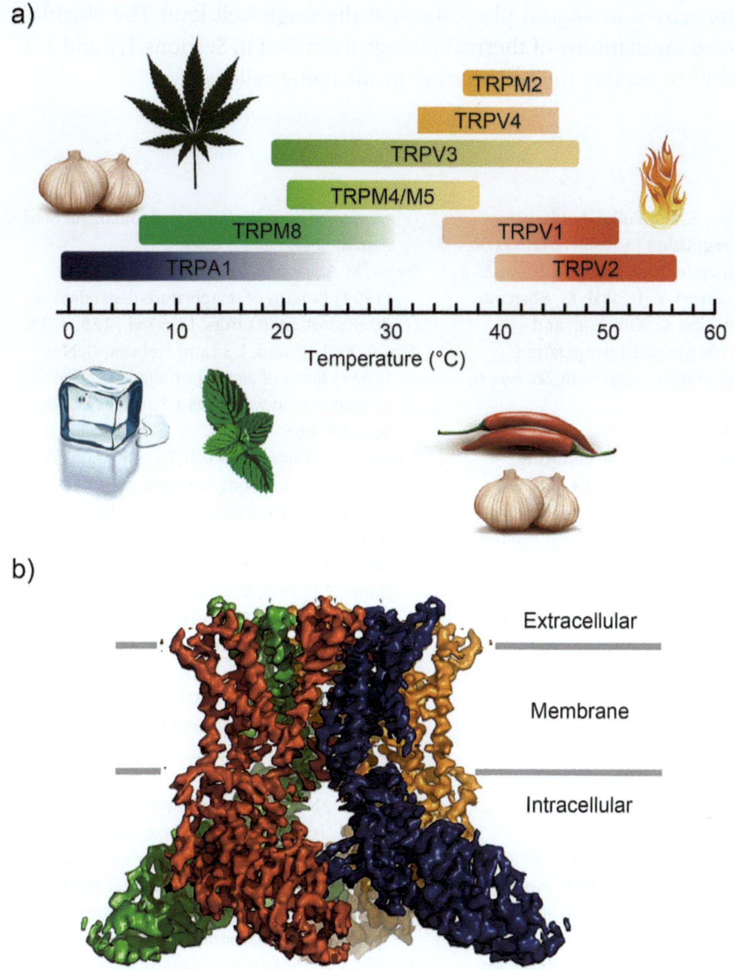

Figure 1.13 Temperature-sensitive TRP channels. (a) Sensitive temperature ranges for open probabilities of TRP channels. The TRP channels are also activated by chemical ligands, including active ingredients such as allicin (in garlic), cannabinoids (in cannabis), menthol (in mint), and capsaicin (in chili pepper). Adapted from Castillo et al. (2018) *Phys. Biol.*, **15**, 021001 / IOP Publishing. (b) Cryo-electron microscopy reconstruction of TRPV4 as a representative. Deng et al. (2018) *Nat. Struct. Mol. Biol.*, **25**, 252–260 / Springer Nature.

1.3
Significance of Intracellular Thermometry

Just as humans maintain a body temperature higher than the environment, so do cells spontaneously produce heat. Heat-induced formation of stress granules in the cytoplasm and expressions of TRP channels in the plasma membrane hint that a detectable temperature variation might take place within a single cell. Accurate intracellular temperature measurements will enable the comparison of temperature-involving biological events between individuals and cells and the monitoring of temperature-related biological phenomena at the single-cell level. The elucidation of profound mechanisms of thermal biology described in Sections 1.1 and 1.2 has motivated us to explore the temperature inside living cells.

References

Baker, F. C., Siboza, F., and Fuller, A. (2020) Temperature regulation in women: effects of the menstrual cycle. *Temperature*, **7**, 226–262.

Black, B., Vishwakarma, V., Dhakal, K., Bhattarai, S., Pradhan, P., Jain, A., Kim, Y.-t., and Mohanty, S. (2016) Spatial temperature gradients guide axonal outgrowth. *Sci. Rep.*, **6**, 29876.

Brack Jr., V., Boyles, J. G., and Cable, T. T. (2022) Warm-blooded mammals: an enduring misconception, *Am. Biol. Teach.*, **84**, 529–534.

Brown, S. A., Zumbrunn, G., Fleury-Olela, F., Preitner, N., and Schibler, U. (2002) Rhythms of mammalian body temperature can sustain peripheral circadian clocks. *Curr. Biol.*, **12**, 1574–1583.

Bull, J. J. and Vogt, R. C. (1979) Temperature-dependent sex determination in turtles. *Science*, **206**, 1186–1188.

Bull, J. J., Vogt, R. C., and McCoy, C. J. (1982) Sex determining temperatures in turtles: a geographic comparison. *Evolution*, **36**, 326–332.

Castillo, K., Diaz-Franulic, I., Canan, J., Gonzalez-Nilo, F., and Latorre, R. (2018) Thermally activated TRP channels: molecular sensors for temperature detection. *Phys. Biol.*, **15**, 021001.

Chang, H. (2004) *Inventing Temperature. Measurement and Scientific Progress*, Oxford University Press, Oxford.

Chen, W. (2019) Thermometry and interpretation of body temperature. *Biomed. Eng. Lett.*, **9**, 3–17.

Childs, C. (2018) Body temperature and clinical thermometry. *Handb. Clin. Neurol.*, **157**, 467–482.

Clapham, D. E. (2003) TRP channels as cellular sensors. *Nature*, **426**, 517–524.

Cossins, A. R. and Bowler, K. (1987) *Temperature Biology of Animals*, Chapman and Hall, London.

Crain, D. A. and Guillette Jr., L. J. (1998) Reptiles as models of contaminant-induced endocrine disruption. *Anim. Reprod. Sci.*, **53**, 77–86.

Ernfors, P., El Manira, A., and Svenningsson, P. (2021) Scientific Background. Discoveries of receptors for temperature and touch. *The Nobel Assembly at Karolinska Institutet*.

Ewert, M. A., Jackson, D. R., and Nelson, C. E. (1994) Patterns of temperature-dependent sex determination in turtles. *J. Exp. Zool.*, **270**, 3–15.

Fox, G. R., Hayward, J. S., and Hobson, G. N. (1979) Effect of alcohol on thermal balance of man in cold water. *Can. J. Physiol. Pharmacol.*, **57**, 860–865.

Geneva, I. I., Cuzzo, B., Fazili, T., and Javaid, W. (2019) Normal body temperature: a systematic review. *Open Forum Infect. Dis.*, **6**, ofz032.

Handrich, Y., Bevan, R. M., Charrassin, J.-B., Butler, P. J., Pütz, K., Woakes, A. J., Lage, J., and Le Maho, Y. (1997) Hypothermia in foraging king penguins. *Nature*, **388**, 64–67.

Harlow, H. J., Lohuis, T., Anderson-Sprecher, R. C., and Beck, T. D. I. (2004) Body surface temperature of hibernating black bears may be related to periodic muscle activity. *J. Mammal.*, **85**, 414–419.

Hirano, K., Nonami, Y., Nakamura, Y., Sato, T., Sato, T., Ishiguro, K.-i., Ogawa, T., and Yoshida, S. (2022) Temperature sensitivity of DNA double-strand break repair underpins heat-induced meiotic failure in mouse spermatogenesis. *Commun. Biol.*, **5**, 504.

Hirsch, S. M., Sundaramoorthy, S., Davies, T., Zhuravlev, Y., Waters, J. C., Shirasu-Hiza, M., Dumont, J., and Canman, J. C. (2018) FLIRT: fast local infrared thermogenetics for subcellular control of protein function. *Nat. Methods*, **15**, 921–923.

Holland, K. N., Brill, R. W., Chang, R. K. C., Sibert, J. R., and Fournier, D. A. (1992) Physiological and behavioural thermoregulation in bigeye tuna (*Thunnus obesus*). *Nature*, **358**, 410–412.

Hossain, M. E., Matsuzaki, K., Katakura, M., Sugimoto, N., Al Mamun, A., Islam, R., Hashimoto, M., and Shido, O. (2017) Direct exposure to mild heat promotes proliferation and neuronal differentiation of neural stem/ progenitor cells in vitro. *PLoS ONE*, **12**, e0190356.

Houdas, Y. and Ring, E. F. J. (1982) *Human Body Temperature. Its Measurement and Regulation*. Springer, New York.

Ito-Inaba, Y., Hida, Y., and Inaba, T. (2009a) What is critical for plant thermogenesis? Differences in mitochondrial activity and protein expression between thermogenic and non-thermogenic skunk cabbages. *Planta*, **231**, 121–130.

Ito-Inaba, Y., Sato, M., Masuko, H., Hida, Y., Toyooka, K., Watanabe, M., and Inaba, T. (2009b) Developmental changes and organelle biogenesis in the reproductive organs of thermogenic skunk cabbage (*Symplocarpus renifolius*). *J. Exp. Bot.*, **60**, 3909–3922.

Ito-Inaba, Y., Hida, Y., Matsumura, H., Masuko, H., Yazu, F., Terauchi, R., Watanabe, M., and Inaba, T. (2012) The gene expression landscape of thermogenic skunk cabbage suggests critical roles for mitochondrial and vacuolar metabolic pathways in the regulation of thermogenesis. *Plant Cell Environ.*, **35**, 554–566.

Jensen, M. P., Allen, C. D., Eguchi, T., Bell, I. P., LaCasella, E. L., Hilton, W. A., Hof, C. A. M., and Dutton, P. H. (2018) Environmental warming and feminization of one of the largest sea turtle populations in the world. *Curr. Biol.*, **28**, 154–159.

Jessen, C. (2001) *Temperature Regulation in Humans and Other Mammals*, Springer, Berlin.

Kamei, Y., Suzuki, M., Watanabe, K., Fujimori, K., Kawasaki, T., Deguchi, T., Yoneda, Y., Todo, T., Takagi, S., Funatsu, T., and Yuba, S. (2009) Infrared laser-mediated gene induction in targeted single cells in vivo. *Nat. Methods*, **6**, 79–81.

Kataoka, N., Shima, Y., Nakajima, K., and Nakamura, K. (2020) A central master driver of psychosocial stress responses in the rat. *Science*, **367**, 1105–1112.

Kedersha, N. L., Gupta, M., Li, W., Miller, I., and Anderson, P. (1999) RNA-binding proteins TIA-1 and TIAR link the phosphorylation of eIF-2α to the assembly of mammalian stress granules. *J. Cell. Biol.*, **147**, 1431–1441.

Khachaturyan, G., Holle, A. W., Ende, K., Frey, C., Schwederski, H. A., Eiseler, T., Paschke, S., Micoulet, A., Spatz, J. P., and Kemkemer, R. (2022) Temperature-sensitive migration dynamics in neutrophil-differentiated HL-60 cells. *Sci. Rep.*, **12**, 7053.

Knight, M. R. and Knight, H. (2012) Low-temperature perception leading to gene expression and cold tolerance in higher plants. *New Phytol.*, **195**, 737–751.

Knutson, R. M. (1979) Plants in heat. *Nat. Hist.*, **88**(3), 42–47.

Kosara, T. (2011) *Hibernation*, Scholastic, New York.

Lang, J. W. and Andrews, H. V. (1994) Temperature-dependent sex determination in crocodilians. *J. Exp. Zool.*, **270**, 28–44.

Lawson, R. N. and Chughtai, M. S. (1963) Breast cancer and body temperature. *Canad. Med. Ass. J.*, **88**, 68–70.

Lee, K. A. (1988) Circadian temperature rhythms in relation to menstrual cycle phase. *J. Biol. Rhythms*, **3**, 255–263.

Lindsey, A. A. (1963) Accuracy of duration temperature summing and its use for *Prunus serrulata*. *Ecology*, **44**, 149–151.

Lowell, B. B. and Spiegelman, B. M. (2000) Towards a molecular understanding of adaptive thermogenesis. *Nature*, **404**, 652–660.

Middleton, W. E. K. (1966) *A History of the Thermometer and Its Uses in Meteorology*, The Johns Hopkins Press, Baltimore.

Mieusset, R. and Bujan, L. (1995) Testicular heating and its possible contributions to male infertility: a review. *Int. J. Androl.*, **18**, 169–184.

Monti, M., Brandt, L., Ikomi-Kumm, J., and Olsson, H. (1986) Microcalorimetric investigation of cell metabolism in tumour cells from patients with non-Hodgkin lymphoma (NHL). *Scand. J. Haematol.*, **36**, 353–357.

Moore, C. R. and Quick, W. J. (1924) The scrotum as a temperature regulator for the testes. *Am. J. Physiol.*, **68**, 70–79.

Nagy, K. A., Odell, D. K., and Seymour, R. S. (1972) Temperature regulation by the inflorescence of philodendron. *Science*, **178**, 1195–1197.

Ohnaka, T., Tochihara, Y., Kubo, M., and Yamaguchi, C. (1995) Physiological and subjective responses to standing showers, sitting showers, and sink baths. *J. Physiol. Anthropol. Appl. Hum. Sci.*, **14**, 235–239.

Patel, D. and Franklin, K. A. (2009) Temperature-regulation of plant architecture. *Plant Signal. Behav.*, **4**, 577–579.

Preußner, M., Goldammer, G., Neumann, A., Haltenhof, T., Rautenstrauch, P., Müller-McNicoll, M., and Heyd, F. (2017) Body temperature cycles control rhythmic alternative splicing in mammals. *Mol. Cell*, **67**, 433–446.

Refinetti, R. and Menaker, M. (1992) The circadian rhythm of body temperature. *Physiol. Behav.*, **51**, 613–637.

Richter, K., Haslbeck, M., and Buchner, J. (2010) The heat shock response: life on the verge of death. *Mol. Cell*, **40**, 253–266.

Roth, J., Rummel, C., Barth, S. W., Gerstberger, R., and Hübschle, T. (2006) Molecular aspects of fever and hyperthermia. *Neurol. Clin.*, **24**, 421–439.

Sato, K. (2014) Body temperature stability achieved by the large body mass of sea turtles. *J. Exp. Biol.*, **217**, 3607–3614.

Sengupta, P. and Garrity, P. (2013) Sensing temperature. *Curr. Biol.*, **23**, R304–R307.

Seymour, R. S. (2001) Biophysics and physiology of temperature regulation in thermogenic flowers. *Biosci. Rep.*, **21**, 223–236.

Shen, Z.-G. and Wang, H.-P. (2014) Molecular players involved in temperature-dependent sex determination and sex differentiation in Teleost fish. *Genet. Sel. Evol.*, **46**, 26.

Shimatani, A., Hoshi, M., Oebisu, N., Takada, N., Ban, Y., and Nakamura, H. (2022) An analysis of tumor-related skin temperature differences in malignant soft-tissue tumors. *Int. J. Clin. Oncol.*, **27**, 234–243.

Tan, C. L. and Knight, Z. A. (2018) Regulation of body temperature by the nervous system. *Neuron*, **98**, 31–48.

Tatsumi, T., Sampei, M., Saito, K., Honda, Y., Okazaki, Y., Arata, N., Narumi, K., Morisaki, N., Ishikawa, T., and Narumi, S. (2020) Age-dependent and seasonal changes in menstrual cycle length and body temperature based on big data. *Obstet. Gynecol.*, **136**, 666–674.

Tøien, Ø., Blake, J., Edgar, D. M., Grahn, D. A., Heller, H. C., and Barnes, B. M. (2011) Hibernation in black bears: independence of metabolic suppression from body temperature. *Science*, **331**, 906–909.

Warner, D. A. and Shine, R. (2008) The adaptive significance of temperature-dependent sex determination in a reptile. *Nature*, **451**, 566–568.

Weber, C., Zhou, Y., Lee, J. G., Looger, L. L., Qian, G., Ge, C., and Capel, B. (2020) Temperature-dependent sex determination is mediated by pSTAT3 repression of *Kdm6b*. *Science*, **368**, 303–306.

Wegner, N. C., Snodgrass, O. E., Dewar, H., and Hyde, J. R. (2015) Whole-body endothermy in a mesopelagic fish, the opah, *Lampris guttatus*. *Science*, **348**, 786–789.

Wigge, P. A. (2013) Ambient temperature signalling in plants. *Curr. Opin. Plant Biol.*, **16**, 661–666.

Zulley, J., Wever, R., and Aschoff, J. (1981) The dependence of onset and duration of sleep on the circadian rhythm of rectal temperature. *Pflügers Arch.*, **391**, 314–318.

2
Fluorescent Molecular Thermometers

Fluorescent molecular thermometers are among the best choices to measure a temperature in small spaces, such as a microfluidic device and a live cell. Fluorescent molecular thermometers, of which fluorescence properties (e.g., intensity, lifetime, and maximum emission wavelength) depend on temperature, are molecular-based, so they are small enough compared to objects intended to be measured. This small size is a tremendous advantage of fluorescent molecular thermometers over other thermometers, such as thermocouples and infrared radiation detectors (thermography). When I started a research project to develop highly sensitive fluorescent molecular thermometers in 2002, only a few examples showed large temperature-dependency in fluorescence properties. Now, dozens of compounds have been reported as fluorescent molecular thermometers, and the recognition of their usefulness in thermometry further accelerates the development of fluorescent molecular thermometers and even other non-fluorescent thermometers with a unique downsizing technique. The latter will be categorized in Chapter 4. Here, I will first briefly summarize the basics of fluorescence and typical fluorescence responses, which will be followed by details of fluorescent molecular thermometers in the order of size, i.e., small organic fluorophores, metal complexes with organic ligands, macromolecules (synthetic polymers, nucleic acids, and proteins), and inorganic substances (quantum dots and other materials). In my opinion, a comprehensive understanding of principles is crucial for reliable experiments (i.e., temperature measurements) using fluorescent molecular thermometers for researchers' own purposes. Fluorescent molecular thermometers that can work only in organic solvents will also be described with their functional mechanisms in this chapter, although the main subject of this textbook is intracellular thermometry. Such thermometers might be able to work under aqueous environments with appropriate structural modifications. Readers can refer to excellent review articles on fluorescent molecular thermometers in general remarks (Appendix 1) or specific categories (as cited in each section if available).

2.1
The Basics of Fluorescence

For a solid understanding of fluorescence phenomena from organic and inorganic molecules, I strongly recommend the two greatest textbooks (Lakowicz, 2006; Valeur and Berberan-Santos, 2012). In the beginning of this chapter, the relaxation processes of excited molecules will be concisely described.

Figure 2.1 is the Jabłoński diagram that summarizes a molecule's excitation and relaxation processes. When a molecule in the ground state (i.e., S_0 state) absorbs light (called "excitation" or "absorption"), an electron transits to an orbital with a higher energy level, and the resultant molecule turns in the excited state. In general, the excitation of a molecule generates a singlet excited molecule with higher energy than the lowest vibrational energy level of the first excited singlet state (i.e., S_1 state). This excitation process occurs within 10^{-15}–10^{-16} seconds. The molecule excited to the S_1 state releases surplus energy as a form of heat and transits to the lowest vibrational energy level of the S_1 state (within 10^{-13} seconds). Then, the excited molecule relaxes to the ground state via the following pathways.

Fluorescence (within 10^{-8}–10^{-9} seconds)
The relaxation pathway to emit light with the energy that corresponds to the difference between S_0 and S_1 states is called "fluorescence". As the excited molecule loses a part of its excitation energy as vibrational energy before fluorescence, the fluorescence energy is always smaller than the absorption energy. Energy is in inverse proportion to wavelength. Thus, the fluorescence wavelength is always longer than the absorption wavelength (Stokes law). Fluorescent molecular thermometers covered in the present textbook involve this radiative process.

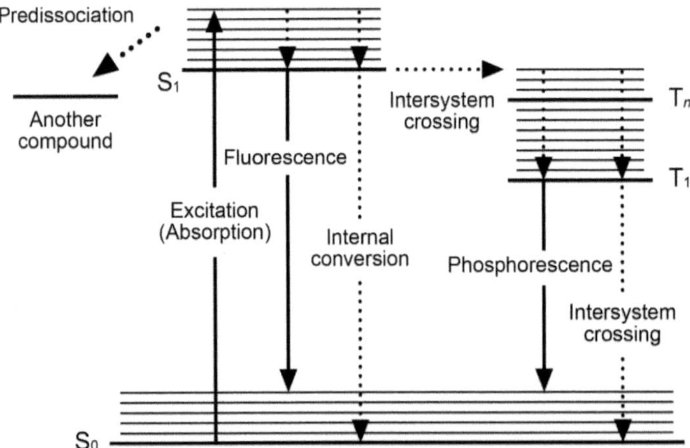

Figure 2.1 Excitation and relaxation pathways of a molecule (Jabłoński diagram). Solid and dotted lines indicate radiative and non-radiative processes, respectively.

Internal conversion (within 10^{-6}–10^{-12} seconds)

"Internal conversion" is the relaxation pathway to lose the energy between S_0 and S_1 states by various molecular vibrations. Internal conversion occurs more efficiently when the energy gap between S_0 and S_1 states is smaller.

Intersystem crossing (within 10^{-6} seconds)

"Intersystem crossing" is the transition from the S_1 state to an excited triplet state (T_n: $n \geq 2$) having a lower energy level. A transition between different spin states is forbidden, but spin–orbit interactions via vibrations enable intersystem crossing. According to El-Sayed's rule, the intersystem crossing is forbidden when the S_1 and T_n states are both $\pi\pi^*$ states or both $n\pi^*$ states (π and n indicate types of electrons. For instance, the $\pi\pi^*$ state means that the excited state results from the electron transition from a π orbital to another π orbital.). On the other hand, when one of the S_1 and T_n states is a $\pi\pi^*$ state, and the other is an $n\pi^*$ state, the intersystem crossing is permitted. Like internal conversion, the smaller the energy difference between the S_1 and T_n states, the more efficiently intersystem crossing will occur. A molecule transitioning from the S_1 state to the T_n state due to intersystem crossing loses excess vibrational energy as heat and relaxes to the lowest excited triplet state (i.e., the T_1 state). Then, it returns to the S_0 state through thermal radiation (within 10^{-2}–10^{-5} seconds), or a process called "phosphorescence" (within 10^{-4}–10^{-1} seconds), which emits light corresponding to the energy difference between the T_1 and S_0 states. Several examples of molecular thermometers show temperature-dependent phosphorescence properties.

Predissociation (within 10^{-9}–10^{-12} seconds)

A photochemical reaction from the S_1 state, such as dissociation of chemical bonds and isomerization, is called "predissociation". Predissociation occurs when potential curves of the S_1 state and the chemically unstable state intersect, and the excitation energy is greater than the energy required for bond dissociation or isomerization of a molecule. Predissociation can occur not only from the S_1 state but also from the T_n state after intersystem crossing.

2.2
Responses of Fluorescent Molecular Thermometers

In general, temperature-dependent fluorescence properties are observed when the fluorescence from the S_1 state is competing with some non-radiative pathway which can be activated by thermal energy (depending on the environmental temperature). In the simplest case (Figure 2.2a), a non-radiative relaxation process competing with fluorescence is accelerated at a higher temperature, decreasing fluorescence intensity. See also Box 2.1 for kinetics on temperature-dependent fluorescence. When a fluorescence spectrum of a molecule varies with changing temperature, as shown in Figure 2.2a, fluorescence lifetime (an index of the relaxation time from the excited state) is also temperature-dependent, i.e., shorter at a higher temperature. It should be noted that fluorescence lifetime is a parameter independent of the concentration of the molecule and the power of the excitation source.

This characteristic is helpful, especially in intracellular imaging. The opposite response, i.e., stronger fluorescence with a longer fluorescence lifetime at a higher temperature, is rare. Therefore, its creation requires a special design of molecules with unusual functional mechanisms. Although the case shown in Figure 2.2b, i.e., the maximum fluorescence wavelength gradually shifts with temperature variation, is easily imagined as a response of fluorescent molecular thermometer, only a few cases (e.g., quantum dots) indicate this type. Instead, a ratiometric response (Figure 2.2c) is observed when a fluorophore has two fluorescent states that are thermally equilibrated: a fluorescence intensity at one wavelength (λ_1) decreases with a change in temperature (from T_1 to T_2), whereas that at another wavelength (λ_2) simultaneously increases.

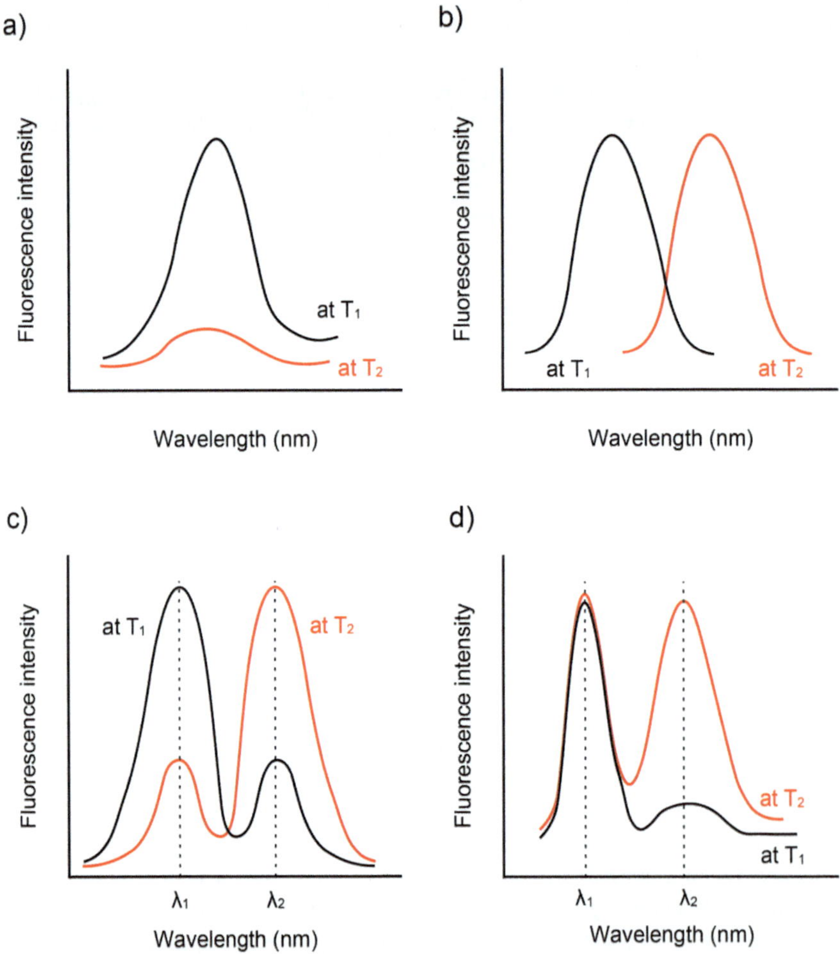

Figure 2.2 Responses of fluorescent molecular thermometers. Model fluorescence spectra are indicated at two temperatures (T_1 and T_2, $T_1 < T_2$). (a)–(d) represent different types of responses.

Moreover, when fluorescent molecular thermometers consist of two fluorophores and temperature-dependent energy transfer between them is involved, the fluorescence spectra also follow that displayed in Figure 2.2c. Similar to the measured parameter fluorescence lifetime, the fluorescence intensity ratio is advantageous over a simple parameter of fluorescence intensity at a single wavelength because the fluorescence intensity ratio is not susceptible to a change in the concentration of a molecule and strength of an excitation source (Lee et al., 2015). Thus, for high practicality, fluorescent molecular thermometers that consist of two fluorophores emitting temperature-dependent and temperature-independent (i.e., reference) fluorescence have been proposed to deliberately obtain a ratiometric response to a temperature variation (Figure 2.2d).

2.3
Small Organic Molecules Involving Intersystem Crossing

9-Methylanthracene (Figure 2.3a) is the simplest example to demonstrate how temperature-dependent fluorescence characteristics are provided by the competition of the radiative fluorescence process with non-radiative processes in a molecule. In 9-methylanthracene, the energy level of an excited triplet state (T_n; n is assumed to be 2) is slightly higher than that of the first excited singlet state (S_1). Thus, the non-radiative intersystem crossing ($S_1 \rightarrow T_n$) competes with the radiative fluorescence process ($S_1 \rightarrow S_0$) with an activation barrier, and the former process is

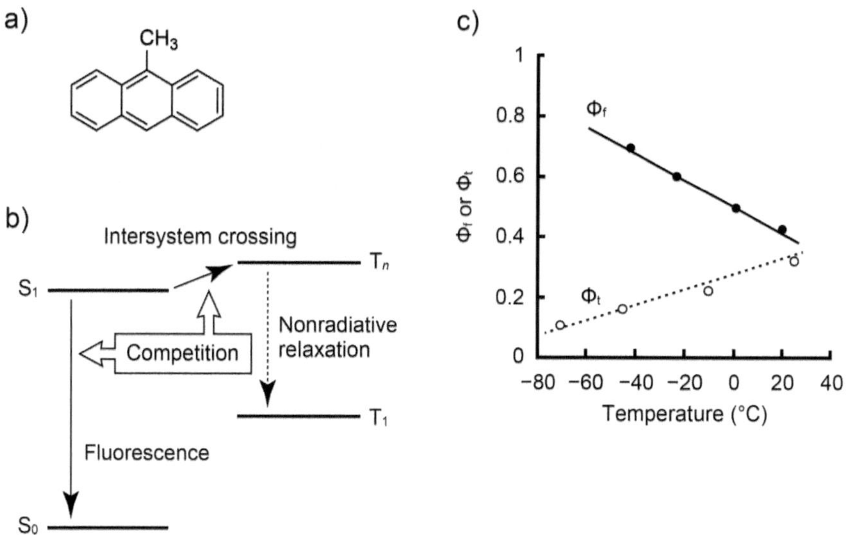

Figure 2.3 9-Methylanthracene. (a) Chemical structure. (b) Energy levels to bear temperature-dependent fluorescence characteristics. (c) Relationships between fluorescence quantum yield (Φ_f) and temperature in ethanol and between the quantum yield of intersystem crossing (Φ_t) and temperature in a lucite acrylic solution. The original data of Φ_f and Φ_t are from Lim et al. (1966) *J. Mol. Spectrosc.*, **19**, 412–420 and Bennett and McCartin (1966) *J. Chem. Phys.*, **44**, 1969–1972, respectively.

accelerated at a higher temperature (Figure 2.3b). As a result, the fluorescence quantum yield of 9-methylanthracene is reduced with the compensation of intersystem crossing by increasing temperature (Figure 2.3c) (Bennett and McCartin, 1966; Lim et al., 1966). The activation energy for intersystem crossing can be evaluated by the Arrhenius plot (see Box 2.1).

Box 2.1 Kinetics of Excited-state Relaxation in a Fluorescent Molecular Thermometer

For an excited fluorescent molecular thermometer in the S_1 state, k_f and k_{nr} denote rate constants of fluorescence and non-radiative relaxation processes, respectively. Suppose intersystem crossing and internal conversion are components of non-radiative relaxation processes. In that case, k_{nr} is the sum of k_{ISC} (a rate constant of intersystem crossing) and k_{IC} (a rate constant of internal conversion), i.e., $k_{nr} = k_{ISC} + k_{IC}$. Then, the fluorescence quantum yield Φ_f can be related to k_f and k_{nr} as Eq. (2.1).

$$\Phi_f = \frac{k_f}{k_f + k_{nr}} \tag{2.1}$$

In general, the fluorescence pathway of a fluorescent molecular thermometer competes with a non-radiative relaxation pathway that is thermally activated. That is, high temperatures increase k_{nr}, decreasing the Φ_f value (and therefore fluorescence intensity too). As the fluorescence lifetime τ_f is defined by Eq. (2.2),

$$\tau_f = \frac{\Phi_f}{k_f} \tag{2.2}$$

so, τ_f is also shortened by increasing temperature.

According to the Arrhenius equation [1], a rate for temperature-dependent non-radiative relaxation pathway from the S_1 state, k_{nr1}, can be represented as

$$k_{nr1} = Ae^{-Ea/RT} \tag{2.3}$$

where A, Ea, R, and T are a pre-exponential factor, activation energy from the S_1 state to a non-radiative state, the gas constant, and the absolute temperature. Temperature-dependent Φ_f of a fluorescent molecular thermometer is correlated to k_f, k_{nr1}, and k_{nr2} (a rate constant for temperature-independent non-radiative relaxation process), as Eq. (2.4).

$$\Phi_f = \frac{k_f}{k_f + k_{nr1} + k_{nr2}} \tag{2.4}$$

Thus, the activation energy from the fluorescent S_1 state to a thermally activated non-radiative state can be evaluated by experimental serial data of the Φ_f value (or relevant parameters such as fluorescence intensity and fluorescence lifetime) and the environmental temperature.

1 Laidler, K. T. (1984) *J. Chem. Educ.*, **61**, 494.

Delayed fluorescence is the emission from the S_1 state after two intersystem crossings, i.e., the first intersystem crossing from the singlet to the triplet state and the second reverse intersystem crossing from the triplet state to the singlet state. As a transition from the triplet excited state to the ground state is generally forbidden, an energy barrier to the reverse intersystem crossing is the cause of temperature-dependent delayed fluorescence. Mlle. Boudin reported fluorescence and phosphorescence of eosin (Figure 2.4a) in a French journal in 1930 (Boudin, 1930), and 30 years later, Parker and Hatchard found the temperature-dependent delayed fluorescence of eosin in a careful photophysical study (Parker and Hatchard, 1961). In ethanol, the phosphorescence intensity of eosin decreases with increasing temperature (phosphorescence quantum yield is 0.0237 and 0.0015 at -196 and $70\,^\circ$C, respectively), while the intensity of delayed fluorescence increases in complement (Figure 2.4b). A plot using temperature-dependent quantum yields of delayed fluorescence and phosphorescence determined the activation energy of the reverse intersystem crossing from the triplet state to the singlet excited state to be approximately 9.7 kcal.

Acridine yellow (Figure 2.4c) in a rigid saccharide glass also showed temperature-dependent delayed fluorescence with the activation energy of reverse intersystem crossing $\Delta E = 2370$ cm^{-1} (Fister et al., 1995). The intensity ratio of delayed fluorescence to phosphorescence of acridine yellow in a saccharide glass increased from 1.0 to 4.0 with increasing temperature from -10 to $22\,^\circ$C.

A member of the fullerene family, C_{70} (Figure 2.4d), is another example that emits temperature-dependent delayed fluorescence (Baleizão and Berberan-Santos, 2006; Baleizão et al., 2007). Figure 2.4e shows temperature-dependent fluorescence spectra of C_{70} in polystyrene film, in which almost all components are ascribed to the delayed fluorescence. The intensity ratio of delayed fluorescence to prompt (normal) fluorescence of C_{70} in polystyrene film at 700 nm is 22, 39, 53, and 70 at 25, 50, 70, and 100 $^\circ$C, respectively. The activation energy of reverse intersystem crossing of C_{70} in polystyrene film was evaluated to be 29 kJ mol^{-1}. The heat-induced enrichment of delayed fluorescence can be visible in a C_{70}-doped poly(tert-butyl methacrylate) film (Figure 2.4f). One of the disadvantages of delayed fluorescence-based molecular thermometers is severe quenching in the presence of oxygen that functions as a quencher of excited triplet states. To avoid using a vacuum or an oxygen-free atmosphere, incorporating polystyrene nanoparticles containing C_{70} into an oxygen-impermeable polyacrylonitrile film is extremely effective (Augusto et al., 2010).

Recent cases include a Zn(II) Schiff base complex, Zn-1, whose delayed fluorescence lifetime in a polystyrene foil is 14.4 and 7.4 ms at 5 and 25 $^\circ$C, respectively (Steinegger and Borisov, 2020). Severe quenching of the delayed fluorescence of Zn-1 by oxygen can be avoided by covering the fluorescent layer with both an off-stoichiometry thiol-ene polymer as an oxygen-consuming layer and a poly(vinylidene chloride-co-acrylonitrile) layer as an oxygen barrier.

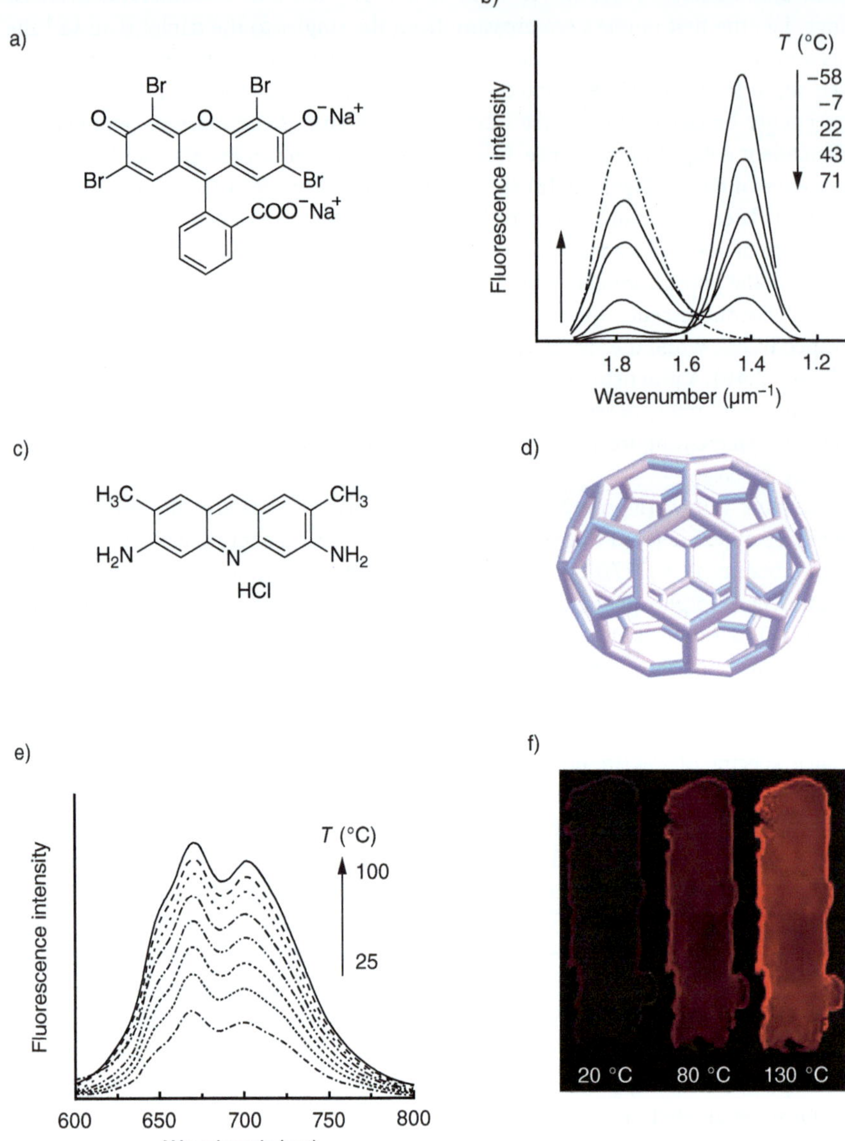

Figure 2.4 Fluorescent molecular thermometers emitting temperature-dependent delayed fluorescence. (a,b) Eosin. (a) Chemical structure. (b) Delayed fluorescence and phosphorescence spectra (solid lines) at −58, −7, 22, 43, and 71 °C and fluorescence spectrum (dotted line) at 22 °C (15 μmol L^{-1}, λ_{ex}: 500 nm) in ethanol (adapted from Parker and Hatchard (1961) *Trans. Faraday Soc.*, **57**, 1894–1904). (c) Chemical structure of acridine yellow. (d–f) C$_{70}$. (d) Chemical structure. (e) Fluorescence spectra in polystyrene (PS) film (λ_{ex}: 470 nm). (f) Visual images in a poly(*tert*-butyl methacrylate) film. Panels (e,f) are adapted from Baleizão et al. (2007) *Chem. Eur. J.*, **13**, 3643–3651.

Zn-1

2.4
Small Organic Molecules with a Rotating Substituent Group

2.4.1
Rhodamine B

Rhodamine B (Figure 2.5a) has a long history as a fluorescent compound. Its structure was identified in 1905 (Noelting and Dziewoński, 1905). As shown in Figure 2.5b, the fluorescence intensity of rhodamine B in water is influenced by the environmental temperature and dramatically decreases when the temperature rises (Karstens and Kobs, 1980). The sensitivity of rhodamine B to a temperature variation is relatively large among well-known fluorescent compounds: e.g., quinine sulfate, which is often used as a standard, shows only a 10% decrease in fluorescence intensity with the temperature change from 10 to 60 °C (Kubin and Fretcher, 1982). The mechanism of the temperature sensitivity of rhodamine B is a heat-induced rotation of a diethylamino substituent group, which was confirmed by the temperature insensitivity of a bridged compound rhodamine 101 (Figure 2.5c, 2.5d). That is, the fluorescence process of rhodamine B in the excited state competes with the non-radiative internal conversion accompanying the rotation of the substituent group (Figure 2.5e) (Snare et al., 1982).

Because of the high sensitivity to temperature variation, the high solubility in water, and the relatively long excitation and emission wavelengths, rhodamine B has been applied for various temperature measurements of non-biological subjects for a long time. It is important to prepare an appropriate calibration curve for accurate temperature measurements (Shah et al., 2009). Table 2.1 summarizes the temperature measurements of non-biological subjects using rhodamine B as a fluorescent molecular thermometer. A couple of cases are introduced below. A microfluidic system is one of the main non-biological targets of fluorescent molecular thermometers. Fluid-temperature mapping in microfluidic systems was successfully conducted with rhodamine B with micrometer spatial resolution and 33 milliseconds temporal resolution (Ross et al., 2001). Control of fluid temperatures in microfluidic systems during reaction and separation is crucial; i.e., temperature

a)

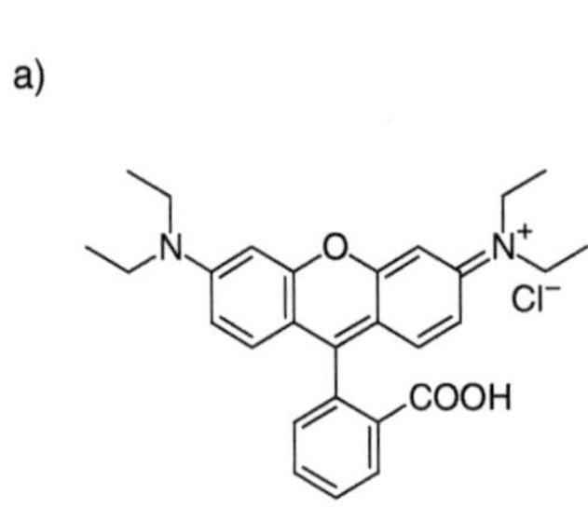

b)

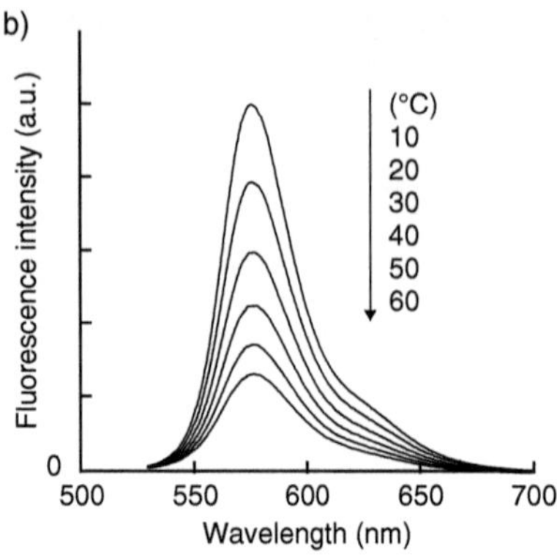

c)

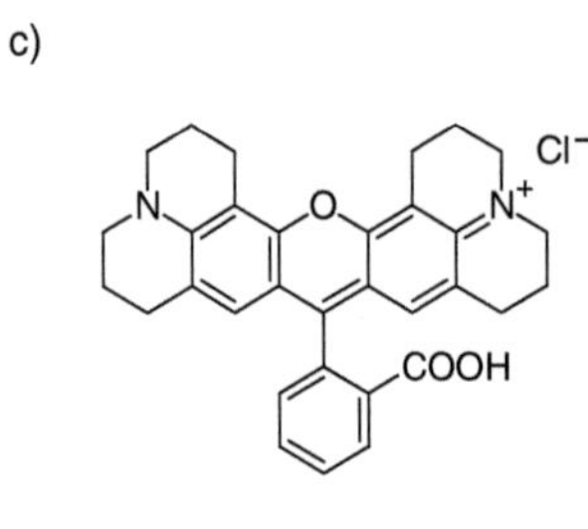

d)

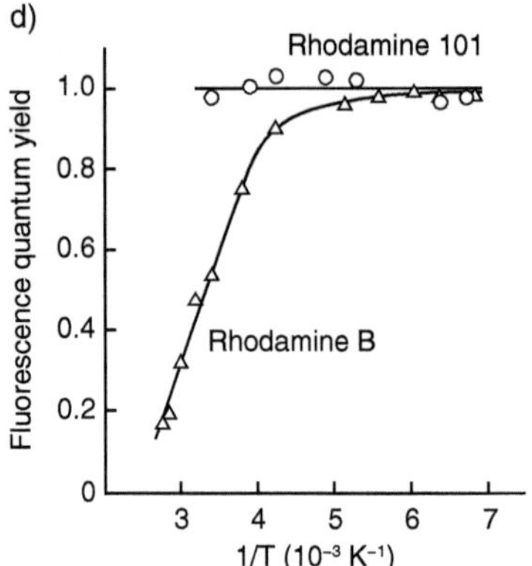

e)

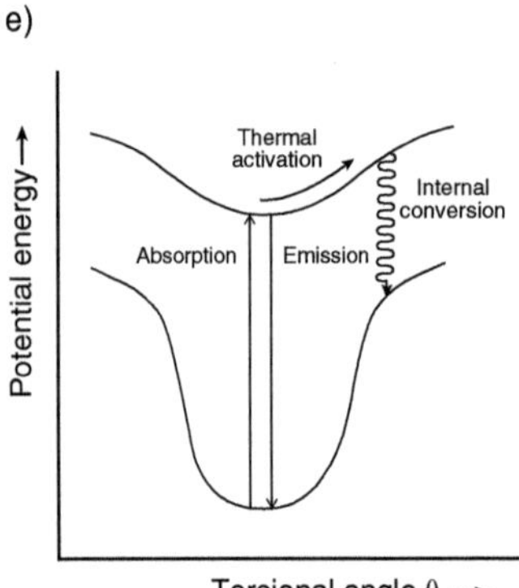

Figure 2.5 Rhodamine B. (a) Chemical structure. (b) Temperature-dependent fluorescence spectra (100 nmol L^{-1}) in water with the excitation at 515 nm (measured in the author's laboratory for this textbook). (c) Chemical structure of a bridged reference compound, rhodamine 101. (d) Relationships between the Φ_f (triangle: rhodamine B, circle: rhodamine 101) and temperature (adapted from Karstens and Kobs (1980) *J. Phys. Chem.*, **84**, 1871–1872). (e) Change in energy levels with the rotation of the diethylamino group.

Table 2.1 Application of rhodamine B in fluorescent thermometry of inanimate objects.

Case	Object	Detection parameter[a]	Temperature range (°C)	Resolution			Ref.
				T (°C)	Spatial	Temporal	
1	Film	FI	20–120	1	2 µm	18 ns	[1]
2	Rectangular box heated by electrical mats	FI ratio of rhodamine B to rhodamine 110	15–40	1.4	0.7 mm		[2]
3	Natural convection flow in a cavity	FI	19.8–55.2	1.7	0.3 mm		[3]
4	Microfluidic circuits	Relative FI normalized at 20 °C	15–90	0.03–3.5	1 µm (2D)	33 ms	[4]
5	Polystyrol microchannel-microheater chip	τ_f	10–60				[5]
6	Poly(dimethylsiloxane) microchip	Relative FI normalized at 23.5 °C	23.5–95	1	∼ 9 µm		[6]
7	Microchannel	τ_f	25–95	1	1 µm (lateral), 10 µm (axial)		[7]
8	Microwave-heated cavity	FI ratio of rhodamine B to rhodamine 110	25–50	1.6	123 µm		[8]
9	Microfluidic pump	FI	23–75				[9]
10	Silicon substrate with a metal wire	FI	30–75	5–10			[10]
11	Dielectrophoretic trap	FI and τ_f	20–50		< µm		[11]
12[b]	Microfluidic chip	FI	20–100	2–6	5 µm	10 ms	[12]

a) FI: fluorescence intensity, τ_f: fluorescence lifetime.
b) Rhodamine B is immobilized on polydimethylsiloxane.

1 Romano, V., Zweig, A. D., Frenz, M., and Weber, H. P. (1989) *Appl. Phys. B*, **49**, 527–533.
2 Sakakibara, J. and Adrian, R. J. (1999) *Exp. Fluids*, **26**, 7–15.
3 Coolen, M. C. J., Kieft, R. N., Rindt, C. C. M., and van Steenhoven, A. A. (1999) *Exp. Fluids*, **27**, 420–426.
4 Ross, D., Gaitan, M., and Locascio, L. E. (2001) *Anal. Chem.*, **73**, 4117–4123.
5 Kitamura, N., Hosoda, Y., Ueno, K., and Iwata, S. (2004) *Anal. Sci.*, **20**, 783–786.
6 Fu, R., Xu, B., and Li, D. (2006) *Int. J. Therm. Sci.*, **45**, 841–847.
7 Benninger, R. K. P., Koç, Y., Hofmann, O., Requejo-Isidro, J., Neil, M. A. A., French, P. M. W., and deMello, A. J. (2006) *Anal. Chem.*, **78**, 2272–2278.
8 Finegan, T., Laibinis, P. E., and Hatton, T. A. (2006) *AIChE J.*, **52**, 2727–2735.
9 Samel, B., Chretien, J., Yue, R., Griss, P., and Stemme, G. (2007) *J. Microelectromech. Syst.*, **16**, 795–801.
10 Löw, P., Kim, B., Takama, N., and Bergaud, C. (2008) *Small*, **4**, 908–914.
11 Gielen, F., Pereira, F., deMello, A. J., and Edel, J. B. (2010) *Anal. Chem.*, **82**, 7509–7514.
12 Wu, J., Kwok, T. Y., Li, X., Cao, W., Wang, Y., Huang, J., Hong, Y., Zhang, D., and Wen, W. (2013) *Sci. Rep.*, **3**, 3321.

gradients in capillaries for electrophoresis have been understood to cause band spreading and, consequently, the reduction of separation efficiency. Therefore, thermometry with rhodamine B can contribute to precise temperature measurements of such a "lab-on-a-chip". Figure 2.6a is a schematic showing the layout of a multi-branched circuit fabricated from poly(methyl methacrylate). When 1000 V was applied to the multi-branched circuit, a temperature gradient analyzed from the fluorescence intensity of rhodamine B was generated near the center channel by Joule heating (Figure 2.6b). The precision of the temperature measurements reached 0.03 °C at the highest. Another example of temperature imaging with rhodamine B is indicated in Figure 2.7. A thin, dried rhodamine B layer monitored the surface temperature of a resistive heater with a nickel wire (Löw et al., 2008).

The fluorescence intensity is the most accessible detection parameter in fluorescence thermometry with fluorescent molecular thermometers. However, the fluorescence intensity occasionally cannot directly correlate to the temperature. For example, it varies when the concentration of a fluorescent molecular thermometer changes or when the power of the excitation source (e.g., laser power) fluctuates. The fluorescence intensity ratio at two different wavelengths is one of the detection parameters that overcome this weakness of fluorescence intensity-based thermometry. Figure 2.8 demonstrates a ratiometric temperature measurement using rhodamine B and rhodamine 110 as a fluorescent molecular thermometer and a temperature-insensitive reference fluorophore, respectively (Sakakibara and Adrian, 1999).

In contrast to rhodamine B, the fluorescence intensity of rhodamine 110 without a bulky diethylamino group is almost constant even if the temperature changes (Figure 2.8a, 2.8b). Thus, the fluorescence intensity ratio of rhodamine B to rhodamine 110 in the mixture of two rhodamine derivatives is temperature-dependent but is greatly influenced by neither the concentration of the mixture nor the power of the excitation source. With the mixed solution of rhodamine B and rhodamine 110, the thermal convection in a rectangular box heated from below by electrical mats was monitored with a random error equivalent to 1.4 K (Figure 2.8c–2.8e).

Fluorescence lifetime is also a useful detection parameter that does not suffer from the changes in concentration of a fluorescent sensor and strength of excitation, although apparatus to measure a fluorescence lifetime, such as a fluorescence lifetime spectrometer and a fluorescence lifetime imaging microscope, are costly. Adopting the fluorescence lifetime of rhodamine B as a temperature-dependent parameter, three-dimensional temperature mapping of a microchannel was carried out (Benninger et al., 2006). As same as fluorescence intensity, the fluorescence lifetime of rhodamine B is sensitive to a temperature variation: it decreases from 1.55 to 0.25 ns with increasing temperature from 25 to 94 °C (Figure 2.9a). Fluorescence lifetime imaging microscopy combined with two-photon excitation enabled three-dimensional temperature imaging of a single microchannel containing a methanolic solution of rhodamine B that moved at a flow rate of 10 $\mu L\,min^{-1}$ (Figure 2.9b). Due to the use of two-photon excitation, rhodamine B could be selectively excited even in a limited space inside a microchannel. As a result, the spatial resolution was improved to ~1 μm in the lateral direction and 10 μm in the axial direction.

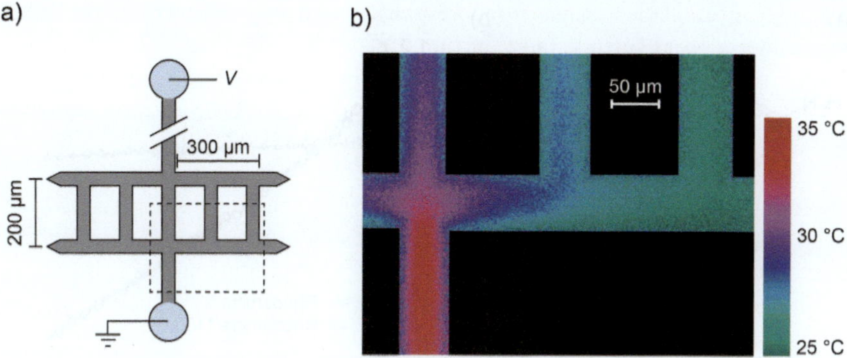

Figure 2.6 Temperature mapping of a microfluidic system with rhodamine B. (a) Schematic diagram of a multi-branched microfluidic system made using poly(methyl methacrylate). (b) Temperature map of the microfluidic system filled with rhodamine B solution (100 µmol L^{-1} in a carbonate buffer (pH 9.4)) (applied voltage: 1 kV). The visual field is the enclosed area by dashed lines in panel (a). Ross et al. (2001) *Anal. Chem.*, **73**, 4117–4123 / American Chemistry Society.

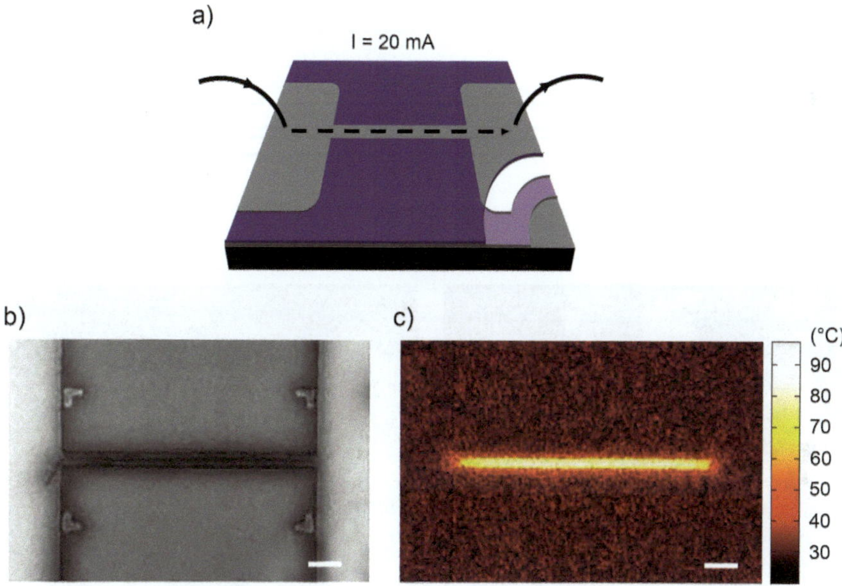

Figure 2.7 Temperature mapping of a resistive heater with a dried layer of rhodamine B. (a) Illustration of a resistive heater composed of a silicon microstructure and a nickel wire. A dried rhodamine B layer was deposited on the top (dark purple) of the silicon substrate. (b) Fluorescence image of rhodamine B on the resistive heater when the current was applied at 20 mA. (c) Corresponding temperature map. Scale bars: 10 µm. Adapted from Löw et al. (2008) *Small*, **4**, 908–914 / JOHN WILEY & SONS, INC.

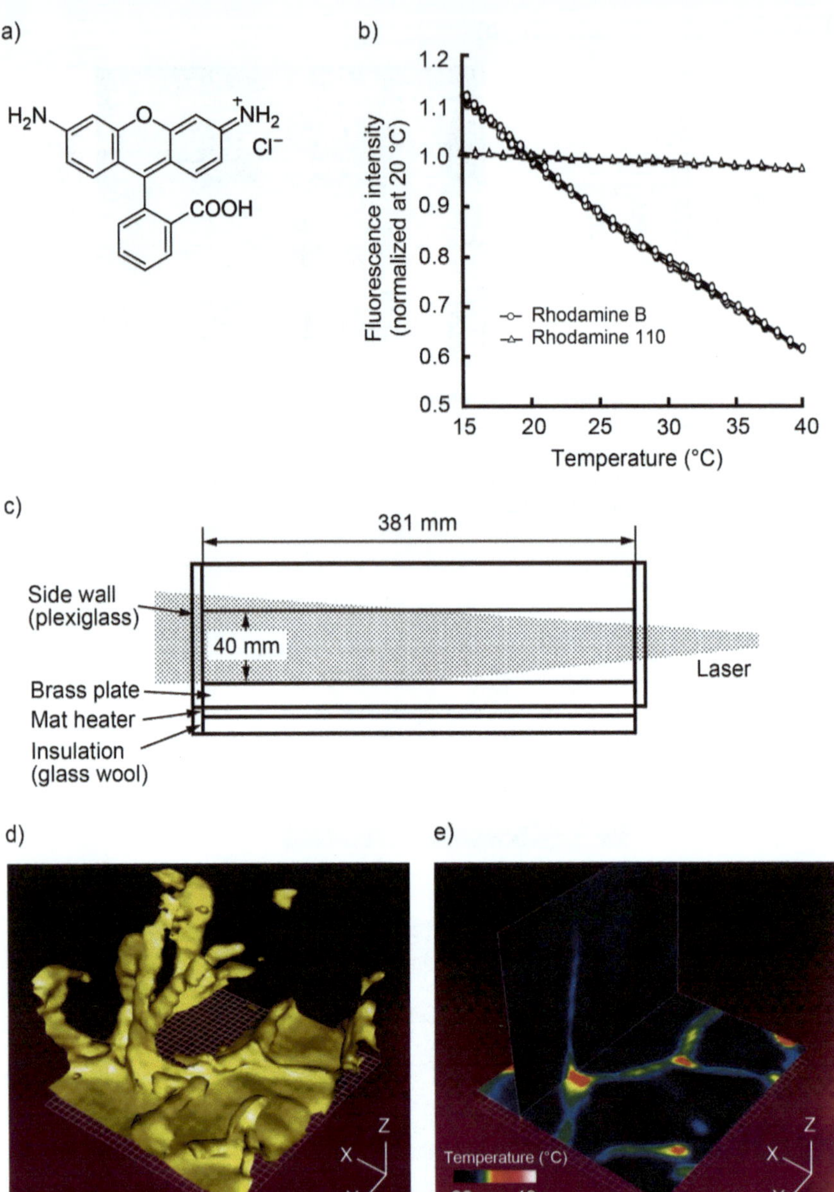

Figure 2.8 Ratiometric temperature measurements by the mixture of rhodamine B and rhodamine 110. (a) Chemical structure of a reference fluorophore, rhodamine 110. (b) Relationships between fluorescence intensity and temperature (λ_{ex}: 488 nm, circle: rhodamine B, triangle: rhodamine 110). (c) Schematic diagram of an apparatus for temperature mapping. (d) The surface of constant temperature in the apparatus. (e) Temperature distribution in the apparatus. Sakakibara and Adrian (1999) *Exp. Fluids*, **26**, 7–15 / Springer Nature.

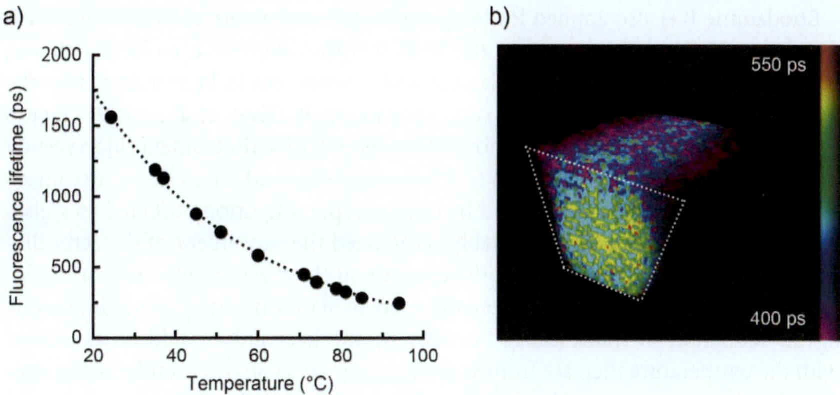

Figure 2.9 Temperature mapping with rhodamine B and fluorescence lifetime imaging microscopy with two-photon excitation. (a) Relationship between fluorescence lifetime of rhodamine B and temperature in methanol. (b) Three-dimensional temperature mapping of a segment (130 × 40 × 100 μm) in a microfluidic system running a methanolic rhodamine B solution (100 μmol L^{-1}) at a flow rate of 10 μL min^{-1}. Benninger et al. (2006) *Anal. Chem.*, **78**, 2272–2278 / American Chemical Society.

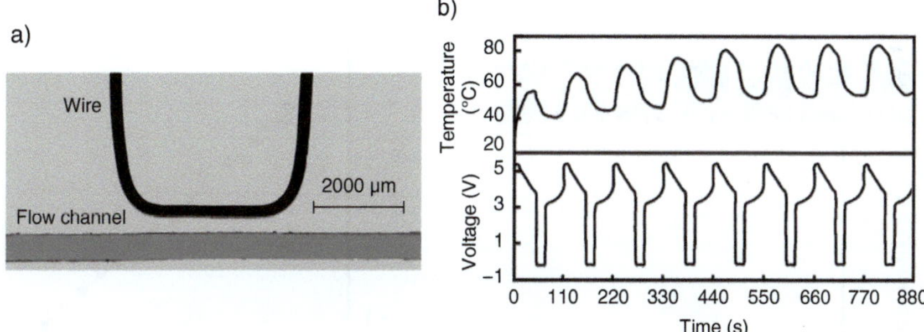

Figure 2.10 Temperature measurement of a microfluidic device, designed for a polymerase chain reaction, by rhodamine B. (a) Image of a microchannel with an embedded resistance (Nickel-Chrome 60) wire. The microfluidic channel was filled with 50 μmol L^{-1} rhodamine B in a sodium carbonate buffer (pH 8.5) (b) Thermal cycling of liquid temperature with a voltage function to the wire. Fu et al. (2006) *Int. J. Therm. Sci.*, **45**, 841–847 / Elsevier.

Besides, the temperature of a poly(dimethylsiloxane) (PDMS) chip designed for polymerase chain reaction (PCR) was monitored with rhodamine B in an aqueous buffer (Fu et al., 2006). PCR to amplify DNA needs precise thermal cycling at different temperature levels, i.e., typically, 94 °C for denaturation, 60 °C for annealing, and 72 °C for extension. A PDMS microfluidic device was built with an embedded resistance nickel-chrome wire (Figure 2.10a), and the configuration and heating/cooling rate were optimized. As a model, thermal cycling for the PDMS microfluidic device at two different temperatures was achieved by controlling the power of a local heater (Figure 2.10b).

Rhodamine B is also applied for temperature measurements of smaller objects at the molecular level. For example, the local temperature near a molecular motor, actomyosin, a complex of actin filament and myosin, could be monitored by the fluorescently labeled actomyosin with rhodamine B (Kato et al., 1999). Figure 2.11a shows the local heat pulse-induced movement of actin filament fluorescently labeled with rhodamine phalloidin (a rhodamine B derivative) on heavy meromyosin (a fragment of myosin prepared by chymotryptic digestion) adhered to a glass surface. A single heat pulse remarkably accelerated the movement of the actin filament, which was accompanied by the decrease in fluorescence intensity of rhodamine phalloidin (Figure 2.11a). From the concept of thermometry using rhodamine B, this reduction in fluorescence intensity by the heat pulse could be associated with the temperature increase from approximately 20 to 60 °C. A similar methodology elucidated the relationship between kinesin velocity along a microtubule and local temperature (Figure 2.11b) (Kawaguchi and Ishiwata., 2001).

Sulforhodamine B (also named Kiton Red) is a sulfonated derivative of rhodamine B and its solubility in water is improved compared to rhodamine B. As the temperature dependency of sulforhodamine B is comparable to that of rhodamine B, sulforhodamine B is also capable of measuring temperature in aqueous media.

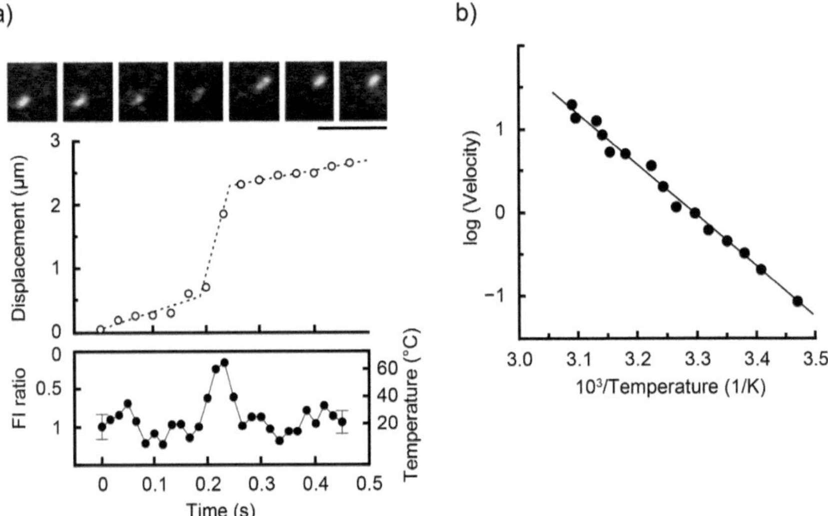

Figure 2.11 Temperature measurements of motor proteins and related compounds by rhodamine B derivatives. (a) Temperature measurements of a sliding actin filament on heavy meromyosin under a single heat pulse. The actin filament was labeled with rhodamine-phalloidin. Fluorescence images (scale bar: 5 µm) (top), the time course of the sliding movement (middle), and the corresponding fluorescence intensity of rhodamine phalloidin (left axis) and calculated temperature (right axis) (bottom) (adapted from Kato et al. (1999) *Proc. Natl. Acad. Sci. USA*, **96**, 9602–9606). (b) Temperature dependence of gliding velocity of kinesin molecules (adapted from Kawaguchi and Ishiwata (2001) *Cell Motil. Cytoskeleton*, **49**, 41–47).

The temperature of a microdroplet that is held and manipulated by optical tweezers in a microfluidic flow with a temperature gradient could be monitored with sulforhodamine B and fluorescence lifetime imaging microscopy (Bennet et al., 2011). The laser tweezer at 780 nm with a maximum output power of 60 mW only induced a temperature increase by 0.1 °C in a microdroplet, which was covered by an oil bubble and contained a sulforhodamine B solution.

Sulforhodamine B

2.4.2
NBD-NHPr

4-n-Propylamino-7-nitro-2,1,3-benzoxadiazole (NBD-NHPr) is a fluorescent compound showing temperature-dependent fluorescence characteristics that originate from the rotation of the substituent n-propylamino group at the 4-position in the excited state (Fery-Forgues et al., 1993). Because of the electron-donating n-propylamino group and the electron-accepting nitro group in NBD-NHPr, its S_1 state is an intramolecular charge transfer (ICT) state. At this excited stage, the fluorescence from the ICT state competes with the process to form a slightly higher non-emissive twisted intramolecular charge transfer (TICT) state with the rotation of the n-propylamino group. Therefore, the fluorescence quantum yield of NBD-NHPr drops with increasing temperature (e.g., $\Phi_f = 0.65$, 0.58, 0.52, and 0.49 at 20, 30, 40, and 50 °C in benzene, respectively). The activation energy from the ICT state to the TICT state in excited NBD-NHPr was evaluated to be 17 kJ mol^{-1} by the Arrhenius plot.

NBD-NHPr

2.4.3
8-Aryl-boron-dipyrromethene Derivatives

Originally, 8-aryl-boron-dipyrromethene (BODIPY) derivatives with a phenyl group at the 8-position were developed as fluorescent molecular viscometers as a non-radiative relaxation process due to the rotation of the aryl group was prevented in highly viscous media (Kuimova et al., 2008). Later, it was recognized that 8-aryl-BODIPY derivatives could also work as fluorescent molecular thermometers because of this rotation of the aryl group, which was accelerated at a higher temperature. For example, the fluorescence quantum yield of PEG-BODIPY is 0.28 and 0.16 at 23 and 49 °C, respectively, in water (Ogle et al., 2019). Since the BODIPY structure is highly hydrophobic, hydrophilic moieties (e.g., ethylene glycol units) are required to increase the compound's solubility in water. To utilize 8-aryl-BODIPY derivatives as fluorescent molecular thermometers, non-specific responses to both viscosity and temperature should be separated (Vyšniauskas et al., 2015). BODIPY3 is one of the answers to this issue: it practically senses only a temperature variation (Vyšniauskas et al., 2021).

PEG-BODIPY

BODIPY3

2.4.4
Triarylborons

In a triarylboron molecule, dipyren-1-yl(2,4,6-triisopropylphenyl)borane (DPTB), a relatively large pyrene group can rotate in 2-methoxyethyl ether to result in temperature-dependent fluorescence properties (Feng et al., 2011). At a low temperature, the excited state of DPTB has a TICT character in which an electron moves from one of the pyrene groups to the electron-deficient boron atom with an empty p orbital. On the other hand, at a high temperature, the TICT character of DPTB is dismissed, and the local excited state emission (LE) character is reinforced.

Consequently, the maximum emission wavelength of DPTB in 2-methoxyethyl ether shifts linearly from 518 to 472 nm with increasing temperature from −50 to 100 °C.

1,1′-(6,6′-((2,4,6-Triisopropylphenyl)boranediyl)bis(pyrene-6,1-diyl))dipyrrolidine (TBBD) is an updated version of DPTB. The maximum emission wavelength of TBBD is bathochromically shifted by 50 nm compared to DPTB. The crosslinked poly(urea-formaldehyde) microcapsules containing TBBD in the mixture of 1,2,3,4-tetrahydronaphthalene and polystyrene were used for temperature imaging of fluids flowing in a T-shaped square tube (Feng et al., 2013). Tuning of a functional temperature range and color could be carried out with the replacement of substituent groups on a boron atom (Liu et al., 2020).

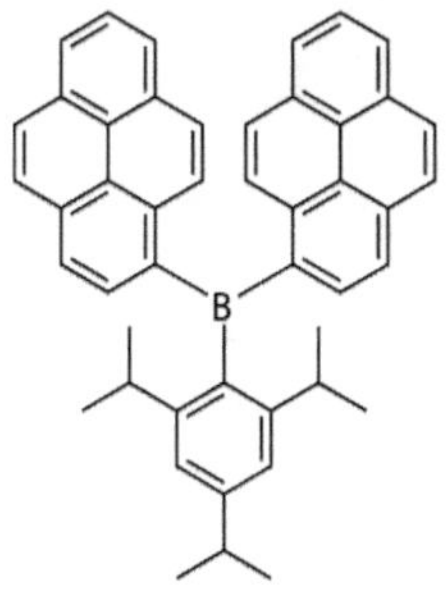

DPTB

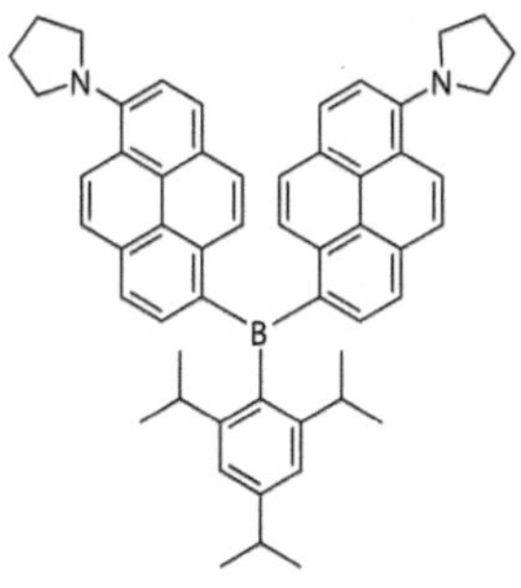

TBBD

2.4.5
Cyanines

Like the rotation of a substituent group described in the previous section, photoisomerization in the excited state can be a reason for the temperature-dependent fluorescence characteristics of a molecule. In 1976, the temperature-dependent fluorescence intensity of 3,3′-diethyl oxadicarbocyanine iodide (DODCI, Figure 2.12a) in ethanol was reported (Rullière, 1976). It was considered that the photoisomerization of DODCI proceeded via the first singlet excited state to a twisted state by thermal activation (Figure 2.12b). From the temperature-dependency of fluorescence quantum yield of DODCI (Figure 2.12c), the activation energy for the photoisomerization was estimated to be 4.8 ± 0.2 kcal M^{-1}. Though not fully confirmed, the photoisomerization of DODCI might be a transformation from the all-trans form to the mono-cis form. Temperature-dependent fluorescence efficiency of a cyanine dye Cy5 can be ascribed to the exact mechanism, i.e., photoisomerization (Widengren and Schwille, 2000).

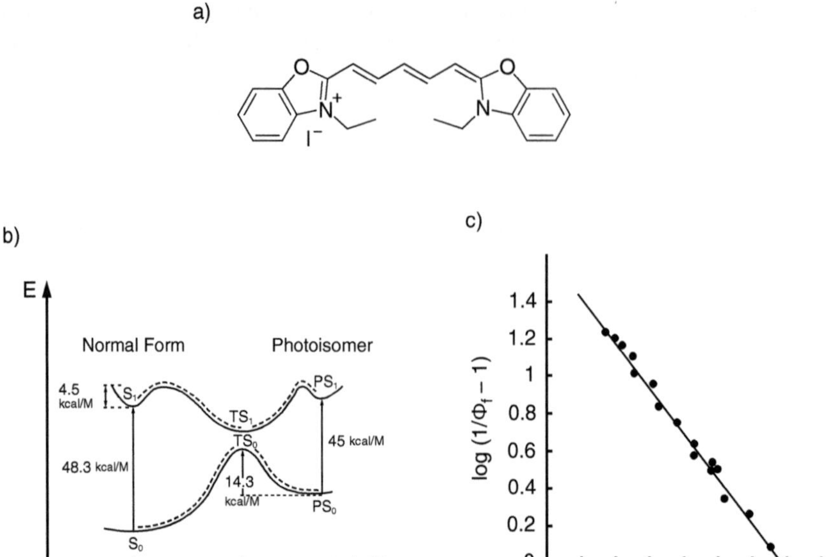

Figure 2.12 3,3′-Diethyl oxadicarbocyanine iodide (DODCI). (a) Chemical structure. (b) Energy diagram for photoisomerization. S$_0$: normal form ground state; S$_1$: normal form singlet excited state; PS$_0$: photoisomer ground state; PS$_1$: photoisomer singlet excited state; TS0: twisted ground state; TS$_1$: twisted singlet excited state. (c) Relationship between Φ_f and temperature in ethanol (Rullière (1976) *Chem. Phys. Lett.*, **43**, 303–308 / Elsevier).

2.5
Reactive Small Organic Molecules

Endo-1 is a fluorescent molecular thermometer based on the retro Diels–Alder reaction and photoinduced electron transfer (Figure 2.13a) (Wang et al., 2005). At a lower temperature, the adduct endo-1 dominates in the equilibrium shown in Figure 2.13a and emits strong fluorescence due to the pyrene structure. On the other hand, at a higher temperature, the retro Diels–Alder reaction proceeds to produce dyad-1, of which fluorescence is quenched by photoinduced electron transfer from the pyrene moiety to the maleimide moiety. Thus, the temperature-dependent fluorescence spectra are recorded with an endo-1 solution (Figure 2.13b). As the Diels–Alder reaction is reversible, the response of endo-1 to a temperature variation can, in principle, be reversible.

A 2H,4H-benzo[1,3]oxazine heterocycle, oxadine-1, has a coumarin chromophore in the structure (Scheme 2.1). In the mixture of acetonitrile and water (1:1, v/v), the

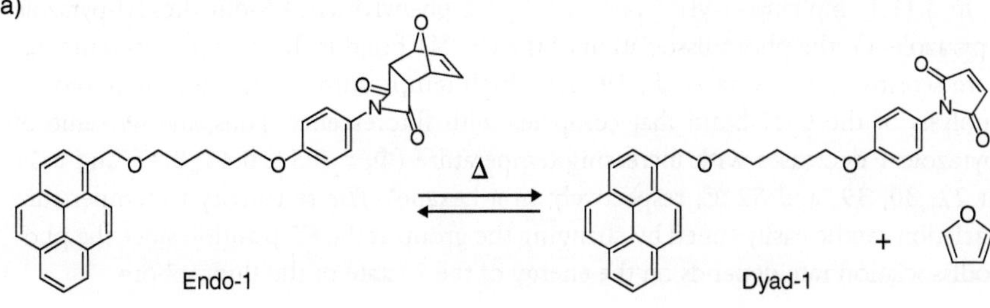

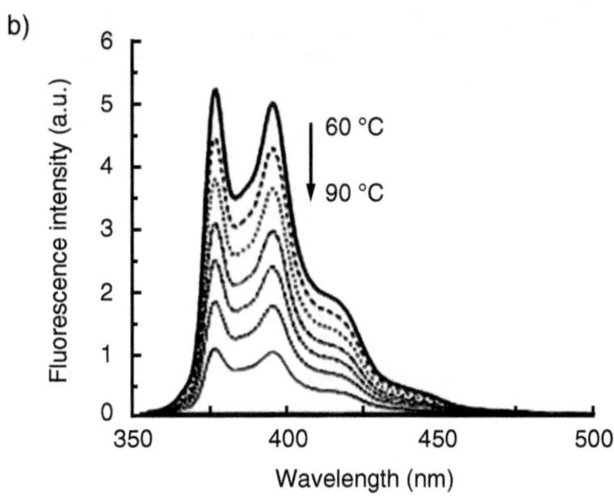

Figure 2.13 Diels–Alder adduct endo-1. (a) Reversible retro Diels–Alder reaction of endo-1. (b) Temperature-dependent fluorescence spectra of endo-1 (10 µmol L^{-1}) in DMF (λ_{ex}: 340 nm) (adapted from Wang et al. (2005) *Tetrahedron Lett.*, **46**, 4609–4612).

Scheme 2.1

structure of oxadine-1 depends on the temperature (Mazza et al., 2019). In a cold solution, oxadine-1 takes an open form in which absorption and emission wavelengths are relatively long due to the extension of coumarin's electronic conjugation over the indolium cation. On the other hand, in a hot solution, oxadine-1 favors a closed form with shorter absorption and emission wavelengths. So, oxadine-1 works as a ratiometric fluorescent molecular thermometer in the temperature range 8–54 °C. A hydrophilic water-soluble version, oxadine-2, which has ethylene glycol units, is also reported (Zheng et al., 2022).

In 3-(1,1′-biphenyl-4-yl)-1-phenyl-5-[(E)-2-phenylvinyl]-4,5-dihydro-1H-pyrazole (pyrazole-1), the photodissociation of the C5–N1 bond is the origin of its temperature sensitivity (de Silva et al., 1995). A high temperature accelerates the photohomolysis of the C–N bond that competes with fluorescence. Thus, the Φ_f value of pyrazole-1 decreases with increasing temperature (Φ_f = 0.52, 0.43, 0.37, and 0.31 at 22, 30, 39, and 52 °C, respectively, in n-hexane). The sensitivity to temperature variation can be easily tuned by changing the group at the C3 position since the photodissociation rate depends on the energy of the S_1 state of the fluorophore.

Oxadine-2

Pyrazole-1

2.6
Viscosity-Sensitive Small Organic Molecules

There are some situations in which a bulk solvent's temperature-dependent characteristics influence a solute molecule's fluorescence behavior. The Φ_f value of 2-(4-pyridyl)-5-(4-N,N-dimethylaminophenyl)-1,3-oxazole (oxazole-1) in ethanol was significantly changed by the increase in temperature, i.e., $\Phi_f \approx$ 1.0, 0.8, 0.5, and 0.4 at −160, −120, −60, and 20 °C, respectively (Khimich et al., 2003). This variation is due to the change in the viscosity of ethanol and the sensitivity of oxazole-1 to the environmental viscosity. The same principle can be found in a viscosity-sensitive BNAP in the mixture of glycerol and water (85:15, v/v) (Xia et al., 2016).

2-Naphthol showed stronger fluorescence at 426 nm, under the excitation at 365 nm, at a higher temperature in the hydrogels that were made from poly(vinyl alcohol), borax ($Na_2B_4O_5(OH)_4 \cdot 8H_2O$), and sodium hydroxide (Lee et al., 2004). In this hydrogel system, the pH value varies from 8.50 to 9.19 by the temperature variation from 30 to 70 °C due to the exothermic reversible didiol-borate complexation. A pH sensor, 2-naphthol, sensed this pH increase to fluoresce strongly at a higher temperature.

A difluoroboron β-diketonate derivative, JBF1 in chloroform shows a temperature-dependent maximum emission wavelength from 445 nm at 67 °C to 592 nm at −196 °C along with a color change from blue to yellow (Wang et al., 2020a). At a higher temperature, JBF1 exists as a monomer in chloroform and emits blue fluorescence. Meanwhile, at a lower temperature, especially below the melting point of chloroform (i.e., −64 °C), JBF1 takes a dimer form to emit yellow fluorescence. A related compound with a moderate sensitivity in the maximum emission wavelength was reported in a different publication (Wang et al., 2020b). Thermochromic fluorescent materials based on the dynamic aggregation/disaggregation of an organic molecule are exceptional.

Oxazole-1

BNAP

2-Naphthol

Poly(vinyl alcohol)

JBF1

2.7
Other Small Organic Molecules

Although detailed mechanisms of temperature dependency are unclear, some organic compounds show temperature-dependent fluorescence properties. Once photophysical studies clarify their functional mechanism, these molecules will be utilized for further applications, including intracellular thermometry.

N,N′-Bis(2,5-di-*tert*-butylphenyl)-3,4,9,10-perylenedicarboximide (BTBP) in methanol showed a slight blue shift in fluorescence spectrum with increasing temperature (Schrum et al., 1994). The variation rate in its maximum fluorescence wavelength was -0.045 nm $°C^{-1}$ between 20 and 60 °C. Although the sensitivity to a temperature variation looked too low to utilize BTBP in real applications, the possibility of fluorescence thermometry with it was discussed in the paper.

The absorption spectrum of fluorescein in a buffer solution (pH $=10$) slightly shifts towards a longer region with increasing temperature. Its maximum absorption wavelength is 489 and 492 nm at 14 and 52 °C, respectively. Thus, when excited at a longer wavelength (e.g., 514 nm), fluorescein emits stronger fluorescence at a higher temperature due to a larger molar extinction coefficient (Coppeta and Rogers, 1998). Fluorescein 27 (2-(2,7-dichloro-6-hydroxy-3-oxo-3H-xanthene-9-yl)benzoic acid) has two chlorine substituent groups on a fluorescein structure. Due to the chlorine groups, fluorescein 27 is more acidic than fluorescein, so fluorescein 27 shows a temperature-dependent absorption spectrum even in neutral water.

Consequently, the fluorescence intensity of fluorescein 27 in water at 84 °C is three times higher than that at 24 °C with the excitation at 532 nm (Sutton et al., 2008). This temperature-dependent fluorescence intensity of fluorescein 27 was well utilized for temperature mapping of a droplet impinging onto a heated wall (Dunand et al., 2012). Ratiometric systems were proposed with rhodamine B or Kiton Red 620 (sulforhodamine B, see Section 2.4.1) that shows the opposite behavior, i.e., weaker fluorescence at a higher temperature. Another research group also discussed a combination of fluorescein 27 and rhodamine B to accomplish accurate ratiometric thermometry (Estrada-Pérez et al., 2011). As a completely different principle, temperature-dependent fluorescence depolarization of fluorescein was proposed for temperature mapping of a fabricated aluminum heater on a glass substrate (Chen et al., 2016).

An open form of a diarylethene derivative 1o shows temperature-dependent fluorescence spectra with the excitation at 326 nm in toluene (Takeuchi et al., 2020). With the increase in temperature from 5 to 95 °C, fluorescence from a locally excited (LE) state at 380 nm decreased, while that from a charge transfer (CT) excited state at 580 nm increased. Activation energy from the LE state to the CT excited state was calculated to be 11.7 kJ mol^{-1}. A white-light-emitting triphenylamine derivative TPA-1 ((Z)-5-(4-(bis(4′-(diphenylamino)-[1,1′-biphenyl]-4-yl)amino)benzylidene)-3-ethyl-2-thioxothiazolidin-4-one) also shows temperature-dependent fluorescence intensity ratio at 428 and 564 nm in tetrahydrofuran in the temperature range −93–57 °C (Ajantha et al., 2021). It is considered that a CT state with a quinonoid configuration having a positive charge on a nitrogen atom in a triphenylamine moiety and a negative charge on an oxygen atom in the ethylrhodanine moiety could be competing with an LE state of TPA-1 in the excited state.

Commercially available fluorescent compounds BD140 and LD688 show temperature-dependent fluorescence properties in organic solvents based on *cis-trans* isomerization in the ground states (Scheme 2.2) (Shen et al., 2021). A computational study revealed that the *cis* forms of both compounds were more stable than the *trans* forms, and the increase in temperature shifted the equilibriums to the right side. Since both *cis* and *trans* forms have inherent absorption and fluorescence properties, BD140 and LD688 worked as fluorescent molecular thermometers.

Doubly boron and nitrogen-containing 2M′2BNM shows a ratiometric response to a temperature variation in chloroform (Shi et al., 2022). As temperature increases from −33 to 47 °C, a fluorescence peak at 450 nm drops while that at 570 nm is enhanced. This temperature-dependent spectral change could be associated with a thermally activated transition from part of the B_{sp^3} state to the B_{sp^2} state.

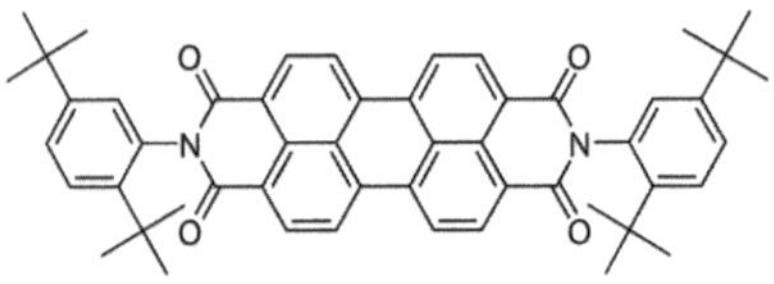

BTBP

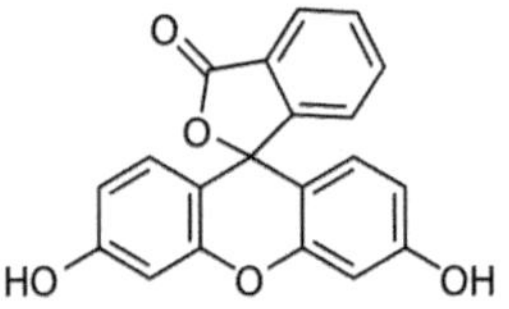

Fluorescein

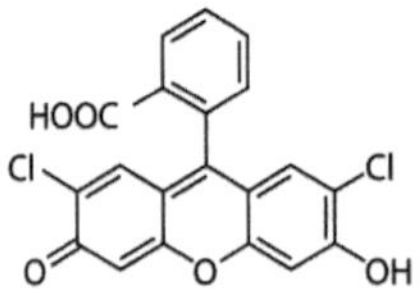

Fluorescein 27

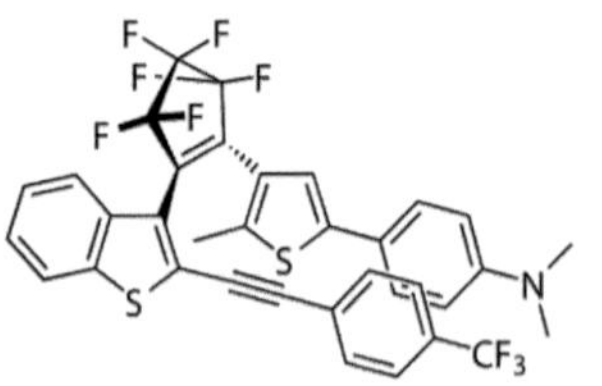

1o

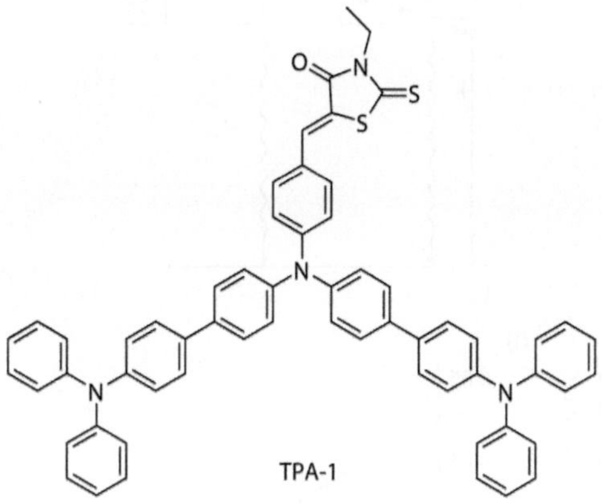
O
N
S
S
N
N
N

TPA-1

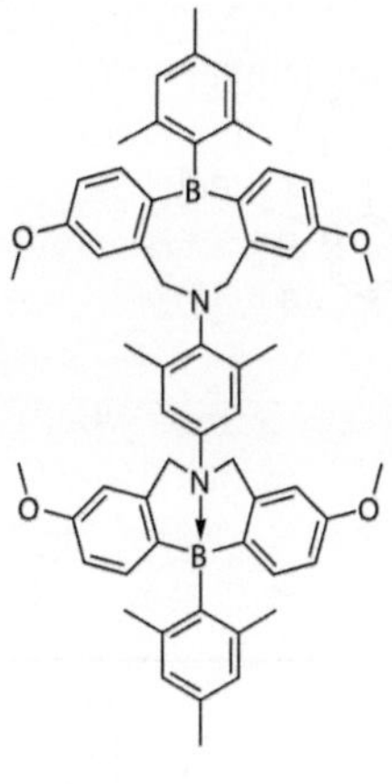
B
O
O
N
O
N
O
B

2M'2BNM

a)

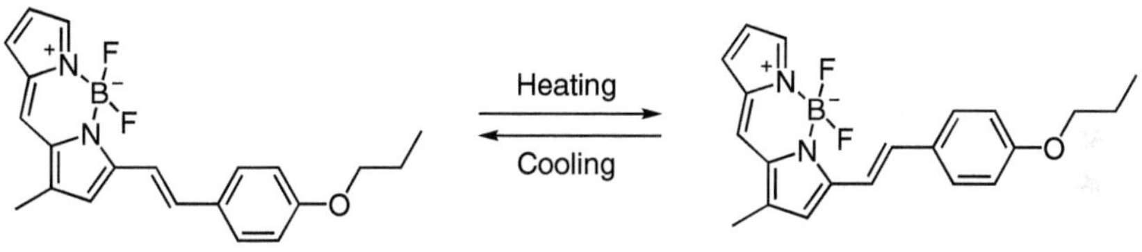
Heating
Cooling

BD140

b)

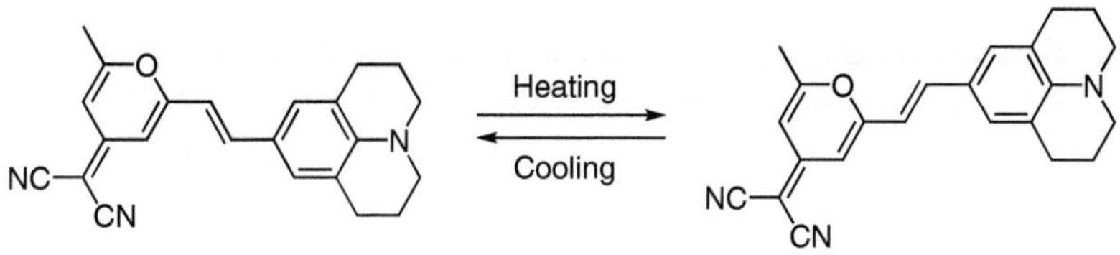
Heating
Cooling

LD688

Scheme 2.2

2.8 Organometallic Complexes

2.8.1 Ruthenium

A tris(2,2′-bipyridyl)ruthenium(II) ($Ru(bpy)_3{}^{2+}$) complex (Figure 2.14a) is a fluorescent molecular thermometer soluble in water (Van Houten and Watts, 1976). The emission of a $Ru(bpy)_3{}^{2+}$ complex in water originates from a triplet metal-to-ligand charge transfer (MLCT) state in which an electron has been transferred from Ru(II) to a 2,2′-bipyridyl moiety. Non-emissive metal-centered (MC) triplet states

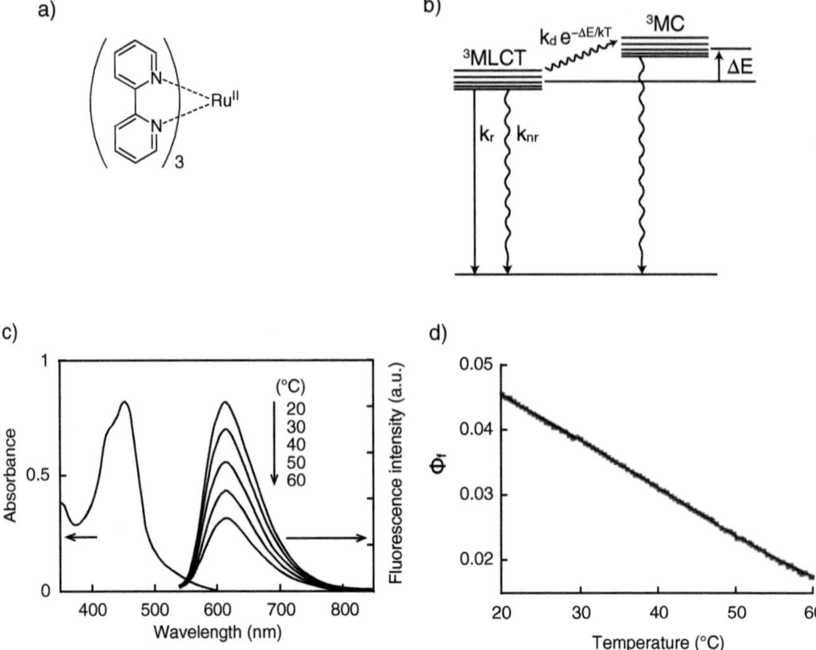

Figure 2.14 Tris(2,2′-bipyridyl)ruthenium(II) ($[Ru(bpy)_3]^{2+}$) complex. (a) Chemical structure. (b) Energy diagram. (c) Absorption and fluorescence spectra of $[Ru(bpy)_3]Cl_2$ solution in water (10 μmol L^{-1}). λ_{ex}: 470 nm. (d) Temperature-dependent Φ_f of $[Ru(bpy)_3]Cl_2$ in water. Panels (c,d) are adapted from Filevich and Etchenique (2006) *Anal. Chem.*, **78**, 7499–7503 / American Chemical Society.

(also called ligand field, LF, states) exist at energies slightly above the MLCT state ($\Delta E \sim 3600\ \text{cm}^{-1}$) (Figure 2.14b). Because the population of an excited Ru(bpy)$_3{}^{2+}$ complex in the MC state becomes feasible by thermal activation, the emission intensity (or the luminescence quantum yield) and the luminescence lifetime are temperature-dependent (Figure 2.14c, 2.14d). Its luminescence quantum yield in water changes from 0.049 to 0.0075 with increasing temperature from 5 to 90 °C. A more recent paper published temperature-dependent luminescence characteristics on a new Ru(II) complex Ru(bpy)$_2$L, in which one 2,2′-bipyridyl ligand is replaced by a 3,5-dicarboxy-2,2′-bipyridyl ligand (Fernando et al., 1996). Interestingly, the fluorescence efficiency of Ru(bpy)$_2$L in water was exceptionally increased with increasing temperature (i.e., $\Phi = 0.0036$, $\tau = 54$ ns at 280 K and $\Phi = 0.0053$, $\tau = 75$ ns at 360 K). This unusual temperature dependence could be attributed to thermally-induced changes occurring within the degree of solvent stabilization of the emitting MLCT state. The effects of ligands on the temperature-dependency of Ru(II) complexes were also systematically investigated (Lumpkin et al., 1990; Bustamante et al., 2018).

Ru(bpy)$_3{}^{2+}$ ion has been utilized in temperature mapping of small objects. An example is shown in Figure 2.15. A 60 cm polyethylene tube with an inside diameter of 0.86 mm for flow-injection analysis was coiled in a spiral, through which 1 mmol L^{-1} [Ru(bpy)$_3$]Cl$_2$ solution at 55 °C flowed from the center. The whole

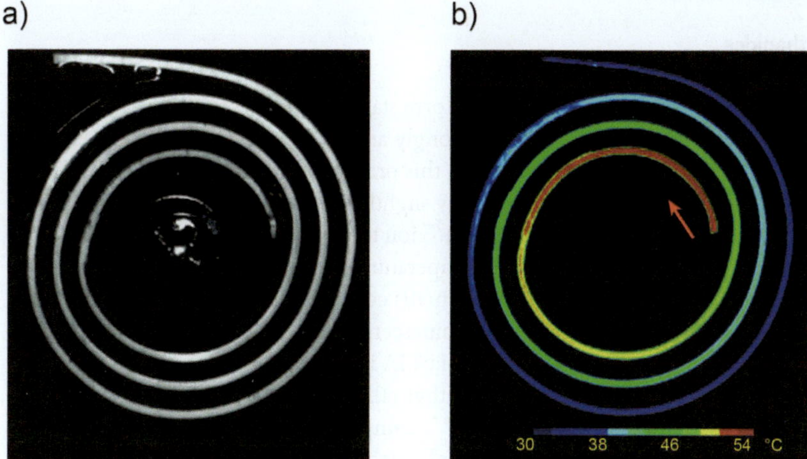

Figure 2.15 Temperature mapping of a flow injection analysis (FIA) reactor by Ru(bpy)$_3^{2+}$ complex. (a) Raw image of the polyethylene reactor with the length of 60 cm. (b) Temperature distribution when hot [Ru(bpy)$_3$]Cl$_2$ solution (1 mmol L^{-1}, 55 °C) flew as indicated by the arrow. Adapted from Filevich and Etchenique (2006) *Anal. Chem.*, **78**, 7499–7503 / American Chemical Society.

spiral was immersed in a cooling bath at 23 °C. By analyzing the emission intensity of [Ru(bpy)$_3$]Cl$_2$, the temperature map of the polyethylene tube was successfully obtained with a temperature resolution of 0.05 °C (Filevich and Etchenique, 2006). Using a microscope with a cooled charge-coupled device (CCD) camera, temperature mapping of a T-shaped microchannel could also be performed with a spatial resolution of 5 μm and a temperature resolution of 0.26 °C (Sato et al., 2003). Combined with microresolution particle image velocimetry, convective heat flux was calculated and compared to heat conduction. It was concluded that in microfluidic devices, the heat flux due to conduction was more significant than convection. In a different case, Ru(bpy)$_3^{2+}$ ion was utilized in three-dimensional temperature mapping of a thick polyacrylate film by two-photon microscopy (Van Keuren et al., 2004). A Ni:chrome wire (diameter: 78.7 μm) was embedded into polyacrylate as a heater, and location-dependent temperature rise by the heater was monitored via two-photon induced fluorescence intensity of a doped Ru(bpy)$_3^{2+}$ ion. From the temperature data obtained, the thermal conductivity of the polyacrylate was calculated to be approximately 0.26 Wm^{-1}K^{-1}.

2.8.2
Lanthanides

Lanthanide ions such as Eu^{3+} and Sm^{3+} form stable complexes with various chelating ligands. Many Eu^{3+} complexes emit strongly among them, and the ligand-centered triplet states are involved as antennas in this process (Crosby et al., 1961). If the triplet T_1 state of a ligand exists at an energy slightly above the emissive state of the Eu^{3+} ion, the energy can transfer from the Eu^{3+} ion to the T_1 state of the ligand with thermal assistance. For example, the temperature-dependent emission intensity of tris(thenoyltrifluoroacetonato)europium(III) complex (EuTTA, Figure 2.16a, 2.16b) is caused by the competition of the luminescence with the energy transfer from the Eu^{3+} ion to the ligand. The Φ_f value of EuTTA is approximately 0.40 at $-200\,°C$ but is 0.06 at $50\,°C$ in the mixture of diethyl ether, isopentane, and ethanol (5:5:2) (Figure 2.16c) (Bhaumik, 1964). Tb^{3+} complexes possess the same functional mechanism. More recent articles on [M⊂branched macrocyclic ligand]$^{3+}$ complexes (M = Eu or Tb) in water (Sabbatini et al., 1994), solid $Eu(thd)_3$ (Berry et al., 1996), nanosized vesicles with an embedded Tb^{3+} complex Tb-1 (Bhuyan and Koenig, 2012), a blend film of $Eu(dbm)_3phen$ and poly(methyl methacrylate) (PMMA) (Lopez et al., 2012), $Eu(TTA)_3Phen$ dispersed in poly(vinyl alcohol) (Shahi et al., 2015), a Eu^{3+} coordination polymer $[Eu(hfa)_3BDPC]_n$ (Kitagawa et al., 2021), Eu-DT in the hybrid nanoparticles of 2-bis(trimethoxysilyl)decane (BTD) and PMMA (Peng et al., 2010a), and Tb^{3+} complexes of thiacalix[4]arenes on polystyrene sulfonate (Zairov et al., 2020) are also available. The case using Eu-DT was further tuned to a ratiometric version by incorporating a reference naphthalimide fluorophore (due to

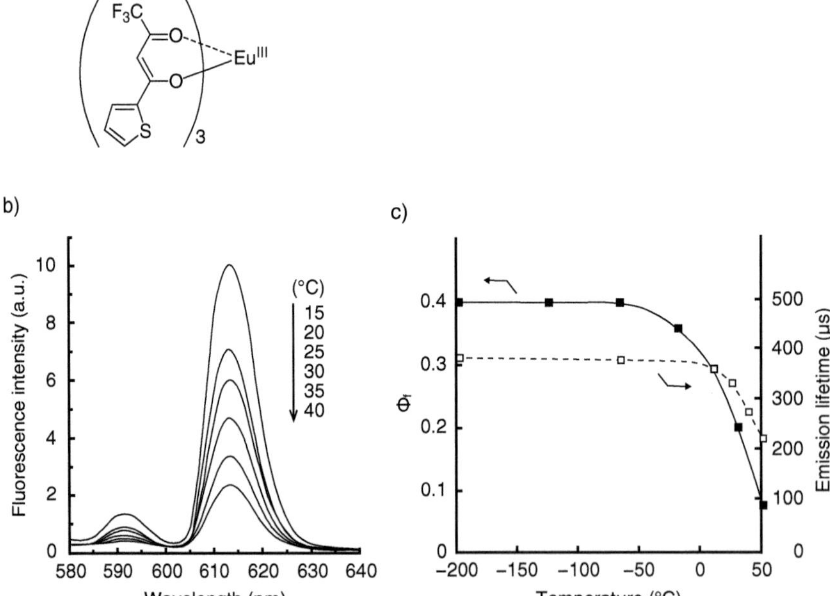

Figure 2.16 Tris(thenoyltrifluoroacetonato)europium(III) complex (EuTTA). (a) Chemical structure. (b) Fluorescence spectra of EuTTA embedded in L-α-lecithin liposomes in a buffered saline solution (pH 7.4) (λ_{ex}: 372 nm) (adapted from Zohar et al. (1998) *Biophys. J.*, **74**, 82–89). (c) Temperature-dependent Φ_f (closed) and emission lifetime (open) in the mixture of diethyl ether, isopentane, and ethanol (5:5:2, v/v/v) (adapted from Bhaumik (1964) *J. Chem. Phys.*, **40**, 3711–3715).

N-octyl-4-(3-aminopropyltrimethoxysilane)-1,8-naphthalimide (OASN)) into the nanoparticles (Peng et al., 2010b).

EuTTA is not stable in water and oxygen-permeable silicone matrices and, therefore, it was applied for temperature mapping as incorporated into suitable polymer films (Basu and Vasantharajan, 2008). A polymer film of perdeutero-poly(methyl methacrylate) doped with EuTTA could continuously monitor the temperature distribution of an integrated circuit with a spatial resolution of 15 μm (Kolodner and Tyson, 1982). Another example of the application is fluorescent temperature imaging of a microheater embedded in a microelectromechanical system (MEMS) (Van Keuren et al., 2005). Figure 2.17a indicates a temperature profile of the microheater with an applied power of 0.5 mW. In this experiment, EuTTA was mixed with PMMA in a chloroform solution and cast onto the heater and substrate. In the temperature map, the pixel intensity is proportional to temperature in the order of 0.1 °C, and the temperature on the heater surface was roughly 50 °C. Figure 2.17b shows the dependence of the surface temperature on the voltage drop across the heater.

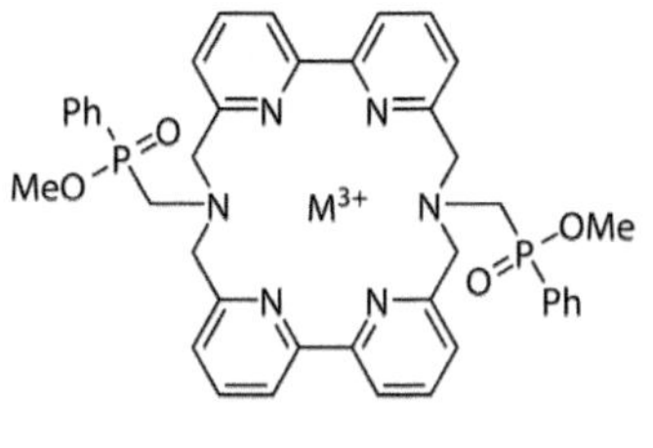

M=branched macrocyclic ligand
(M = Eu or Tb)

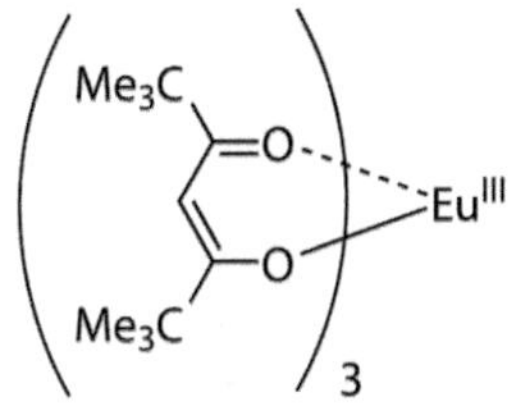

Eu(thd)$_3$

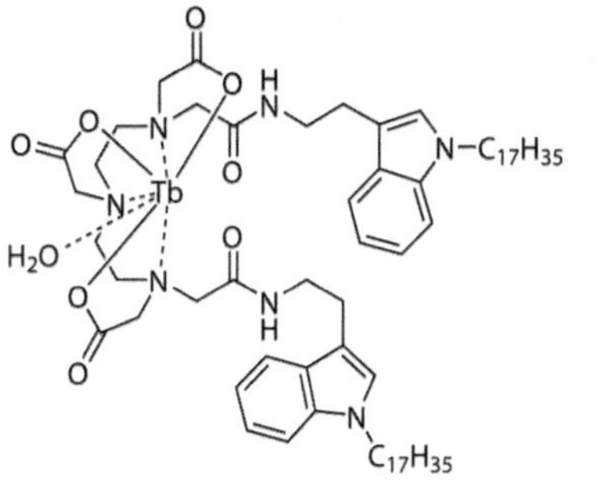

Tb-1

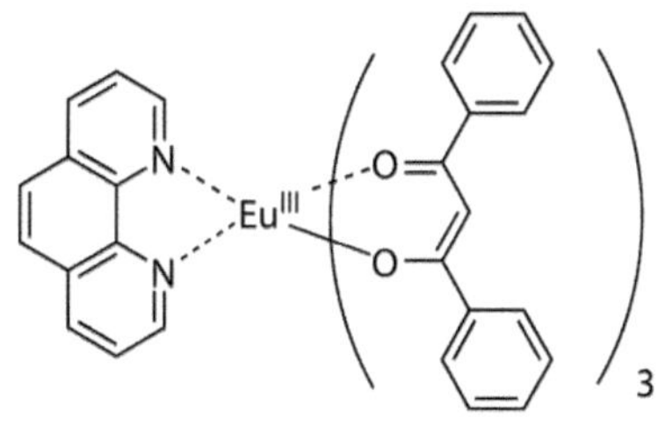

Eu(dbm)$_3$phen

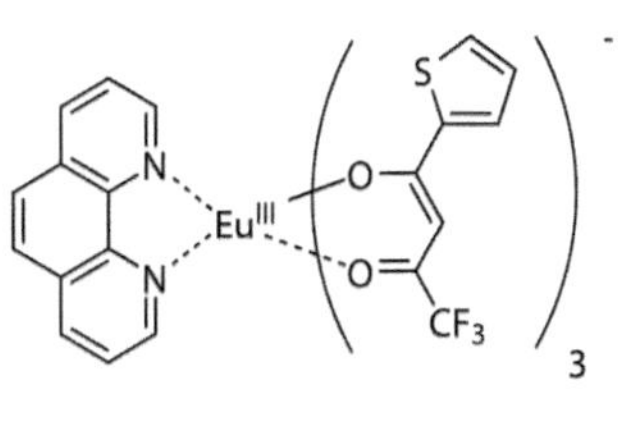

Eu(TTA)$_3$Phen

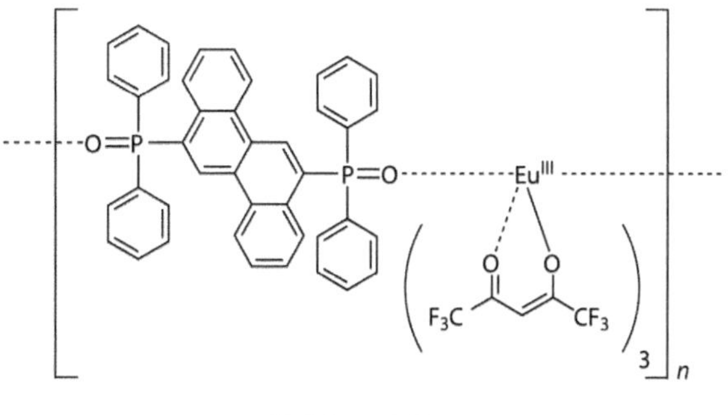

[Eu(hfa)$_3$BDPC]$_n$

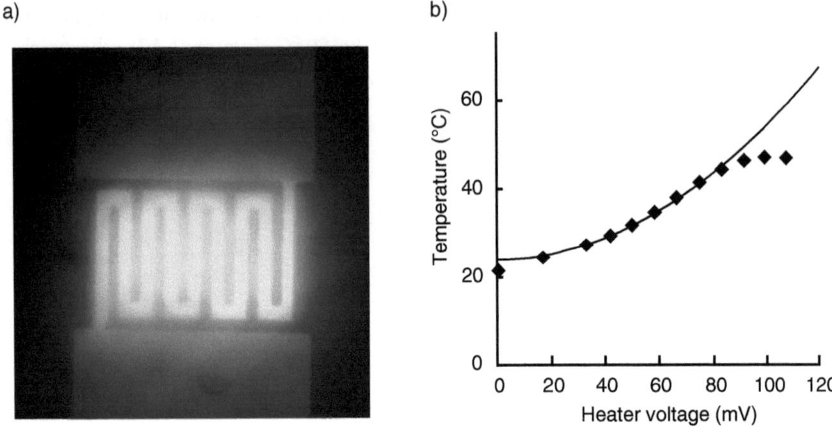

Figure 2.17 Temperature measurement of a microheater by EuTTA in poly(methyl methacrylate). (a) The temperature profile of the microheater, in which pixel intensity is proportional to temperature. (b) Relationship between the surface temperature and the applied voltage across the microheater. Adapted from Van Keuren et al. (2005) *Sens. Mater.*, **17**, 1–6 / Myu Scientific Publishing.

2.8.3
Porphyrins

2,3,7,8,12,13,17,18-Octaethyl-21H,23H-porphyrin platinum(II) (PtOEP) shows temperature-dependent fluorescence spectra in a polystyrene film with the excitation at 380 nm. The fluorescence intensity ratio of PtOEP at 540 nm and 650 nm increases from 0.16 to 0.58 with increasing temperature from 17 to 47 °C (Lupton, 2002). A later study clarified the fluorescence properties of a different metal complex, 5,10,15,20-tetraphenyl-21H,23H-porphyrin copper(II) (CuTPP) (Fujino et al., 2009). CuTPP in polystyrene fiber revealed three fluorescence peaks at 653, 720, and 780 nm. With increasing temperature, the fluorescence intensity at 653 nm was almost constant while that at 780 nm gradually decreased. Thus, the fluorescence intensity ratio at 780 nm and 653 nm decreased with increasing temperature (i.e., 0.62 at 25 °C to 0.39 at 70 °C). The temperature-dependent fluorescence lifetime of CuTPP at 780 nm (i.e., 11 μs at 75 K and 0.89 μs at 300 K) revealed that the activation energy for some non-radiative transition process was 0.08 meV. In benzene, metal-free tetraphenylporphyrin (TPP) and its zinc complex (ZnTPP) also show a temperature-dependent fluorescence intensity ratio at two different wavelengths (651 and 719 nm for TPP; 598 and 646 nm for ZnTPP) (Kolesnikov et al., 2019). Tautomerization of TPP in the excited state (Uttamlal and Holmes-Smith, 2008) was proposed as a mechanism of temperature dependency. In the same way, water-soluble porphyrin derivatives, i.e., (*meso*-mono(4-pyridyl)-triphenylporphyrinato)phosphorus(V) bromide (MPyPP(OH)$_2$) and *meso*-tetrasulfonatophenylporphyrinate tetrasodium salt (TSPH$_2$) were developed as fluorescent molecular thermometers (Kolesnikov et al., 2021).

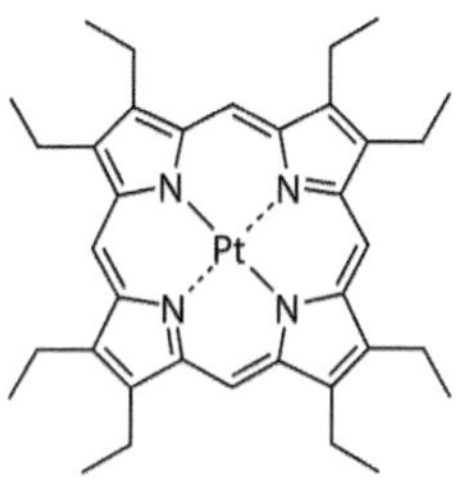

PtOEP

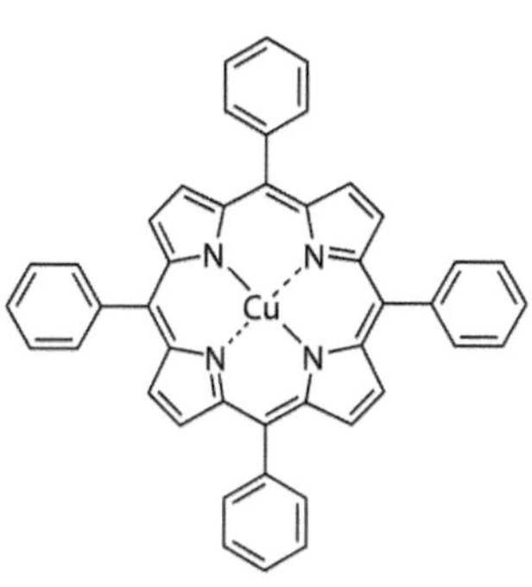

CuTPP

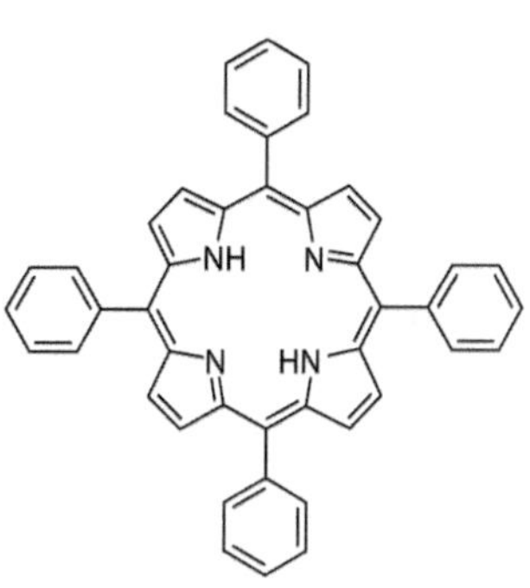

TPP

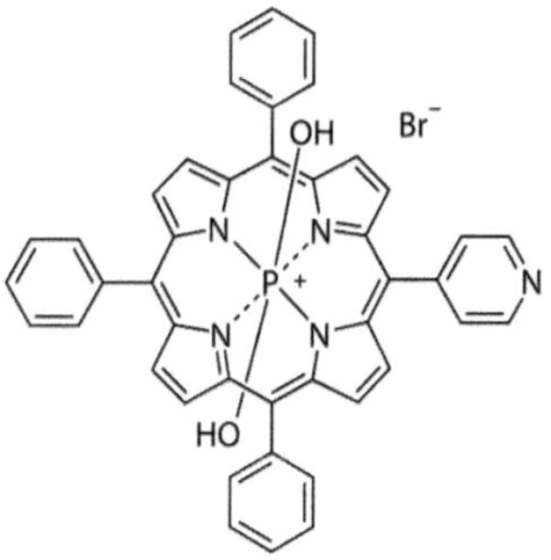

MPyPP(OH)$_2$

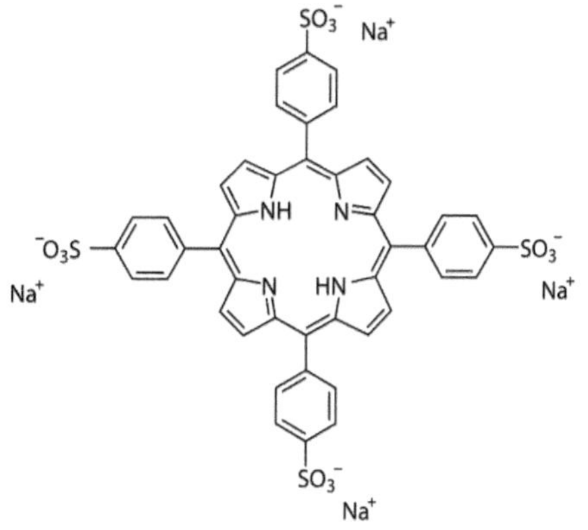

TSPH$_2$

2.8.4
Others

A tetraazamacrocycle with a naphthalene fluorophore (azaNap, Figure 2.18a) forms a complex with a Ni^{2+} ion. Interestingly, the Ni^{2+} complex of azaNap gives an equilibrium mixture of high- and low-spin states in acetonitrile,

$$\left[Ni(azaNap)S_2\right]^{2+} \rightleftarrows \left[Ni(azaNap)\right]^{2+} + 2S \tag{2.1}$$

    high-spin octahedral   low-spin square-planer

where S represents acetonitrile (solvent). In this equilibrium, the conversion from the high-spin state to the low-spin state is endothermic; hence, the proportion of the high-spin state becomes lower at high temperatures. The fluorescence of naphthalene is quenched by the Ni^{2+}-catching macrocyclic moiety in both states, but the quenching efficiencies are not identical (the quenching mechanism has been suggested to be an energy transfer). So, the fluorescence intensity of the azaNap-Ni^{2+} complex becomes temperature-dependent (Engeser et al., 1999). Figure 2.18b shows the fluorescence spectra of the azaNap-Ni^{2+} complex at different temperatures and the relationship between its fluorescence intensity and temperature. This fluorescent molecular thermometer revealed an interesting behavior: the fluorescence intensity increased at higher temperatures, whereas most fluorescent molecular thermometers displayed the opposite behavior. Unfortunately, the ClO_4^- salt of azaNap-Ni^{2+} was not soluble in water even at a concentration as low as 100 nmol L^{-1} owing to the hydrophobicity of naphthalene moiety.

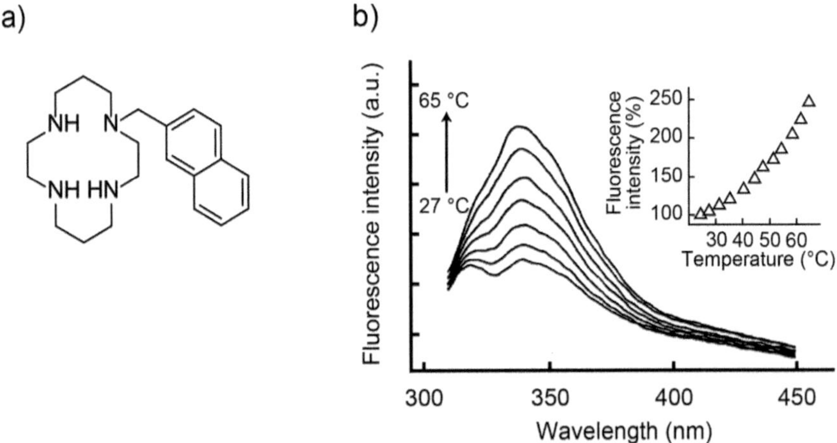

Figure 2.18 Nickel(II) tetraazamacrocyclic complex, $[Ni(azaNap)]^{2+}$. (a) Chemical structure of azaNap. (b) Temperature-dependent fluorescence spectra of $[Ni(azaNap)](ClO_4)_2$ (10 µmol L^{-1}) in acetonitrile and the corresponding relationship between fluorescence intensity and temperature (inset). Engeser et al. (1999) *Chem. Commun.*, 1191–1192 / Royal Society of chemistry.

A Pt(II) complex-based ionic liquid, [C₂mim][Pt(CN)₂(ptpy)] (C₂mim: 1-ethyl-3-methylimidazolium ion, Hptpy: 2-(p-tolyl)-pyridine) shows a fluorescence color change with a temperature variation from -196 to $25\,°C$ (Ogawa et al., 2015). With increasing temperature, the emission from the monomeric species of [C₂mim][Pt(CN)₂(ptpy)] at 530 nm decreases, and that from the aggregated species at 570 nm becomes predominant. In [C₂mim][Pt(CN)₂(ptpy)], the transition to the triplet metal–metal-to-ligand charge transfer (^{3}MMLCT) state of the dimerized Pt(II) complex competes with the monomeric emission. In a later paper, the same research group reported other ionic liquids based on cyclometalated Pt(II) complexes (i.e., [C₄mim][Pt(CN)₂(mpi)] and [C₄mim][Pt(CN)₂(bzq)], C4mim: 1-methyl-3-butyl-imidazolium ion; H₂mpi: 1-methyl-3-phenyl-1H-imidazolium; Hbzq: benzo[h]quinoline) that exhibited thermochromic behavior in different color regions (Ogawa et al., 2018).

The chromium(III)-based complex, [Cr(ddpd)₂]$^{3+}$, shows dual emission (^{2}E emission band at 775 nm and ^{2}T₁ emission band at 738 nm) in the near-infrared region. Because the energy difference of these two emissive states is sufficiently small (i.e., 7.7 kJ mol^{-1}) for thermal activation, a temperature-dependent ratiometric emission signal (i.e., the ^{2}E emission diminishes while the ^{2}T₁ emission increases) is observed with increasing temperature from 5 to $85\,°C$ in aqueous solution (Otto et al., 2017).

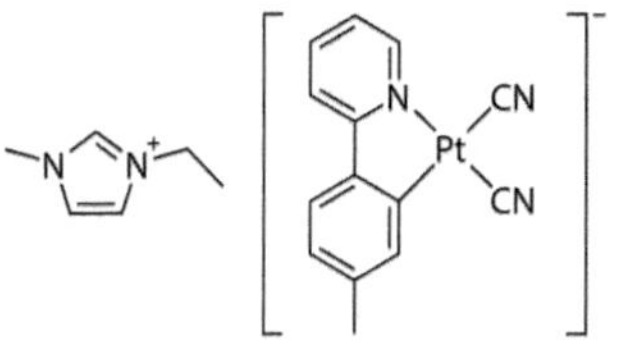

[C₂mim][Pt(CN)₂(ptpy)]

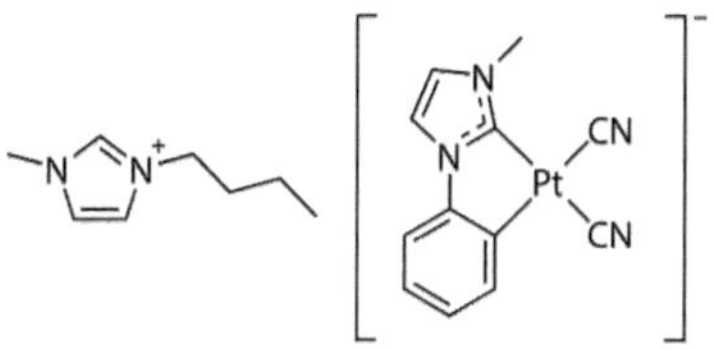

[C₄mim][Pt(CN)₂(mpi)]

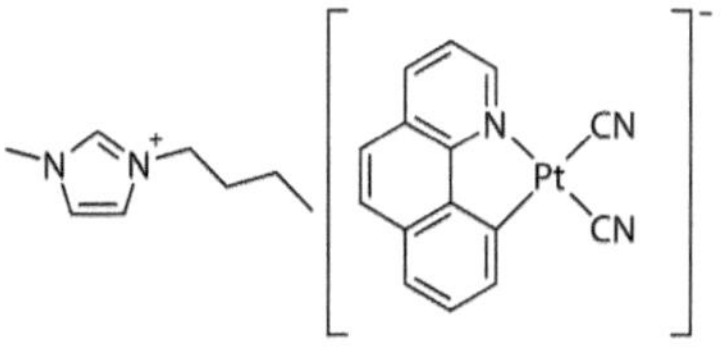

[C₄mim][Pt(CN)₂(bzq)]

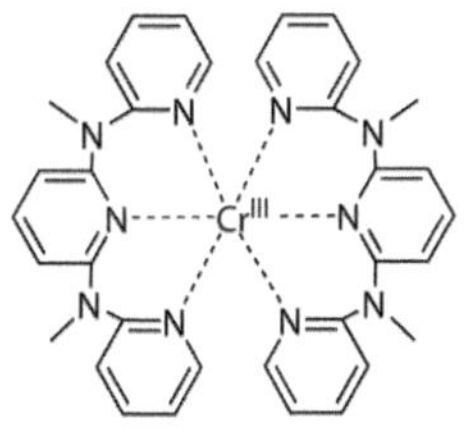

[Cr(ddpd)₂]$^{3+}$

2.9
Excimers and Exciplexes

The fluorescent molecular thermometers presented in the former sections are mainly based on the competition of fluorescence with another non-emissive process within a molecule. In this section, I will describe fluorescent molecular thermometers in which two different fluorescent states showing different florescence properties equilibrate.

An excimer (excited-state dimer) forms from the same fluorophores as the following equilibrium (Eq. (2.2))

$$M* + M \rightleftarrows E* \tag{2.2}$$

where M*, M, and E* represent the excited monomer, another in the ground state, and the excimer, respectively. Emission from the excimer is generally broadened at a longer wavelength than monomer emission. Because the complexation of two fluorophores is required for excimer emission, an intramolecular excimer, in which a linker connects two fluorophores, needs less concentration to obtain excimer emission.

1,3-Bis-(1'-pyrenyl)-propane (PYPYP) owns two pyrenyl units connected by a propyl linker (Figure 2.19a). Pyrene is the most well-known fluorophore to show excimer emission, and the propyl linker isolates two pyrenyl units at their ground state. Figure 2.19b displays normalized fluorescence spectra of PYPYP in hexadecane (Gossage and Melton, 1987). Above 110 °C, the ratio of excimer emission to monomer emission decreases with increasing temperature (Figure 2.19c) because the excimer formation in Eq. (2.2) is an exothermic process. Below 100 °C, the rate of excimer formation is not sufficiently fast for equilibrium to be obtained within the fluorescence lifetime of the monomer. It is somewhat dependent on the viscosity of hexadecane. The linear relationship above 110 °C in Figure 2.19c corresponds to the accuracy of 0.7 °C in temperature determination with a detection error assumed to be 1%. Temperature-dependent monomer/excimer emission of PYPYP was also observed in an ionic liquid, 1-butyl-1-methylpyrrolidinium bis(trifluoromethylsulfonyl)imide ([C$_4$mpy] [Tf$_2$N]) (Baker et al., 2003). The emission intensity ratio of a pyrene excimer at 476 nm to a pyrene monomer at 376 nm in [C$_4$mpy][Tf$_2$N] increased from 0.1 to 1.1 with increasing temperature from 25 to 140 °C, and this response to temperature was completely reversible. The viscosity of [C$_4$mpy][Tf$_2$N] significantly reduces at a higher temperature, facilitating pyrene excimer formation.

N-(1-Pyrenylmethyl)-1-pyreneacetamide (PMPAA) and N-(1-pyrenylmethyl)-1-pyrenebutanamide (PMPBA) are analogs of PYPYP and radiate temperature-dependent monomer/excimer emission in organic solvents (Lou et al., 1997). The emission intensity ratio of a pyrene excimer at 421 nm to a pyrene monomer at 376 nm of PMPAA in dodecane decreases from 4.2 to 1.7 with increasing temperature from 20 to 100 °C, while that of an excimer at 448 nm to a monomer at 396 nm of PMPBA in dodecane changes from 0.79 to 0.26 with the same temperature variation. A kinetic analysis using the natural logarithm of emission intensity ratio and the reciprocal number of temperature determined the excimer stabilization energies of PMPAA and PMPBA to be 10.4 kJ mol^{-1} and 13.1 kJ mol^{-1}, respectively. The precisions of temperature measurement using PMPAA and PMPBA were evaluated to be 0.7–0.8 °C. A more recent case of temperature-dependent monomer/excimer emission is found with a perylene derivative DEH-PDI (N,N'-di(2-ethylhexyl)-3,4,9,10-perylenetetracarboxylic diimide) in methylcyclohexane in the temperature range 10–70 °C (Liang et al., 2019).

The above cases used organic solvents to produce temperature-dependent monomer/excimer emission of fluorescent molecular thermometers. In contrast, a

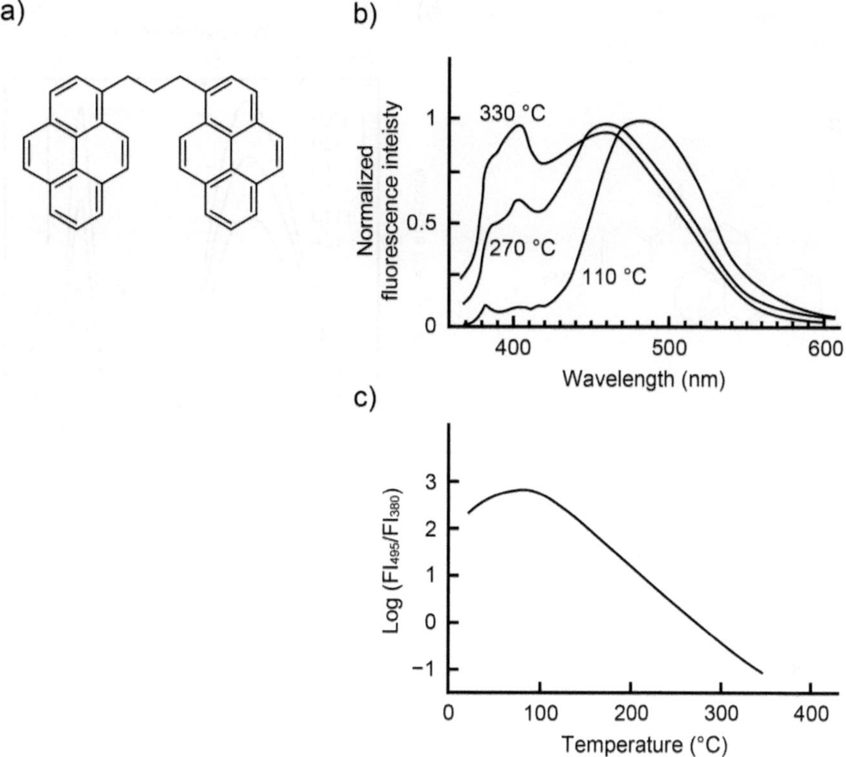

Figure 2.19 1,3-Bis-(1′-pyrenyl)-propane (PYPYP). (a) Chemical structure. (b) Normalized fluorescence spectra in hexadecane ([PYPYP] = 38 µmol L^{-1}, λ_{ex}: 345 nm). (c) Temperature-dependent fluorescence intensity ratio at 495 and 380 nm (Fl$_{495}$/Fl$_{380}$) in hexadecane. Reproduced with permission of Gossage and Melton (1987) *Appl. Opt.*, **26**, 2256–2259 / The Optical Society.

tripodal polyamine receptor containing three naphthalene fluorophores **L** can be dissolved in water. The emission intensity ratio of a naphthalene excimer at 399 nm to a naphthalene monomer at 324 nm is temperature-dependent (1.3 at 5 °C and 0.34 at 50 °C) at pH = 1, in which almost all the amino moieties are protonated (Albelda et al., 2003). When pH exceeds 6.5, excimer and monomer emissions dramatically weaken, probably due to the photoinduced electron transfer from free nitrogen atoms to naphthalene fluorophores. Thus, the function of **L** as a fluorescent molecular thermometer is recognizable only under acidic environments.

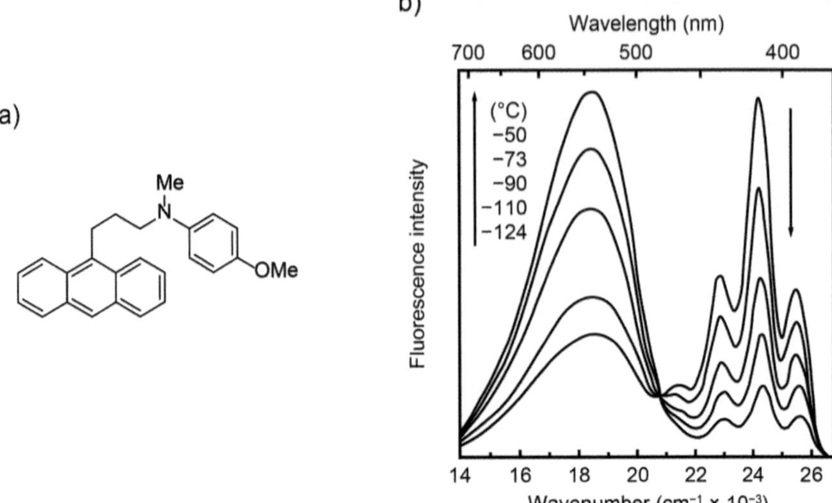

Figure 2.20 1-(*N-p*-Anisyl-*N*-methyl)-amino-3-(9-anthryl)-propane (AMAAP). (a) Chemical structure. (b) Fluorescence spectra in the degassed mixture of methylcyclohexane and isopentane (3:1) (250 µmol L^{-1}, λ$_{ex}$: 360 nm). Pragst et al. (1977) *Chem. Phys. Lett.*, **48**, 36–39 / Elsevier.

A complex of a fluorophore and another structure in the excited state is called an exciplex (excited-state complex). Intramolecular exciplex also contributes to temperature-dependent fluorescence characteristics. For example, 1-(N-p-anisyl-N-methyl)-amino-3-(9-anthryl)-propane (AMAAP) shows temperature-dependent fluorescence spectra (Figure 2.20) (Pragst et al., 1977). The relative emission intensity from the exciplex (the intramolecular complex of anthracene and anisidine moieties, 540 nm) to that from anthracene (413 nm) increases with increasing temperature. From a relationship between the quantum yield ratio of exciplex

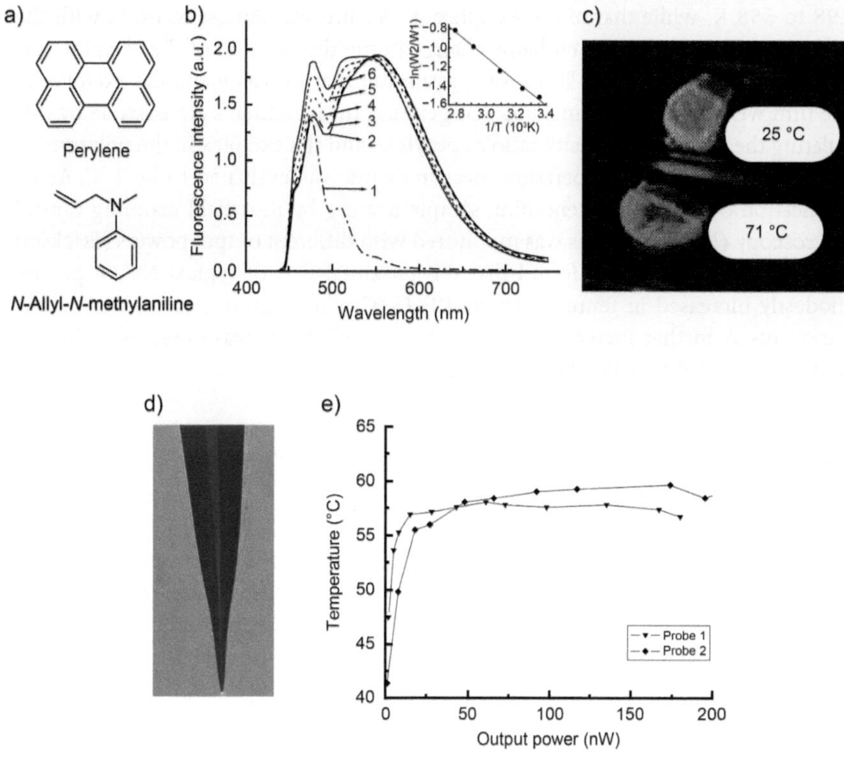

Figure 2.21 Exciplex of perylene and *N*-methylaniline encapsulated in a polystyrene film. (a) Chemical structure of perylene and a monomer *N*-allyl-*N*-methylaniline. (b) Temperature-dependent emission spectra of a polystyrene film composed of 0.017 wt% of perylene and 18.96 wt% of *N*-allyl-*N*-methylaniline (λ_{ex}: 386 nm). 1: The emission spectrum of a film of perylene alone at 298 K; 2: 298 K; 3: 304 K; 4: 323 K; 5: 343 K; 6: 358 K; W2/W1: emission ratio of the exciplex to monomer.). (c) Fluorescence image at two different temperatures. Panels (b,c) are adapted from Chandrasekharan et al. (2001) *J. Am. Chem. Soc.*, **123**, 9898–9899 / American Chemical Society. (d,e) Application for temperature measurement of aluminum-coated fiber optic probes in near-field scanning optical microscopy. (d) Phase contrast image of a probe. (e) Temperature as the function of output power. Panels (d,e) are adapted from Erickson et al. (2005) *Appl. Phys. Lett.*, **87**, 201102 / AIP Publishing.

emission to anthracene emission of AMAAP and the reciprocal of temperature, the activation energy for the exciplex formation was evaluated to be 2.9 kcal mol^{-1}.

In the same way as intramolecular excimers and exciplexes, intermolecular excimers and exciplexes are adopted as a basis of fluorescent molecular thermometers. However, such a thermometer is not unimolecular. Compared with intramolecular types, an impractically high concentration of molecules is required to form an intermolecular excimer or exciplex (Murray and Melton, 1985). Nevertheless, a polystyrene film, in which perylene and N-allyl-N-methylaniline (NA) were encapsulated and anchored, respectively, to allow the formation of an intermolecular exciplex functioned as a fluorescence thermometer (Figure 2.21a–2.21c) (Chandrasekharan and Kelly, 2001). As displayed in Figure 2.21b, the emission intensity due to perylene at 475 nm increased with the temperature increase from

298 to 358 K, while that of the exciplex at 551 nm was almost constant with the temperature variation. The enthalpy change in the dissociation of the exciplex was calculated to be 2.6 ± 0.2 kcal mol^{-1}, assuming that perylene and the exciplex in the film were in equilibrium. Thus, temperature measurements are possible by calculating the emission intensity ratio of perylene and the exciplex in the polystyrene film. The accuracy of temperature measurements was evaluated to be 1 °C. As an application of this polystyrene film, sample heating by near-field scanning optical microscopy (NSOM) probes was monitored with different output powers (Erickson and Dunn, 2005). It was found that the proximal end of typical NSOM probes modestly increased in temperature to 40–45 °C when output powers were a few nanowatts. A further increase in temperature to 55–65 °C was observed at higher output powers of 50 nW or greater (Figure 2.21d, 2.21e).

2.10
Host–guest Interactions

The fluorescence intensity of NapCD, a β-cyclodextrin (CD) derivative bearing a naphthalene fluorophore, decreases with increasing temperature (Figure 2.22) (Nakashima et al., 2001). In a HEPES (N-(2-hydroxyethyl)piperazine-N′-ethanesulfonic acid) buffer (pH = 7), its intensity at 380 nm at 80 °C is approximately 18% of that at 10 °C. The mechanism of this temperature sensitivity was unclear. However, it was considered to relate to the change in the relative position of the naphthalene moiety to the β-CD cavity. Experimental data suggested that the naphthalene moiety was always located outside of the β-CD cavity in the range of 10–80 °C, and the distance between the naphthalene moiety and the β-CD cavity

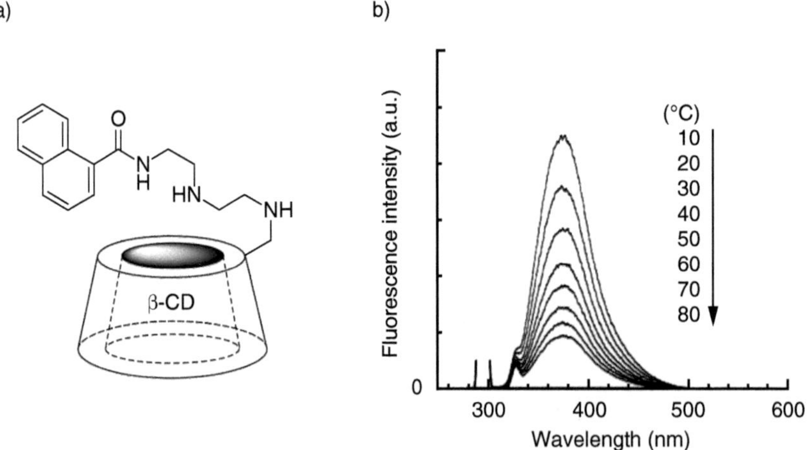

Figure 2.22 Naphthalene-linked β-cyclodextrin (NapCD). (a) Chemical structure. (b) Fluorescence spectra in a HEPES buffer (pH 7) (12.5 µmol L^{-1}, λ_{ex}: 295 nm). Adapted from Nakashima et al. (2001) *J. Chem. Soc., Perkin Trans. 2*, 2096–2103.

increased with the increase in temperature. In general, the emission efficiency of the fluorophore depends on the local environment.

Although relatively high concentrations are required for host and guest molecules, an intermolecular host–guest complex is also a candidate for molecular-based fluorescent thermometers. For instance, the mixture of 4-(N,N-dimethylamino)benzonitrile (DMABN) and β-CD shows temperature-dependent fluorescence spectra in water (Figure 2.23) (Figueroa et al., 1998). In water, DMABN can emit fluorescence from two different excited states; the Frank–Condon excited state and the twisted internal charge-transfer state (El Baraka et al., 1994). In the mixture of 10 μmol L^{-1} DMABN and 3 mmol L^{-1} β-CD in water, the fluorescence of DMABN from the Frank–Condon state at 350 nm was predominant at a low temperature. In contrast, fluorescence from the twisted internal charge-transfer state at 530 nm was relatively increased at a high temperature (Figure 2.23b). The mechanism of fluorescence responses of the mixture of DMABN and β-CD was not fully understood, but it was considered that the equilibrium (DMABN + β-CD $\leftrightarrows$ DMABN-β-CD complex) shifted to the left at a higher temperature and a more significant fraction of DMABN was transferred to the outer aqueous phase where the solvents quenched DMABN. The temperature-dependent fluorescence spectra of the mixture of DMABN and β-CD were equivalent to the temperature resolution of 2.5 K.

A related case is the mixture of 1-bromonaphthalene (1-BrNp), a derivative of β-CD, and selected alcohol. In the original paper, temperature-dependent phosphorescence of 1-BrNp was observed in the mixture of 1-BrNp, mono glucosyl-β-CD, and cyclohexanol (Thomson and Maynes, 2001). Later, the temperature-dependent phosphorescence lifetime of 1-BrNp in the mixture of 1-BrNp, maltosyl-β-CD (Mβ-CD), and alcohols (not specified) was applied for temperature mapping of a water droplet (Huang and Hu, 2007). As displayed in Figure 2.23c, the phosphorescence lifetime of 1-BrNp in the mixture shortens from 3.4 to 1.5 ms with increasing temperature from 22 to 36 °C. Temperature mapping of a water droplet containing the mixture was conducted by calculating the phosphorescence lifetime of 1-BrNp by comparing two phosphorescence images acquired at two different timings (Figure 2.23d, 2.23e). The same research group published serial papers including applications by the mixture of 1-BrNp, a derivative of CD, and alcohol; simultaneous temperature and velocity measurement in a pulsed jet flow (Hu and Koochesfahani, 2003), temperature measurement in the wake of a heated cylinder (Hu et al., 2006), and simultaneous measurements of transient temperature, droplet size, and flying velocity of in-flight droplets (Li et al., 2015). A combination of a different phosphorophore and CD (i.e., 6-bromo-2-naphthol and α-cyclodextrin) was also reported from another laboratory (Brewster et al., 2001).

1-BrNp   6-Bromo-2-naphthol

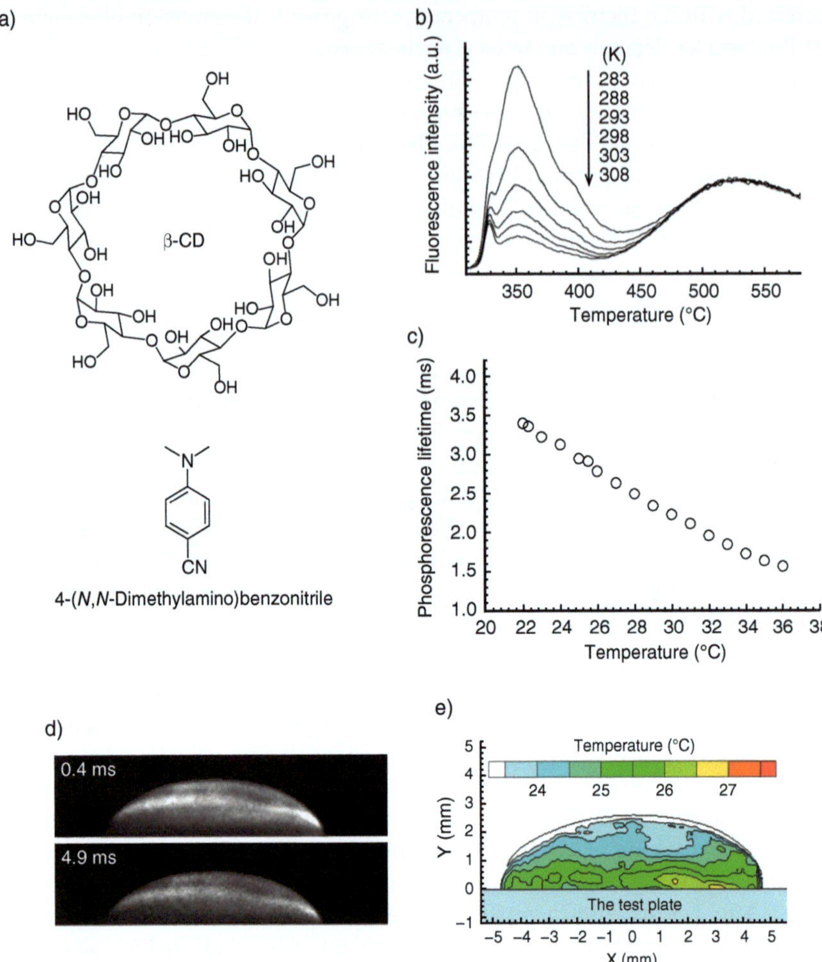

Figure 2.23 Intermolecular host–guest complexes. (a,b) A mixture of 4-(*N*,*N*-dimethylamino) benzonitrile (DMABN) and β-cyclodextrin (β-CD). (a) Chemical structures. (b) Fluorescence spectra of the mixture ([DMABN] = 10 µmol L^{-1}, [β-CD] = 3 mmol L^{-1}, λ_{ex}: 295 nm) in water. Panels (a,b) are adapted from Figueroa et al. (1998) *Anal. Chem.*, **70**, 3974–3977. (c-e). A mixture of 1-bromonaphthalene, maltosyl-modified β-cyclodextrin, and alcohol. (c) Relationship between phosphorescence lifetime of 1-bromonaphthalene in the mixture and temperature in water. (d) Phosphorescence images of a water droplet containing the mixture. Images were acquired for calculating phosphorescence lifetime at the indicated times after the excitation by a laser pulse. (e) Temperature distribution of the droplet. Panels (c–e) [Reprinted/Adapted] with permission from [Huang and Hu (2007) *Opt. Lett.*, **32**, 3534–3536] © The Optical Society.

A calix[4]arene derivative CA-1 is functionalized with a pair of Förster resonance energy transfer (FRET) donor and acceptor fluorophores (coumarin 343 and NBD, respectively). In chloroform, CA-1 showed a temperature-dependence of fluorescence intensity ratio at 515 and 465 nm upon the excitation at 438 nm in the temperature range from −33 to 27 °C (Bardi et al., 2019). With increasing

temperature, the calixarene scaffold opens, and the distance between two fluorophores increases to reduce the FRET efficiency.

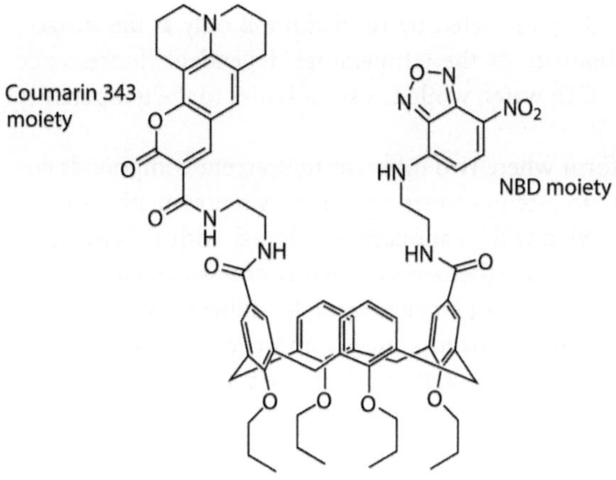

2.11
Synthetic Polymers

2.11.1
Use of Polymers as a Medium for Fluorescent Molecular Thermometers

Sometimes, a fluorescent molecular thermometer is encapsulated into a polymer at
the expense of spatial resolution and a homogeneous nature. However, there are at
least two great merits in using a polymer as a medium for a fluorescent molecular
thermometer: high robustness and capability of functional integration. For robustness, for instance, incorporating a fluorescent molecular thermometer, rhodamine
B (see Section 2.4.1), into poly(ethylene glycol) diacrylate gel improved the stability of the dye toward photobleaching remarkably (Hashim et al., 2019). In a related
case, co-encapsulation of rhodamine B and a reference compound rhodamine 6G
into mesoporous silica nanoparticles (diameter: 50 nm) permitted efficient FRET
from rhodamine 6G to rhodamine B to give temperature-dependent fluorescence
intensity ratio (Kalaparthi et al., 2021).

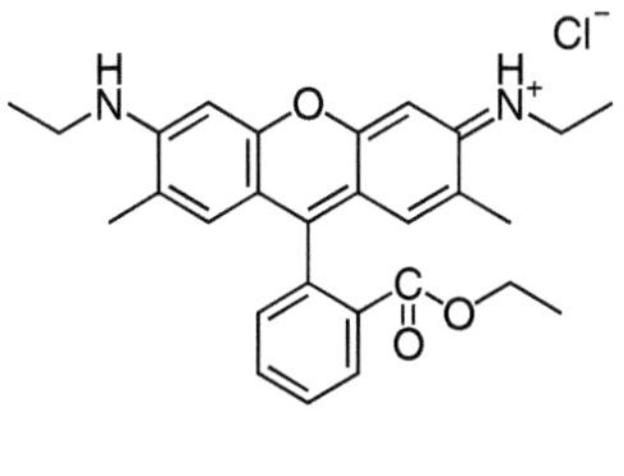

A French group investigated the effects of fluorescent labeling methods on the function of polystyrene latex particles (diameter ~ 900 nm) (Soleilhac et al., 2016). In the comparison between particles uniformly doped with a fluorescent molecular thermometer, rhodamine B and those coated by rhodamine B only at the surface, the former showed a better linearity of the temperature-dependent fluorescence intensity in the range 30–100 °C in water, while the sensitivities to the temperature variation were equivalent.

Nanoparticles can be a platform where two different fluorescent compounds co-exist for ratiometric responses to a temperature variation. A representative case is poly(vinylidene chloride-co-acrylonitrile) nanoparticles doped with a fluorescent thermometer Eu(BA)$_3$Phen (tris(benzoylacetonato)mono(phenanthroline)europium(III)) and a reference 4,4'-bis(2-benzoxazolyl) stilbene (Wang et al., 2011). The containment in polymer nanoparticles improved the stability of Eu(BA)$_3$Phen. A relatively large variation (-3.4% °C^{-1}) in the fluorescence intensity ratio at 611 and 412 nm was observed.

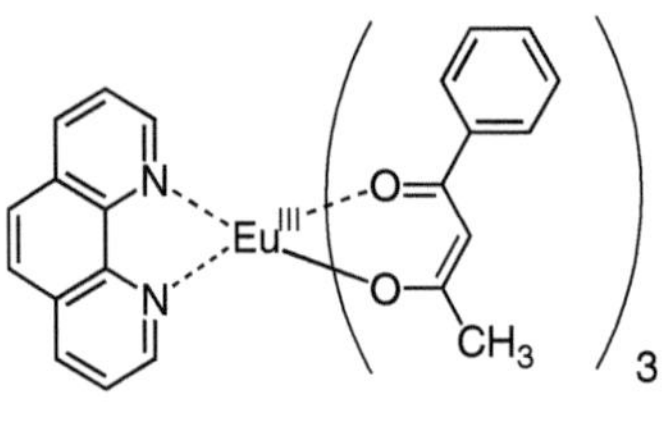

Eu(BA)$_3$ Phen

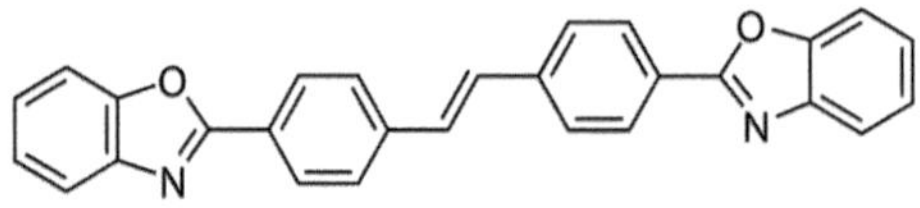

4,4'-Bis(2-benzoxazolyl)stilbene

Nanocapsules with a crosslinked polymethacrylate nanoshell (diameter: 180 $\pm$ 20 nm) were proposed as materials in which a fluorescent molecular thermometer, indocyanine green in PBS (phosphate-buffered saline) was contained (Zhegalova et al., 2014). Temperature-dependency of indocyanine green's fluorescence is due to photoisomerization (as described in Section 2.4.5). Co-encapsulation of an additional fluorescent compound as a reference (e.g., rhodamine 640 identical to rhodamine 101, see Figure 2.5c) and indocyanine green yielded a ratiometric fluorescent thermometer with emission at 640 and 800 nm.

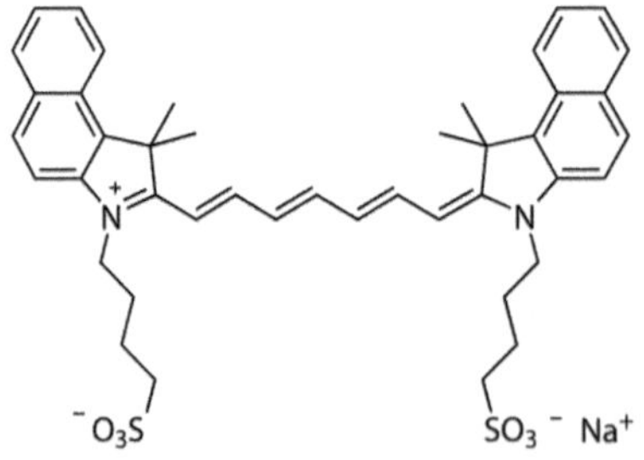

Indocyanine green

The fluorescence properties of indenoquinacridone (IQA) are determined by its dissociation/aggregation status. Significant heat-induced fluorescence enhancement was observed when IQA was co-encapsulated with tetradecanol or hexadecanol into polymer matrixes (Zhang et al., 2021) and SiO_2 nanoparticles (Zhang et al., 2020). Adopting PDMS as a polymer matrix, IQA was almost entirely quenched by aggregation with intermolecular hydrogen bondings at 25 °C. On the other hand, it emitted strong yellow fluorescence at 570 nm (with the excitation at 500 nm) at a temperature above 42 °C, where IQA was dissociated with hydrogen bondings with tetradecanol.

A bipyridine-based ligand BPPA (4,4′-([2,2′-bipyridine]-4,4′-diyl)bis(N,N-diphenylaniline)) fluoresces itself at 463 nm in dichloromethane, and its Zn^{2+} complex (Zn-BPPA) emits at 565 nm in the same solvent (Ma et al., 2020). An efficient ICT process presumably causes this shift in emission wavelength derived from the complexation with Zn^{2+} ion. Furthermore, when the Zn^{2+}-BPPA complex is incorporated into PEG8000, it displayed temperature-dependent fluorescence, i.e., yellow at a low temperature (25 °C) due to the complex and blue at a high temperature (65 °C) due to the dissociated ligand BPPA.

A "threshold" fluorescent thermometer change its fluorescence properties when an environmental temperature exceeds the temperature set previously. The response of a threshold fluorescent thermometer is irreversible, such as a clinical thermometer. An appropriate selection of a cyano-substituted oligo(p-phenylene vinylene) and a glassy amorphous polymer brought individual responses of a threshold fluorescent thermometer (Crenshaw et al., 2007). For example, C18-RG in poly(butyl methacrylate) (glass transition temperature: 23 °C) emitted green fluorescence when kept at 0 °C, whereas it emitted orange fluorescence with the annealing at 60 °C. The heat-induced color change was caused by aggregation of C18-RG in the softened polymer vehicle. Afterward, different combinations of a fluorophore and a polymer for higher operating temperatures (130–200 °C) were reported (Sing et al., 2009).

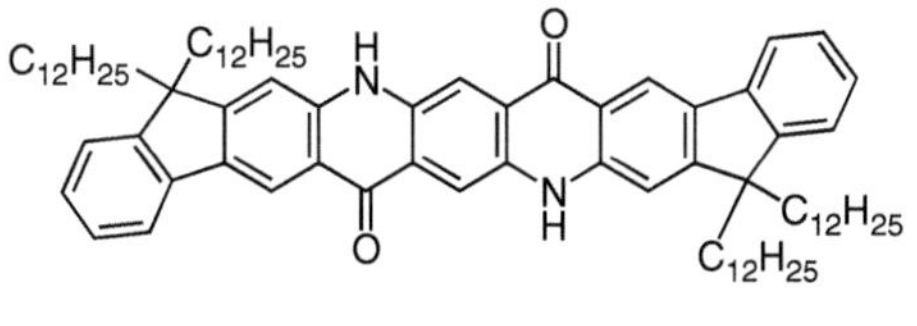

IQA

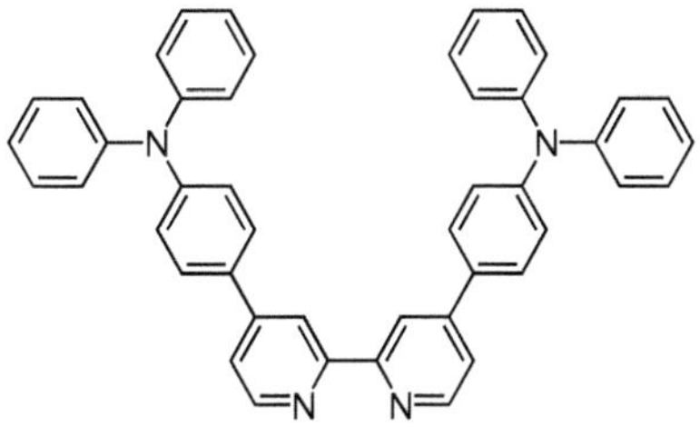

BPPA

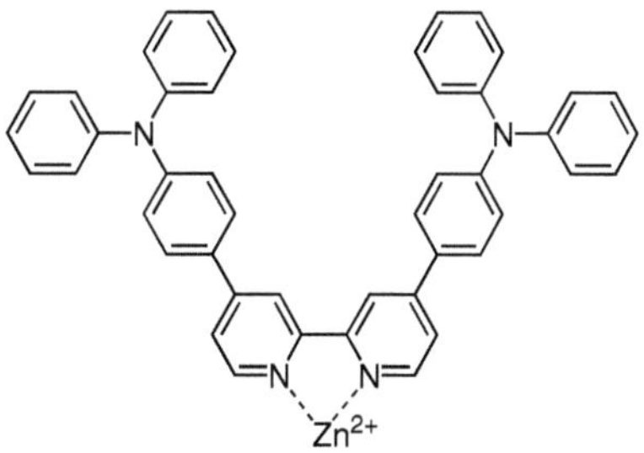

Zn-BPPA

C18-RG

2.11.2
Polymers of Fluorescent Monomer Units

Fluorescent poly[(9,9-dioctylfluorenyl-2,7-diyl)-*co*-(1,4-benzo-{2,1′,3}-thiadiazole)] (PFBT) nanoparticles or poly[{9,9-dioctyl-2,7-divinylene-fluorenylene}-*alt*-*co*-{2-methoxy-5-(2-ethylhexyloxy)-1,4-phenylene}] (PFPV) was used for the construction of ratiometric fluorescent thermometers by mixing with rhodamine B-labeled polystyrene (PS-RhB) (Ye et al., 2011). As described in Section 2.4.1, the rhodamine B moiety in PS-RhB exhibited temperature-dependent fluorescence intensity at approximately 570 nm. On the other hand, PFBT or PFPV was fluorescent at 540 nm and could also transfer a part of the excited energy to the temperature-sensitive rhodamine B structure. Thus, the PFBT or PFPV nanoparticles containing PS-RhB showed a temperature-dependent fluorescence intensity ratio in an aqueous solution.

In chlorobenzene, the π-conjugated polymer having oligophenylenevinylene (OPV) chromophores and polymethylene backbones (poly(OPV-DD)) shows blue fluorescence (λ_{em} = 405 nm), while its Eu^{3+} complex bound to the carboxylic groups emits red fluorescence (λ_{em} = 615 nm) via ligand-to-metal energy transfer (Balamurugan et al., 2013). When the temperature increased, fluorescence from Eu^{3+} was significantly quenched due to non-radiative deactivation from its $^5D_0$ excited state. The ligand-to-metal energy transfer was incomplete in other solvents such as DMF and DMSO, and fluorescence from poly(OPV-DD) remained.

Poly(phenylene ethynylene) carboxylate (PPE-CO$_2$-7) takes two conformations in an aqueous solution containing an amphiphilic polyvinylpyrrolidone (PVP). At a low temperature, PPE-CO$_2$-7 takes an aggregated form by π–π stacking and radiates an excimer emission. In contrast, it disaggregates into individual polymer chains at a high temperature, producing a monomer emission. PVP shifts this equilibrium of PPE-CO$_2$-7 to the less aggregated state and optimizes the fluorescence response to a temperature variation. The mixture of PPE-CO$_2$-7 and PVP shows a ratiometric response to a temperature variation. The fluorescence intensity ratio at 450 nm (due to the monomer) and 520 nm (due to the excimer) changes with a temperature variation from 20 to 70 °C in 10 mmol L^{-1} HEPES (4-(2-hydroxyethyl)piperazine-1-ethanesulfonic acid) and 150 mmol L^{-1} NaCl buffer (pH 7.0) (Darwish et al., 2016a, 2016b). The combination of PPE-CO$_2$-7 and PVP is valid in agarose or agar hydrogel, which can be conveniently utilized in temperature imaging of rigid subjects (Darwish et al., 2017).

A thin film of poly[1-(trimethylsilyl)phenyl-2-phenylacetylene] (PTMSDPA) shows temperature-dependent fluorescence properties (Kwak et al., 2006). The fluorescence of the PTMSDPA film at 530 nm with the excitation at 430 nm significantly decreased to 54% upon heating from 25 to 95 °C. The increase in temperature causes a slight molecular perturbation in the main chain and significant exciton deconfinement, which induces fluorescence quenching.

PFBT

PFPV

Poly(OPV-DD)

PPE-CO$_2$-7

PVP

PTMSDPA

2.11.3

Fluorescent Polymeric Thermometer Based on the Combination of a Thermo-responsive Polymer and an Environment-sensitive Fluorophore

The development of fluorescent polymeric thermometers based on the combination of a thermo-responsive polymer and an environment-sensitive fluorophore is the most outstanding achievement in the author's research group. Functionality in aqueous media and high sensitivity to temperature variation are significant advantages of fluorescent polymeric thermometers. This superiority of fluorescent polymeric thermometers has accelerated the further development of various types of fluorescent polymeric thermometers in many laboratories and enabled intracellular thermometry as a consequence. Because of space limitations, these venerable works cannot be introduced in the main text in detail. However, readers can access all the information in Appendices 2–5: Appendix 2 summarizes fluorescent polymeric thermometers. Appendix 3 lists fluorescent nanogel thermometers. Appendix 4 digests non-covalent systems consisting of a polymer and a fluorophore. Appendix 5 introduces fluorescent polymeric thermometer-based logic gates that simultaneously sense multiple factors (one is temperature). This section is an updated version of the author's expansive review article published in *Chemical Communications* in 2017 (Uchiyama et al., 2017) and is written subjectively. Related review articles on fluorescent polymeric thermometers are also available from other laboratories and can be referred to (Pietsch et al., 2011; Qiao et al., 2016).

2.11.3.1 **Background and Design Concept**

My focus on the development of fluorescent polymeric thermometers based on the combination of a thermo-responsive polymer and an environment-sensitive fluorophore started in 2002 with the following inspiration: (i) synthetic fluorescent probes for various kinds of ions and molecules had been successfully developed (de Silva et al., 1997; Prodi et al., 2011; Daly et al., 2015) and often applied for intracellular fluorescence imaging (Domaille et al., 2008; Chan et al., 2012; Yang et al., 2013). In contrast, there were few fluorescent sensors to monitor intracellular physical parameters. (ii) Some organic molecules and inorganic complexes show temperature-dependent fluorescence, as described in the previous sections. However, their sensitivities to temperature variations were not sufficient for distinguishing a difference of 1 °C. Thus, I considered that improving the sensitivity of fluorescent molecular thermometers should be addressed. (iii) Properties of a thermo-responsive polymer, poly(N-isopropylacrylamide) (PNIPAM), were stunning (Schild, 1992; Halperin et al., 2015). PNIPAM is soluble in water below 32 °C (this temperature is defined as the lower critical solution temperature, LCST) by hydration of amido groups in the NIPAM units with solvent water molecules. In contrast, the solution undergoes a phase transition and becomes colloidal when the temperature exceeds 32 °C. In the latter case, the hydrogen bonds between the NIPAM units and water are broken, and PNIPAM takes a globular form due to hydrophobic interactions by the isopropyl groups of the NIPAM units.

The above factors inspired me to design fluorescent polymeric thermometers based on a new design concept (Figure 2.24). Because the sensitivity of fluorescent

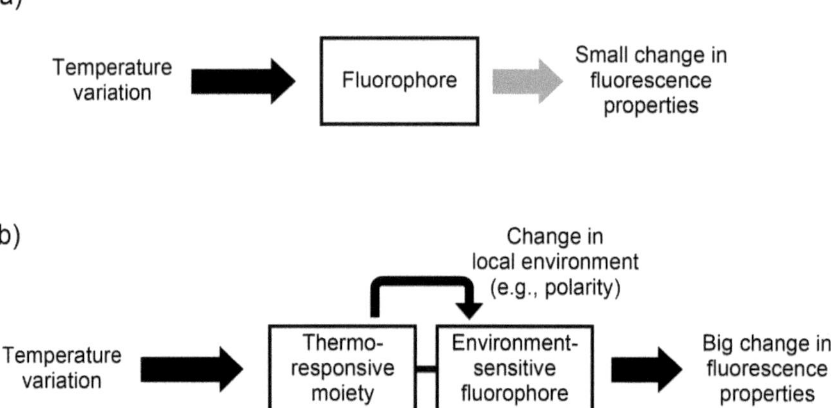

Figure 2.24 Design concept of fluorescent polymeric thermometers. (a) The temperature has a relatively small effect on the relaxation process of a fluorophore. (b) Therefore, novel fluorescent thermometers include a thermo-responsive moiety to sense a temperature variation in addition to the fluorophore. When the thermo-responsive moiety encounters a temperature variation, the microenvironment near the thermo-responsive moiety changes, accompanied by its three-dimensional structural transformation. Finally, the environment-sensitive fluorophore in a fluorescent thermometer outputs a signal by sensing the environmental variation. These schemes were originally published in Uchiyama (2019) *J. Synth. Org. Chem., Jpn.*, **77**, 1116 / The Society of Synthetic Organic Chemistry.

compounds to a temperature variation is generally low (Figure 2.24a), I decided to separate the requisite roles in fluorescent thermometers, i.e., sensing a temperature variation and signaling the variation. As shown in Figure 2.24b, novel fluorescent polymeric thermometers consisted of thermo-responsive and environment-sensitive fluorescent units. The unit ratio of environment-sensitive fluorescent units to thermo-responsive units should be kept low (usually from 1:1000 to 1:100) to not interfere with the original sensitivity of the thermo-responsive units. In my research projects, benzofurazan (2,1,3-benzoxadiazole) was selected as an environment-sensitive fluorophore due to its high sensitivity to polarity and the hydrogen bonding ability of the surroundings. While nitrobenzofurazan (NBD) is a more common fluorophore, N,N-dimethylaminosulfonylbenzofurazan (DBD) was chosen in my serial studies due to the higher sensitivity to the environment (see Box 2.2).

It should be noted that before the launch of my projects, thermo-responsive PNIPAMs labeled with a fluorophore such as pyrene (Ringsdorf et al., 1991), 9-(4-N,N-dimethylaminophenyl)phenanthrene (Iwai et al., 1998), and acenaphthylene (Chee et al., 2001) was reported. From the current viewpoint, these copolymers are also fluorescent polymeric thermometers. Nevertheless, these works aimed to investigate the change in local environments of thermo-responsive polymers with a temperature variation, and there was no mention of fluorescent thermometers. These copolymers exhibited only moderate sensitivity to temperature variation due to the characteristics of the adopted fluorophores.

Box 2.2   Benzofurazan (2,1,3-benzoxadiazole) as an Environment-sensitive Fluorophore

Benzofurazan (2,1,3-benzoxadiazole) is a unique structure that can generate fluorescence. Non-substituted benzofurazan does not emit fluorescence. However, when we introduce appropriate substituent groups (typically a pair of electron-donating and electron-accepting groups) into the 4- and 7- positions of the benzofurazan structure, most compounds emit fluorescence [1]. Interestingly, the maximum absorption and emission wavelengths depend on the substituent groups. For instance, NBD-NHPr (described in Section 2.4.2) has a relatively long emission wavelength (> 500 nm) even when compared to other fluorescent 10π systems (e.g., naphthalene). In addition, the fluorescence properties of 4,7-disubstituted benzofurazan are influenced by considerable solvent effects [2]. In a more hydrophilic solvent, its maximum emission wavelength is bathochromically shifted, and fluorescence quantum yield is significantly decreased. Fluorescence characteristics of model compounds, NBD-NHPr and DBD-IA (N,2-dimethyl-N-(2-{methyl[7-(dimethylsulfamoyl)-2,1,3-benzoxadiazol-4-yl]amino}ethyl)propenamide, [3]) in the mixture of 1,4-dioxane (Dx) and water are summarized in Table B2.2.1. Although NBD (nitro-benzofurazan) derivatives are likely more widespread than DBD (dimethylsulfamoyl-benzofurazan) derivatives, the latter show more considerable variations, especially in terms of the fluorescence quantum yield, by solvent effects. Thus, the DBD fluorophore was selected to be incorporated into thermo-responsive fluorescent polymeric thermometers in my

Table B2.2.1 Fluorescence properties of NBD-NHPr and DBD-IA in the mixture of 1,4-dioxane (Dx) and water at 20 °C.

Dx (%)[a]	NBD-NHPr [Ref.1]				DBD-IA[b]			
	λ_{ab} (nm)	λ_{em} (nm)	Φ_f	τ_f (ns)	λ_{ab} (nm)	λ_{em} (nm)	Φ_f	τ_f (ns)
100	451	523	0.67	10.2	444	555	0.91	19.0
90	462	536	0.49	8.6	447	570	0.63	14.7
75	467	542	0.32	6.2				
70					450	583	0.37	9.5
50	474	550	0.15	3.7	452	593	0.19	5.9[c]
25	479	557	0.07	2.0				
20					453	607	0.05	2.4[d]
0	482	566	0.027	1.0	452	616	0.027	1.1

a) The proportions of Dx/water were based on volume.
b) Unpublished data collected in the author's laboratory. Experimental conditions are the same to those used in Gota, C., Uchiyama, S., Yoshihara, T., Tobita, S., and Ohwada, T. (2008) *J. Phys. Chem. B*, **112**, 2829–2836.
c) Average fluorescence lifetime of double components ($\tau_1 = 4.02$ ns (29%) and $\tau_2 = 6.66$ ns (71%)).
d) Average fluorescence lifetime of double components ($\tau_1 = 2.04$ ns (89%) and $\tau_2 = 5.25$ ns (11%)).
1 Fery-Forgues, S., Fayet, J.-P., and Lopez, A. (1993) *J. Photochem. Photobiol. A*, **70**, 229–243.

research projects. Besides fluorescent polymeric thermometers, the NBD or DBD structure serves as an environmental-sensitive fluorophore in stimulus-responsive macromolecules for sensing pH [4] and glutamine [5].

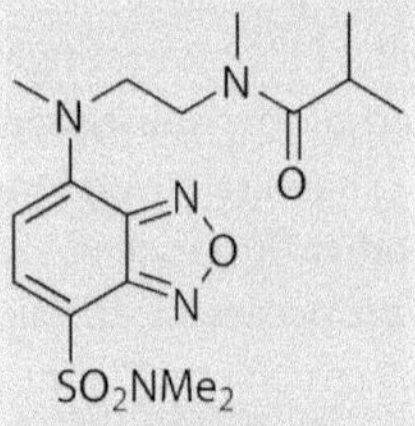

DBD-IA

1 Uchiyama, S., Santa, T., Fukushima, T., Homma, H., and Imai, K. (1998) *J. Chem. Soc. Perkin Trans. 2*, 2165–2173.

2 Uchiyama, S., Santa, T., and Imai, K. (1999) *J. Chem. Soc. Perkin Trans. 2*, 2525–2532.

3 Gota, C., Uchiyama, S., Yoshihara, T., Tobita, S., and Ohwada, T. (2008) *J. Phys. Chem. B*, **112**, 2829–2836.

4 Uchiyama, S. and Makino, Y. (2009) *Chem. Commun.*, 2646–2648.

5 Wada, A., Mie, M., Aizawa, M., Lahoud, P., Cass, A. E. G., and Kobatake, K. (2003) *J. Am. Chem. Soc.*, **125**, 16228–16234.

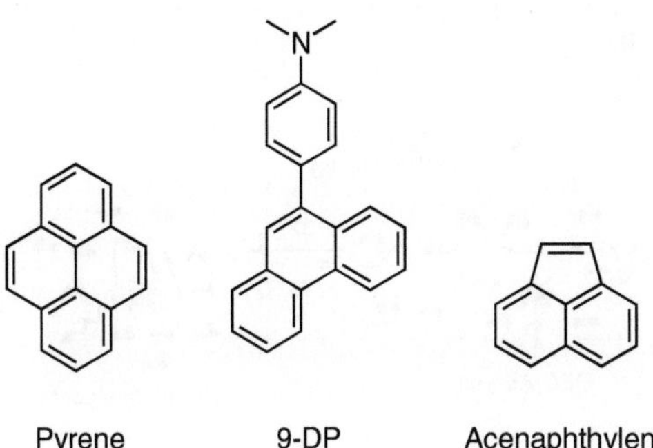

2.11.3.2
First Fluorescent Polymeric Thermometer

Based on the design concept described above, I prepared the first fluorescent polymeric thermometer, poly(NIPAM-*co*-DBD-AE) which consisted of thermo-responsive NIPAM units and fluorescent DBD-AE (4-N-(2-acryloyloxyethyl)-N-methylamino-7-N,N-dimethylaminosulfonyl-2,1,3-benzoxadiazole) units (weight-average molecular weight Mw = 39 400, number-average molecular weight Mn = 27 600, Mw/Mn = 1.43) (Figure 2.25a) (Uchiyama et al., 2003). The synthetic procedure for fluorescent polymeric thermometers is generally straightforward:

1) Monomers and radical initiator (normally AIBN (2,2′-azobisisobutyronitrile)) are mixed in a solvent such as 1,4-dioxane or N,N-dimethylformamide.
2) Dissolved oxygen is removed by bubbling argon or nitrogen gas (or freeze-pump-thaw).
3) The solution is left to stand at 60 °C for 4–16 hours.
4) The fluorescent polymeric thermometer obtained is purified by reprecipitation or dialysis.

The functional mechanism of poly(NIPAM-*co*-DBD-AE) is displayed in Figure 2.25b. Below its LCST, poly(NIPAM-*co*-DBD-AE) takes an extended form due to hydration, and water molecules quench the fluorescent DBD-AE units. On the other hand, above the LCST, poly(NIPAM-*co*-DBD-AE) undergoes a phase transition and takes a globular form by hydrophobic interaction, where the water molecules are repelled by the NIPAM units and the DBD-AE units strongly fluoresce. Figure 2.25c shows the fluorescence spectra of poly(NIPAM-*co*-DBD-AE) in water with increasing temperature. As expected, its fluorescence intensity dramatically increased at approximately 32 °C, the phase transition temperature of PNIPAM (Figure 2.25d, 2.25e). The average variation in the fluorescence intensity of poly(NIPAM-*co*-DBD-AE) was 38% °C^{-1} from 29 to 37 °C, the highest among fluorescent molecular thermometers ever reported.

The application of poly(NIPAM-*co*-DBD-AE) for temperature mapping of a microfluidic device presented the usefulness of its fluorescence lifetime in thermometry (Graham et al., 2010). Like fluorescence intensity, the fluorescence lifetime of poly(NIPAM-*co*-DBD-AE) is temperature-dependent around the LCST of PNIPAM (Figure 2.26a). Temperature imaging of a microfluidic device under a temperature gradient was successfully carried out with temperature resolution better than 0.1 °C in a detection volume of a few hundred femtolitres when applying a fluorescence

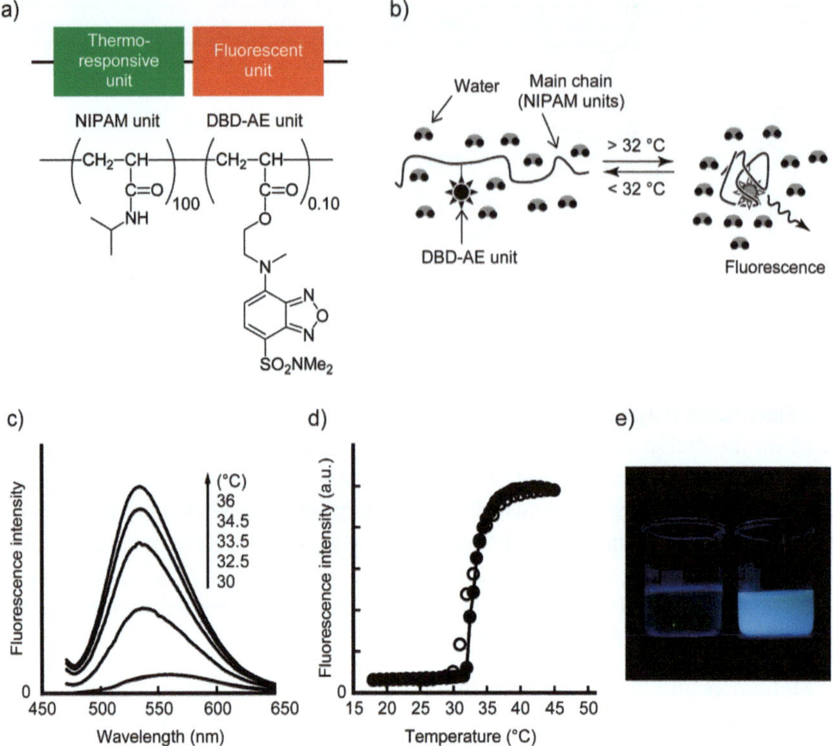

Figure 2.25 Thermo-responsive NIPAM polymer labeled with an environment-sensitive fluorescent unit DBD-AE. (a) Chemical structure of poly(NIPAM-*co*-DBD-AE). (b) The functional mechanism in water. (c) Fluorescence spectra in water (0.01 w/v%, λ_{ex}: 444 nm). (d) Relationship between fluorescence intensity and temperature (closed: heating, open: cooling). (e) Photograph of the aqueous solutions (left: cold, right: hot) under irradiation of UV light. Panels (c,d) are adapted from Uchiyama et al. (2003) *Anal. Chem.*, **75**, 5926–5935.

lifetime imaging microscope for picturing temperature-dependent fluorescence lifetime of poly(NIPAM-*co*-DBD-AE) (Figure 2.26b). Meanwhile, the fluorescence image of the same subject was severely distorted (Figure 2.26c). Because the fluorescence intensity is influenced by the concentration of a fluorescent sensor and an optical setup, such as the strength of an excitation source, selecting a detection parameter free from these influences is necessary for accurate thermometry and temperature imaging with fluorescent thermometers.

Another application can be found in the simultaneous monitoring of flow velocity and temperature in the wake of a heated cylinder by poly(NIPAM-*co*-NBD-AE) in which the environment-sensitive fluorescent units are changed from DBD-AE units to NBD-AE units (Cellini et al., 2017). Less sensitive compared to poly(NIPAM-*co*-DBD-AE), but still a considerable fluorescence enhancement (*ca* 210%) was observed in poly(NIPAM-*co*-NBD-AE) with the increase in temperature from 28 to 40 °C. In this work, poly(NIPAM-*co*-NBD-AE) in water was encapsulated into optically transparent poly(dimethylsiloxane) (PDMS) particles for particle image velocimetry.

A photophysical study analyzing the temperature-dependent fluorescence lifetime of poly(NIPAM-*co*-DBD-AA) (Mw = 112 000, Mn = 36 400, Mw/Mn = 3.08) revealed the detailed role of fluorescent benzofurazan units in the function as a fluorescent

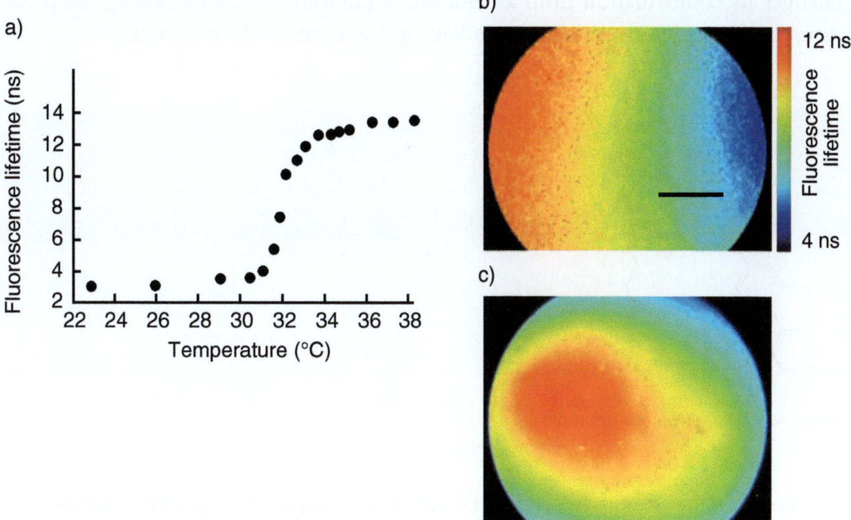

Figure 2.26 Temperature mapping of a microfluidic chamber with poly(NIPAM-*co*-DBD-AE) and fluorescence lifetime imaging microscopy. (a) Relationship between the average fluorescence lifetime of poly(NIPAM-*co*-DBD-AE) and temperature in water. The average fluorescence lifetime was calculated from fitted multiexponential fluorescence decay functions. (b) Fluorescence lifetime image of poly(NIPAM-*co*-DBD-AE) in a microfluidic chamber with a temperature decrease gradient from left to right. Scale bar: 100 μm. (c) Fluorescence image corresponding to (b). Graham et al. (2010) *Lab Chip*, **10**, 1267–1273 / Royal Society of chemistry.

Poly(NIPAM-*co*-NBD-AE)

Poly(NIPAM-*co*-DBD-AA)

polymeric thermometer (Gota et al., 2008). In the study, a new amide-type fluorescent monomer, DBD-AA (N-{2-[(7-N,N-dimethylaminosulfonyl)-2,1,3-benzoxadiazol-4-yl](methyl)amino}ethyl-N-methylacrylamide) was used because it is more resistant to hydrolysis than an ester-type DBD-AE while both monomers include the same fluorophore. The DBD-AA units of poly(NIPAM-*co*-DBD-AA) not only sense the decrease in local polarity involved in the heat-induced phase transition of NIPAM units but also concurrently lose the hydrogen bondings at the nitrogen atoms of the benzofurazan ring and the dimethylsulfamoyl substituent group with water molecules. It should be noted that the fluorescent polymeric thermometer poly(NIPAM-*co*-DBD-AA) was independently applied for ultrasound-switchable fluorescence imaging (see Box 2.3).

Although the function is only observed in organic solvents, another fluorescent polymeric thermometer based on a thermo-responsive polymer has been reported (Nomura et al., 2002). A pyrene-labeled poly(N-propargylamide) in chloroform changes its conformation from a helix into a random coil by increasing temperature, resulting in temperature-dependent pyrene excimer fluorescence.

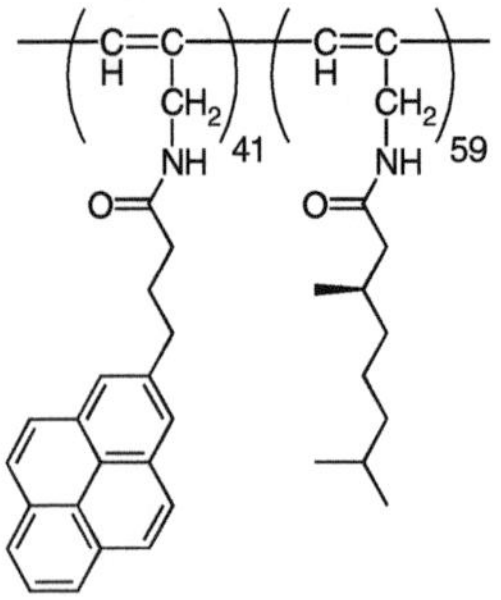

Pyrene-labeled poly(*N*-propargylamide)

Box 2.3   Ultrasound-switchable Fluorescence Imaging Using a Fluorescent Polymeric Thermometer

One of the promising applications of fluorescent polymeric thermometers based on a thermo-responsive polymer is ultrasound-switchable fluorescence imaging with high resolution in deep turbid media, which has been exclusively led by Yuan's research group at The University of Texas at Arlington. Generally, photon diffusion dramatically reduces the spatial resolution of fluorescence imaging in deep optically turbid media (e.g., tissue). The ultrasound-mediated fluorescence technique has been expected to overcome this limitation of fluorescence imaging because ultrasound-mediated fluorescence signal induced by a highly focused ultrasound beam can be detected to quantify the fluorescence information in the small ultrasound focal volume. In ultrasound-switchable fluorescence imaging, a fluorescent polymeric thermometer can be used as a contrast agent that switches on the fluorescence when the temperature increase induced by a highly intensified and focused ultrasound exceeds the lower critical solution temperature (LCST) [1]. Figure B2.3.1 indicates a schematic diagram of the measurement system in the first report on ultrasound-switchable fluorescence imaging using a fluorescent polymeric thermometer, in which poly(NIPAM-*co*-DBD-AA) was used as a fluorescent polymeric thermometer whose fluorescence intensity changes according to ultrasound irradiation. As a model sample, a small silicon tube (inner diameter: 180 μm) labeled with poly(NIPAM-*co*-DBD-AA) was inserted into a silicone phantom mimicking a blood vessel (inner diameter: 787 μm). Figure B2.3.2 shows a photon distribution due to ultrasound-switchable fluorescence along a lateral direction of the silicon tube. Ultrasound-switchable fluorescence could be detected only at the location where the temperature was 32 °C or higher. The length that afforded high ultrasound-switchable fluorescence was 260 μm which was longer than the actual sample tube (180 μm). Nevertheless, ultrasound-switchable fluorescence can image the small fluorescent tube with a diameter of 180 μm at a depth of 20 mm in an optically turbid medium.

The quality of an ultrasound-switchable fluorescence image depends on the properties of a temperature-sensitive contrast agent. The research group continuously reported improved polymers. For example, a PNIPAM nanoparticle encapsulating indocyanine green (ICG, see Section 2.11.1) fluoresces with long

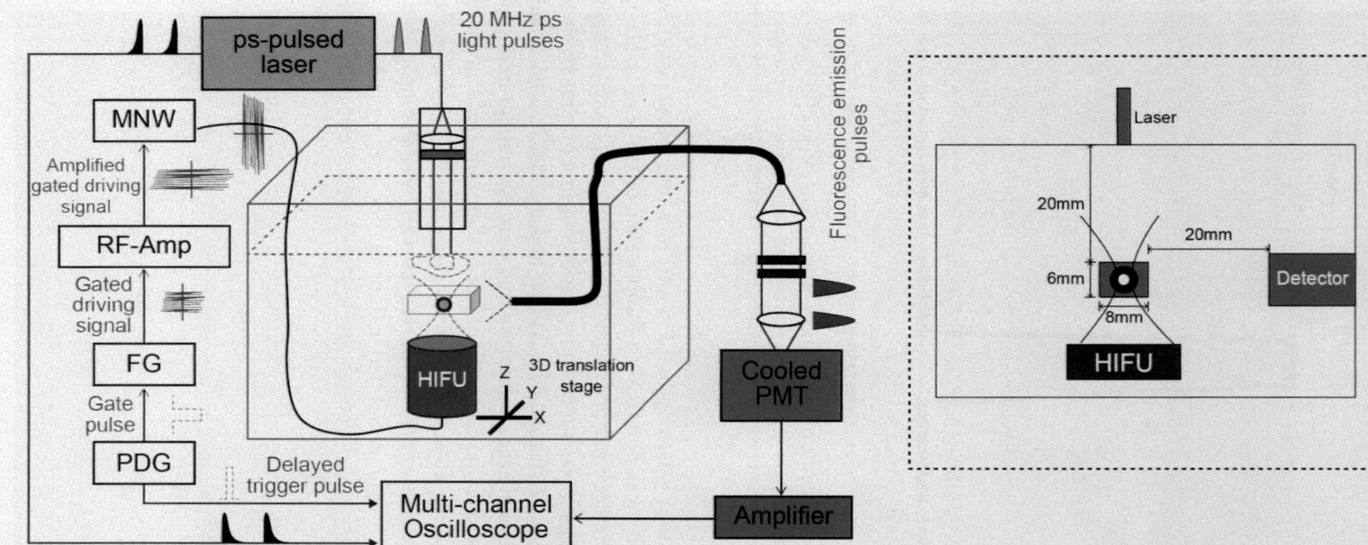

Figure B2.3.1 Schematic diagram of ultrasound-switchable fluorescence measurements. PDG: pulse-delay generator; FG: function generator; RF-Amp: radio frequency power amplifier; MNW: matching network; ps: pico-second; HIFU: high intensity focused ultrasound; PMT: photomultiplier tube. Inside the dotted square: the phantom, laser, and detector configuration. Adapted from Yuan, B. et al. (2012) *Appl. Phys. Lett.*, **101**, 033703.

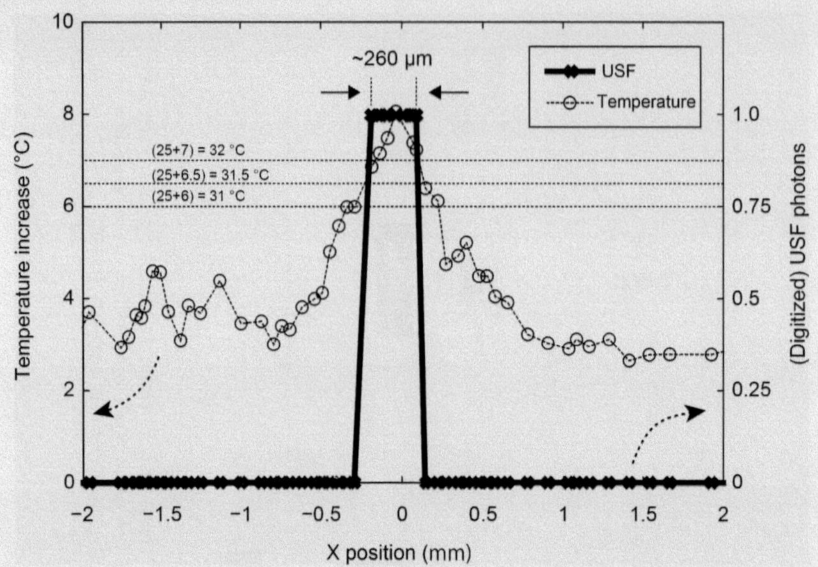

Figure B2.3.2 Photon distribution of ultrasound-switchable fluorescence (USF) along a lateral direction. The left and right vertical axes represent the temperature increase and the digitalized photons of ultrasound-switchable fluorescence. Adapted from Yuan, B. et al. (2012) *Appl. Phys. Lett.*, **101**, 033703.

excitation and emission wavelengths (808 and 830 nm, respectively), which can minimize interferences due to auto-fluorescence of tissues [2]. Using the ICG-encapsulated PNIPAM nanoparticles, the first *in vivo* ultrasound-switchable fluorescence imaging was successfully carried out in a porcine heart tissue and mouse breast tumor (Figure B2.3.3) [3]. In a mouse, ICG-encapsulated PNIPAM nanoparticles injected via the tail vein were distributed in the spleen. A different type of molecules-based fluorescent thermometer, Pluronic F127 nanocapsules containing an aza-BODIPY derivative, ADP(CA)$_2$ (λ_{ex} and λ_{em} in dichloromethane are 683 and 717 nm, respectively), was also proposed as a contrast agent for ultrasound-switchable fluorescence imaging [4].

1 Yuan, B., Uchiyama, S., Liu, Y., Nguyen, K. T., and Alexandrakis, G. (2012) *Appl. Phys. Lett.*, **101**, 033703.

2 Yu, S., Cheng, B., Yao, T., Xu, C., Nguyen, K., Hong, Y., and Yuan, B. (2016) *Sci. Rep.*, **6**, 35942.

3 Yao, T., Yu, S., Liu, Y., and Yuan, B. (2019) *Sci. Rep.*, **9**, 9855.

4 Cheng, B., Bandi, V., Wei, M.-Y., Pei, Y., D'Souza, F., Nguyen, K. T., Hong, Y., and Yuan, B. (2016) PLoS ONE, **11**, e0165963.

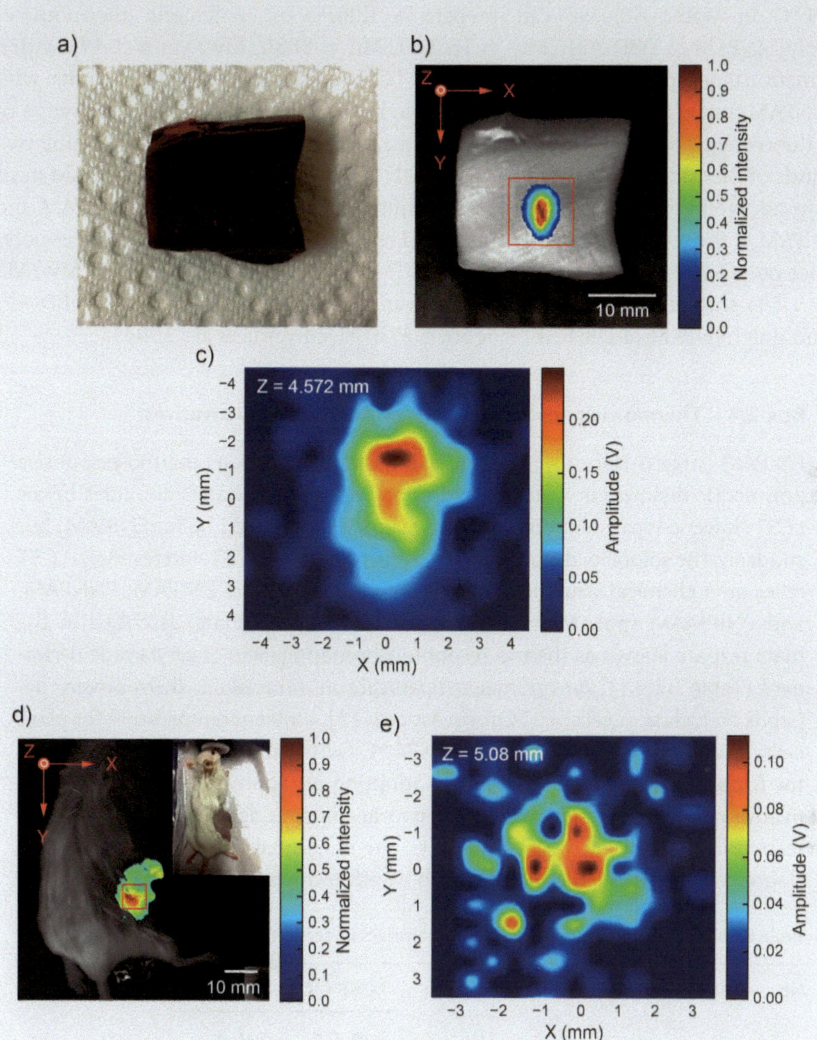

Figure B2.3.3 *In vivo* ultrasound-switchable fluorescence imaging. (a–c) Heart tissue sample. (a) A photograph. (b) Distribution of ICG-nanoparticles (λ_{ex}: 808 nm, λ_{em}: 830 nm). The red square represents the ultrasound-switchable fluorescence scan area. (c) Ultrasound-switchable fluorescence image. (d,e) Mouse with a breast tumor. (d) A photograph and a pseudo-colored fluorescence intensity image of the ICG-encapsulated PNIPAM nanoparticles locally injected in the tumor (λ_{ex}: 808 nm, λ_{em}: 830 nm). The red square represents the ultrasound-switchable fluorescence scan area. (e) Ultrasound-switchable fluorescence image. Adapted from Yao, T. et al. (2019) *Sci. Rep.*, **9**, 9855.

2.11.3.3 Modulation of the Functional Temperature Range of Fluorescent Polymeric Thermometers

Although NIPAM is the most successful thermo-responsive polymer, other homopolymers of acrylamide derivatives also own phase transition temperatures, leading to a three-dimensional structural transformation in water (Box 2.4). For instance, poly(N-n-propylacrylamide) (PNNPAM) undergoes a phase transition at 21 °C in water. So, we can prepare a fluorescent polymeric thermometer poly(NNPAM-*co*-DBD-AE) (Mw = 13 100, Mn = 8830, Mw/Mn = 1.49) with a functional temperature range around 21 °C by replacing the NIPAM units with NNPAM units (Figure 2.27). More amazingly, the functional temperature range of a fluorescent polymeric thermometer can be more precisely tuned by selecting two kinds of thermo-responsive units. When a 1:1 mixture of NNPAM and NIPAM units was adopted as the thermo-responsive units (i.e., the molar ratio of NNPAM and NIPAM in the feed was 1:1), the functional temperature range of a resultant copolymer poly(NNPAM-*co*-NIPAM-*co*-DBD-AE) (Mw = 26 900, Mn = 17 500, Mw/Mn = 1.54) shifted to approximately the mean of those of poly(NIPAM-*co*-DBD-AE) and poly(NNPAM-*co*-DBD-AE) (Figure 2.27b) (Uchiyama et al., 2004).

Box 2.4 Thermo-responsive Polymers of Acrylamide Derivatives

PNIPAM (poly(*N*-isopropylacrylamide)) is the most famous thermo-responsive polymer. It dissolves in water by hydration with solvent water molecules below LCST (lower critical solution temperature; approximately 32 °C for PNIPAM) but suddenly, the solubility drops by phase transition above LCST. Interestingly, LCST relies on a chemical structure, and many examples besides PNIPAM, PNNPAM, and PNIPMAM (poly(*N*-isopropylmethacrylamide)) that are described in the main text are known as thermo-responsive homopolymers of acrylamide derivatives (Table B2.4.1). An optimal temperature in intracellular thermometry depends on a class of cells; 20 °C for yeast cells [1], ambient temperature for plant cells, ~37 °C for mammalian cells, 70–75 °C for thermophiles [2], and > 80 °C for hyperthermophiles. Therefore, by adopting an appropriate acrylamide derivative or a suitable combination of two acrylamide derivatives as monomers responsible for thermo-responsive units, the functional temperature range of a fluorescent polymeric thermometer can be arbitrarily tuned for subject cells.

Table B2.4.1 Solution properties of homopolymers of acrylamide derivatives in water.

No.	Polymer	LCST (°C)	
		in Ref. 1	in Ref. 2
1	Poly(N-isopropylacrylamide)	30.9	32
2	Poly(N-n-propylacrylamide)	21.5	
3	Poly(N-isopropylmethacrylamide)	44.0	
4	Poly(N-tert-butylacrylamide)		(insoluble)[a]
5	Poly(N-acryloylpiperidine)	5.5	
6	Poly(N-methyl-N-n-propylacrylamide)	19.8	

No.	Polymer	LCST (°C)	
		in Ref. 1	in Ref. 2
7	Poly(N-methyl-N-isopropylacrylamide)	22.3	
8	Poly(N-n-propylmethacrylamide)	28.0	
9	Poly(N,N-diethylacrylamide)	32.0	32
10	Poly(N-cyclopropylacrylamide)	45.5	
11	Poly(N-ethylmethacrylamide)	50.0	
12	Poly(N-methyl-N-ethylacrylamide)	56.0	
13	Poly(N-acryloylpyrrolidine)	56.0	
14	Poly(N-cyclopropylmethacrylamide)	59.0	
15	Poly(N-ethylacrylamide)	72.0	82
16	Poly(acrylamide)		(soluble)[b]
17	Poly(N,N-dimethylacrylamide)		(soluble)[b]

a) Insoluble in the whole range of 0–100 °C.
b) Soluble in the whole range of 0–100 °C.
1 Ito, S. (1989) *Kobunshi Ronbunshu*, **46**, 437–443.
2 Liu, H.Y. and Zhu, X.X. (1999) *Polymer*, **40**, 6985–6900.

1 Yoshida, S., Imoto, J., Minato, T., Oouchi, R., Sugihara, M., Imai, T., Ishiguro, T., Mizutani, S., Tomita, M., Soga, T., and Yoshimoto, H. (2008) *Appl. Environ. Microbiol.*, **74**, 2787–2796.

2 Brock, T. D. and Freeze, H. (1969) *J. Bacteriol.*, **98**, 289–297.

2.11.3.4 Introduction of Ionic Units to Improve Solubility and Accuracy as a Fluorescent Thermometer

The apparent shortcoming of the first fluorescent polymeric thermometer poly(NIPAM-*co*-DBD-AE) is the hysteresis, i.e., different fluorescence responses at the same temperature in heating and cooling cycles, as previously shown in Figure 2.25d. In practical use as a fluorescent thermometer, this hysteresis significantly decreases the accuracy of thermometry. Because the hysteresis of PNIPAM is derived from interpolymeric aggregation, ionic SPA (3-sulfopropyl acrylate) units were introduced into a fluorescent polymeric thermometer to increase its hydrophilicity (Gota et al., 2007). Figure 2.28a shows the chemical structure of poly(NIPAM-*co*-SPA-*co*-DBD-AA) (Mw = 48 700, Mn = 30 000, Mw/Mn = 1.62) and its responses to temperature variation. Because of the presence of ionic SPA units, the response curves upon heating and cooling were identical (Figure 2.28b), confirming that the ionic SPA units improved accuracy. Furthermore, the functional temperature range of poly(NIPAM-*co*-SPA-*co*-DBD-AA) broadened and shifted to a higher temperature region compared to poly(NIPAM-*co*-DBD-AA) because of the increased hydrophilicity of the polymer. The increase in the hydrophilicity of fluorescent polymeric thermometers is essential for applying them in living cells, as polymeric thermometers aggregate more easily in an intracellular space under high ionic strength than in water. Thus, ionic units in fluorescent polymeric thermometers can effectively prevent aggregation in cellular environments.

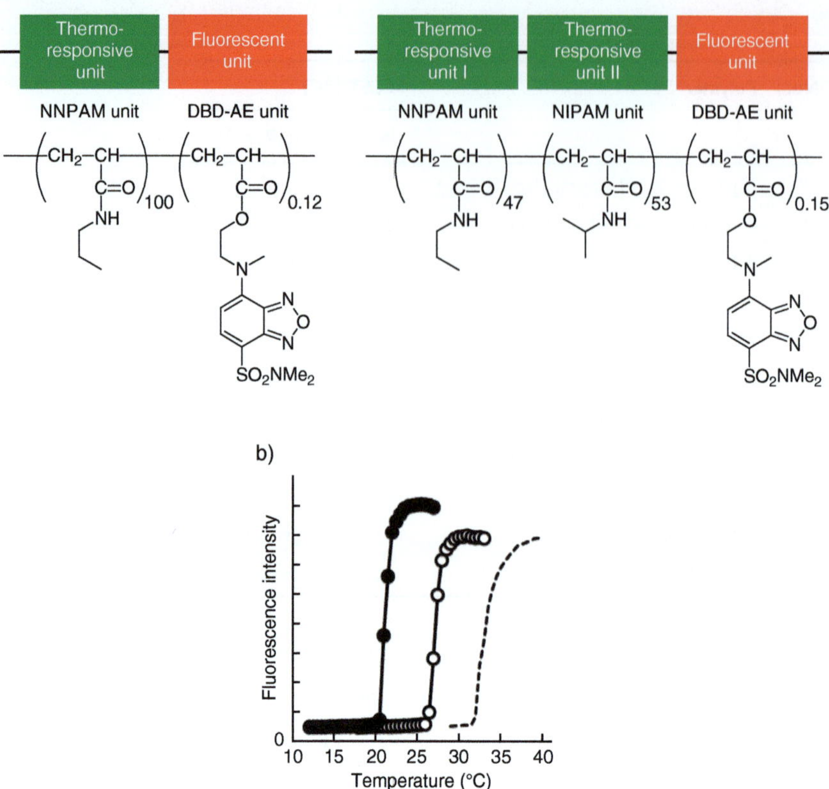

Figure 2.27 Modulation of the functional temperature range of fluorescent thermoresponsive copolymers. (a) Chemical structures of poly(NNPAM-*co*-DBD-AE) and poly(NNPAM-*co*-NIPAM-*co*-DBD-AE). (b) Relationships between the fluorescence intensity and temperature in water (0.01 w/v%, λ_{ex}: 444 nm, closed circle: poly(NNPAM-*co*-DBD-AE), open circle: poly(NNPAM-*co*-NIPAM-*co*-DBD-AE), dotted line: poly(NIPAM-*co*-DBD-AE)). Adapted from Uchiyama et al. (2004) *Anal. Chem.*, **76**, 1793–1798.

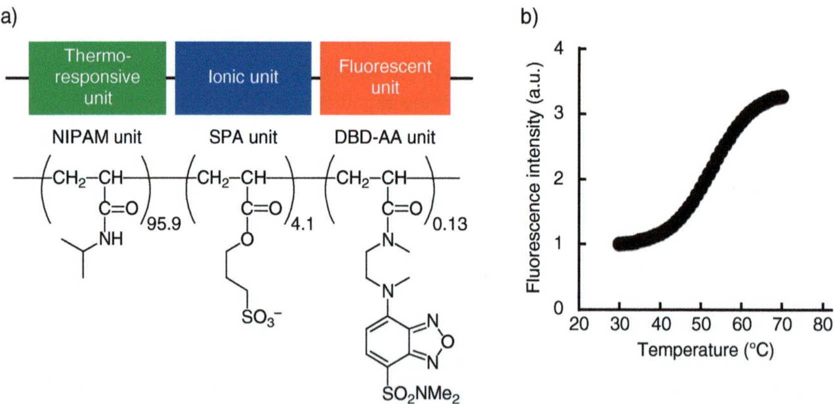

Figure 2.28 Improvement of accuracy in a fluorescent polymeric thermometer by introducing ionic units. (a) Chemical structure of poly(NIPAM-*co*-SPA-*co*-DBD-AA). (b) Relationship between the fluorescence intensity of poly(NIPAM-*co*-SPA-*co*-DBD-AA) (0.01 w/v%, λ_{ex}: 456 nm, λ_{em}: 558 nm) and temperature in water. Response data upon heating (closed circle) was fully overlapped with that upon cooling (open circle). Adapted from Gota et al. (2007) *Analyst*, **132**, 121–126.

2.11.3.5 Gelation

One of the morphological modifications of fluorescent polymeric thermometers is the gelation. The use of crosslinker MBAM (N,N′-methylenebisacrylamide) in polymerization resulted in the formation of a three-dimensional nanogel with a diameter in sub-micrometer scale. The gelation brings structural robustness, while the spatial resolution in thermometry with a fluorescent nanogel thermometer is inferior to that with a fluorescent polymeric thermometer due to its large size. The functional mechanism of a fluorescent nanogel thermometer composed of NIPAM, DBD-AE, and MBAM units is comparable to that of poly(NIPAM-*co*-DBD-AE) with a linear structure. Below a phase transition temperature, called the "volume phase transition temperature" for a thermo-responsive nanogel, the fluorescent nanogel thermometer swells by including water molecules in the interior, where the environment-sensitive DBD-AE units are quenched. On the other hand, above the volume phase transition temperature, the fluorescent nanogel thermometer shrinks by repelling water molecules from the interior, where the DBD-AE units emit strong fluorescence (Figure 2.29a). As demonstrated in Figure 2.29, a fluorescent nanogel thermometer

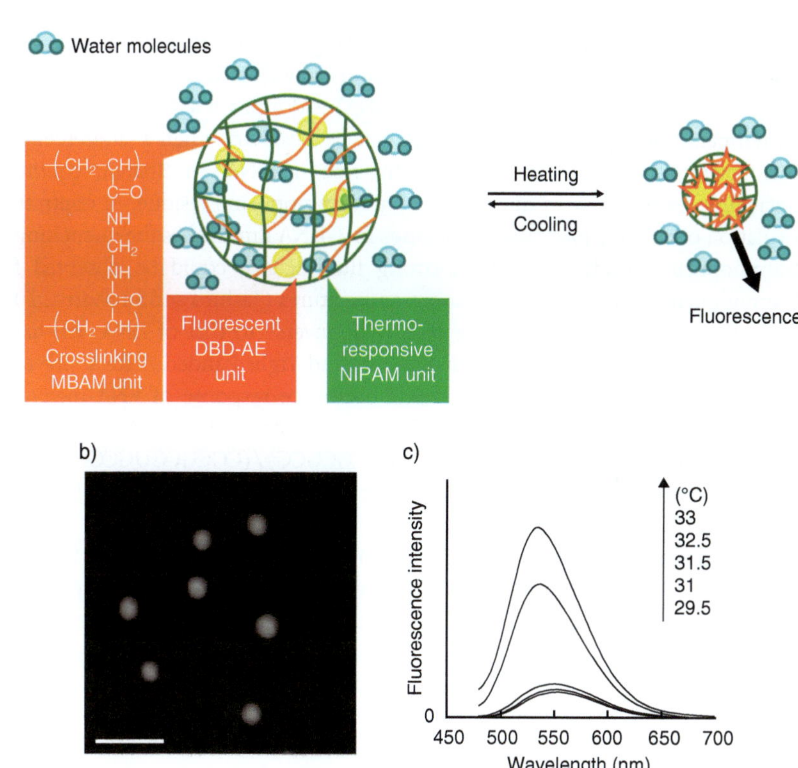

Figure 2.29 Fluorescent nanogel thermometer composed of thermo-responsive units, environment-sensitive fluorescent DBD-AE units, and crosslinking MBAM units. (a) Schematic diagram of the structure and function. (b) AFM (atomic force microscopy) image of a fluorescent nanogel thermometer with thermo-responsive NNPAM units. Scale bar: 500 nm. (c) Fluorescence spectra of a fluorescent nanogel thermometer with thermo-responsive NIPAM units in water (0.1 w/v%, λ_{ex}: 444 nm). Panels (b,c) are adapted from Iwai et al. (2005) *J. Mater. Chem.*, **15**, 2796–2800 / Royal Society of Chemistry.

based on a thermo-responsive unit (e.g., NNPAM, NIPAM, or NIPMAM (N-isopropylmethacrylamide)) and an environment-sensitive fluorescent DBD-AE unit dramatically starts to fluoresce when the temperature exceeds the volume phase transition temperature of the thermo-responsive units (Iwai et al., 2005).

2.12
Nucleic Acids (DNA and RNA)

DNA is perhaps the most notable macromolecule in biology and biotechnology, and it has also been exploited as a thermo-responsive moiety in fluorescent molecular thermometers. For instance, the DNA oligomer labeled with a rhodamine G dye (5′-RhG-GATGATGAGAAGAAC-3′, RhG: rhodamine-G dye, G: guanine, A: adenine; T: thymine, C: cytosine) shows temperature-dependent fluorescence lifetime in the range 15–35 °C (Figure 2.30a) (Jeon et al., 2003). The temperature-dependency of the fluorescence lifetime of the DNA oligomer is due to its conformation changes with a temperature variation, and the quenching efficiency by guanine and thymine toward the fluorophore increases at a higher temperature.

DNA and RNA convert between a right-handed helix (A form for RNA and B form for DNA) and a left-handed helix (Z form) (Figure 2.30b). The B–Z transition of DNA can be controlled by changing temperature: the Z form (Z-DNA) is preferred at lower temperatures, while the B form (B-DNA) is dominant at higher temperatures. So, temperature-dependent fluorescence can be observed when a fluorescent adenine analog, 2-aminopurine (Ap), is inserted in DNA. In B-DNA, continuous π-stackings cause efficient quenching of a 2-aminopurine residue. In contrast, the formation of discrete four-base π-stackings in Z-DNA diminishes the quenching of a 2-aminopurine residue. Therefore, strong fluorescence could be observed from 2-aminopurine-contaiing DNA at lower temperatures (Tashiro and Sugiyama, 2003). Interestingly, the thermal response of RNA is inverse to that of DNA: left-handed Z-RNA is preferred in higher temperatures, and right-handed A-RNA prevails in lower temperatures. Consequently, the fluorescence response of RNA including a 2-aminopurine residue is inverse, too. The fluorescence response of an RNA duplex containing an Ap residue (r(GCGCGC-Ap-CGCGCG)/(CGCGCGUGCGCGC)) to a temperature variation in aqueous solution is indicated in Figure 2.30c (Tashiro and Sugiyama, 2005).

A *molecular beacon* is a designed DNA hairpin structure first reported in 1996 (Tyagi and Kramer, 1996). Molecular beacons have a characteristic stem-loop structure terminated by a fluorophore-quencher pair or a fluorescence donor-fluorescence acceptor pair. We can find fascinating utilization of molecular beacons in gene screening, development of biosensors, construction of biochips, detection of single-nucleotide polymorphisms (SNPs), and intracellular messenger-RNA (mRNA) monitoring (Wang et al., 2009). A molecular beacon based on L-DNA, the enantiomeric form of natural D-DNA, is an excellent format for developing fluorescent

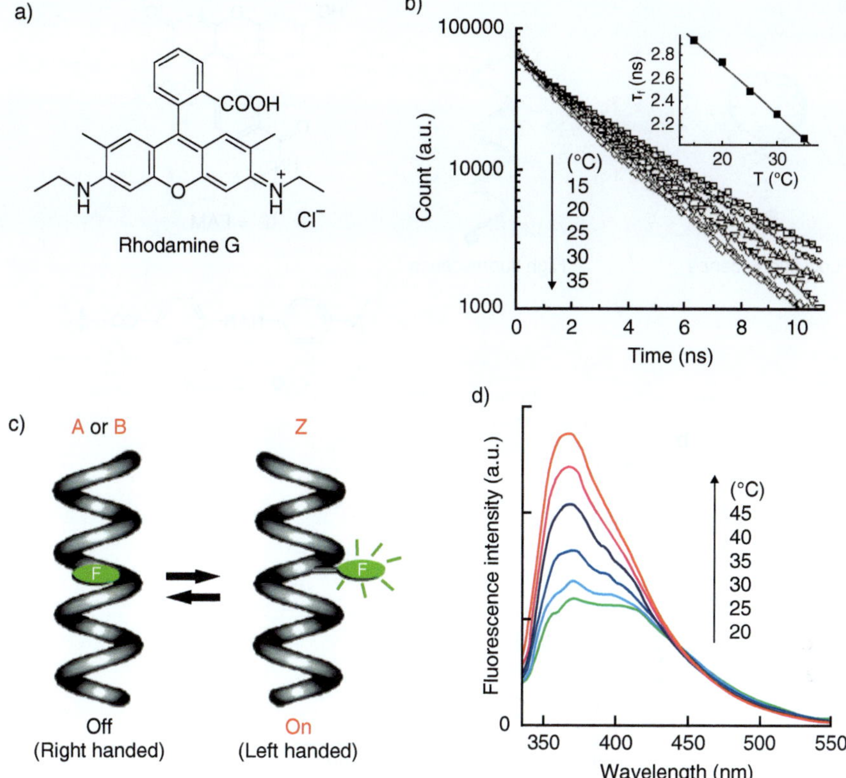

Figure 2.30 Nucleic acid-based fluorescent thermometers (the sequences are indicated in the main text). (a,b) Utilization of quenching ability of guanine and thymine toward rhodamine G-labeled DNA. (a) Chemical structure of rhodamine G. (b) The temperature-dependent fluorescence lifetime of rhodamine G-labeled DNA (2 µmol L^{-1}) in water (adapted from Jeon et al. (2003) *J. Am. Chem. Soc.*, **125**, 9908–9909). (c,d) Utilization of temperature-regulated transitions of DNA and RNA between B (or A) form and Z form. (c) Functional mechanism. (d) Fluorescence spectra of an RNA duplex containing 2-aminopurine residue in a 6 mol L^{-1} NaClO$_4$ solution (adapted from Tashiro and Sugiyama (2005) *J. Am. Chem. Soc.*, **127**, 2094–2097).

molecular thermometers (Ke et al., 2012). An L-DNA-based molecular beacon, FAM-CGAG TTT TTT TTT TTT TTT CTCG-Dabcyl is a hairpin-structured dual-labeled oligonucleotide. As indicated in the function scheme in Figure 2.31a, the distance between the fluorophore (FAM) and the quencher (Dabcyl) varies with temperature. The fluorescence intensity of the L-DNA-based molecular beacon in a PBS buffer (pH 7.4) increased 3.5 times with a temperature increase from 20 to 55 °C (Figure 2.31b). Utilizing non-natural L-DNA is crucial in intracellular thermometry, as a D-DNA-based molecular beacon should not exhibit any responses, likely due to its rapid digestion by endogenous nucleases. Afterward, molecular beacons conjugated to gold nanoparticles were proposed (Ebrahimi et al., 2014). The work

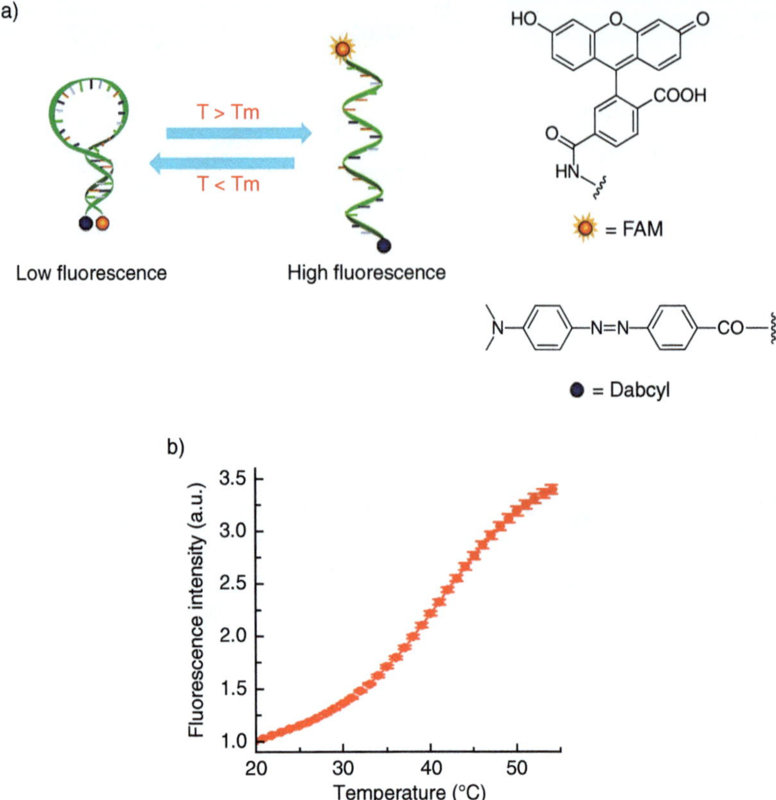

Figure 2.31 Nucleic acid-based fluorescent thermometers (continued). Molecular beacon with FAM and Dabcyl end groups. (a) Functional mechanism. Tm stands for melting temperature. (b) Temperature-dependent fluorescence intensity of the molecular beacon in a PBS buffer (pH 7.4) with Ke et al. (2012) *J. Am. Chem. Soc.*, **134**, 18908–18911 / American Chemical Society.

used gold nanoparticles as quenchers and carriers of molecular beacons. The flexibility in the design of molecular beacons enabled easy tuning of operating temperature ranges. Also, control of temperature sensitivity was demonstrated with structural modifications in molecular beacons (Gareau et al., 2016). A more intricate molecular beacon (called "self-assembled DNA tetrahedron") with a FRET system between 6′-carboxyfluorescein (FAM; λ_{em} = 522 nm) and tetramethyl rhodamine (TAMRA; λ_{em} = 580 nm) was reported as a ratiometric fluorescent molecular thermometer that functioned from 27 to 54 °C in a Tris (tris(hydroxymethyl)aminomethane) buffer (pH 7.4) with the high-temperature resolution (0.1–1 °C) (Xie et al., 2017). Another molecular beacon labeled with fluorophores (Cy5 and perylene, see Section 2.4.5 and 2.9, respectively) and a quencher (BHQ3: Black Hole Quencher®-3) is also available (Lee et al., 2019).

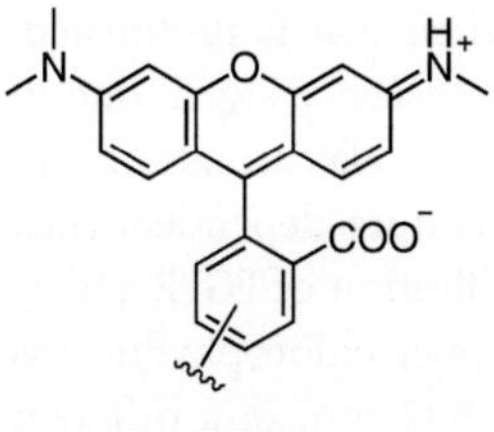

TAMRA

2.13
Peptides

The temperature-induced conformational change was also observed in peptide-1 (Scheme 2.3), which contains an α-aminoisobutyric acid residue (Aib), a methoxynaphthalene-containing residue (Don, as an electron donating group), and a piperidone-containing residue (Acc, as an electron accepting group) (Hungerford et al., 1996). The peptide in acetonitrile at a given temperature emits fluorescence with two different lifetimes. These two lifetimes originate from two equilibrating conformations (α-helix and 3_{10}-helix) of peptide-1, in which the distances between Don and Acc, in other words, efficiencies of photoinduced electron transfer between these residues, are not identical. Therefore, the proportion of fluorescence with each lifetime is temperature-dependent.

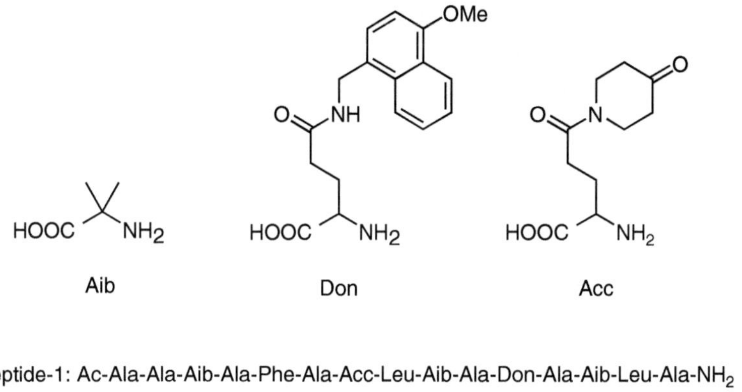

Aib

Don

Acc

Peptide-1: Ac-Ala-Ala-Aib-Ala-Phe-Ala-Acc-Leu-Aib-Ala-Don-Ala-Aib-Leu-Ala-NH$_2$

Scheme 2.3

2.14
Fluorescent Proteins

An early case used enhanced green fluorescent protein (EGFP) that can alternate between two states with different fluorescent properties (Wong et al., 2007). When the hydroxyl group of Tyrosine-66 on the chromophore is deprotonated, EGFP fluoresces upon 488 nm excitation. On the other hand, when that is protonated, EGFP is mostly non-fluorescent. The fluctuations in the fluorescence signal due to the reversible protonation are called blinking. Under pH = 5, the relaxation time (τ_B) of the blinking associated with protonation is strongly temperature-dependent (i.e., τ_B = 160 and 15 µs at 10 and 55 °C, respectively). As an application of EGFP, a temperature increase of a liquid sample containing sodium copper chlorophyllin was monitored under continuous radiation by a laser diode at 637 nm. In a different study, the fluorescence intensity itself of EGFP at 510 nm indicated moderate sensitivity to a temperature variation, i.e., fluorescence intensity at 70 °C is 44% of that at 20 °C (Toca-Herrera et al., 2006). In an exceptional case, individual *Escherichia coli* over-expressing EGFP was used as a fluorescent thermometer to measure the temperature of the bacterial culture in a Fabry–Perot optofluidic planar microcavity (Lahoz et al., 2017). Nevertheless, the fluorescence intensity of EGFP was also remarkably affected by pH variation. This low selectivity of responses of EGFP would prevent its use for real applications. Although photophysical mechanisms were unclear, temperature-dependent fluorescence spectra of EGFP and red fluorescent protein (RFP, DsRed1) in buffer solutions (pH 7.5) (Figure 2.32) were reported in work on infrared laser-mediated gene induction (Kamei et al., 2009). Temperature-dependent fluorescence intensities of red fluorescent proteins (mRFP1 (-1.27% $°C^{-1}$), mRFP1-P63A (-1.26% $°C^{-1}$), and mRFP1-P63A[(4R)-FP] (-0.77% $°C^{-1}$)) were also reported after expressed in *E. coli* (Deepankumar et al., 2015). In 2022, temperature-sensitive fluorescence spectra of a red fluorescent protein, mPlum, were reported (Lyu et al., 2022). Due to the interconversion between direct hydrogen bonding and water-mediated hydrogen bonding between the chromophore acylimine carbonyl and the Glu16 side chain, the emission wavelength of a dominant mPlum form in a pH 8.0 Tris buffer shifted from 630 nm to 650 nm with the increase in temperature from -33 °C to 22 °C (upon excitation at 555 nm).

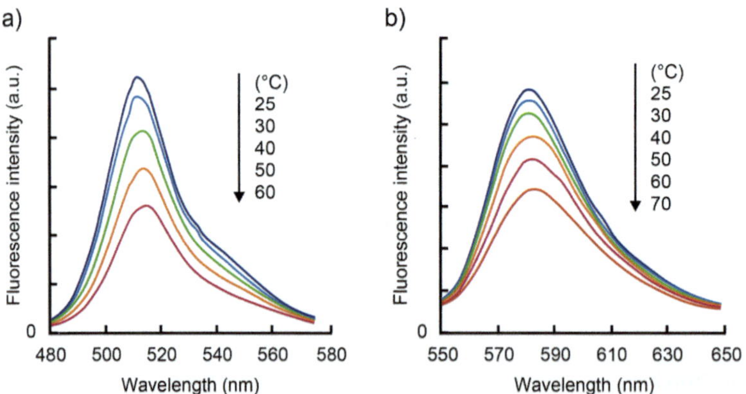

Figure 2.32 Fluorescent proteins. Temperature-dependent fluorescence spectra of (a) EGFP (1 µg mL^{-1}, λ_{ex}: 473 nm) and (b) RFP (DsRed1) (1.2 µg mL^{-1}, λ_{ex}: 532 nm) in HEPES buffer solutions (pH 7.5). Adapted from Kamei et al. (2009) *Nat. Methods*, **6**, 79–81

2.15
Non-covalent Systems Based on Thermo-responsive Self-assembly and an Environment-sensitive Fluorophore

As well as covalent macromolecular systems described above, non-covalent designed systems, including supramolecular assembly, can create temperature-dependent fluorescence signals. Thermo-sensitive liposomes of phosphatidylcholines release cargos when the environmental temperature exceeds the thresholds that depend on the length of alkyl chains (n) in phosphatidylcholines (31, 39, and 47 °C for $n = 15$, 16, and 17, respectively). If fluorescent calcein is encapsulated in the liposome, it is self-quenched below a threshold temperature. Once the environmental temperature increases over the threshold, calcein is released from the liposome to start emitting fluorescence (Sou et al., 2016). For non-covalent systems, reversibility and independency of function in complex situations like living cells are the most critical issue to be addressed. Later, the same research group reported the combination of phosphatidylcholines and a far-red fluorescent dye SQR22 (Sou et al., 2019). In the liposome solution, SQR22 was incorporated into the lipid bilayer membranes of phosphatidylcholines. At a cold temperature, the fluorescence of SQR22 incorporated into the bilayer was almost negligible due to aggregation-caused quenching. With increasing temperature, the lipid bilayer membrane transitioned from a rigid gel state to a more flexible liquid crystalline state in which SQR22 favored being a monomer to strongly emit red fluorescence at 664 nm under the excitation at 365 nm. Though showing a considerable hysteresis (i.e., different fluorescence responses to a temperature variation upon heating and cooling), the reversibility in response was confirmed. The thermo-responsive household unsalted butter and aggregation-induced emission (AIE) dye were proposed to construct a temperature-sensing system (Gao et al., 2020).

Calcein

Phosphatidylcholine

SQR22

A dual-responsive organic fluorophore TICT@AIE responds to the molecular state of aggregation and environmental polarity. When TICT@AIE is encapsulated into a natural phase-change material composed of lauric acid and stearic acid, external temperature information can be correlated with the fluorescence properties of TICT@AIE (Xue et al., 2021). At a low temperature, the phase-change material takes a solid state where the maximum emission wavelength of TICT@AIE is red-shifted with a dramatic decrease in fluorescence intensity because of the self-aggregation. At a high temperature, the material prefers a liquid phase where TICT@AIE disaggregates and adopts a more planar conformation by a hydrophobic microenvironment imposed by the long alkyl chains of fatty acids to emit strong fluorescence at a shorter emission wavelength. The functional temperature range and color can be tuned by the composition of fatty acids and the concentration of TICT@AIE, respectively.

The most straightforward phase transition is the solid–liquid transition, e.g., ice and water. A recent report shows that an AIE luminophore TPE-octane suddenly starts to fluoresce when the temperature decreases blow thresholds depending on solvents: -64, -29, and $-95\,°C$ in octane, decane, and hexane, respectively (Guan et al., 2022). This fluorescence switching is because the non-radiative intramolecular motion of TPE-octane is effectively restricted below the freezing points of solvents.

TICT@AIE

TPE-octane

As described in later Section 2.16.5, inorganic lanthanide-based upconversion nanoparticles (UCNPs) are typical for observing fluorescence signals with multiple-photon excitation at a wavelength longer than an emission wavelength. Recently, it has been found that organic compound-based nanoparticles also show the upconversion phenomenon involving triplet-triplet annihilation (Zhou et al., 2015). In the presence of a triplet sensitizer, a fluorescent compound at the triplet excited state is produced by an energy transfer. Then, two fluorescent molecules at the triplet excited state generate one fluorescent molecule at the singlet excited state (i.e., triplet-triplet annihilation) to emit fluorescence. Because highly dense fluorescent compounds and sensitizers are required for this triplet-triplet annihilation upconversion phenomenon, polymer matrices are used to enclose them. Also, triplet-triplet annihilation-based upconversion nanoparticles can function only under anoxic conditions because oxygen atoms quench triplet states of compounds. The first example of triplet-triplet annihilation-based upconversion nanoparticles for temperature sensing attempted to utilize the sensitivity to oxygen concentration as it depended on the temperature in living cells (Iyisan et al., 2020). Then, a thermo-responsive F127 (poly(ethylene oxide)-poly(propylene oxide)-poly(ethylene oxide)) micelle was reported as a suitable nanoarchitecture for triplet-triplet annihilation-based upconversion (Li et al., 2021), where a sensitizer PtOEP (see Section 2.8.3) and a fluorophore p-DHMPA (9,10-di(p-hydroxymethylphenyl)anthracene) were contained (Figure 2.33a). In aqueous solution, F127 micelles show a thermo-responsive swelling/shrinking: average hydrodynamic diameter was 24.59 and 19.54 nm at 30 and 60 °C, respectively. Thus, the efficiency of triplet-triplet annihilation involving PtOEP and p-DHMPA is higher at 60 °C, where the distance between the compounds in F127 micelles becomes closer compared to 30 °C. With increasing temperature, the upconverted fluorescence of p-DHMPA is enhanced while its phosphorescence is diminished (Figure 2.33b, 2.33c). Interestingly, F127 micelles shielded upconversion fluorescence from oxygen quenching without special treatments.

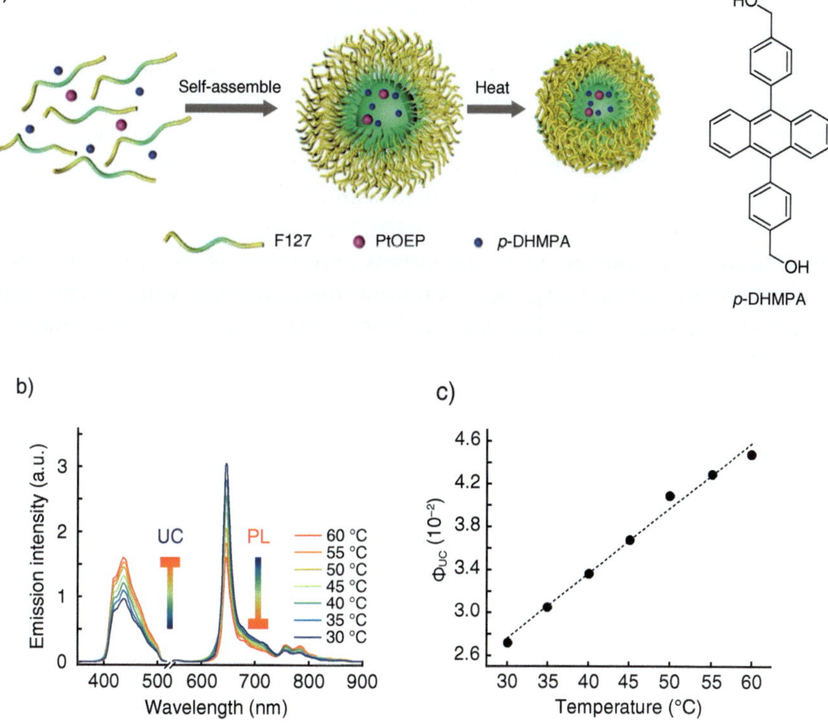

Figure 2.33 Triplet-triplet annihilation (TTA) upconversion nanomicelles. (a) Schematic illustration of thermo-responsive TTA upconversion F127 nanomicelles containing PtOEP and p-DHMPA. (b) Temperature-dependent emission spectra of the TTA upconversion F127 nanomicelles containing PtOEP (10 μmol L^{-1}) and p-DHMPA (250 μmol L^{-1}) in water (λ_{ex}: 532 nm, UC: upconversion fluorescence, PL: phosphorescence). (c) The temperature-dependent quantum yield of upconversion fluorescence (Φ_{UC}) of the TTA upconversion F127 nanomicelles in water. Adapted from Li et al. (2021) *Angew. Chem. Int. Ed.*, **60**, 26725–26733 / JOHN WILEY & SONS, INC.

2.16
Inorganic Nanomaterials

Atomic species and small inorganic materials can achieve the thermometer function in a similar fashion to organic molecule-based fluorescent thermometers. For example, it has been known for a long time that the precious stone ruby has fluorescence characteristics that are affected by temperature (Nelson and Sturge, 1965; Anghel et al., 1995). This fluorescence behavior is caused by two competing relaxation pathways in its main optical component, Cr^{3+} ion. In the early stages of a research field on fluorescent thermometry, it was often considered that inorganic nanomaterials were not compatible with the field, especially in intracellular thermometry, because of relatively large size, potential cytotoxicity, and low dispersibility in aqueous environments in comparison with molecule-based fluorescent thermometers. Nevertheless, researchers' substantial efforts have raised the reputation of inorganic materials with numerous reports relevant to methods for the preparation, temperature-sensitive fluorescence properties in aqueous solution, and applications including intracellular thermometry. In this section, I will summarize inorganic nanomaterials, whose fluorescence properties are temperature-sensitive in aqueous solution. In addition, hybrid nanomaterials, e.g., polymeric nanoparticles containing temperature-sensitive fluorescent inorganic materials, will also be picked up in a later subsection. Hundreds of research papers have been published on a solid inorganic material that shows temperature-dependent fluorescence. Furthermore, some solid inorganic materials were applied for temperature mapping. For example, temperature imaging of a printed circuit board was performed with Zn_2GeO_4:Mn^{2+} with a temperature resolution of 0.68 °C (Chi et al., 2019). However, I consider that the use of fluorescent thermometers in a solid state is beyond the scope of this textbook and, therefore, solid inorganic fluorescent thermometers will be concisely touched on in Appendix 6. Lastly, a gas phase example is given: the temperature-dependent fluorescence of nitric oxide from various rotational levels. This feature helps measure the temperature of flames seeded with nitric oxide (Tamura et al., 1998).

2.16.1
Quantum Dots

Quantum dots (also called "Q-dots") are spherical inorganic substances typically with a 2–10 nm diameter and are expected as novel fluorescent semiconductor materials. Due to high emission quantum yields, high molar extinction coefficients, photoluminescence spectra with a narrow width, and high resistance to chemical decomposition, no wonder quantum dots have been exploited in live cell imaging, in vivo imaging, and even diagnostics (Medintz et al., 2005; Michalet et al., 2005). Quantum dots consist of less than 100 000 atoms and, strictly speaking, are not fluorescent "molecular" thermometers. Nevertheless, quantum dots are sufficiently small and highly fluorescent, so the utilization of quantum dots in fluorescent thermometry of tiny spaces is quite reasonable. The excitation and emission wavelengths of quantum dots depend on the particle size, implying that the control of their size can easily tune the color of quantum dots. Moreover, it is advantageous that quantum dots are relatively resistant to photobleaching compared to organic molecule-based fluorescent molecular thermometers. The initial case used CdTe quantum dots in the dried state (Wang et al., 2002). Afterward, emission properties of CdTe quantum dots (diameter: $\leq$ 7.8 nm) dispersed in water were thoroughly examined under two-photon optical excitation (Maestro et al., 2011). Among various quantum dots, CdSe quantum dots showed highly temperature-dependent emission characteristics in the visible light wavelength region. However, CdSe quantum dots are not very suitable for temperature measurements due to the instability toward oxidation and the low emission efficiency. Thus, CdSe quantum dots were coated by ZnS to improve these weaknesses, and the resultant CdSe/ZnS quantum dots' emission properties were investigated in detail to use as fluorescent thermometers (Walker et al., 2003). Figure 2.34a shows temperature-dependent emission spectra of CdSe/ZnS quantum dots with a 5.0–5.5 nm diameter in a poly(lauryl methacrylate) matrix. With increasing temperature from −173 to 42 °C (from 100 to 315 K), the emission intensity of CdSe/ZnS quantum dots decreased to approximately 9% with a red shift of maximum emission wavelength. Cretí et al. also investigated the emission properties of CdSe/ZnS quantum dots in a polystyrene matrix (Figure 2.34b) and concluded that the primary non-radiative process, which competes with emission, is a thermal escape from the quantum dots assisted by scattering with four longitudinal-optical-phonons (Valerini et al., 2005). For thermometry adopting the temperature-dependent emission wavelength as a detection parameter, CdTe/CdSe colloidal heteronanocrystals with a relatively constant (i.e., temperature-independent) emission quantum yield were proposed (Chin et al., 2007). Ratiometric responses to a temperature variation have been also proposed by using plural lumophores such as $Zn_{1-x}Mn_xSe/ZnCdSe$ core/shell nanocrystals (Vlaskin et al., 2010), $Zn_{1-x}Mn_xSe/ZnS/CdS/ZnS$ multilayer nanocrystals (McLaurin et al., 2011), water-soluble D-penicillamine-passivated Mn^{2+}-doped (CdSSe)ZnS (core)shell nanocrystals (Hsia et al., 2011), and Mn-doped CdS–ZnS core-shell nanocrystals (Park et al., 2013).

We can find several applications of quantum dots for temperature measurements of non-biological subjects. An early attempt used CdSe quantum dots, in which a maximum emission wavelength was adopted as a temperature-dependent but concentration and excitation strength-independent parameter (Li et al., 2007). As shown in Figure 2.34c, 2.34d, CdSe quantum dots in the dried state measured the local temperature of a microelectromechanical system (MEMS) aluminum heater with dimensions of $1200 \times 40 \times 0.1$ μm^3. In a different study, the temperature of a microreactor (300 μm in width and 200 μm in depth) was correlated to emission properties of CdSe/ZnSe/ZnS core/shell-type semiconductor nanocrystals dispersed in octadecene in the channel (Lee et al., 2007). Mn-doped CdS-ZnS nanocrystals were employed in ratiometric temperature imaging of a

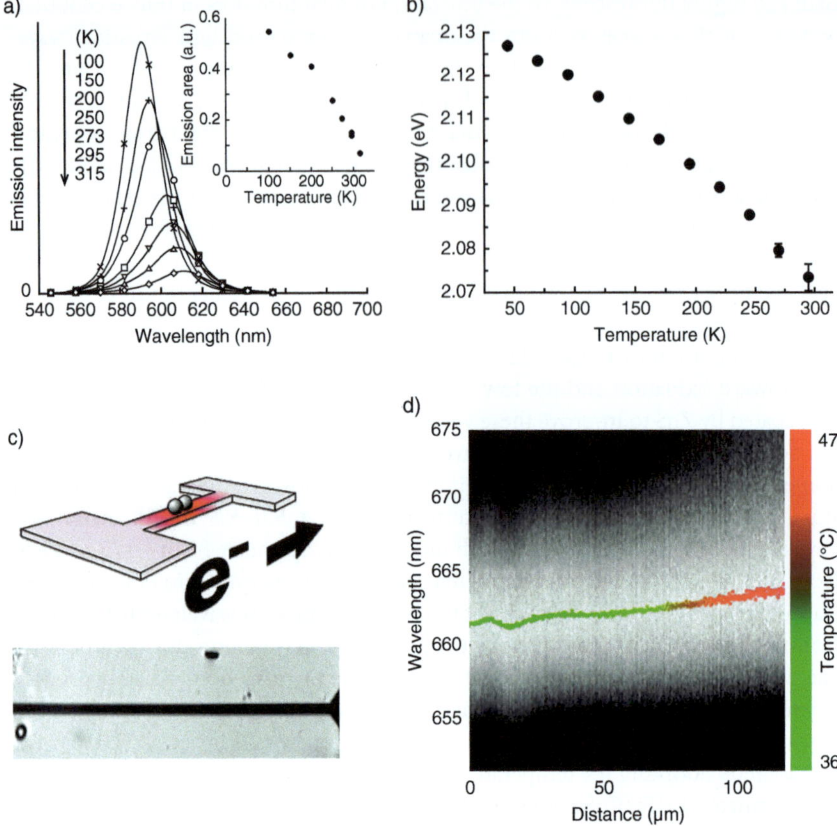

Figure 2.34 Quantum dots (Q-dots). (a) Emission spectra of CdSe/ZnS Q-dots in a poly(lauryl methacrylate) matrix (λ_{ex}: 480 nm) and corresponding temperature-dependent emission area (inset) Walker et al. (2003) *Appl. Phys. Lett.*, **83**, 3555–3557 / AIP Publishing. (b) Relationship between the emission peak energy of CdSe/ZnS Q-dots in a polystyrene film and temperature (λ_{ex}: 380 nm) (adapted from Valerini et al. (2005) *Phys. Rev. B*, **71**, 235409). (c,d) Temperature measurement of an aluminum microheater by CdSe Q-dots. (c) Schematic diagram (top) and wild image (bottom) of the microheater. (d) Temperature profile calculated from the maximum emission wavelength of CdSe Q-dots. Panels (c,d) are adapted from Li et al. (2007) *Nano Lett.*, **7**, 3102–3105 / American Chemical Society.

cryo-cooling device with a temperature gradient from −196 to −13 °C (Park et al., 2013). ZnO microcrystals could be used for temperature mapping of a carbon nanotube mat in a wide temperature range (−190–200 °C) with an accuracy of 0.1 °C (Shinde and Nanda, 2013). Very recently, a thick hollow core fiber filling an octane solution of CdZnSe/ZnSe/ZnS quantum dots was proposed for real-time monitoring of human thermal activities (Xu et al., 2023). Biological applications, including intracellular thermometry, have been performed with selected quantum dots (see Chapter 3).

Remarkably, approximately 2 nm sized CdTe quantum dots adhered to myosin (bovine cardiac or rabbit skeletal muscle myosin) were capable of directly measuring heat which was released in the hydrolysis of ATP by myosin motors (Laha et al., 2017). The spatial and thermal resolutions of this thermometry were 80 nm and 1 mK, respectively. Muscle efficiency during work could be evaluated by monitoring heat loss (i.e., temperature increase) by the CdTe quantum dots-attached myosin motors. It was found that rabbit skeletal myosin was more efficient than that of bovine cardiac.

2.16.2
Gold Nanoparticles

Gold nanoparticles are emissive nanomaterials. Although gold nanoparticles in an early case only indicate low photoluminescence quantum yield (approximately 10^{-5}–10^{-4}) (Wilcoxon et al., 1998), an advanced method for preparing gold nanoparticles (diameter: 1–2 nm) with a surface ligand, 1-octadecylmercaptane improved it to 0.166 in chloroform at room temperature (Bomm et al., 2012). The gold nanoparticles showed strong thermo-sensitive photoluminescence at around 620 nm with an excitation at 320 nm probably due to heat-induced non-radiative recombinations of electrons and holes: photoluminescence quantum yield increased to 0.286 at $-7\,°C$ and decreased to 0.112 at $36\,°C$. Aimed at intracellular thermometry, temperature-sensitive gold nanoclusters stabilized by lipoic acid (Shang et al., 2013) or by 6-aza-2-thiothymine and L-arginine were also reported (Kundu et al., 2019). The surface-modified gold nanoclusters showed excellent temperature-dependency in emission intensity or emission lifetime in an aqueous environment. For instance, with increasing temperature from 14 to $43\,°C$, the intensity-weighted average emission lifetime of the lipoic acid-capped gold nanoclusters with the excitation at 470 nm and the emission of 690 ± 35 nm was reduced from 970 to 670 ns in fixed HeLa (human epithelial carcinoma) cells in PBS (Shang et al., 2013).

2.16.3
Carbon Dots

Carbon dots (also called "C-dots") are carbon element-based nanomaterials that were reported in 2004 for the first time (Xu et al., 2004). Carbon dots can be prepared from natural materials such as bacteria, plants, and vegetables and chemical compounds such as glucose and citric acid with the benefit of laser ablation, microwave irradiation, hydrothermal treatment, sonication, etc. In addition to the low toxicity and low environmentally hazardous nature compared to metal element-based nanomaterials, carbon dots show bright fluorescence in both solution (or dispersion) and the solid state (Sun et al., 2006). Within a decade, various reports on intracellular fluorescence imaging with carbon dots have been stored (Lin et al., 2022).

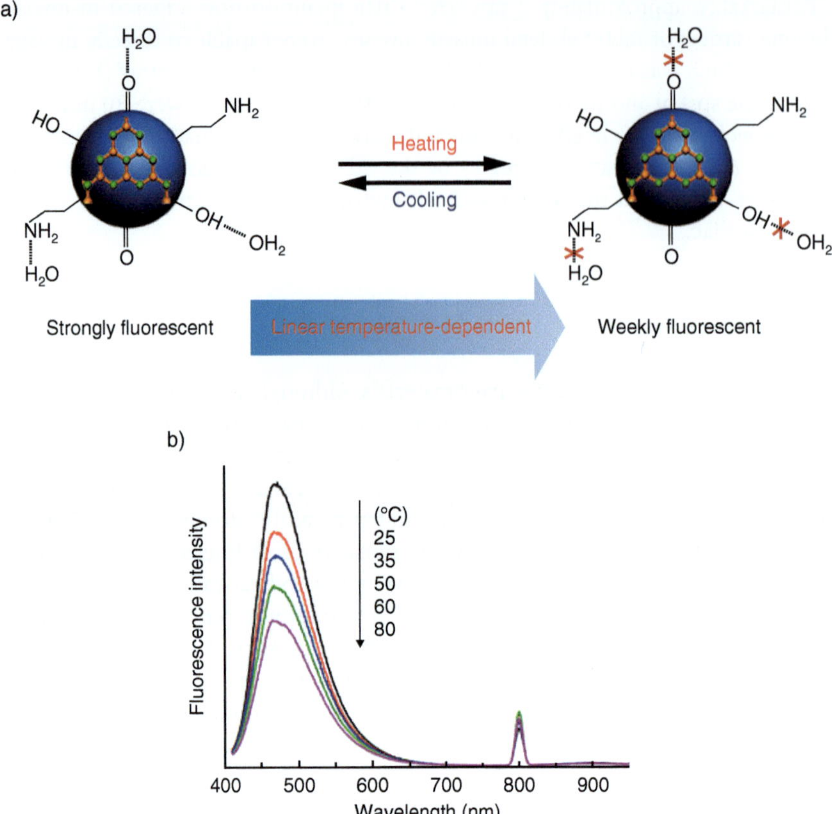

Figure 2.35 N-Doped carbon dots. (a) Schematic diagram of functional mechanism in water. (b) Fluorescence spectra in water (λ_{ex}: 400 nm). Yang et al. (2015) *ACS Appl. Mater. Interfaces*, **7**, 27324–27330 / American Chemical Society.

The first example of carbon dots which function as fluorescent thermometers, is summarized in Figure 2.35 (Yang et al., 2015). In the study, N-doped carbon dots with a diameter of 5 nm were prepared from C_3N_4 and ethanediamine by refluxing. The fluorescence intensity of N-doped carbon dots in water was higher than in ethanol. Furthermore, the fluorescence intensity of N-doped carbon dots in water dropped when the –C=O groups of their surface were removed by the reduction with $NaBH_4$. Therefore, it can be concluded that the hydrogen bondings are required for the fluorescence of N-doped carbon dots and are reasons for their temperature-dependency in water. In cold water, N-doped carbon dots fluoresce strongly because of the hydrogen bondings with solvent water molecules, whereas the breaking of hydrogen bondings at a higher temperature diminishes the fluorescence of N-doped carbon dots (Figure 2.35a, 2.35b).

Until now, various kinds of carbon dots have been reported as fluorescent thermometers, as summarized in Table 2.2. It should be noted that most of the reported carbon dots were introduced in live or fixed cells to check the function as fluorescent thermometers in a biological milieu and the low toxicity to cells. Nevertheless, there have yet to be reports on biological studies using intracellular thermometry with carbon dots.

2.16.4
Fluorescent Nanodiamond

Fluorescent nanodiamond is a novel nanomaterial boosted by a striking paper published in 2005 (Yu et al., 2005). Then, rapid progress is found in the preparation of surface-modified fluorescent nanodiamonds as well as its applications in cell labeling and bioimaging with advantages in the inherent biocompatibility and robustness (Mochalin et al., 2012; Hsiao et al., 2016). Naturally, thermometry is a field in the territory of fluorescent nanodiamonds (Sotoma et al., 2018). The presence of nitrogen-vacancy (NV) centers (a nitrogen atom next to vacancy) in nanodiamonds leads to strong fluorescence, and two types of NV centers are known: neutral (NV^0) and negatively charged (NV^-) centers. These two forms have different emission spectra, and especially the latter (NV^-) center is important for the temperature-dependent emission of fluorescent nanodiamonds (Figure 2.36a). The NV^- center absorbs light at 550 nm for the $^3A \rightarrow {}^3E$ transition (Figure 2.36b, 2.36c), and its emission band peak at 685 nm with a high fluorescence quantum yield ($\Phi_f \sim 1$, Figure 2.36d). As indicated in Figure 2.36e, careful observation of the emission spectrum of the NV^- center of fluorescent nanodiamond dispersed in water could lead to the fact that the zero-phonon line (ZPL) around 637 nm (corresponding to the $^3A_2 \rightarrow {}^3E$ transition) was red-shifted as the temperature increased from 28 to 75 °C (Tsai et al., 2017). Nevertheless, the sensitivity of the ZPL of the NV^- center of fluorescent nanodiamonds to a temperature variation is not high, so this property has rarely been utilized for fluorescence thermometry. Recently, the emission lifetime of the NV^- center of fluorescent nanodiamonds was proposed as a temperature-dependent parameter but is not widespread yet (Bommidi and Pickel, 2021). Instead, a more sensitive temperature-dependent zero-field splitting parameter (D), initially reported in 2010 (Acosta et al., 2010), has been proposed as a principle of fluorescence thermometry using the NV^- center of fluorescent nanodiamond. The electron spins in the ground state of NV^- centers can be optically polarized, provided by the intersystem crossing in the excited state, enabling optical readout of their spin states ($m_s = 0$ and ± 1) at the single molecular level by using optically detected magnetic resonance (ODMR) techniques (Figure 2.36f, 2.36g) (Hsiao et al., 2016). In 2013, three independent research groups published fluorescence thermometry using fluorescent nanodiamonds and the ODMR technique one after another (Kucsko et al., 2013; Neumann et al., 2013; Toyli et al., 2013). It was insisted that this method's temperature resolution reached a mK level. Furthermore, it was reported that a Monte Carlo simulation could improve the accuracy of the ODMR temperature measurements with fluorescent

Table 2.2 Carbon dots as fluorescent thermometers functioning in water.

Entry	Material	λ_{ex}(nm)	λ_{em}(nm)	Parameter to be measured[a]	Functional temperature range (°C)	Parameter variation	Remarks	Ref.
1	C3N4	400	475	FI	20–80	1–0.49	N-Doped carbon dots	[1]
2	Citric acid	367	430, (605 for gold nanocluster)	FI ratio	20–75	0.72–2.39	Ratiometric with gold nanoclusters	[2]
3	p-Phenylenediamine	365	615	FI	10–80	1–1.75		[3]
4	Citric acid, L-cysteine	355	421	τ_f	2–80	11–5.3 (ns)	N,S-co-Doped carbon dots	[4]
5	Bovine serum albumin	405	482, (586 for rhodamine B)	FI ratio	5–50	1.49–1.21	Labeled by rhodamine B	[5]
6	Glutathione, formamide	405	460, 680	Peak area ratio	5–60	1–3.5		[6]
7	Tyrosine, urea	410	518	FI	25–80	1–0.69	N-Doped carbon dots, also sensing Co^{2+} ion	[7]
8	*Dendrobium officinale*, ethylenediamine	362	448 (for C-dots), 660 (for gold)	FI ratio	5–75	1.98–1.15	N-Doped carbon dots, hybrid with gold nanoparticles, also sensing tyrosine	[8]
9	Citric acid, thionine	500	650	FI	4–80	1–1.9		[9]
10	Reduced graphene oxide	475	950	FI	25–45	1–0.93		[10]
11	Glucosamine	400	500	FI	25–49	1–0.81		[10]

12	Phloroglucinol	405	521, 560	Peak area ratio	33–47	2.66–1.47	Fluorescence spectra were deconvoluted	[11]
13	N,N-Diethyl-p-phenylenediamine	405	505, 630	Peak area ratio	34–46	2.55–2.10	Fluorescence spectra were deconvoluted	[11]
14	Trisodium citrate, L-cysteine	360	445	FI	25–70	1–0.40	N,S-co-Doped carbon dots	[12]

a) FI: fluorescence intensity; FI ratio: fluorescence intensity ratio. τ_f: fluorescence lifetime.

1 Yang, Y., Kong, W., Li, H., Liu, J., Yang, M., Huang, H., Liu, Y., Wang, Z., Wang, Z., Sham, T.-K., Zhong, J., Wang, C., Liu, Z., Lee, S.-T., and Kang, Z. (2015) *ACS Appl. Mater. Interfaces*, **7**, 27324–27330.

2 Wang, C., Lin, H., Xu, Z., Huang, Y., Humphrey, M. G., and Zhang, C. (2016) *ACS Appl. Mater. Interfaces*, **8**, 6621–6628.

3 Wang, C., Jiang, K., Wu, Q., Wu, J., and Zhang, C. (2016) *Chem. Eur. J.*, **22**, 14475–14479.

4 Kalytchuk, S., Poláková, K., Wang, Y., Froning, J. P., Cepe, K., Rogach, A. L., and Zbořil, R. (2017) *ACS Nano*, **11**, 1432–1442.

5 Wei, L., Ma, Y., Shi, X., Wang, Y., Su, X., Yu, C., Xiang, S., Xiao, L., and Chen, B. (2017) *J. Mater. Chem. B*, **5**, 3383–3390.

6 Macairan, J.-R., Jaunky, D. B., Piekny, A., and Naccache, R. (2019) *Nanoscale Adv.*, **1**, 105–113.

7 Shi, L., Chang, D., Zhang, G., Zhang, C., Zhang, Y., Dong, C., Chu, L., and Shuang, S. (2019) *RSC Adv.*, **9**, 41361–41367.

8 Hua, J., Mu, Z., Hua, P., Wang, M., and Qin, K. (2020) *Talanta*, **219**, 121279.

9 Xu, Y., Yang, Y., Lin, S., and Xiao, L. (2020) *Anal. Chem.*, **92**, 15632–15638.

10 Lee, B. H., McKinney, R. L., Hasan, Md. T., and Naumov, A. V. (2021) *Materials*, **14**, 616.

11 Han, Y., Liu, Y., Zhao, H., Vomiero, A., and Li, R. (2021) *J. Mater. Chem. B*, **9**, 4111–4119.

12 Khan, W. U., Qin, L., Alam, A., Zhou, P., Peng, Y., and Wang, Y. (2021) *ACS Appl. Bio Mater.*, **4**, 5786–5796.

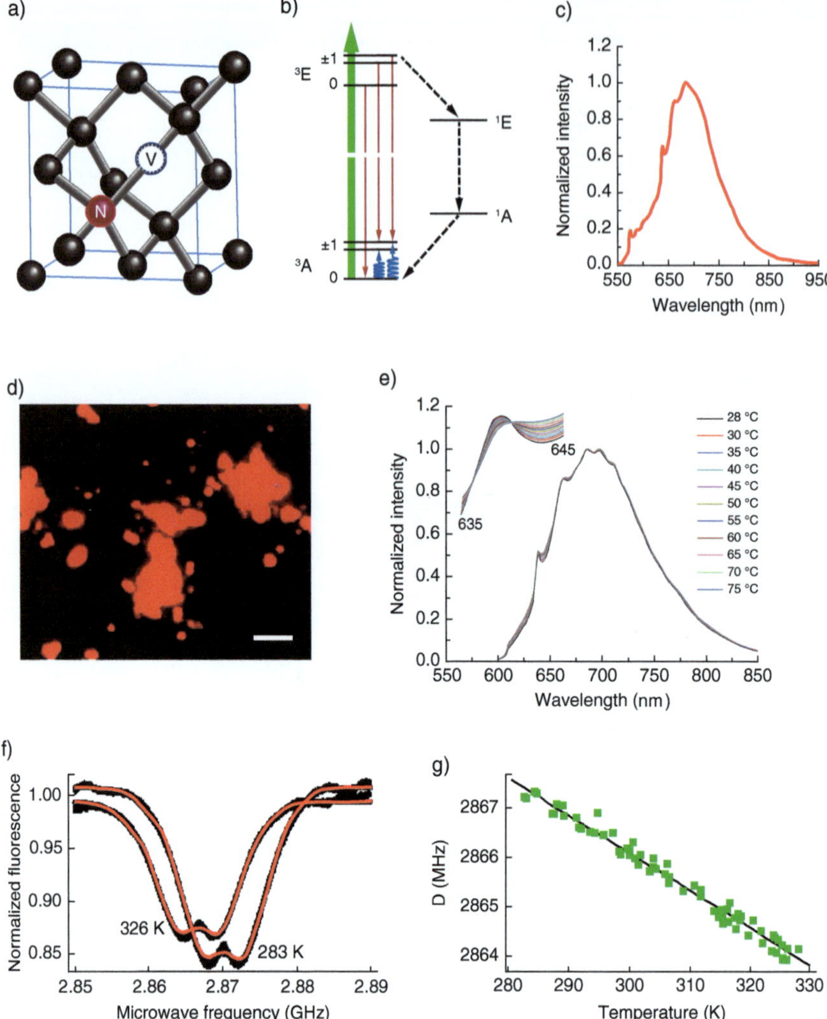

Figure 2.36 Fluorescent nanodiamonds. (a) Structure of the NV⁻ center in nanodiamonds. Black spheres, a red sphere, and a blue circle represent carbon atoms, a nitrogen atom, and vacancy, respectively. (b) Energy level diagram of the NV⁻ center in nanodiamonds. The green and red arrows indicate excitation and fluorescence emission, respectively. (c) Fluorescence spectrum (λ_{ex}: 532 nm). Panels (a–c) are Hsiao et al. (2016) *Acc. Chem. Res.*, **49**, 400–407 / American Chemical Society. (d) Fluorescence image under the excitation by a mercury vapor lamp (λ_{ex}: 510–560 nm). Scale bar: 10 µm. Adapted from Yu et al. (2005) *J. Am. Chem. Soc.*, **127**, 17604–17605 / American Chemical Society. (e) Temperature-dependent fluorescence spectra of fluorescent nanodiamonds in water (λ_{ex}: 594 nm). Inset: Enlarged view of the temperature-induced shift of the zero phonon lines. Tsai et al. (2017) *Angew. Chem. Int. Ed.*, **56**, 3025–3030 / John Wiley & Sons. (f) Optically detected magnetic resonance (ODMR) spectra of fluorescent nanodiamonds at 283 and 326 K. (g) Temperature-dependent axial zero-field splitting (ZFS) parameter. Panels (f and g) are Acosta et al. (2010) *Phys. Rev. Lett.*, **104**, 070801 / American Physical Society.

nanodiamonds by determining an optimal frequency sweep range (Yanagi et al., 2020). In addition to temperature measurements at local places near a microstructured loop-gap resonator (heater) (Neumann et al., 2013), intracellular thermometry with fluorescent nanodiamonds started to emerge (Kucsko et al., 2013). Readers can refer to several reports concerning technical improvement in the ODMR measurements and limitations in thermometry using fluorescent nanodiamonds, including intracellular thermometry (Simpson et al., 2017; Yukawa et al., 2020; An et al., 2021; Nishimura et al., 2021).

Recently, a temperature-dependent emission wavelength of "silicon-vacancy" centers in fluorescent nanodiamonds was also reported (Romshin et al., 2021). In water, upon the excitation at 473 nm, the maximum emission wavelength of the silicon-vacancy centers of fluorescent nanodiamond shifted from 738.09 to 738.36 nm with the increase in temperature from 22 to 50 °C. In a different research group, silicon-vacancy nanodiamonds were coated by a protein-derived biopolymer (i.e., based on human serum albumin and polyethylene glycol (PEG)) to enhance colloidal stability in an aqueous environment. This modification of silicon-vacancy nanodiamonds was suitable for their utilization in the intracellular environment, but, at the same time, it was recognized that there was a considerable deviation in the maximum emission wavelength in each nanocrystal (Liu et al., 2022).

2.16.5
Upconversion Nanoparticles

Upconversion nanoparticles (UCNPs), which stands typically for lanthanide-doped UCNPs, are relatively new materials that emit multiple narrow emission peaks in the visible spectral range by excitation in the near-infrared (NIR) region (i.e., 800–1000 nm). Uniquely, lanthanide-doped UCNPs can combine two or more low-energy photons to generate a single high-energy photon by an anti-Stokes process. In addition to the high photostability, the NIR excitation is a significant advantage of lanthanide-doped UCNPs because it can substantially reduce possible interference from biological tissues (e.g., light scattering) in bioimaging (Han et al., 2014; Wen et al., 2018). It should be noted that readers can consult review articles on UCNPs specific to thermometry (Fischer et al., 2011; Ansari et al., 2021).

The earliest examples of UCNPs as fluorescent thermometers were CaF_2:Er^{3+},Yb^{3+} nanoparticles that could be excited at 920 nm to show a temperature-dependent emission intensity ratio at 522 and 538 nm, and CaF_2:Tm^{3+},Yb^{3+} nanoparticles that could be excited at the same wavelength to show it at 790 and 800 nm in dispersion in PBS (Dong et al., 2011). As a later related case, hexagonal $NaYF_4$ nanocrystals doped with Yb^{3+} as the sensitizer and Er^{3+} as the activator displayed multiple emission lines that respond differently to a temperature variation. UCNPs of $NaYF_4$:Yb, Er covered by an inactive shell of $NaYF_4$ could be dispersed in water after their surface ligand was exchanged to citric acid. Under the excitation at 980 nm, the emission intensity ratio of UCNPs at 541 nm ($^4S_{3/2} \rightarrow {}^4I_{15/2}$ transition) and 523 nm ($^2H_{11/2} \rightarrow {}^4I_{15/2}$ transition) in water allowed for distinguishing a temperature difference of less than 0.5 °C in the physiological range (i.e., 20–45 °C) (Sedlmeier et al., 2012). A control of excitation wavelength to escape from tissue heating by infrared excitation at 980 nm was proposed using core-shell-shell UCNPs (β-NaYF$_4$:40%Yb, 2%Er@NaYF$_4$:20%Yb@NaNdF$_4$:10%Yb) that showed a temperature-dependent emission intensity ratio with the excitation at 806 nm (Green et al., 2018). Besides, NaGdF$_4$:Yb^{3+}:Er^{3+} UCNPs coated with a EuTTA-containing silica shell (Nigoghossian et al., 2017) have been reported. In order to improve the sensitivity of ratiometric responses of UCNPs, a better combination of two doped lumophores and a matrix material was discussed (Li et al., 2019). Notably, NaYF$_4$:Yb/Er@NaYF$_4$ core-shell UCNPs (~23 nm in diameter) were used to determine the instantaneous ballistic velocity of Brownian motion in a fluid (Brites et al., 2016). By employing the UCNPs as Brownian nanocrystals, a temperature variation during the measurement of instantaneous ballistic velocity could be precisely monitored.

2.16.6
Others

In contrast to UCNPs described in the previous section, NaYF$_4$-based emissive nanoparticles that do not utilize upconversion are also reported. In the temperature range 25–46 °C, β-NaYF$_4$ nanoparticles co-doped with Yb^{3+}, Ho^{3+}, and Er^{3+} (NaYF$_4$:Yb^{3+}, Ho^{3+}, Er^{3+} nanoparticles) display a temperature-dependent emission ratio at 1150 nm ($^5I_6 \rightarrow {}^5I_8$ transition of Ho^{3+}) and 1550 nm ($^4I_{13/2} \rightarrow {}^4I_{15/2}$ transition of Er^{3+}) in cyclohexane under the excitation at 980 nm (Sekiyama et al., 2018). The nanoparticles with long emission wavelengths are intended to avoid the scattering and absorption of light by biological tissues. The function of the nanoparticles in a silicone composite was also evaluated at 25 and 46 °C. The same research group reported temperature imaging by measuring the fluorescence lifetime of NaYF$_4$ co-doped with Nd^{3+} and Yb^{3+} at 1000 nm upon the excitation at 808 nm (Chihara et al., 2019).

A Cu$_5$ metal cluster is a metal nanoparticle composed of five copper atoms bound to three highly conjugated dianionic cationic ligands (EtNC(S)PPh$_2$NPPh$_2$C(S)NEt)$^-$. The Cu$_5$ metal cluster shows red emission at around 650 nm in mixed alcohol or dichloromethane when excited at 405 nm, and the emission quantum yield and excited-state lifetime decrease by about two orders of magnitude between −45 and 45 °C (Cauzzi et al., 2012). Experimental and computational results suggested that emissive S$_1$–S$_6$ states competed with the non-emissive S$_7$ state with the activation energy of 9.7 kcal mol^{-1} (in dichloromethane). Copper nanoclusters prepared with glutathione as a protective layer could disperse in water and showed temperature-dependent emission intensity at 605 ± 5 nm with excitation at 400 nm (Wang et al., 2015). The emission intensity of the copper nanoclusters in water decreased with an increase in temperature, and that at 80 °C was only 11% of that at 15 °C.

Later research indicated that the copper nanoclusters stabilized using glutathione also responded to Cr^{6+} ions in an aqueous solution and a temperature variation (Kong et al., 2016).

Recently, europium (Eu^{3+})-doped silicon nanoparticles (Eu@SiNPs; diameter: ~4.8 nm) were reported as a ratiometric fluorescent thermometer that could work in a cellular environment (Wang et al., 2020c). Eu@SiNPs showed both blue (455 nm, due to SiNPs) and red (620 nm, due to Eu^{3+} ion) emissions, and their intensity ratio was temperature-dependent from 25 to 70 °C in PBS.

Cu$_5$ metal cluster
(only one ligand is shown for clarity)

2.17
Hybrid Nanomaterials

A fluorescent DNA-templated silver nanocluster (DNA-AgNC) consists of a few silver atoms that are stabilized by a single-stranded DNA scaffold. Temperature-dependent fluorescence properties of DNA-AgNC (DNA sequence: 5'-TTCCCACCCACCCCGGCCC-3') were reported (Cerretani et al., 2017). At 5 °C, the DNA-AgNC in $10\,mmol\,L^{-1}$ NH$_4$OAc solution emitted strong fluorescence (Φ_f = 0.88) at 638 nm by the excitation at 561 nm. When the temperature increased to 40 °C, fluorescence intensity moderately decreased (Φ_f = 0.70) with a slight bathochromic shift in maximum emission wavelength (642 nm). The non-radiative process that competes with fluorescence at a high temperature is still unclear.

An early example is based on the temperature-dependent fashion of Eu^{3+}/Tb^{3+} emissions (Brites et al., 2010, 2013). The fluorescent thermometer consists of [Eu(btfa)$_3$(MeOH)(bpeta)] and [Tb(btfa)$_3$(MeOH)(bpeta)] embedded into organic–inorganic hybrid nanoclusters, in which a maghemite (γ-Fe$_2$O$_3$) magnetic core (hydrodynamic size of 21 nm) is coated with a tetraethyl orthosilicate/aminopropyltriethoxysilane organosilica shell. Responses due to the difference in temperature sensitivity between Eu^{3+} ($^5D_0 \rightarrow {}^7F_2$ emission at 612 nm) and Tb^{3+} ($^5D_4 \rightarrow {}^7F_5$ emission at 545 nm) can be observed. A di-ureasil film version co-doped with Eu^{3+} and Tb^{3+} ions was proposed for temperature measurement of an integrated circuit. Review articles on lanthanide-based hybrid materials are also available (Brites et al., 2011, 2019; Rocha et al., 2016)

A novel type of thermo-sensitive nanohybrid, CD-MSN@UP38, consists of a carbon dot-based mesoporous silica nanoparticle (as a core) and thermo-responsive poly(N-vinylimidazole-co-1-vinyl-2-(hydroxymethyl)-imidazole) (UPVH as a shell). In the presence of cell lysis in an aqueous solution, environment-sensitive fluorescent benzofurazan (4-benzylamino-7-nitro-2,1,3-benzoxadiazole, BBD)-loaded CD-MSN@UP38 functions as an irreversible, threshold-type fluorescent thermometer (Dong et al., 2016). At approximately 36 °C, fluorescent benzofurazan is released from CD-MSN@UP38 because thermo-responsive UPVH with an upper critical solution temperature (UCST) is swollen. Then, the environment-sensitive hydrophobic fluorescent benzofurazan released is taken in hydrophobic biomolecules (e.g., proteins) to fluoresce strongly.

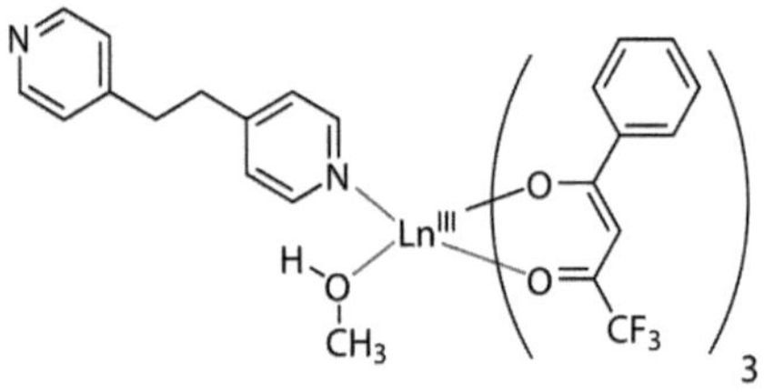

[Ln(btfa)$_3$(MeOH)(bpeta)]
(Ln = Eu or Tb)

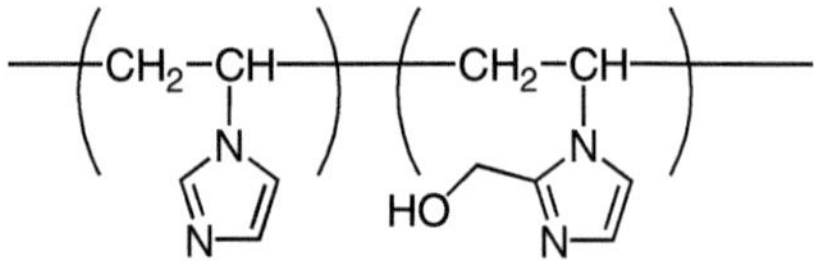

UPVH

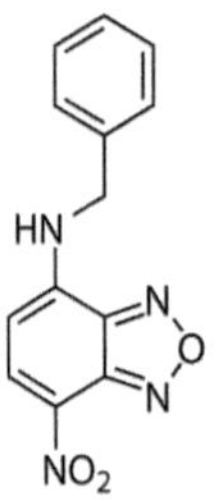

BBD

References

Acosta, V. M., Bauch, E., Ledbetter, M. P., Waxman, A., Bouchard, L.-S., and Budker, D. (2010) Temperature dependence of the nitrogen-vacancy magnetic resonance in diamond. *Phys. Rev. Lett.*, **104**, 070801.

Ajantha, J., Yuvaraj, P., Karuppusamy, M., and Easwaramoorthi, S. (2021) Single-molecule white-light-emitting starburst donor-acceptor triphenylamine derivatives and their application as ratiometric luminescent molecular thermometers. *Chem. Eur. J.*, **27**, 11319–11325.

Albelda, M. T., García-España, E., Gil, L., Lima, J. C., Lodeiro, C., de Melo, J. S., Melo, M. J., Parola, A. J., Pina, F., and Soriano, C. (2003) Intramolecular excimer formation in a tripodal polyamine receptor containing three naphthalene fluorophores. *J. Phys. Chem. B*, **107**, 6573–6578.

An, H., Yin, Z., Mitchell, C., Semnani, A., Hajrasouliha, A. R., and Hosseini, M. (2021) Nanodiamond ensemble-based temperature measurement in living cells and its limitations. *Meas. Sci. Technol.*, **32**, 015701.

Anghel, F., Iliescu, C., Grattan, K. T. V., Palmer, A. W., and Zhang, Z. Y. (1995) Fluorescent-based lifetime measurement thermometer for use at subroom temperature (200–300 K). *Rev. Sci. Instrum.*, **66**, 2611–2614.

Ansari, A. A., Parchur, A. K., Nazeeruddin, M. K., and Tavakoli, M. M. (2021) Luminescent lanthanide nanocomposites in thermometry: chemistry of dopant ions and host matrices. *Coord. Chem. Rev.*, **444**, 214040.

Augusto, V., Baleizão, C., Berberan-Santos, M. N., and Farinha, J. P. S. (2010) Oxygen-proof fluorescence temperature sensing with pristine C_{70} encapsulated in polymer nanoparticles. *J. Mater. Chem.*, **20**, 1192–1197.

Baker, G. A., Baker, S. N., and McCleskey, T. M. (2003) Noncontact two-color luminescence thermometry based on intramolecular luminophore cyclization within an ionic liquid. *Chem. Commun.*, 2932–2933.

Balamurugan, A., Reddy, M. L. P., and Jayakannan, M. (2013) π-Conjugated polymer-Eu^{3+} complexes: versatile luminescent molecular probes for temperature sensing. *J. Mater. Chem. A*, **1**, 2256–2266.

Baleizão, C. and Berberan-Santos, M. N. (2006) A molecular thermometer based on the delayed fluorescence of C_{70} dispersed in a polystyrene film. *J. Fluoresc.*, **16**, 215–219.

Baleizão, C., Nagl, S., Borisov, S. M., Schäferling, M., Wolfbeis, O. S., and Berberan-Santos, M. N. (2007) An optical thermometer based on the delayed fluorescence of C_{70}. *Chem. Eur. J.*, **13**, 3643–3651.

Bardi, B., Tosi, I., Faroldi, F., Baldini, L., Sansone, F., Sissa, C., and Terenziani, F. (2019) A calixarene-based fluorescent ratiometric temperature probe. *Chem. Commun.*, **55**, 8098–8101.

Basu, B. B. J. and Vasantharajan, N. (2008) Temperature dependence of the luminescence lifetime of a europium complex immobilized in different polymer matrices. *J. Lumin.*, **128**, 1701–1708.

Bennett, R. G. and McCartin, P. J. (1966) Radiationless deactivation of the fluorescent state of substituted anthracenes. *J. Chem. Phys.*, **44**, 1969–1972.

Bennet, M. A., Richardson, P. R., Arlt, J., McCarthy, A., Buller, G. S., and Jones, A. C. (2011) Optically trapped microsensors for microfluidic temperature measurement by fluorescence lifetime imaging microscopy. *Lab Chip*, **11**, 3821–3828.

Benninger, R. K. P., Koç, Y., Hofmann, O., Requejo-Isidro, J., Neil, M. A. A., French, P. M. W., and deMello, A. J. (2006) Quantitative 3D mapping of fluidic temperatures within microchannel networks using fluorescence lifetime imaging. *Anal. Chem.*, **78**, 2272–2278.

Berry, M. T., May, P. S., and Xu, H. (1996) Temperature dependence of the Eu^{3+} $^5D_0$ lifetime in europium tris(2,2,6,6-tetramethyl-3,5-heptanedionato). *J. Phys. Chem.*, **100**, 9216–9222.

Bhaumik, M. L. (1964) Quenching and temperature dependence of fluorescence in rare-earth chelates. *J. Chem. Phys.*, **40**, 3711–3715.

Bhuyan, M. and Koenig, B. (2012) Temperature responsive phosphorescent small unilamellar vesicles. *Chem. Commun.*, **48**, 7489–7491.

Bomm, J., Günter, C., and Stumpe, J. (2012) Synthesis and optical characterization of thermosensitive, luminescent gold nanodots. *J. Phys. Chem. C*, **116**, 81–85.

Bommidi, D. K. and Pickel, A. D. (2021) Temperature-dependent excited state lifetimes of nitrogen vacancy centers in individual nanodiamonds. *Appl. Phys. Lett.*, **119**, 254103.

Boudin, S. (1930) Phosphorescence des solutions glycériques d'éosine influence des iodures. *J. Chim. Phys.*, **27**, 285–290.

Brewster, R. E., Kidd, M. J., and Schuh, M. D. (2001) Optical thermometer based on the stability of a phosphorescent 6-bromo-2-naphthol/α-cyclodextrin$_2$ ternary complex. *Chem. Commun.*, 1134–1135.

Brites, C. D. S., Lima, P. P., Silva, N. J. O., Millán, A., Amaral, V. S., Palacio, F., and Carlos, L. D. (2010) A luminescent molecular thermometer for long-term absolute temperature measurements at the nanoscale. *Adv. Mater.*, **22**, 4499–4504.

Brites, C. D. S., Lima, P. P., Silva, N. J. O., Millán, A., Amaral, V. S., Palacio, F., and Carlos, L. D. (2011) Lanthanide-based luminescent molecular thermometers. *New J. Chem.*, **35**, 1177–1183.

Brites, C. D. S., Lima, P. P., Silva, N. J. O., Millán, A., Amaral, V. S., Palacio, F., and Carlos, L. D. (2013) Thermometry at the nanoscale using lanthanide-containing organic-inorganic hybrid materials. *J. Lumines.*, **133**, 230–232.

Brites, C. D. S., Xie, X., Debasu, M. L., Qin, X., Chen, R., Huang, W., Rocha, J., Liu, X., and Carlos, L. D. (2016) Instantaneous ballistic velocity of suspended Brownian nanocrystals measured by upconversion nanothermometry. *Nat. Nanotechnol.*, **11**, 851–856.

Brites, C. D. S., Balabhadra, S., and Carlos, L. D. (2019) Lanthanide-based thermometers: at the cutting-edge of luminescence thermometry. *Adv. Optical Mater.*, **7**, 1801239.

Bustamante, N., Ielasi, G., Bedoya, M., and Orellana, G. (2018) Optimization of temperature sensing with polymer-embedded luminescent Ru(II) complexes. *Polymers*, **10**, 234.

Cauzzi, D., Pattacini, R., Delferro, M., Dini, F., Di Natale, C., Paolesse, R., Bonacchi, S., Montalti, M., Zaccheroni, N., Calvaresi, M., Zerbetto, F., and Prodi, L. (2012) Temperature-dependent fluorescence of Cu$_5$ metal clusters: a molecular thermometer. *Angew. Chem. Int. Ed.*, **51**, 9662–9665.

Cellini, F., Peterson, S. D., and Porfiri, M. (2017) Flow velocity and temperature sensing using thermosensitive fluorescent polymer seed particles in water. *Int. J. Smart Nano Mater.*, **8**, 232–252.

Cerretani, C., Carro-Temboury, M. R., Krause, S., Bogh, S. A., and Vosch, T. (2017) Temperature dependent excited state relaxation of a red emitting DNA-templated silver nanocluster. *Chem. Commun.*, **53**, 12556–12559.

Chan, J., Dodani, S. C., and Chang, C. J. (2012) Reaction-based small-molecule fluorescent probes for chemoselective bioimaging. *Nat. Chem.*, **4**, 973–984.

Chandrasekharan, N. and Kelly, L. A. (2001) A dual fluorescence temperature sensor based on perylene/exciplex interconversion. *J. Am. Chem. Soc.*, **123**, 9898–9899.

Chee, C. K., Rimmer, S., Shaw, D. A., Soutar, I., and Swanson, L. (2001) Manipulating the thermoresponsive behavior of poly(N-isopropylacrylamide). 1. On the conformational behavior of a series of N-isopropylacrylamide-styrene statistical copolymers. *Macromolecules*, **34**, 7544–7549.

Chen, C., Du, Z., Wang, J., and Pan, L. (2016) Temperature mapping using molecular diffusion based fluorescence thermometry via simultaneous imaging of two numerical apertures. *Opt. Express*, **24**, 26599–26611.

Chi, F., Jiang, B., Zhao, Z., Chen, Y., Wei, X., Duan, C., Yin, M., and Xu, W. (2019) Multimodal temperature sensing using Zn$_2$GeO$_4$:Mn^{2+} phosphor as highly sensitive luminescent thermometer. *Sens. Actuators: B. Chem.*, **296**, 126640.

Chihara, T., Umezawa, M., Miyata, K., Sekiyama, S., Hosokawa, N., Okubo, K., Kamimura, M., and Soga, K. (2019) Biological deep temperature imaging with fluorescence lifetime of rare-earth-doped ceramics particles in the second NIR biological window. (2019). *Sci. Rep.*, **9**, 12806.

Chin, P. T. K., de Mello Donegá, C., van Bavel, S. S., Meskers, S. C. J., Sommerdijk, N. A. J. M., and Janssen, R. A. J. (2007) Highly luminescent CdTe/CdSe colloidal heteronanocrystals with temperature-dependent emission color. *J. Am. Chem. Soc.*, **129**, 14880–14886.

Coppeta, J. and Rogers, C. (1998) Dual emission laser induced fluorescence for direct planar scalar behavior measurements. *Exp. Fluids*, **25**, 1–15.

Crenshaw, B. R., Kunzelman, J., Sing, C. E., Ander, C., and Weder, C. (2007) Threshold temperature sensors with tunable properties. *Macromol. Chem. Phys.*, **208**, 572–580.

Crosby, G. A., Whan, R. E., and Alire, R. M. (1961) Intramolecular energy transfer in rare earth chelates. Role of the triplet state. *J. Chem. Phys.*, **34**, 743–748.

Daly, B., Ling, J., and de Silva, A. P. (2015) Current developments in fluorescent PET (photoinduced electron transfer) sensors and switches. *Chem. Soc. Rev.*, **44**, 4203–4211.

Darwish, G. H., Koubeissi, A., Shoker, T., Shaheen, S. A., and Karam, P. (2016a) Turning the heat on conjugated polyelectrolytes: an off-on ratiometric nanothermometer. *Chem. Commun.*, **52**, 823–826.

Darwish, G. H., Abouzeid, J., and Karam, P. (2016b) Tunable nanothermometer based on short poly(phenylene ethynylene). *RSC Adv.*, **6**, 67002–67010.

Darwish, G. H., Fakih, H. H., and Karam, P. (2017) Temperature mapping in hydrogel matrices using unmodified digital camera. *J. Phys. Chem. B*, **121**, 1033–1040.

Deepankumar, K., Nadarajan, S. P., Bae, D.-H., Baek, K.-H., Choi, K.-Y., and Yun, H. (2015) Temperature sensing using red fluorescent protein. *Biotechnol. Bioprocess Eng.*, **20**, 67–72.

de Silva, A. P., Gunaratne, H. Q. N., Jayasekera, K. R., O'Callaghan, S., and Sandanayake, K. R. A. S. (1995) Temperature dependent fluorescence of 'tunable fluorophore-fissile bond' systems based on 1-phenyl, 3-aryl Δ^2-pyrazolines as a means of quantitating photofission processes. *Chem. Lett.*, 123–124.

de Silva, A. P., Gunaratne, H. Q. N., Gunnlaugsson, T., Huxley, A. J. M., McCoy, C. P., Rademacher, J. T., and Rice, T. E. (1997) Signaling recognition events with fluorescent sensors and switches. *Chem. Rev.*, **97**, 1515–1566.

Domaille, D. W., Que, E. L., and Chang, C. J. (2008) Synthetic fluorescent sensors for studying the cell biology of metals. *Nat. Chem. Biol.*, **4**, 168–175.

Dong, N.-N., Pedroni, M., Piccinelli, F., Conti, G., Sbarbati, A., Ramírez-Hernández, J. E., Maestro, L. M., Iglesias-de la Cruz, M. C., Sanz-Rodriguez, F., Juarranz, A., Chen, F., Vetrone, F., Capobianco, J. A., Solé, J. G., Bettinelli, M., Jaque, D., and Speghini, A. (2011) NIR-to-NIR two-photon excited $CaF_2:Tm^{3+},Yb^{3+}$ nanoparticles: multifunctional nanoprobes for highly penetrating fluorescence bio-imaging. *ACS Nano*, **5**, 8665–8671.

Dong, F., Zheng, T., Zhu, R., Wang, S., and Tian, Y. (2016) An engineered thermo-sensitive nanohybrid particle for accurate temperature sensing at the single-cell level and biologically controlled thermal therapy. *J. Mater. Chem. B*, **4**, 7681–7688.

Dunand, P., Castanet, G., and Lemoine, F. (2012) A two-color planar LIF technique to map the temperature of droplets impinging onto a heated wall. *Exp. Fluids*, **52**, 843–856.

Ebrahimi, S., Akhlaghi, Y., Kompany-Zareh, M., and Rinnan, Å (2014) Nucleic acid based fluorescent nanothermometers. *ACS Nano*, **8**, 10372–10382.

El Baraka, M., García, R., and Quiñones, E. (1994) A study of the inclusion complexes of β-cyclodextrin with three electronic states of 4-(N,N-dimethylamino)benzonitrile. *J. Photochem. Photobiol. A: Chem.*, **79**, 181–187.

Engeser, M., Fabbrizzi, L., Licchelli, M., and Sacchi, D. (1999) A fluorescent molecular thermometer based on the nickel(II) high-spin/low-spin interconversion. *Chem. Commun.*, 1191–1192.

Erickson, E. S. and Dunn, R. C. (2005) Sample heating in near-field scanning optical microscopy. *Appl. Phys. Lett.*, **87**, 201102.

Estrada-Pérez, C. E., Hassan, Y. A., and Tan, S. (2011) Experimental characterization of temperature sensitive dyes for laser induced fluorescence thermometry. *Rev. Sci. Instrum.*, **82**, 074901.

Feng, J., Tian, K., Hu, D., Wang, S., Li, S., Zeng, Y., Li, Y., and Yang, G. (2011) A triarylboron-based fluorescent thermometer: sensitive over a wide temperature range. *Angew. Chem. Int. Ed.*, **50**, 8072–8076.

Feng, J., Xiong, L., Wang, S., Li, S., Li, Y., and Yang, G. (2013) Fluorescent temperature sensing using triarylboron compounds and microcapsules for detection of a wide temperature range on the micro- and macroscale. *Adv. Funct. Mater.*, **23**, 340–345.

Fernando, S. R. L., Maharoof, U. S. M., Deshayes, K. D., Kinstle, T. H., and Ogawa, M. Y. (1996) A negative activation energy for luminescence decay: specific solvation effects on the emission properties bis(2,2'-bipyridine)(3,5-dicarboxy-2,2'-bipyridine) ruthenium(II) chloride. *J. Am. Chem. Soc.*, **118**, 5783–5790.

Fery-Forgues, S., Fayet, J.-P., and Lopez, A. (1993) Drastic changes in the fluorescence properties of NBD probes with the polarity of the medium: involvement of a TICT state? *J. Photochem. Photobiol. A: Chem.*, **70**, 229–243.

Figueroa, I. D., El Baraka, M., Quiñones, E., and Rosario, O. (1998) A fluorescent temperature probe based on the association between the excited states of 4-(N,N-dimethylamino) benzonitrile and β-cyclodextrin. *Anal. Chem.*, **70**, 3974–3977.

Filevich, O. and Etchenique, R. (2006) 1D and 2D temperature imaging with a fluorescent ruthenium complex. *Anal. Chem.*, **78**, 7499–7503.

Fischer, L. H., Harms, G. S., and Wolfbeis, O. S. (2011) Upconverting nanoparticles for nanoscale thermometry. *Angew. Chem. Int. Ed.*, **50**, 4546–4551.

Fister, J. C. III, Rank, D., and Harris, J. M. (1995) Delayed fluorescence optical thermometry. *Anal. Chem.*, **67**, 4269–4275.

Fu, R., Xu, B., and Li, D. (2006) Study of the temperature field in microchannels of a PDMS chip with embedded local heater using temperature-dependent fluorescent dye. *Int. J. Therm. Sci.*, **45**, 841–847.

Fujino, N., Ichikawa, M., Koyama, T., and Taniguchi, Y. (2009) A molecular thermometer based on luminescence of copper(II) tetraphenylporphyrin. *Thin Solid Films*, **518**, 563–566.

Gao, H., Kam, C., Chou, T. Y., Wu, M.-Y., Zhao, X., and Chen, S. (2020) A simple yet effective AIE-based fluorescent nanothermometer for temperature mapping in living cells using fluorescence lifetime imaging microscopy. *Nanoscale Horiz.*, **5**, 488–494.

Gareau, D., Desrosiers, A., and Vallée-Bélisle, A. (2016) Programmable quantitative DNA nanothermometers. *Nano Lett.*, **16**, 3976–3981.

Gossage, H. E. and Melton, L. A. (1987) Fluorescence thermometers using intramolecular exciplexes. *Appl. Opt.*, **26**, 2256–2259.

Gota, C., Uchiyama, S., and Ohwada, T. (2007) Accurate fluorescent polymeric thermometers containing an ionic component. *Analyst*, **132**, 121–126.

Gota, C., Uchiyama, S., Yoshihara, T., Tobita, S., and Ohwada, T. (2008) Temperature-dependent fluorescence lifetime of a fluorescent polymeric thermometer, poly(N-isopropylacrylamide), labeled by polarity and hydrogen bonding sensitive 4-sulfamoyl-7-aminobenzofurazan. *J. Phys. Chem. B*, **112**, 2829–2836.

Graham, E. M., Iwai, K., Uchiyama, S., de Silva, A. P., Magennis, S. W., and Jones, A. C. (2010) Quantitative mapping of aqueous microfluidic temperature with sub-degree resolution using fluorescence lifetime imaging microscopy. *Lab Chip*, **10**, 1267–1273.

Green, K., Huang, K., Pan, H., Han, G., and Lim, S. F. (2018) Optical temperature sensing with infrared excited upconversion nanoparticles. *Front. Chem.*, **6**, 416.

Guan, W., Yao, R., Tang, X., and Lu, C. (2022) Design of a temperature-independent luminescent probe for visualization of ice-to-liquid transition at −129 °C. *J. Phys. Chem. C*, **126**, 7556–7563.

Halperin, A., Kröger, M., and Winnik, F. M. (2015) Poly(N-isopropylacrylamide) phase diagrams: fifty years of research. *Angew. Chem. Int. Ed.*, **54**, 15342–15367.

Han, S., Deng, R., Xie, X., and Liu, X. (2014) Enhancing luminescence in lanthanide-doped upconversion nanoparticles. *Angew. Chem. Int. Ed.*, **53**, 11702–11715.

Hashim, H., Maruyama, H., Akita, Y., and Arai, F. (2019) Hydrogel fluorescence microsensor with fluorescence recovery for prolonged stable temperature measurements. *Sensors*, **19**, 5247.

Hsia, C.-H., Wuttig, A., and Yang, H. (2011) An accessible approach to preparing water-soluble Mn^{2+}-doped (CdSSe)ZnS (core)shell nanocrystals for ratiometric temperature sensing. *ACS Nano*, **5**, 9511–9522.

Hsiao, W. W.-W., Hui, Y. Y., Tsai, P.-C., and Chang, H.-C. (2016) Fluorescent nanodiamond: a versatile tool for long-term cell tracking, super-resolution imaging, and nanoscale temperature sensing. *Acc. Chem. Res.*, **49**, 400–407.

Hu, H. and Koochesfahani, M. M. (2003) A novel technique for quantitative temperature mapping in liquid by measuring the lifetime of laser induced phosphorescence. *J. Vis.*, **6**, 143–153.

Hu, H., Koochesfahani, M., and Lum, C. (2006) Molecular tagging thermometry with adjustable temperature sensitivity. *Exp. Fluids*, **40**, 753–763.

Huang, D. and Hu, H. (2007) Molecular tagging thermometry for transient temperature mapping within a water droplet. *Opt. Lett.*, **32**, 3534–3536.

Hungerford, G., Martinez-Insua, M., Birch, D. J. S., and Moore, B. D. (1996) A reversible transition between an α-helix and a 3_{10}-helix in a fluorescence-labeled peptide. *Angew. Chem. Int. Ed. Engl.*, **35**, 326–329.

Iwai, K., Matsumoto, N., Niki, M., and Yamamoto, M. (1998) Fluorescence probe studies of thermosensitive N-isopropylacrylamide copolymers in aqueous solutions. *Mol. Cryst. Liq. Cryst.*, **315**, 53–58.

Iwai, K., Matsumura, Y., Uchiyama, S., and de Silva, A. P. (2005) Development of fluorescent microgel thermometers based on thermo-responsive polymers and their modulation of sensitivity range. *J. Mater. Chem.*, **15**, 2796–2800.

Iyisan, B., Thiramanas, R., Nazarova, N., Avlasevich, Y., Mailänder, V., Baluschev, S., and Landfester, K. (2020) Temperature sensing in cells using polymeric upconversion nanocapsules. *Biomacromolecules*, **21**, 4469–4478.

Jeon, S., Turner, J., and Granick, S. (2003) Noncontact temperature measurement in microliter-sized volumes using fluorescent-labeled DNA oligomers. *J. Am. Chem. Soc.*, **125**, 9908–9909.

Kalaparthi, V., Peng, B., Peerzade, S. A. M. A., Palantavida, S., Maloy, B., Dokukin, M. E., and Sokolov, I. (2021) Ultrabright fluorescent nanothermometers. *Nanoscale Adv.*, **3**, 5090–5101.

Kamei, Y., Suzuki, M., Watanabe, K., Fujimori, K., Kawasaki, T., Deguchi, T., Yoneda, Y., Todo, T., Takagi, S., Funatsu, T., and Yuba, S. (2009) Infrared laser-mediated gene induction in targeted single cells in vivo. *Nat. Methods*, **6**, 79–81.

Karstens, T. and Kobs, K. (1980) Rhodamine B and rhodamine 101 as reference substances for fluorescence quantum yield measurements. *J. Phys. Chem.*, **84**, 1871–1872.

Kato, H., Nishizaka, T., Iga, T., Kinoshita, K. Jr., and Ishiwata, S. (1999) Imaging of thermal activation of actomyosin motors. *Proc. Natl. Acad. Sci. USA*, **96**, 9602–9606.

Kawaguchi, K. and Ishiwata, S. (2001) Thermal activation of single kinesin molecules with temperature pulse microscopy. *Cell Motil. Cytoskeleton*, **49**, 41–47.

Ke, G., Wang, C., Ge, Y., Zheng, N., Zhu, Z., and Tang, C. J. (2012) L-DNA molecular beacon: a safe, stable, and accurate intracellular nano-thermometer for temperature sensing in living cells. *J. Am. Chem. Soc.*, **134**, 18908–18911.

Khimich, M. N., Volchkov, V. V., and Uzhinov, B. M. (2003) Fluorescence study of excited-state relaxation processes of 2-pyridyl-5-aryloxazoles. *J. Fluoresc.*, **13**, 301–305.

Kitagawa, Y., Kumagai, M., da Rosa, P. P. F., Fushimi, K., and Hasegawa, Y. (2021) Long-range LMCT coupling in Eu^{III} coordination polymers for an effective molecular luminescent thermometer. *Chem. Eur. J.*, **27**, 264–269.

Kolesnikov, I. E., Kalinichev, A. A., Kurochkin, M. A., Kolesnikov, E. Y., and Lähderanta, E. (2019) Porphyrins as efficient ratiometric and lifetime-based contactless optical thermometers. *Mater. Des.*, **184**, 108188.

Kolesnikov, I. E., Kurochkin, M. A., Meshkov, I. N., Akasov, R. A., Kalinichev, A. A., Kolesnikov, E. Y., Gorbunova, Y. G., and Lähderanta, E. (2021) Water-soluble multimode fluorescent thermometers based on porphyrins photosensitizers. *Mater. Des.*, **203**, 109613.

Kolodner, P. and Tyson, J. A. (1982) Microscopic fluorescent imaging of surface temperature profiles with 0.01 °C resolution. *Appl. Phys. Lett.*, **40**, 782–784.

Kong, L., Chu, X., Liu, W., Yao, Y., Zhu, P., and Ling, X. (2016) Glutathione-directed synthesis of Cr(VI)- and temperature-responsive fluorescent copper nanoclusters and their applications in cellular imaging. *New J. Chem.*, **40**, 4744–4750.

Kundu, S., Mukherjee, D., Maiti, T. K., and Sarkar, N. (2019) Highly luminescent thermoresponsive green emitting gold nanoclusters for intracellular nanothermometry and cellular imaging: a dual function optical probe. *ACS Appl. Bio Mater.*, **2**, 2078–2091.

Kubin, R. F. and Fretcher, A. N. (1982) Fluorescence quantum yields of some rhodamine dyes. *J. Lumines*, **27**, 455–462.

Kucsko, G., Maurer, P. C., Yao, N. Y., Kubo, M., Noh, H. J., Lo, P. K., Park, H., and Lukin, M. D. (2013) Nanometre-scale thermometry in a living cell. *Nature*, **500**, 54–58.

Kuimova, M. K., Yahioglu, G., Levitt, J. A., and Suhling, K. (2008) Molecular rotor measures viscosity of live cells via fluorescence lifetime imaging. *J. Am. Chem. Soc.*, **130**, 6672–6673.

Kwak, G., Fukao, S., Fujiki, M., Sakaguchi, T., and Masuda, T. (2006) Temperature-dependent, static, and dynamic fluorescence properties of disubstituted acetylene polymer films. *Chem. Mater.*, **18**, 2081–2085.

Laha, S. S., Naik, A. R., Kuhn, E. R., Alvarez, M., Sujkowski, A., Wessells, R. J., and Jena, B. P. (2017) Nanothermometry measure of muscle efficiency. *Nano Lett.*, **17**, 1262–1268.

Lahoz, F., Martín, I. R., Walo, D., Freire, R., Gil-Rostra, J., Yubero, F., and Gonzalez-Elipe, A. R. (2017) Enhanced green fluorescent protein in optofluidic Fabry-Perot microcavity to detect laser induced temperature changes in a bacterial culture. *Appl. Phys. Lett.*, **111**, 111103.

Lakowicz, J. R. (2006) *Principles of Fluorescence Spectroscopy*, 3rd edn. Springer, New York.

Lee, S. M., Chung, W. Y., Kim, J. K., and Suh, D. H. (2004) A novel fluorescence temperature sensor based on a surfactant-free PVA/borax/2-naphthol hydrogel network system. *J. Appl. Polym. Sci.*, **93**, 2114–2118.

Lee, C.-G., Uehara, M., Yamaguchi, Y., Nakamura, H., and Maeda, H. (2007) Micro-space synthesis of core-shell-type semiconductor nanocrystals for thermosensing. *Bull. Chem. Soc. Jpn.*, **80**, 794–796.

Lee, M. H., Kim, J. S., and Sessler, J. L. (2015) Small molecule-based ratiometric fluorescence probes for cations, anions, and biomolecules. *Chem. Soc. Rev.*, **44**, 4185–4191.

Lee, M.-H., Lin, H.-Y., and Yang, C.-N. (2019) A DNA-based two-way thermometer to report high and low temperatures. *Anal. Chim. Acta*, **1081**, 176–183.

Li, S., Zhang, K., Yang, J.-M., Lin, L., and Yang, H. (2007) Single quantum dots as local temperature markers. *Nano Lett.*, **7**, 3102–3105.

Li, H., Chen, F., and Hu, H. (2015) Simultaneous measurements of droplet size, flying velocity and transient temperature of in-flight droplets by using a molecular tagging technique. *Exp. Fluids*, **56**, 194.

Li, P., Jia, M., Liu, G., Zhang, A., Sun, Z., and Fu, Z. (2019) Investigation on the fluorescence intensity ratio sensing thermometry based on nonthermally coupled levels. *ACS Appl. Bio Mater.*, **2**, 1732–1739.

Li, L., Zhang, C., Xu, L., Ye, C., Chen, S., Wang, X., and Song, Y. (2021) Luminescence ratiometric nanothermometry regulated by tailoring annihilators of triplet–triplet annihilation upconversion nanomicelles. *Angew. Chem. Int. Ed.*, **60**, 26725–26733.

Liang, S., Wang, Y., Wu, X., Chen, M., Mu, L., She, G., and Shi, W. (2019) An ultrasensitive ratiometric fluorescent thermometer based on frustrated static excimers in the physiological temperature range. *Chem. Commun.*, **55**, 3509–3512.

Lim, E. C., Laposa, J. D., and Yu, J. M. H. (1966) Temperature dependence of intersystem crossing in substituted anthracenes. *J. Mol. Spectrosc.*, **19**, 412–420.

Lin, F., Jia, C., and Wu, F.-G. (2022) Carbon dots for intracellular sensing. *Small Struct.*, **3**, 2200033.

Liu, X., Liu, J., Zhou, H., Yan, M., Liu, C., Guo, X., Xie, J., Li, S., and Yang, G. (2020) Ratiometric dual fluorescence tridurylboron thermometers with tunable measurement ranges and colors. *Talanta*, **210**, 120630.

Liu, W., Alam, M. N. A., Liu, Y., Agafonov, V. N., Qi, H., Koynov, K., Davydov, V. A., Uzbekov, R., Kaiser, U., Lasser, T., Jelezko, F., Ermakova, A., and Weil, T. (2022) Silicon-vacancy nanodiamonds as high performance near-infrared emitters for live-cell dual-color imaging and thermometry. *Nano Lett.*, **22**, 2881–2888.

Lopez, I. S., Mendonça, A. L., Fernandes, M., de Zea Bermudez, V., Morgado, J., Del Pozo, G., Romero, B., and Cabanillas-Gonzalez, J. (2012) Europium complex-based thermochromic sensor for integration in plastic optical fibers. *Opt. Mater.*, **34**, 1447–1450.

Lou, J., Hatton, T. A., and Laibinis, P. E. (1997) Fluorescent probes for monitoring temperature in organic solvents. *Anal. Chem.*, **69**, 1262–1264.

Löw, P., Kim, B., Takama, N., and Bergaud, C. (2008) High-spatial-resolution surface-temperature mapping using fluorescent thermometry. *Small*, **4**, 908–914.

Lumpkin, R. S., Kober, E. M., Worl, L. A., Murtaza, Z., and Meyer, T. J. (1990) Metal-to-ligand charge-transfer (MLCT) photochemistry. Experimental evidence for the participation of a higher lying MLCT state in polypyridyl complexes of ruthenium (II) and osmium (II). *J. Phys. Chem.*, **94**, 239–243.

Lupton, J. M. (2002) A molecular thermometer based on long-lived emission from platinum octaethyl porphyrin. *Appl. Phys. Lett.*, **81**, 2478–2480.

Lyu, T., Sohn, S. H., Jimenez, R., and Joo, T. (2022) Temperature-dependent fluorescence of mPlum fluorescent protein from 295 to 20 K. *J. Phys. Chem. B*, **126**, 2337–2344.

Ma, Y., Dong, Y., Liu, S., She, P., Lu, J., Liu, S., Huang, W., and Zhao, Q. (2020) Chameleon-like thermochromic luminescent materials with controllable response behaviors for multilevel security printing. *Adv. Optical Mater.*, **8**, 1901687.

Maestro, L. M., Jacinto, C., Silva, U. R., Ventrone, F., Capobianco, J. A., Jaque, D., and Solé, J. G. (2011) CdTe quantum dots as nanothermometers: towards highly sensitive thermal imaging. *Small*, **7**, 1774–1778.

Mazza, M. M. A., Cardano, F., Cusido, J., Baker, J. D., Giordani, S., and Raymo, F. M. (2019) Ratiometric temperature sensing with fluorescent thermochromic switches. *Chem. Commun.*, **55**, 1112–1115.

McLaurin, E. J., Vlaskin, V. A., and Gamelin, D. R. (2011) Water-soluble dual-emitting nanocrystals for ratiometric optical thermometry. *J. Am. Chem. Soc.*, **133**, 14978–14980.

Medintz, I. L., Uyeda, H. T., Goldman, E. R., and Mattoussi, H. (2005) Quantum dot bioconjugates for imaging, labelling and sensing. *Nat. Mater.*, **4**, 435–446.

Michalet, X., Pinaud, F. F., Bentolila, L. A., Tsay, J. M., Doose, S., Li, J. J., Sundaresan, G., Wu, A. M., Gambhir, S. S., and Weiss, S. (2005) Quantum dots for live cells, in vivo imaging, and diagnostics. *Science*, **307**, 538–544.

Mochalin, V. N., Shenderova, O., Ho, D., and Gogotsi, Y. (2012) The properties and applications of nanodiamonds. *Nat. Nanotechnol.*, **7**, 11–23.

Murray, A. M. and Melton, L. A. (1985) Fluorescence methods for determination of temperature in fuel sprays. *Appl. Opt.*, **24**, 2783–2787.

Nakashima, H., Takenaka, Y., Higashi, M., and Yoshida, N. (2001) Fluorescent behaviour in host–guest interactions. Part 2. Thermal and pH-dependent sensing properties of two geometric isomers of fluorescent amino-β-cyclodextrin derivatives. *J. Chem. Soc. Perkin Trans. 2*, 2096–2103.

Nelson, D. F. and Sturge, M. D. (1965) Relation between absorption and emission in the region of the R lines of ruby. *Phys. Rev.*, **137**, A1117–A1130.

Neumann, P., Jakobi, I., Dolde, F., Burk, C., Reuter, R., Waldherr, G., Honert, J., Wolf, T., Brunner, A., Shim, J. H., Suter, D., Sumiya, H., Isoya, J., and Wrachtrup, J. (2013) High-precision nanoscale temperature sensing using single defects in diamond. *Nano Lett.*, **13**, 2738–2742.

Nigoghossian, K., Messaddeq, Y., Boudreau, D., and Ribeiro, S. J. L. (2017) UV and temperature-sensing based on $NaGdF_4$:Yb^{3+}:Er^{3+}@SiO_2–Eu(tta)$_3$. *ACS Omega*, **2**, 2065–2071.

Nishimura, Y., Oshimi, K., Umehara, Y., Kumon, Y., Miyaji, K., Yukawa, H., Shikano, Y., Matsubara, T., Fujiwara, M., Baba, Y., and Teki, Y. (2021) Wide-field fluorescent nanodiamond spin measurements toward real-time large-area intracellular thermometry. *Sci. Rep.*, **11**, 4248.

Noelting, E. and Dziewoński, K. (1905) Zur kenntniss der rhodamine. *Ber. Dtsch. Chem. Ges.*, **38**, 3516–3527.

Nomura, R., Yamada, K., and Masuda, T. (2002) A chromophore-labeled poly(N-propargylamide): a new strategy for a stimuli-responsive conjugated polymer. *Chem. Commun.*, 478–479.

Ogawa, T., Yoshida, M., Ohara, H., Kobayashi, A., and Kato M. (2015) A dual-emissive ionic liquid based on an anionic platinum(II) complex. *Chem. Commun.*, **51**, 13377–13380.

Ogawa, T., Sameera, W. M. C., Yoshida, M., Kobayashi, A., and Kato, M. (2018) Luminescent ionic liquids based on cyclometalated platinum(II) complexes exhibiting thermochromic behaviour in different colour regions. *Dalton Trans.*, **47**, 5589–5594.

Ogle, M. M., McWilliams, A. D. S., Ware, M. J., Curley, S. A., Corr, S. J., and Martí, A. A. (2019) Sensing temperature in vitro and in cells using a BODIPY molecular probe. *J. Phys. Chem. B*, **123**, 7282–7289.

Otto, S., Scholz, N., Behnke, T., Resch-Genger, U., and Heinze, K. (2017) Thermo-chromium: a contactless optical molecular thermometer. *Chem. Eur. J.*, **23**, 12131–12135.

Park, Y., Koo, C., Chen, H.-Y., Han, A., and Son, D. H. (2013) Ratiometric temperature imaging using environment-insensitive luminescence of Mn-doped core-shell nanocrystals. *Nanoscale*, **5**, 4944–4950.

Parker, C. A. and Hatchard, C. G. (1961) Triplet-singlet emission in fluid solutions. *Trans. Faraday Soc.*, **57**, 1894–1904.

Peng, H., Stich, M. I. J., Yu, J., Sun, L.-n., Fischer, L. H., and Wolfbeis, O. S. (2010a) Luminescent europium(III) nanoparticles for sensing and imaging of temperature in the physiological range. *Adv. Mater.*, **22**, 716–719.

Peng, H.-S., Huang, S.-H., and Wolfbeis, O. S. (2010b) Ratiometric fluorescence nanoparticles for sensing temperature. *J. Nanopart. Res.*, **12**, 2729–2733.

Pietsch, C., Schubert, U. S., and Hoogenboom, R. (2011) Aqueous polymeric sensors based on temperature-induced polymer phase transitions and solvatochromic dyes. *Chem. Commun.*, **47**, 8750–8765.

Pragst, F., Hamann, H.-J., Teuchner, K., Naether, M., Becker, W., and Daehne, S. (1977) Steady state and laser spectroscopic investigations on the intramolecular exciplex formation of 1-(N-p-anisyl-N-methyl)-amino-3-anthryl-(9)-propane. *Chem. Phys. Lett.*, **48**, 36–39.

Prodi, L., Montalti, M., and Zaccheroni, N. (eds.) (2011) *Luminescence Applied in Sensor Science. Top. Curr. Chem.*, vol. **300**, Springer, Heidelberg.

Qiao, J., Mu, X., and Qi, L. (2016) Construction of fluorescent polymeric nano-thermometers for intracellular temperature imaging: a review. *Biosens. Bioelectron.*, **85**, 403–413.

Ringsdorf, H., Venzmer, J., and Winnik, F. M. (1991) Fluorescence studies of hydrophobically modified poly(N-isopropylacrylamide). *Macromolecules*, **24**, 1678–1686.

Rocha, J., Brites, C. D. S., and Carlos, L. D. (2016) Lanthanide organic framework luminescent thermometers. *Chem. Eur. J.*, **22**, 14782–14795.

Romshin, A. M., Zeeb, V., Martyanov, A. K., Kudryavtsev, O. S., Pasternak, D. G., Sedov, V. S., Ralchenko, V. G., Sinogeykin, A. G., and Vlasov, I. I. (2021) A new approach to precise mapping of local temperature fields in submicrometer aqueous volumes. *Sci. Rep.*, **11**, 14228.

Ross, D., Gaitan, M., and Locascio, L. E. (2001) Temperature measurement in microfluidic systems using a temperature-dependent fluorescent dye. *Anal. Chem.*, **73**, 4117–4123.

Rullière, C. (1976) Laser action and photoisomerisation of 3,3'-diethyl oxadicarbocyanine iodide (DODCI): influence of temperature and concentration. *Chem. Phys. Lett.*, **43**, 303–308.

Sabbatini, N., Guardigli, M., Bolletta, F., Manet, I., and Ziessel, R. (1994) Luminescent Eu^{3+} and Tb^{3+} complexes of a branched macrocyclic ligand incorporating 2,2'-bipyridine in the macrocycle and phosphinate esters in the side arms. *Angew. Chem. Int. Ed. Engl.*, **33**, 1501–1503.

Sakakibara, J. and Adrian, R. J. (1999) Whole field measurement of temperature in water using two-color laser induced fluorescence. *Exp. Fluids*, **26**, 7–15.

Sato, Y., Irisawa, G., Ishizuka, M., Hishida, K., and Maeda, M. (2003) Visualization of convective mixing in microchannel by fluorescence imaging. *Meas. Sci. Technol.*, **14**, 114–121.

Schild, H. G. (1992) Poly(N-isopropylacrylamide): experiment, theory and application. *Prog. Polym. Sci.*, **17**, 163–249.

Schrum, K. F., Williams, A. M., Haerther, S. A., and Ben-Amotz, D. (1994) Molecular fluorescence thermometry. *Anal. Chem.*, **66**, 2788–2790.

Sedlmeier, A., Achatz, D. E., Fischer, L. H., Gorris, H. H., and Wolfbeis, O. S. (2012) Photon upconverting nanoparticles for luminescent sensing of temperature. *Nanoscale*, **4**, 7090–7096.

Sekiyama, S., Umezawa, M., Kuraoka, S., Ube, T., Kamimura, M., and Soga, K. (2018) Temperature sensing of deep abdominal region in mice by using over-1000 nm near-infrared luminescence of rare-earth-doped NaYF$_4$ nanothermometer. *Sci. Rep.*, **8**, 16979.

Shah, J. J., Gaitan, M., and Geist, J. (2009) Generalized temperature measurement equations for rhodamine B dye solution and its application to microfluidics. *Anal. Chem.*, **81**, 8260–8263.

Shahi, P. K., Singh, A. K., Singh, S. K., Rai, S. B., and Ullrich, B. (2015) Revelation of the technological versatility of the Eu(TTA)$_3$Phen complex by demonstrating energy harvesting, ultraviolet light detection, temperature sensing, and laser applications. *ACS Appl. Mater. Interfaces*, **7**, 18231–18239.

Shang, L., Stockmar, F., Azadfar, N., and Nienhaus, G. U. (2013) Intracellular thermometry by using fluorescent gold nanoclusters. *Angew. Chem. Int. Ed.*, **52**, 11154–11157.

Shen, T., Wu, X., Tan, D., Xu, Z., and Liu, X. (2021) Thermal equilibria between conformers enable highly reliable single-fluorophore ratiometric thermometers. *Analyst*, **146**, 4219–4225.

Shi, Y., Li, C., Ma, H., Cao, Z., Liu, K., Yin, X., Wang, N., and Chen, P. (2022) Two-in-one approach toward white-light emissions of dimeric B/N Lewis pairs by tuning the ortho-substitution effect. *Org. Lett.*, **24**, 5497–5502.

Shinde, S. L. and Nanda, K. K. (2013) Wide-range temperature sensing using highly sensitive green-luminescent ZnO and PMMA-ZnO film as a non-contact optical probe. *Angew. Chem. Int. Ed.*, **52**, 11325–11328.

Simpson, D. A., Morrisroe, E., McCoey, J. M., Lombard, A. H., Mendis, D. C., Treussart, F., Hall, L. T., Petrou, S., and Hollenberg, L. C. L. (2017) Non-neurotoxic nanodiamond probes for intraneuronal temperature mapping. *ACS Nano*, **11**, 12077–12086.

Sing, C. E., Kunzelman, J., and Weder, C. (2009) Time-temperature indicators for high temperature applications. *J. Mater. Chem.*, **19**, 104–110.

Snare, M. J., Treloar, F. E., Ghiggino, K. P., and Thistlethwaite, P. J. (1982) The photophysics of rhodamine B. *J. Photochem.*, **18**, 335–346.

Soleilhac, A., Girod, M., Dugourd, P., Burdin, B., Parvole, J., Dugas, P.-Y., Bayard, F., Lacôte, E., Bourgeat-Lami, E., and Antoine, R. (2016) Temperature response of rhodamine B-doped latex particles. From solution to single particles. *Langmuir*, **32**, 4052–4058.

Sou, K., Chan, L. Y., and Lee, C.-L. K. (2016) Temperature tracking in a three-dimensional matrix using thermosensitive liposome platform. *ACS Sens.*, **1**, 650–655.

Sou, K., Chan, L. Y., Arai, S., and Lee, C.-L. K. (2019) Highly cooperative fluorescence switching of self-assembled squaraine dye at tunable threshold temperatures using thermosensitive nanovesicles for optical sensing and imaging. *Sci. Rep.*, **9**, 17991.

Sotoma, S., Epperla, C. P., and Chang, H.-C. (2018) Diamond nanothermometry. *ChemNanoMat*, **2**, 15–27.

Steinegger, A. and Borisov, S. M. (2020) Zn(II) Schiff bases: bright TADF emitters for self-referenced decay time-based optical temperature sensing. *ACS Omega*, **5**, 7729–7737.

Sun, Y.-P., Zhou, B., Lin, Y., Wang, W., Fernando, K. A. S., Pathak, P., Meziani, M. J., Harruff, B. A., Wang, X., Wang, H., Luo, P. G., Yang, H., Kose, M. E., Chen, B., Veca, L. M., and Xie, S.-Y. (2006) Quantum-sized carbon dots for bright and colorful photoluminescence. *J. Am. Chem. Soc.*, **128**, 7756–7757.

Sutton, J. A., Fisher, B. T., and Fleming, J. W. (2008) A laser-induced fluorescence measurement for aqueous fluid flows with improved temperature sensitivity. *Exp. Fluids*, **45**, 869–881.

Takeuchi, S., Nakagawa, T., and Yokoyama, Y. (2020) A thermoresponsive fluorophore based on a photochromic diarylethene having donor-acceptor moieties. *Chem. Commun.*, **56**, 6492–6494.

Tamura, M., Luque, J., Harrington, J. E., Berg, P. A., Smith, G. P., Jeffries, J. B., and Crosley, D. R. (1998) Laser-induced fluorescence of seeded nitric oxide as a flame thermometer. *Appl. Phys. B*, **66**, 503–510.

Tashiro, R. and Sugiyama, H. (2003) A nanothermometer based on the different π stackings of B- and Z-DNA. *Angew. Chem. Int. Ed.*, **42**, 6018–6020.

Tashiro, R. and Sugiyama, H. (2005) Biomolecule-based switching devices that respond inversely to thermal stimuli. *J. Am. Chem. Soc.*, **127**, 2094–2097.

Thomson, S. L. and Maynes, D. (2001) Spatially resolved temperature measurements in a liquid using laser induced phosphorescence. *J. Fluids Eng.*, **123**, 293–302.

Toca-Herrera, J. L., Küpcü, S., Diederichs, V., Moncayo, G., Pum, D., and Sleytr, U. B. (2006) Fluorescence emission properties of S-layer enhanced green fluorescent fusion protein as a function of temperature, pH conditions, and guanidine hydrochloride concentration. *Biomacromolecules*, **7**, 3298–3301.

Toyli, D. M., de las Casas, C. F., Christle, D. J., Dobrovitski, V. V., and Awschalom, D. D. (2013) Fluorescence thermometry enhanced by the quantum coherence of single spins in diamond. *Proc. Natl. Acad. Sci. USA,*, **10**, 8417–8421.

Tsai, P.-C., Epperla, C. P., Huang, J.-S., Chen, O. Y., Wu, C.-C., and Chang, H.-C. (2017) Measuring nanoscale thermostability of cell membranes with single gold-diamond nanohybrids. *Angew. Chem. Int. Ed.*, **56**, 3025–3030.

Tyagi, S. and Kramer, F. R. (1996) Molecular beacons: probes that fluoresce upon hybridization. *Nat. Biotechnol.*, **14**, 303–308.

Uchiyama, S., Matsumura, Y., de Silva, A. P., and Iwai, K. (2003) Fluorescent molecular thermometers based on polymers showing temperature-induced phase transitions and labeled with polarity-responsive benzofurazans. *Anal. Chem.*, **75**, 5926–5935.

Uchiyama, S., Matsumura, Y., de Silva, A. P., and Iwai, K. (2004) Modulation of the sensitive temperature range of fluorescent molecular thermometers based on thermoresponsive polymers. *Anal. Chem.*, **76**, 1793–1798.

Uchiyama, S., Gota, C., Tsuji, T., and Inada, N. (2017) Intracellular temperature measurements with fluorescent polymeric thermometers. *Chem. Commun.*, **53**, 10976–10992.

Uttamlal, M. and Holmes-Smith, A. S. (2008) The excitation wavelength dependent fluorescence of porphyrins. *Chem. Phys. Lett.*, **454**, 223–228.

Valerini, D., Cretí, A., Lomascolo, M., Manna, L., Cingolani, R., and Anni, M. (2005) Temperature dependence of the photoluminescence properties of colloidal CdSe/ZnS core/shell quantum dots embedded in a polystyrene matrix. *Phys. Rev. B*, **71**, 235409.

Valeur, B. and Berberan-Santos, M. N. (2012) *Molecular fluorescence. Principles and Applications*, 2nd edn. Wiley-VCH, Weinheim.

Van Houten, J. and Watts, R. J. (1976) Temperature dependence of the photophysical and photochemical properties of the tris(2,2′-bipyridyl)ruthenium(II) ion in aqueous solution. *J. Am. Chem. Soc.*, **98**, 4853–4858.

Van Keuren, E., Littlejohn, D., and Schrof, W. (2004) Three-dimensional thermal imaging using two-photon microscopy. *J. Phys. D: Appl. Phys.*, **37**, 2938–2943.

Van Keuren, E., Cheng, M., Albertini, O., Luo, C., Currie, J., and Paranjape, M. (2005) Temperature profiles of microheaters using fluorescence microthermal imaging. *Sens. Mater.*, **17**, 1–6.

Vlaskin, V. A., Janssen, N., van Rijssel, J., Beaulac, R., and Gamelin, D. R. (2010) Tunable dual emission in doped semiconductor nanocrystals. *Nano Lett.*, **10**, 3670–3674.

Vyšniauskas, A., Qurashi, M., Gallop, N., Balaz, M., Anderson, H. L., and Kuimova, M. K. (2015) Unravelling the effect of temperature on viscosity-sensitive fluorescent molecular rotors. *Chem. Sci.*, **6**, 5773–5778.

Vyšniauskas, A., Cornell, B., Sherin, P. S., Maleckaitė, K., Kubánková, M., Izquierdo, M. A., Vu, T. T., Volkova, Y. A., Budynina, E. M., Molteni, C., and Kuimova, M. K. (2021) Cyclopropyl substituents transform the viscosity-sensitive BODIPY molecular rotor into a temperature sensor. *ACS Sens.*, **6**, 2158–2167.

Walker, G. W., Sundar, V. C., Rudzinski, C. M., Wun, A. W., Bawendi, M. G., and Nocera, D. G. (2003) Quantum-dot optical temperature probes. *Appl. Phys. Lett.*, **83**, 3555–3557.

Wang, S., Westcott, S., and Chen, W. (2002) Nanoparticle luminescence thermometry. *J. Phys. Chem. B*, **106**, 11203–11209.

Wang, Z., Zhang, D., and Zhu, D. (2005) Synthesis and thermal modulation of the fluorescence of a pyrene-arylmaleimide dyad and its Diels–Alder adduct. *Tetrahedron Lett.*, **46**, 4609–4612.

Wang, K., Tang, Z., Yang, C. J., Kim, Y., Fang, X., Li, W., Wu, Y., Medley, C. D., Cao, Z., Li, J., Colon, P., Lin, H., and Tan, W. (2009) Molecular engineering of DNA: molecular beacons. *Angew. Chem. Int. Ed.*, **48**, 856–870.

Wang, X.-d., Song, X.-h., He, C.-y., Yang, C. J., Chen, G., and Chen, X. (2011) Preparation of reversible colorimetric temperature nanosensors and their application in quantitative two-dimensional thermo-imaging. *Anal. Chem.*, **83**, 2434–2437.

Wang, C., Ling, L., Yao, Y., and Song, Q. (2015) One-step synthesis of fluorescent smart thermo-responsive copper clusters: a potential nanothermometer in living cells. *Nano Res.*, **8**, 1975–1986.

Wang, J.-X., Zhang, T.-S., Zhu, X., Li, C.-X., Dong, L., Cui, G., and Yang, Q.-Z. (2020a) Organic thermometers based on aggregation of difluoroboron β-diketonate chromophores. *J. Phys. Chem. A*, **124**, 10082–10089.

Wang, J.-X., Yu, Y.-S., Niu, L.-Y., Zou, B., Wang, K., and Yang, Q.-Z. (2020b) A difluoroboron β-diketonate based thermometer with temperature-dependent emission wavelength. *Chem. Commun.*, **56**, 6269–6272.

Wang, J., Jiang, A., Wang, J., Song, B., and He, Y. (2020c) Dual-emission fluorescent silicon nanoparticle-based nanothermometer for ratiometric detection of intracellular temperature in living cells. *Faraday Discuss.*, **222**, 122–134.

Wen, S., Zhou, J., Zheng, K., Bednarkiewicz, A., Liu, X., and Jin, D. (2018) Advances in highly doped upconversion nanoparticles. *Nat. Commun.*, **9**, 2415.

Wong, F. H. C., Banks, D. S., Abu-Arish, A., and Fradin, C. (2007) A molecular thermometer based on fluorescent protein blinking. *J. Am. Chem. Soc.*, **129**, 10302–10303.

Widengren, J. and Schwille, P. (2000) Characterization of photoinduced isomerization and back-isomerization of the cyanine dye Cy5 by fluorescence correlation spectroscopy. *J. Phys. Chem. A*, **104**, 6416–6428.

Wilcoxon, J. P., Martin, J. E., Parsapour, F., Wiedenman, B., and Kelley, D. F. (1998) Photoluminescence from nanosize gold clusters. *J. Chem. Phys.*, **108**, 9137–9143.

Xia, T., Wang, L., Qu, Y., Rui, Y., Cao, J., Hu, Y., Yang, J., Wu, J. and Xu, J. (2016) A thermoresponsive fluorescent rotor based on a hinged naphthalimide for a viscometer and a viscosity-related thermometer. *J. Mater. Chem. C*, **4**, 5696–5701.

Xie, N., Huang, J., Yang, X., He, X., Liu, J., Huang, J., Fang, H., and Wang, K. (2017) Scallop-inspired DNA nanomachine: a ratiometric nanothermometer for intracellular temperature sensing. *Anal. Chem.*, **89**, 12115–12122.

Xu, X., Ray, R., Gu, Y., Ploehn, H. J., Gearheart, L., Raker, K., and Scrivens, W. A. (2004) Electrophoretic analysis and purification of fluorescent single-walled carbon nanotube fragments. *J. Am. Chem. Soc.*, **126**, 12736–12737.

Xu, W., Qu, J., Liu, Y., Bai, J., Li, Y., and Qu, S. (2023) A calibration-free fiber sensor based on CdZnSe/ZnSe/ZnS quantum dots for real-time monitoring of human thermal activities. *Measurement*, **206**, 112315.

Xue, K., Wang, C., Wang, J., Lv, S., Hao, B., Zhu, C., and Tang, B. Z. (2021) A sensitive and reliable organic fluorescent nanothermometer for noninvasive temperature sensing. *J. Am. Chem. Soc.*, **143**, 14147–14157.

Yanagi, T., Kaminaga, K., Kada, W., Hanaizumi, O., and Igarashi, R. (2020) Optimization of wide-field ODMR measurements using fluorescent nanodiamonds to improve temperature determination accuracy. *Nanomaterials*, **10**, 2282.

Yang, Y., Zhao, Q., Feng, W., and Li, F. (2013) Luminescent chemodosimeters for bioimaging. *Chem. Rev.*, **113**, 192–270.

Yang, Y., Kong, W., Li, H., Liu, J., Yang, M., Huang, H., Liu, Y., Wang, Z., Wang, Z., Sham, T.-K., Zhong, J., Wang, C., Liu, Z., Lee, S.-T., and Kang, Z. (2015) Fluorescent N-doped carbon dots as in vitro and in vivo nanothermometer. *ACS Appl. Mater. Interfaces*, **7**, 27324–27330.

Ye, F., Wu, C., Jin, Y., Chan, Y.-H., Zhang, X., and Chiu, D. T. (2011) Ratiometric temperature sensing with semiconducting polymer dots. *J. Am. Chem. Soc.*, **133**, 8146–8149.

Yu, S.-J., Kang, M.-W., Chang, H.-C., Chen, K.-M., and Yu, Y.-C. (2005) Bright fluorescent nanodiamonds: no photobleaching and low cytotoxicity. *J. Am. Chem. Soc.*, **127**, 17604–17605.

Yukawa, H., Fujiwara, M., Kobayashi, K., Kumon, Y., Miyaji, K., Nishimura, Y., Oshimi, K., Umehara, Y., Teki, Y., Iwasaki, T., Hatano, M., Hashimoto, H., and Baba, Y. (2020) A quantum thermometric sensing and analysis system using fluorescent nanodiamonds for the evaluation of living stem cell functions according to intracellular temperature. *Nanoscale Adv.*, **2**, 1859–1868.

Zairov, R. R., Dovzhenko, A. P., Sapunova, A. S., Voloshina, A. D., Sarkanich, K. A., Daminova, A. G., Nizameev, I. R., Lapaev, D. V., Sudakova, S. N., Podyachev, S. N., Petrov, K. A., Vomiero, A., and Mustafina, A. R. (2020) Terbium (III)- thiacalix[4]arene nanosensor for highly sensitive intracellular monitoring of temperature changes within the 303–313 K range. *Sci. Rep.*, **10**, 20541.

Zhang, W., Ji, X., Peng, B.-J., Che, S., Ge, F., Liu, W., Al-Hashimi, M., Wang, C., and Fang, L. (2020) High-performance thermoresponsive dual-output dye system for smart textile application. *Adv. Funct. Mater.*, **30**, 1906463.

Zhang, W., Ji, X., Al-Hashimi, M., Wang, C., and Fang, L. (2021) Feasible fabrication and textile application of polymer composites featuring dual optical thermoresponses. *Chem. Eng. J.*, **419**, 129553.

Zhegalova, N. G., Dergunov, S. A., Wang, S. T., Pinkhassik, E., and Berezin, M. Y. (2014) Design of fluorescent nanocapsules as ratiometric nanothermometers. *Chem. Eur. J.*, **20**, 10292–10297.

Zheng, Y., Meana, Y., Mazza, M. M. A., Baker, J. D., Minnett, P. J., and Raymo, F. M. (2022) Fluorescence switching for temperature sensing in water. *J. Am. Chem. Soc.*, **144**, 4759–4763.

Zhou, J., Liu, Q., Feng, W., Sun, Y., and Li, F. (2015) Upconversion luminescent materials: advances and applications. *Chem. Rev.*, **115**, 395–465.

3
Intracellular Thermometry with Fluorescent Molecular Thermometers

A living cell is the most attractive place for a fluorescent molecular thermometer to work. Fluorescent molecular thermometers that can stably exist in aqueous media and show high sensitivity to a temperature variation are candidates for intracellular thermometry. Until now, various fluorescent molecular thermometers that can measure intracellular temperature have been developed and used in research. Expertise clarified by fluorescent molecular thermometers includes the confirmation of old beliefs, such as hot mitochondria and thermogenesis by uncoupling there (Nakamura and Matsuoka, 1978), at a single-cell level. In addition, rather unexpected results on intracellular temperature profiles have been published. Every week, we can find a new paper concerning intracellular temperature measured by fluorescent molecular thermometers. In this chapter, I systematically summarize the biologically significant results of intracellular thermometry with fluorescent molecular thermometers, which follows a brief presentation of early attempts in intracellular thermometry and procedures for introducing fluorescent molecular thermometers into living cells and their cytotoxicity assessments. Scientific papers published up to July 2023 are included in this chapter. Besides this book, some well-organized review articles and book chapters specifically on intracellular thermometry by fluorescent molecular thermometers are recommended to read (see Appendix 1 for a list).

3.1
Early Attempts for Cellular Thermometry

3.1.1
Polarity-sensitive Fluorophores Located in Membranes

The first case utilized fluorescent membrane probes, 6-dodecanoyl-2-dimethylami-nonaphthalene (laurdan) and 2-(12-N-NBD-amino)dodecanoyl-1-hexadecanoyl-sn-glycero-3-phosphocholine (NBD-PC) (Figure 3.1a) (Chapman et al., 1995). Both fluorescent compounds were localized on the cellular membrane of CHO (Chinese hamster ovary) cells upon incubation of the cells in a solution containing the fluorescent compound. A heat-induced gel–liquid crystalline phase transition of the cell membrane changes the bilayer's permeability toward water molecules. As laurdan is sensitive to environmental polarity, its temperature-dependent fluorescence spectra were obtained when localized in the cell membrane of CHO cells (Figure 3.1b). The generalized polarization (G.P.) value, calculated as the ratio $(FI_G - FI_L)/(FI_G + FI_L)$ where FI_G and FI_L are the laurdan fluorescence intensities measured at 440 nm (for the gel phase) and 490 nm (for the liquid–crystalline phase), respectively, was linearly correlated with the temperature of CHO cells (Figure 3.1c). On the other hand, the rotation of an amino substituent group is the origin of the temperature sensitivity of NBD-PC in the cell membrane of CHO cells. The fluorescence lifetime of NBD-PC localized in the cell membrane of CHO cells decreased from 4.2 to 2.4

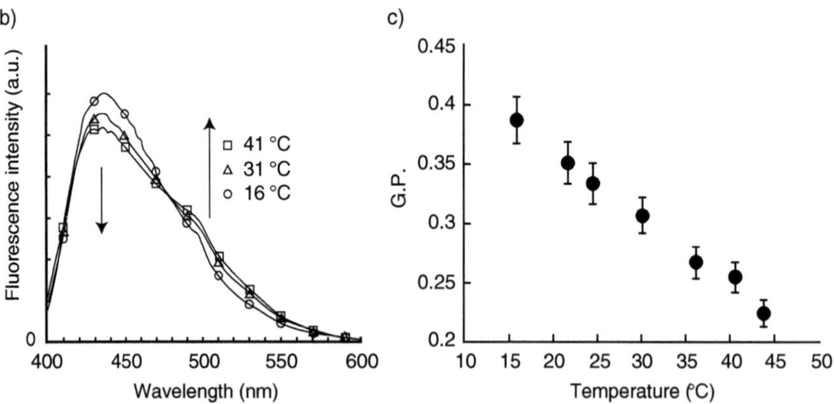

Figure 3.1 Early work on cellular thermometry with fluorescent membrane probes. (a) Chemical structures of laurdan and NBD-PC. (b) Fluorescence spectra of laurdan in CHO cells at 16 °C (circle), 31 °C (triangle), and 41 °C (square). (c) Relationship between the generalized polarization (G.P.) value (see text) and temperature of CHO cells. Adapted from Chapman et al. (1995) *Photochem. Photobiol.*, **62**, 416–425.

ns with the temperature variation from 25 to 42 °C. In a later study, laurdan was utilized for temperature measurements of CHO cells confined by infrared (IR) optical tweezers (Liu et al., 1995). A 1.064 µm optical tweezer with applied power at 200 mW increased the temperature of a CHO cell by 2.2 °C. The heating rate with the IR optical tweezers was evaluated to be 1.15 ± 0.25 °C per 100 mW.

3.1.2
Europium(III) Complex Located in Membranes

The second case applied europium(III) thenoyltrifluoroacetonate (EuTTA). EuTTA itself shows temperature-dependent fluorescence characteristics as described in Section 2.8.2. While EuTTA is easily decomposed in an aqueous environment, it can stably exist when localized in membranes. So, hydrophobic EuTTA stained the cell membrane of CHO cells, and the temperature change of CHO cells was monitored after the treatment with acetylcholine, which activates metabotropic m1-muscarinic receptors (Zohar et al., 1998). Figure 3.2 displays qualitative temperature maps of CHO cells before and after the chemical stimulation by analyzing the fluorescence characteristics of EuTTA. A bi-phasic heat wave was observed, but it was blocked by pre-treatment with a muscarinic receptor antagonist, atropine. Interestingly, atropine alone evoked a mono-phasic heat wave in CHO cells.

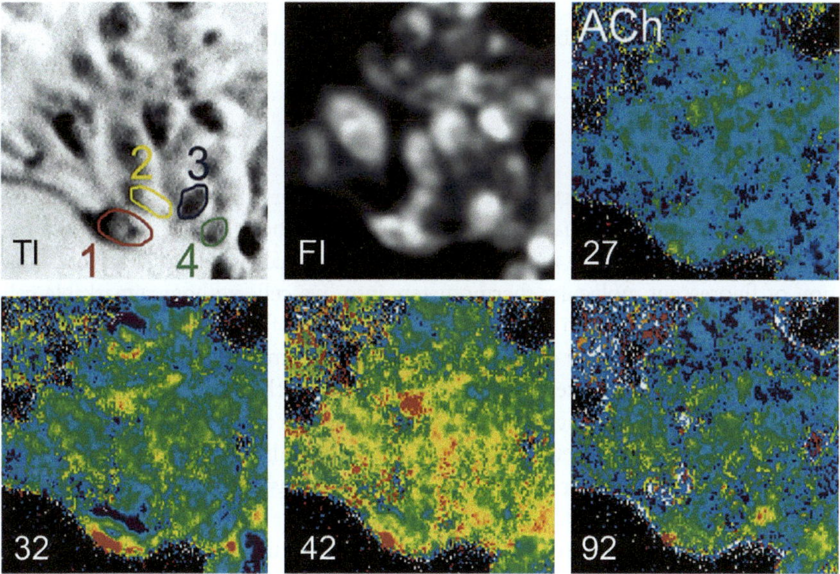

Figure 3.2 Early work on cellular thermometry with EuTTA. Imaging of the heat wave in the cluster of CHO cells after the stimulation with acetylcholine (ACh). Transmitted light (TI), fluorescence (FI), and pseudo-color images. In the transmitted light image, four cells are indicated with circles. The numbers shown on the pseudo-color images are frame numbers (frame interval: 1 s before frame 36, 5 s after frame 36). Acetylcholine was added at frame 27. In the pseudo-color images, cold and warm areas are represented by blue and red, respectively. Adapted from Zohar et al. (1998) *Biophys. J.*, **74**, 82–89 / with permission from ELSEVIER.

A more indirect utilization of EuTTA for cellular thermometry was the containment of EuTTA solution in dimethyl sulfoxide into a glass micropipette. Temperature variation of a single HeLa (human epithelial carcinoma) cell accompanying ionomycin-induced calcium ion influx from an extracellular space was monitored with the EuTTA solution-filled micropipette (Suzuki et al., 2007).

3.2
Introduction of Fluorescent Molecular Thermometers into a Living Cell

When scientists got a highly sensitive fluorescent molecular thermometer functioning in aqueous media, the next obstacle in intracellular temperature measurements was its introduction into target living cells. Some fluorescent molecular thermometers can luckily enter living cells by endocytosis when living cells are incubated in an appropriate medium containing the fluorescent sensor. However, utilizing this bioprocess generally takes longer than a few hours and encounters difficulty dispersing a fluorescent molecular thermometer inside a living cell. In this section, I will outline how to introduce fluorescent molecular thermometers into cells, giving examples. It should be stressed that once a fluorescent molecular thermometer enters living cells, the confirmation of its response to a temperature variation under intracellular circumstances is necessary. Usually, this is done by changing the temperature of a culture medium or heating a local spot with an external heat source (e.g., IR (infrared) laser). In performing intracellular thermometry, selecting a suitable intracellular delivery method is required.

3.2.1
Microinjection

Among various physical methods for intracellular delivery (Stewart et al., 2016), the microinjection technique has been well-established. Although a dedicated apparatus and excellent user skill are required, microinjection enables the efficient introduction of fluorescent molecular thermometers into live cells without remarkable cell damage. As summarized in Figure 3.3, the first case of intracellular thermometry adopted the microinjection of fluorescent nanogel thermometers (Figure 3.3a) into COS7 (African green monkey kidney) cells in my research group (Gota et al., 2009). The hydrophilic fluorescent nanogel thermometer with abundant sulfate groups on the surface, which were specially designed for intracellular thermometry to disperse without aggregation under the high ionic strength inside cells, emitted stronger fluorescence as dots when the temperature of the culture medium increased (Figure 3.3b, 3.3c). The temperature resolution of the hydrophilic fluorescent nanogel thermometers in intracellular thermometry was 0.29–0.50 °C over the range 27–33 °C.

3.2.2
Penetration of Silicon Nanowires

In the first report on intracellular thermometry with fluorescent nanodiamonds, surface-modified vertical silicon nanowires were used for the introduction of the nitrogen-vacancy centers in diamond nanocrystals into a WS1 (human skin

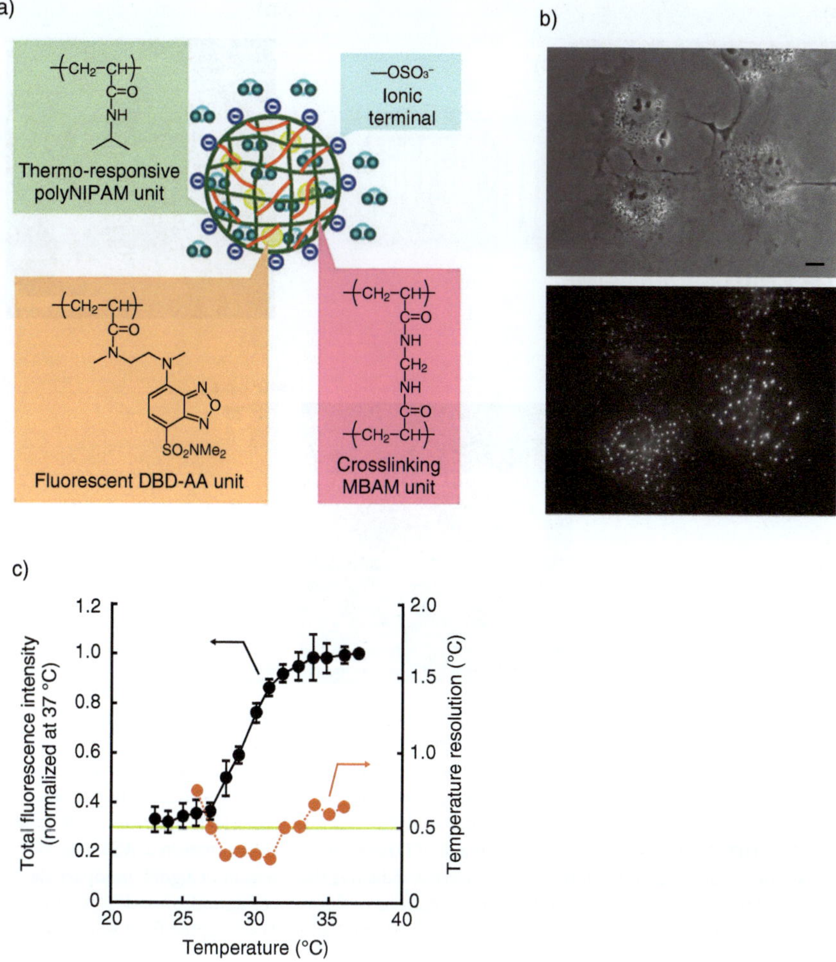

Figure 3.3 Intracellular thermometry with microinjected fluorescent nanogel thermometers. (a) Schematic diagram of the structure of a fluorescent nanogel thermometer. (b) Phase-contrast (upper) and fluorescence (bottom) images of COS7 cells containing fluorescent nanogel thermometers introduced by microinjection. (c) Relationship between the total fluorescence intensity of fluorescent nanogel thermometers in a single COS7 cell and the temperature of culture media (left axis) and the corresponding temperature resolution (right axis). Scale bar: 10 μm. Gota et al. (2009) *J. Am. Chem. Soc.*, **131**, 2766–2767 / American Chemical Society.

fibroblast) cell (Kucsko et al., 2013). This method relies on the ability of silicon nanowires to penetrate a plasma membrane and subsequently release surface-bound functional materials directly into the cytosol (Figure 3.4a–3.4c) (Shalek et al., 2010). In the first example, fluorescent nanodiamonds and gold nanoparticles were delivered into a WS1 cell as thermometers and laser irradiation-activated local heaters (Figure 3.4d).

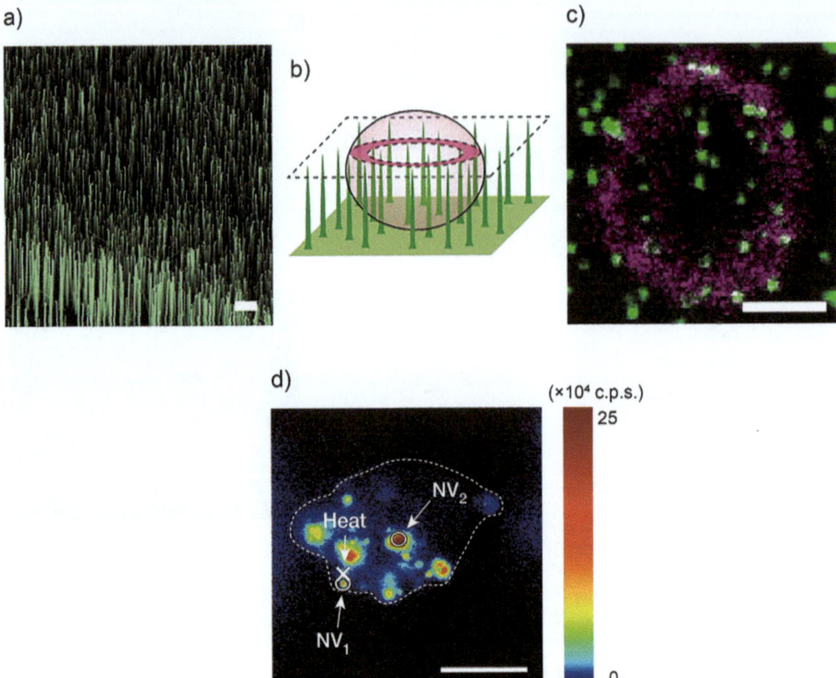

Figure 3.4 Intracellular thermometry with fluorescent nanodiamonds introduced by silicon nanowire-assisted delivery. (a) Scanning electron micrograph of vertical silicon nanowires fabricated via chemical vapor deposition. Scale bar: 1 μm. (b) Schematic rendering of a cell (pink) penetrated by silicon nanowires (green). (c) Confocal microscope image corresponding to the dashed plane in panel (b). Scale bar: 10 μm. Panels (a–c) are Shalek et al. (2010) *Proc. Natl. Acad. Sci. USA*, **107**, 1870–1875. (d) Confocal image of fluorescent nanodiamonds in a WS1 cell. The cell shape is outlined with a dotted line. The cross indicates the position of a gold nanoparticle subjected to heating. The circles indicate the locations of fluorescent nanodiamonds (NV₁ and NV₂) used to measure the temperature. Scale bar: 10 μm. Panel (d) is adapted from Kucsko et al. (2013) *Nature*, **500**, 54–58 / Springer Nature.

3.2.3
Use of Optical Tweezers

Polystyrene particles (diameter: 1 μm) containing a fluorescent molecular thermometer, rhodamine B, and an infrared absorbing dye, FDN-002 (Yamada Chemical, the chemical structure is not disclosed), could be injected into MDCK (Madin–Darby canine kidney) cells using optical tweezers: The polystyrene microsensor was trapped by a 1064 nm laser and transported to the plasma membrane of a target cell. Then, it was heated using an 808 nm laser for 10 seconds to move inside the cytoplasm (Maruyama et al., 2020).

3.2.4
Use of Transfection Reagents

In contrast to the physical methods described above, chemical methods do not need costly apparatus and advanced skills. Initially, the method for delivering fluorescent sensors into cells applies transfection reagents originally developed for transporting a gene's DNA into living cells (Behr, 1993). Transfection reagents have a cationic nature because the electrostatic attraction between the cargo and the anionic plasma membrane is the first delivery process. Lipofectamine 2000 (or simply "lipofectamine") is a transfection reagent that is composed of DOSPA (2,3-dioleyloxy-N-[2-(sperminecarboxamido)ethyl]-N,N-dimethyl-1-propanaminium hydrogen chloride salt) and DOPE (1,2-dioleoyl-*sn*-glycero-3-phosphoethanolamine) (3:1, w/w) (Figure 3.5a) (Yang and Huang, 1998). A fluorescent molecular thermometer based on an L-DNA molecular beacon (Section 2.12) was delivered into HeLa (human epithelial carcinoma) cells by incubation with lipofectamine 2000 for 4 hours (Ke et al., 2012). Although not mentioned in the original paper, an L-DNA molecular beacon-based fluorescent molecular thermometer was likely localized only in the nucleus (Figure 3.5b). There, the temperature-dependent response was confirmed (Figure 3.5c).

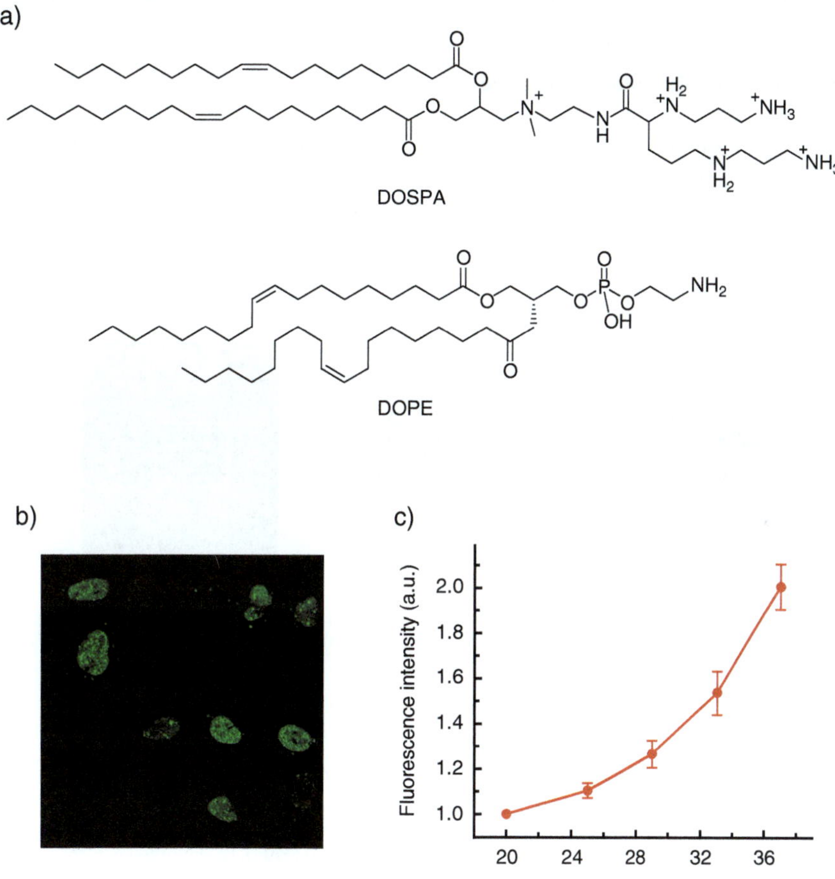

Figure 3.5 Delivery of a molecular beacon-based fluorescent thermometer into HeLa cells by lipofectamine 2000. (a) Chemical structures of the components (DOSPA and DOPE) of lipofectamine 2000. (b) Confocal microscopy image of HeLa cells transfected with a molecular beacon-based fluorescent thermometer at 37 °C. (C) Relationship between the fluorescence intensity of HeLa cells transfected with the fluorescent thermometer and the temperature. The fluorescence intensity is normalized at 20 °C. Panels (b,c) are adapted from Ke et al. (2012) *J. Am. Chem. Soc.*, **134**, 18908–18911 / American Chemical Society.

3.2.5
Use of a Synthetic Cationic Polymer

As scientists deepened their understanding of the importance of a cationic structure for intracellular delivery, another motif, a synthetic cationic polymer, was proposed as a transfection reagent (Godbey et al., 1999). Based on this idea, nanoparticles containing temperature-sensitive EuTTA were introduced in a HeLa cell by coating with poly(allylamine hydrochloride) (PAH) (Figure 3.6) (Oyama et al., 2012). The intracellular delivery of the nanoparticles required 1 hour at 37 °C. A similar concept can be found in the delivery of dual-emitting quantum dot/quantum rod-based fluorescent thermometers in HeLa cells (Figure 3.7a) (Albers et al., 2012). A cationic copolymer with ammonium moieties in the side chains assisted the entry of quantum dot/quantum rod-based thermometers into HeLa cells to bring out intracellular temperature-dependent fluorescence characteristics (Figure 3.7b).

3.2.6
Use of a Cationic Biomacromolecule

Once the high utility of cationic molecules was recognized in intracellular delivery, it is reasonable to screen cationic biomolecules as efficient transporters of fluorescent sensors into living cells. Cationic biomacromolecules are expected to be less

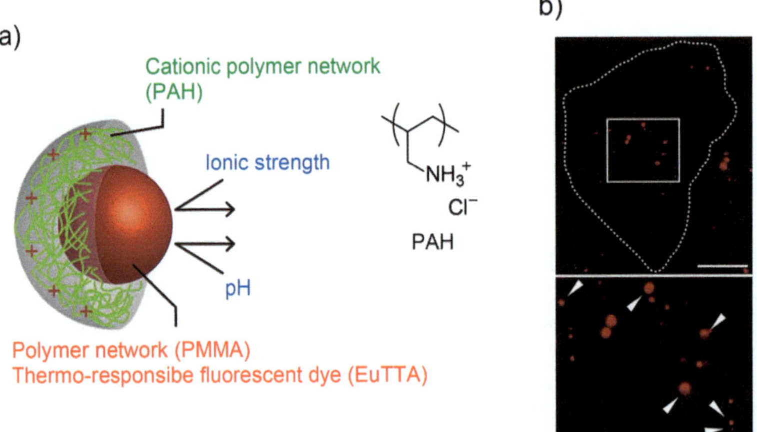

Figure 3.6 Cationic polymer-assisted intracellular delivery of a fluorescent nanoparticle thermometer into HeLa cells. (a) Schematic illustration of a thermometer with a core-shell structure. A poly(methyl methacrylate) (PMMA) core containing temperature-sensitive EuTTA is coated with a cationic polymer poly(allylamine hydrochloride) (PAH). (b) Fluorescence images of EuTTA in the thermometer delivered into a HeLa cell. The dashed line in the upper panel indicates the outline of a HeLa cell. The area surrounded by a white square in the upper panel is enlarged in the lower panel. White arrowheads in the lower panel indicate the nanoparticle thermometers enclosed in either endosomes or lysosomes. Scale bar: 20 μm. Oyama et al. (2012) *Lab Chip*, **12**, 1591–1593 / Royal Society of Chemistry.

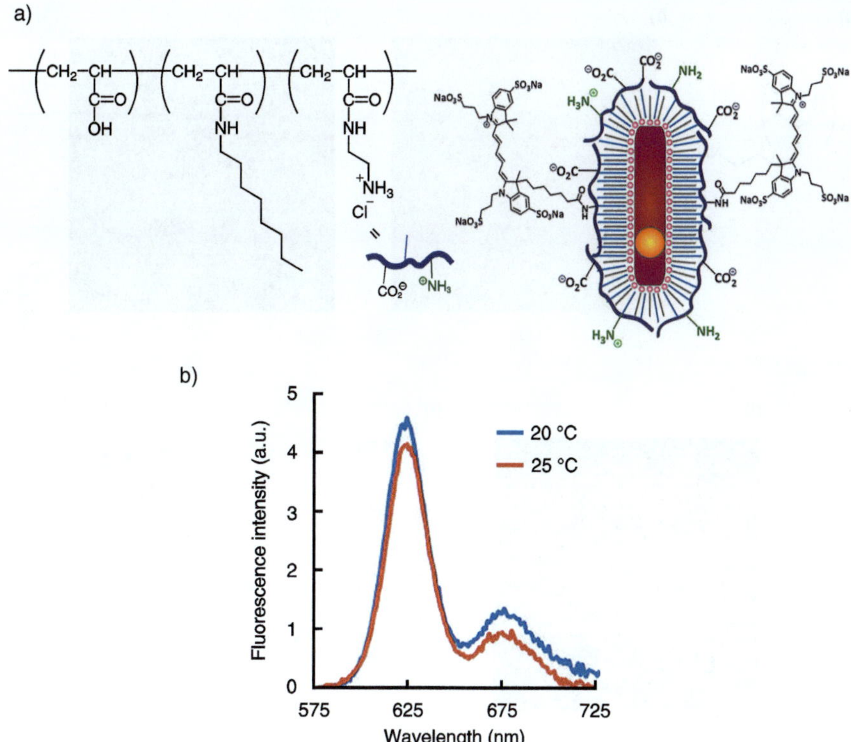

Figure 3.7 Introduction of dual-emitting quantum dot/quantum rod-based fluorescent thermometers into HeLa cells with assistance from an amphiphilic random copolymer having alkylammonium moieties. (a) Schematic structure of quantum dot (orange circle)/quantum rod (purple rectangle)-based fluorescent thermometers covered by an amphiphilic random copolymer. (b) Temperature-dependent fluorescence spectra of quantum dot/quantum rod-based fluorescent thermometers in HeLa cells at 20 (blue) and 25 (red) °C. Albers et al. (2012) *J. Am. Chem. Soc.*, **134**, 9565–9568. Reproduced with the permission of American Chemical Society.

cytotoxic compared to synthetic cationic polymers. Poly-L-lysine (Figure 3.8a) exists as a cationic biomacromolecule by protonation in a neutral aqueous solution. Silica nanoparticles doped with the fluorescent thermometer $Ru(bpy)_3^{2+}$ were successfully delivered to HepG2 (human hepatocellular carcinoma) cells by further coating with poly-L-lysine and incubation for 24 hours (Figure 3.8b) (Yang et al., 2014).

In 1998, it was found that the human immunodeficiency virus type 1 (HIV-1) transactivating transcriptional activator (TAT) protein can penetrate cells in a receptor-independent mechanism and activate HIV-1 specific target genes (Frankel and Pabo, 1988; Green and Loewenstein, 1988). The ability of HIV-TAT protein to enter cells is due to the TAT protein transduction domain, an arginine-rich region comprising residues 48–60 (i.e., GRKKRRQRRRPPQ) (Vivès et al., 1997). Now, numberless protein transduction domains (also called cell-penetrating peptides) are known (Copolovici et al., 2014). A ratiometric fluorescent thermometer, rhodamine B-conjugated carbon

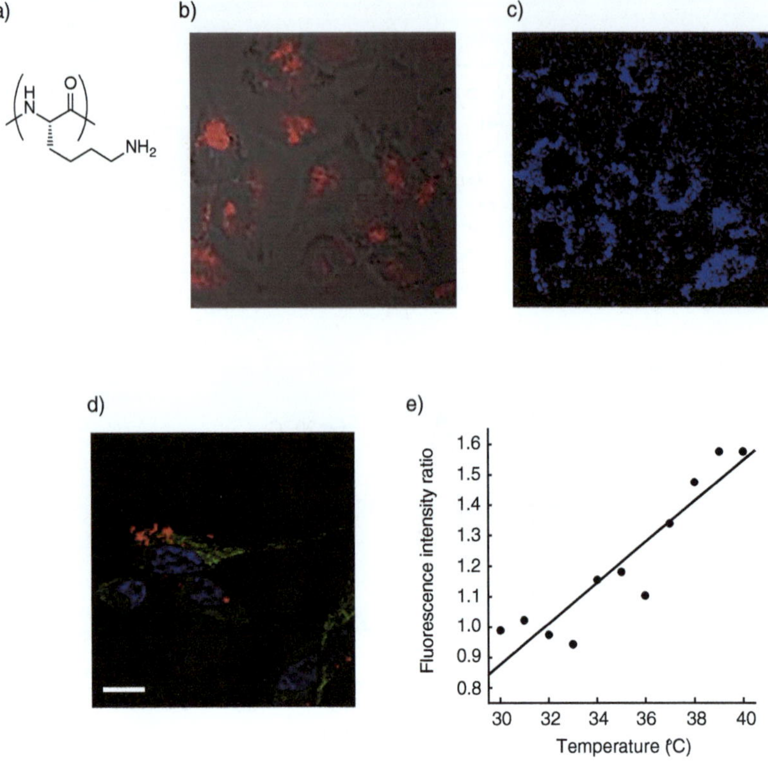

Figure 3.8 Intracellular delivery of fluorescent thermometers into live cells with the assistance of biomacromolecules. (a) Chemical structure of poly-L-lysine. (b) Introduction of Ru(bpy)$_3^{2+}$-doped silica nanoparticles (red) coated with poly-L-lysine into HepG2 cells. The scale was not indicated in the original report. Panel (b) is adapted from Yang et al. (2014) *Microchim. Acta,* **181**, 743–749 / Springer Nature. (c) Insertion of rhodamine B-conjugated carbon dots with support of a cell-penetrating peptide, human immunodeficiency virus 1 transacting activator of transcription peptide (HIV-I TAT). The scale was not indicated in the original report. Panel (c) is adapted from Wei et al. (2017) *J. Mater. Chem. B,* **5**, 3383–3390 / Royal Society of Chemistry. (d) Merged fluorescence image of Hoechst 33342 (blue, nucleus), MitoTracker Green FM (green, mitochondria), and quantum dots (red) in SH-SY5Y cells. Quantum dots were introduced into the cells using a Qtracker™ cell labeling kit. Scale bar: 20 μm. (e) Relationship between the fluorescence intensity ratio of a single quantum dot in an SH-SY5Y cell and temperature. The fluorescence intensity ratio was calculated by dividing the fluorescence intensity in 650–670 nm by that in 630–650 nm. Panels (d,e) are adapted from Tanimoto et al. (2016) *Sci. Rep.,* **6**, 22071 / Springer Nature / CCBY 4.0 / Public domain.

dot was delivered to HeLa cells by incubation for 2 hours after the surface of the probe was further modified with the HIV-1 TAT peptide (Figure 3.8c) (Wei et al., 2017).

Commercially available Qtracker™ cell labeling kits used for the introduction of quantum dots into living cells most likely employ some cell-penetrating peptides such as polyarginine: an example is shown in Figure 3.8d (Tanimoto et al., 2016). Quantum dots are introduced into a human-derived neuronal cell line SH-SY5Y by incubation with a Qtracker™ 655 cell labeling kit at 37 °C for 60 min, followed by checking the fluorescence response to a temperature variation (Figure 3.8e).

3.2.7
Cationic Fluorescent Polymeric Thermometers

Followed by the first intracellular thermometry with fluorescent nanogel thermometers microinjected into COS7 cells, a thermo-responsive polymer-based fluorescent polymeric thermometer with the ability to enter living cells spontaneously was developed in the author's laboratory (Tsuji et al., 2013). The original motivation for us to develop such a microinjection-free polymeric thermometer was the intracellular thermometry of yeast cells. Although technical progress was reported in microinjection in yeast cells (Riveline and Nurse, 2009), it cannot be routinely applied to yeast cells because the cell wall prevents the ingression of a glass needle. As a polymer motif facilitates functional integration in a molecule, a cationic fluorescent polymeric thermometer with the ability to enter live cells could be straightforwardly designed (Figure 3.9a). In addition to thermo-responsive NNPAM (N-n-propylacrylamide) units and environment-sensitive fluorescent DBD-AA

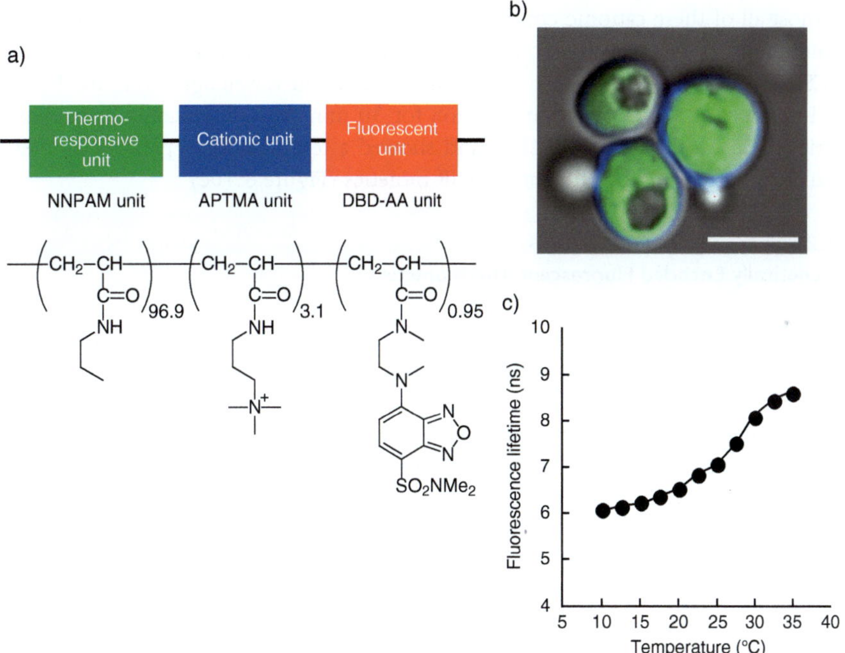

Figure 3.9 Cationic fluorescent polymeric thermometer with the ability to enter living cells. (a) Chemical structure. (b) Merged image of differential interference contrast of yeast cells and confocal fluorescence of Calcofluor white M2R (cell wall, blue) and the cationic fluorescent polymeric thermometer (green) in the yeast cells. Scale bar: 2 μm. (c) The function of the cationic fluorescent polymeric thermometer in yeast cells. λ_{ex}: 456 nm, λ_{em}: 565 nm. Adapted from Tsuji et al. (2013) *Anal. Chem.*, **85**, 9815–9823 / American Chemical Society.

(N-{2-[(7-N,N-dimethylaminosulfonyl])-2,1,3-benzoxadiazol-4-yl}(methyl) amino}ethyl-N-methylacrylamide) units, cationic APTMA ((3-acrylamidopropyl) trimethylammonium) units are included in the cationic fluorescent polymeric thermometer, which spontaneously entered into the yeast *Saccharomyces cerevisiae* strain SYT001 by treatment at 25 °C for 20 min in 1.2 mol L^{-1} sorbitol solution (Figure 3.9b). A supplementary explanation is that the cationic APTMA units also increase the hydrophilicity of the fluorescent polymeric thermometer not to aggregate inside cells under high ionic strength. After the introduction into yeast cells, the cationic fluorescent polymeric thermometer stably dispersed in the cytoplasm to show a temperature-dependent fluorescence lifetime (Figure 3.9c). In the same paper (Tsuji et al., 2013), spontaneous entry of the cationic fluorescent polymeric thermometer was also observed in the mammalian MOLT-4 (human acute lymphoblastic leukemia) and HEK293T (human embryonic kidney) cells.

3.2.8
Small Cationic Organic Molecules

Only occasionally, small organic fluorescent compounds, which have a cationic moiety (e.g., quaternary ammonium unit) and a relatively hydrophobic moiety (including a fluorophore), can enter live cells. Due to the electrostatic attraction, almost all of these cationic compounds tend to be localized in negatively charged mitochondria. Mito-RTP is composed of two cationic rhodamine-like structures (CS NIR and RhB) with different maximum emission wavelengths (Figure 3.10a) (Homma et al., 2015). After incubation for 30 min, Mito-RTP stained mitochondria in HeLa cells (Figure 3.10b) and showed a temperature-dependent fluorescence intensity ratio of RhB and CS NIR moieties (Figure 3.10c).

3.2.9
Genetically Encoded Fluorescent Thermometers

Inspired by the high biocompatibility of the green fluorescent protein (GFP) (Tsien, 1998), a large number of genetically encoded fluorescent sensors have been developed for monitoring chemical species and physical parameters (Wang et al., 2023). For temperature measuring, some genetically encoded fluorescent thermometers have been reported. Once a gene for a genetically encoded fluorescent thermometer is determined, cells expressing it can be prepared in a few hours to days. As the temperature sensitivity of the original GFP was not large enough, the adoption of its fluorescence polarization anisotropy, which reflected depolarization of fluorescence as a temperature-dependent parameter, was discussed for intracellular thermometry of HeLa and U-87 MG (human glioblastoma-astrocytoma) cells overexpressing GFP (Donner et al., 2012). A specially designed GFP was used to improve the sensitivity to temperature variation. Figure 3.11a displays the intracellular thermometry of HeLa cells with a genetically encoded fluorescent thermometer pair, gTEMP, that consists of two fluorescent proteins, temperature-sensitive Sirius and reference mT-Sapphire (Nakano et al., 2017). gTEMP is uniformly expressed in the cytoplasm and the nucleus. With increasing temperature, the fluorescence intensity ratio of mT-Sapphire (λ_{em}: 502.5–537.5 nm) to Sirius (λ_{em}: 417–477 nm) in the cytoplasm of HeLa cells linearly increased (Figure 3.11b).

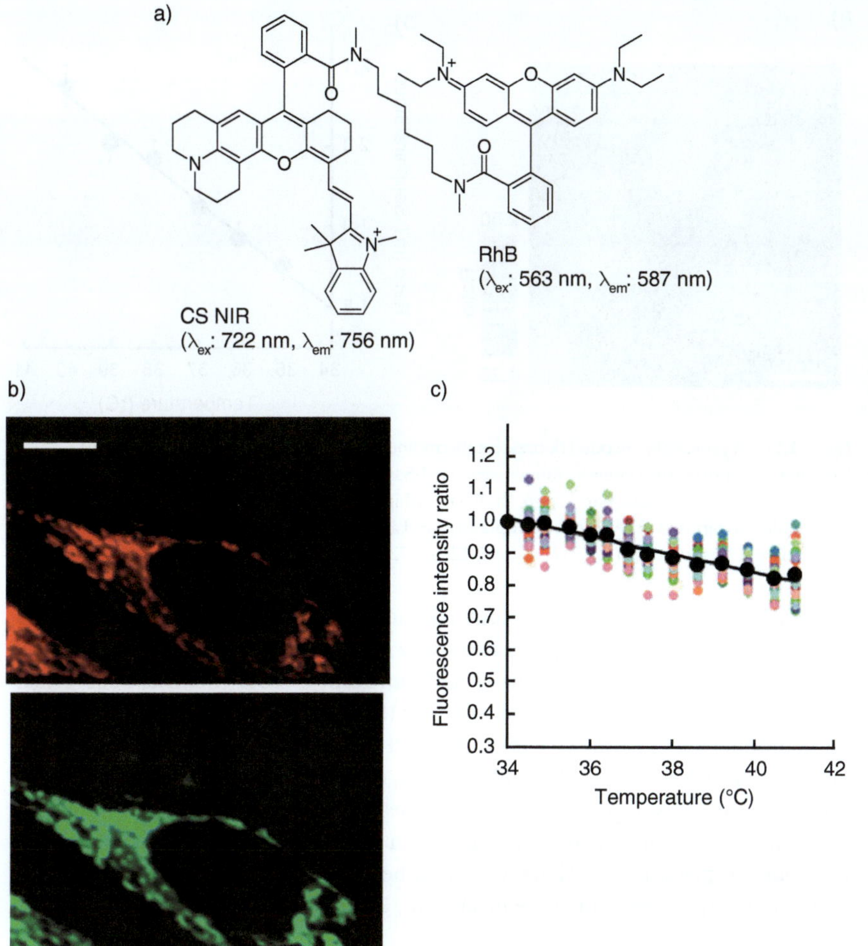

Figure 3.10 A fluorescent molecular thermometer Mito-RTP with the ability to enter live cells and localize at mitochondria. (a) Chemical structure of Mito-RTP composed of two rhodamine-like structures (CS NIR and RhB). (b) Fluorescence images of Mito-RTP (upper) and Mito-Tracker green (lower) in HeLa cells. Scale bar: 5 µm. (c) Temperature-dependent fluorescence intensity ratio of the RhB moiety to the CS NIR moiety of Mito-RTP in HeLa cells (black: average, colored: individual cell) . Adapted from Homma et al. (2015) *Chem. Commun.*, **51**, 6194–6197 / Royal Society of Chemistry.

3.3
Cytotoxicity Assessment

Once a fluorescent probe has stained the subject cells, its cytotoxicity is the next important issue. Cytotoxicity depends on the type of fluorescent probes, cell lines, and even experimental conditions. Therefore, it is desirable to check the cytotoxicity of a fluorescent molecular thermometer in each experimental setup, even though the original report on its development confirmed no toxicity. Researchers who intend to use a fluorescent molecular thermometer must remind themselves that its cytotoxicity might influence the regular temperature of cells. Although not valid for supporting low cytotoxicity in a scientific paper, careful observation of

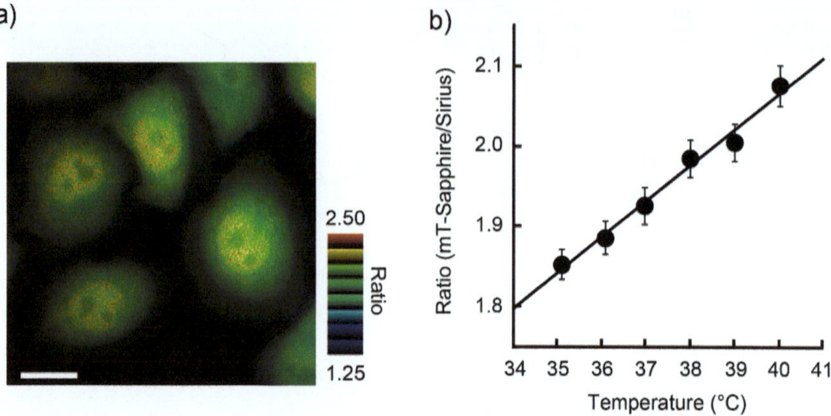

Figure 3.11 A genetically encoded fluorescent thermometer gTEMP composed of Sirius and mT-Sapphire. (a) Fluorescence intensity ratio image of mT-Sapphire to Sirius in gTEMP ubiquitously expressed in HeLa cells. Scale bar: 20 μm. (b) Relationship between the fluorescence intensity ratio of mT-Sapphire to Sirius in gTEMP in the cytoplasm of HeLa cells (n = 13) and the medium temperature. Adapted from Nakano et al. (2017) *PLoS ONE*, **12**, e0172344 / PLOS / CC BY 4.0.

labeled cells in terms of morphology and adhesion (if the cells are adherent) is essential to the cytotoxicity assessment of a fluorescent molecular thermometer. The bias of scientists is occasionally far from reality: most chemists believe that GFPs are almost safe for biological cells, but the fact is that toxic hydrogen peroxide is produced every time GFP is synthesized inside cells (Craggs, 2009). In addition to a fluorescent molecular thermometer, excitation light (especially UV light) can harm live cells. Typical methods for assessing the cytotoxicity of fluorescent molecular thermometers are summarized in Table 3.1. Brief explanations of the methods are given below. There are a number of prominent review publications on cytotoxicity assessments (Lewinski et al., 2008; Soenen et al., 2011).

3.3.1
Propidium Iodide Assay

Propidium iodide (PI) intercalates with DNA and RNA, and this binding significantly enhances the fluorescence of PI due to the hindrance of a non-radiative proton transfer process (Samanta et al., 2012). If live cells are incubated with PI in an isotonic solution, the plasma membrane prevents the entry of PI into the cytoplasm. On the other hand, when cells are dead, PI can move into the nucleus, and strong fluorescence is observed by intercalation with DNA (λ_{ex}: 488 nm, λ_{em}: 575–670 nm) (Krishan, 1975).

Propidium iodide (PI)

Table 3.1 Cytotoxicity assessment for fluorescent molecular thermometers

Name	Principle and remarks	Representative ref.
PI assay	Nuclei of dead cells are stained by an intercalator, propidium iodide (PI) while those of live cells are not.	[1]
MTT assay	MTT is cleaved by intracellular reductase to form blue formazan.	[2–4]
WST-1 assay	Water-soluble tetrazolium salt, WST-1 is reduced by NADH to produce the corresponding water-soluble yellow formazan.	[5]
CCK-8 assay	Highly water-soluble tetrazolium salt, WST-8, is reduced by NADH to produce the corresponding highly water-soluble orange formazan.	[6,7]
Methylene blue assay	Enzymes contained in viable cells reduce methylene blue to a colorless compound. Generally used in cytotoxicity assessment toward yeast cells.	[8]
Total ATP assay	In the presence of luciferase, O_2, and Mg^{2+}, ATP converts luciferin to oxyluciferin with the emission of bioluminescence.	[9]
ROS measurements	Reactive oxygen species (ROS) generated in live cells can be quantified by a fluorescent probe $CM\text{-}H_2DCFDA$.	[10]

1 Hayashi, T., Fukuda, N., Uchiyama, S., and Inada, N. (2015) *PLoS ONE*, **10**, e0117677.
2 Dong, N.-N., Pedroni, M., Piccinelli, F., Conti, G., Sbarbati, A., Ramírez-Hernández, J. E., Maestro, L. M., Iglesias-de la Cruz, M. C., Sanz-Rodriguez, F., Juarranz, A., Chen, F., Vetrone, F., Capobianco, J. A., Solé, J. G., Bettinelli, M., Jaque, D., and Speghini, A. (2011) *ACS Nano*, **5**, 8665–8671.
3 Yang, L., Peng, H.-S., Ding, H., You, F.-T., Hou, L.-L., and Teng, F. (2014) *Microchim. Acta*, **181**, 743–749.
4 Chen, Z., Zhang, K. Y., Tong, X., Liu, Y., Hu, C., Liu, S., Yu, Q., Zhao, Q., and Huang, W. (2016) *Adv. Funct. Mater.*, **26**, 4386–4396.
5 Homma, M., Takei, Y., Murata, A., Inoue, T., and Takeoka, S. (2015) *Chem. Commun.*, **51**, 6194–6197.
6 Qiao, J., Qi, L., Shen, Y., Zhao, L., Qi, C., Shangguan, D., Mao, L., and Chen, Y. (2012) *J. Mater. Chem.*, **22**, 11543–11549.
7 Qiao, J., Hwang, Y.-H., Chen, C.-F., Qi, L., Dong, P., Mu, X.-Y., and Kim, D.-P. (2015) *Anal. Chem.*, **87**, 10535–10541.
8 Tsuji, T., Yoshida, S., Yoshida, A., and Uchiyama, S. (2013) *Anal. Chem.*, **85**, 9815–9823.
9 Jenkins, J., Borisov, S. M., Papkovsky, D. B., and Dmitriev, R. I. (2016) *Anal. Chem.*, **88**, 10566–10572.
10 Kalytchuk, S., Poláková, K., Wang, Y., Froning, J. P., Cepe, K., Rogach, A. L., and Zbořil, R. (2017) *ACS Nano*, **11**, 1432–1442.

3.3.2
MTT Assay

A pale yellow substrate, 3-(4,5-dimethylthiazol-2-yl)-2,5-diphenyl tetrazolium bromide (MTT, Scheme 3.1) produces a dark blue formazan product with absorption at 570 nm when incubated with live cells. The cleavage of MTT is proceeded by the intracellular succinate-tetrazolium reductase system (Slater et al., 1963). Because MTT is cleaved by all living cells and the amount of the product formazan is directly proportional to a cell number, this assay can be used to measure cytotoxicity, proliferation, or activation (Mosmann, 1983).

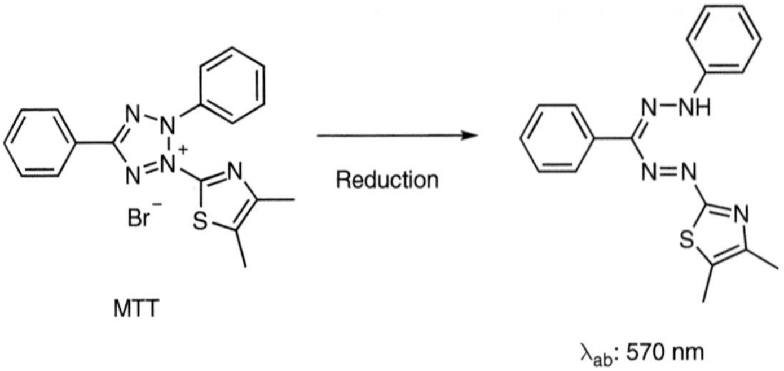

Scheme 3.1

3.3.3
WST-1 Assay

A disadvantage of MTT (Section 3.3.2) is that the formazan product is insoluble in water. So, an extra step to dissolve the formazan product by the addition of alcohol (e.g., ethanol) is needed in the MTT assay. 2-(4-Iodophenyl)-3-(4-nitrophenyl)-5-(2,4-disulfophenyl)-2H-tetrazolium, monosodium salt (WST-1, Scheme 3.2) is the improved reagent in the solubility issue (Ishiyama et al., 1993). Due to two sulfonate groups, the formazan product of WST-1 is highly water-soluble. The maximum absorption wavelength of the formazan product of WST-1 was hypsochromically shifted to 438 nm in compensation for the high solubility.

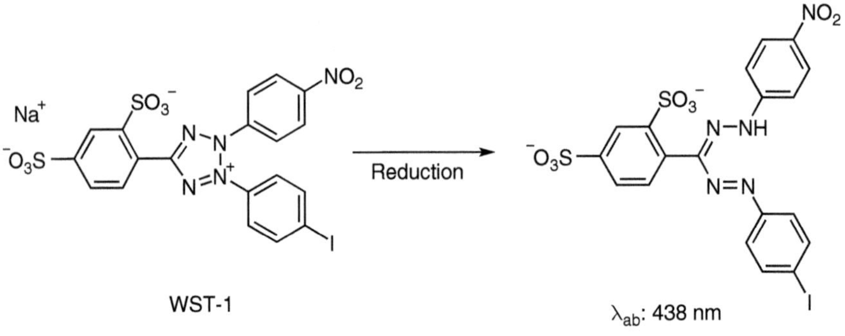

Scheme 3.2

3.3.4
CCK-8 Assay

CCK-8 (cell counting kit-8) assay has been established by a reagent company, Dojindo Laboratories, as a finalized method for assessing cell viability or cytotoxicity with a tetrazolium salt (Ishiyama et al., 1997). CCK-8 assay uses 4-[3-(2-methoxy-4-nitrophenyl)-2-(4-nitrophenyl)-2H-5-tetrazolio]-1,3-benzene disulfonate sodium salt (WST-8, Scheme 3.3) as a tetrazolium salt. WST-8 is reduced by a reduced form of nicotinamide adenine dinucleotide (NADH) produced by dehydrogenases inside living cells to yield the corresponding formazan dye ($\lambda_{ab} = 460$

nm) in the presence of an electron mediator, 1-methoxy-5-methylphenazinium methosulfate. WST-8 is superior to other tetrazolium salts, including WST-1, regarding chemical stability in aqueous solution, sensitivity, and reactivity. A linear colorimetric response was obtained in proportion to the number of viable cells, thus useful for evaluating cytotoxicity.

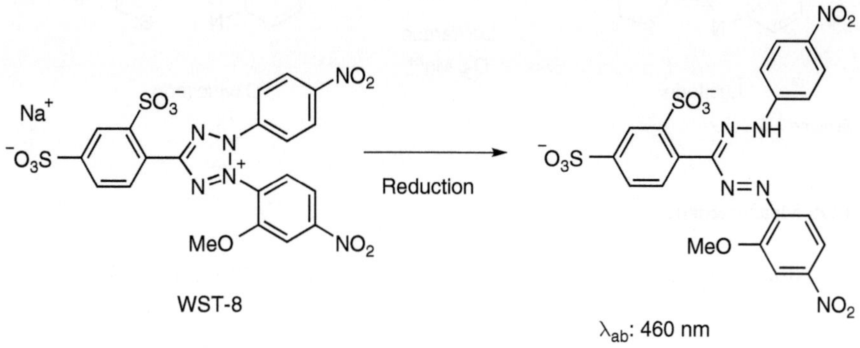

Scheme 3.3

3.3.5
Methylene Blue Assay

For cytotoxicity assessment of yeast cells, methylene blue assay is the most typical method (European Brewery Convention, 1977; Painting and Kirsop, 1990). Living yeast cells contain an enzyme to reduce blue "methylene blue" to colorless leucomethylene blue (Scheme 3.4). When yeast cells are suspended in a methylene blue solution, it penetrates both living and dead cells but is reduced only by living cells, and dead cells are stained blue. Thus, the cytotoxicity of fluorescent thermometers can be evaluated by counting dead blue cells stained with methylene blue.

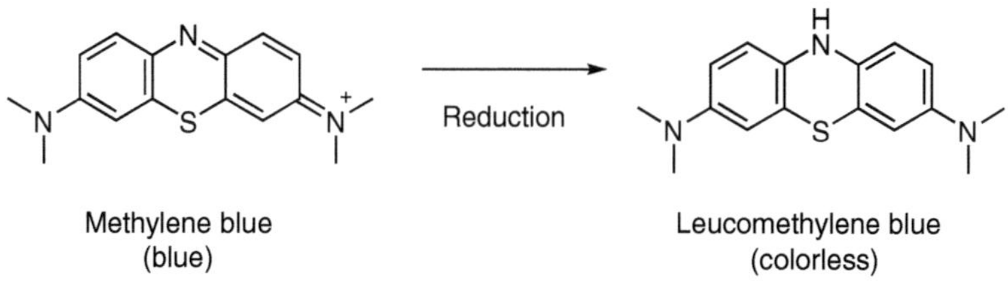

Scheme 3.4

3.3.6
ATP Assay

Apart from a matter of life or death, there are some methodologies to evaluate the cytotoxicity of fluorescent molecular thermometers for live cells. The first is the total adenosine triphosphate (ATP) assay (Crouch et al., 1993), in which intracellular ATP produced in mitochondria is quantified based on the bioluminescence in the oxidation of luciferin in the presence of luciferase, Mg^{2+} ion, and oxygen (Scheme 3.5) following cell lysis (Zhdanov et al., 2011). However, it should be noted that oxidative phosphorylation and glycolysis compensate for each other in mitochondria when either pathway to synthesize ATP is disturbed. That is, a constant level of ATP does not entirely mean no damage (or influence) to mitochondrial functions.

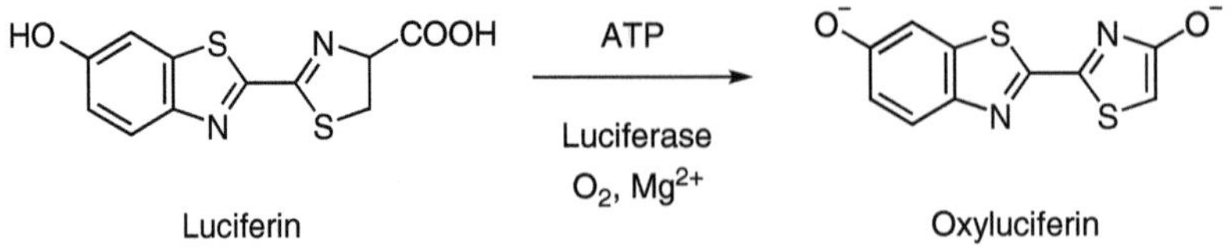

Scheme 3.5

3.3.7
ROS Measurements

Occasionally, quantification of reactive oxygen species (ROS) was performed to evaluate oxidative stresses inflicted upon living cells. For this purpose, 5-(and-6)-chloromethyl-2′,7′-dichlorodihydrofluorescein diacetate (CM-H$_2$DCFDA, Scheme 3.6) is quite applicable: CM-H$_2$DCFDA freely passes the plasma membrane due to increased hydrophobicity by plural acetyl groups, is hydrolyzed to the carboxylate anion form to reside in the cytoplasm, and is oxidized to form a strongly fluorescent fluorescein product (λ_{ex}: 518 nm; λ_{em}: 534 nm) (Oparka et al., 2016). The chloromethyl group of CM-H$_2$DCFDA can react with a thiol group of proteins to prevent a passive leakage from the cell.

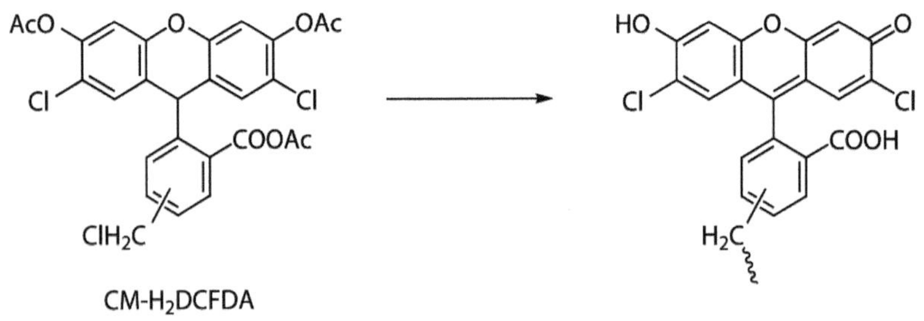

Scheme 3.6

3.4
Practice of Intracellular Thermometry with Fluorescent Molecular Thermometers

3.4.1
Temperature Distribution inside a Living Cell

The development of useful fluorescent molecular thermometers enables intracellular temperature mapping to visualize temperature distribution. The following organelles were reported to be warmer than the cytoplasm.

3.4.1.1 Nucleus
In the first intracellular temperature mapping, an anionic fluorescent polymeric thermometer composed of thermo-responsive NNPAM units, ionic 3-sulfopropyl acrylate

(SPA) units, and fluorescent DBD-AA units (Figure 3.12a) was microinjected into COS7 cells to disperse it in a whole cell and its temperature-dependent fluorescence lifetime was observed with fluorescence lifetime imaging microscopy (FLIM) (Okabe et al., 2012). The most notable feature of intracellular temperature mapping using the fluorescent polymeric thermometer is the difference in fluorescence lifetime between the nucleus and the cytoplasm (Figure 3.12b). In a representative cell, the temperature of the nucleus, calculated from fluorescence lifetime, was higher than that of the cytoplasm (Figure 3.12c). The average temperature difference between the nucleus and the cytoplasm in COS7 cells was 0.96 °C (cell number was 62) (Figure 3.12d). The temperature gap was markedly cell cycle-dependent (see later Section 3.4.2). It was discussed that the origin of the temperature difference between the nucleus and the cytoplasm was higher activity in the nucleus in DNA replication, transcription, and RNA processing besides structural separation by the nuclear membrane.

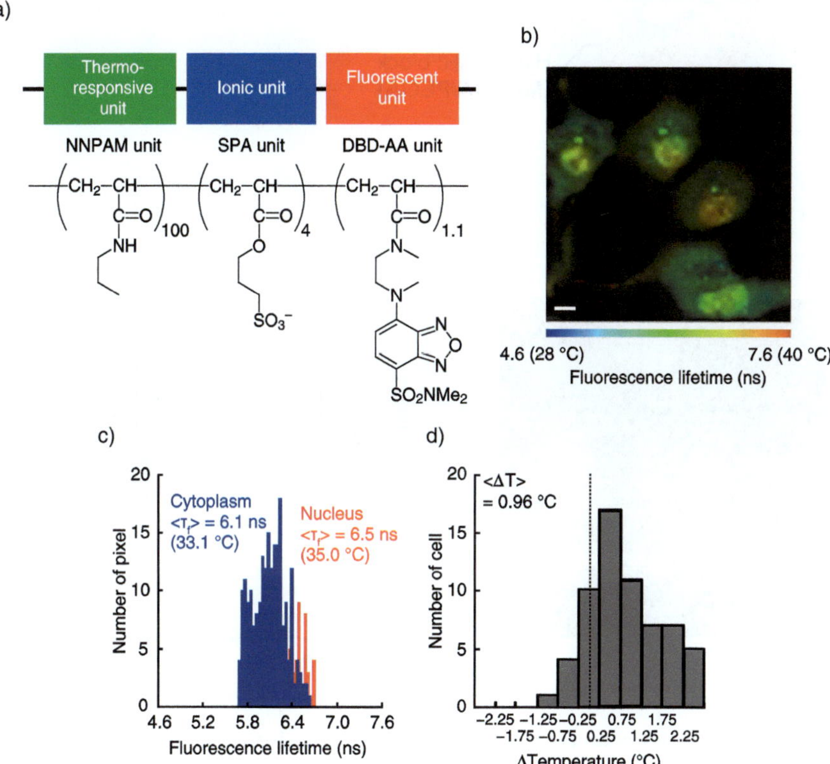

Figure 3.12 Intracellular temperature mapping with a fluorescent polymeric thermometer visualizes a temperature difference between the nucleus and the cytoplasm in live cells. (a) Chemical structure of the fluorescent polymeric thermometer microinjected into COS7 cells. (b) Fluorescence lifetime image of the fluorescent polymeric thermometer in COS7 cells. The temperature of the medium was maintained at 30 °C. Scale bar: 10 μm. (c) Histograms of the fluorescence lifetime in the nucleus and the cytoplasm in a representative cell (the leftmost cell in the panel (b)). <Δτf> is an average of the histogram. (d) Histogram of the temperature difference between the nucleus and the cytoplasm (n = 62). ΔTemperature was calculated by subtracting the average temperature of the cytoplasm from that of the nucleus. <ΔT> is an average of the histogram. Adapted from Adapted from Okabe et al. (2012) *Nat. Commun.*, **3**, 705 / Springer Nature / CCBY 3.0 / Public domain.

Intracellular temperature mapping using luminescent Ln^{3+} (Ln = Sm, Eu)-bearing polymeric micellar probes illustrated the temperature discrepancy between the nucleolus and the rest of the cell (Piñol et al., 2020). Figure 3.13a is a schematic diagram of the lanthanide-bearing polymeric micellar probes, formed as a self-organized structure consisting of a single block copolymer of hydrophobic and hydrophilic units. A fluorescence image of MDA-MB-468 (human breast metastatic adenocarcinoma) cells incubated with the lanthanide-bearing polymeric micellar

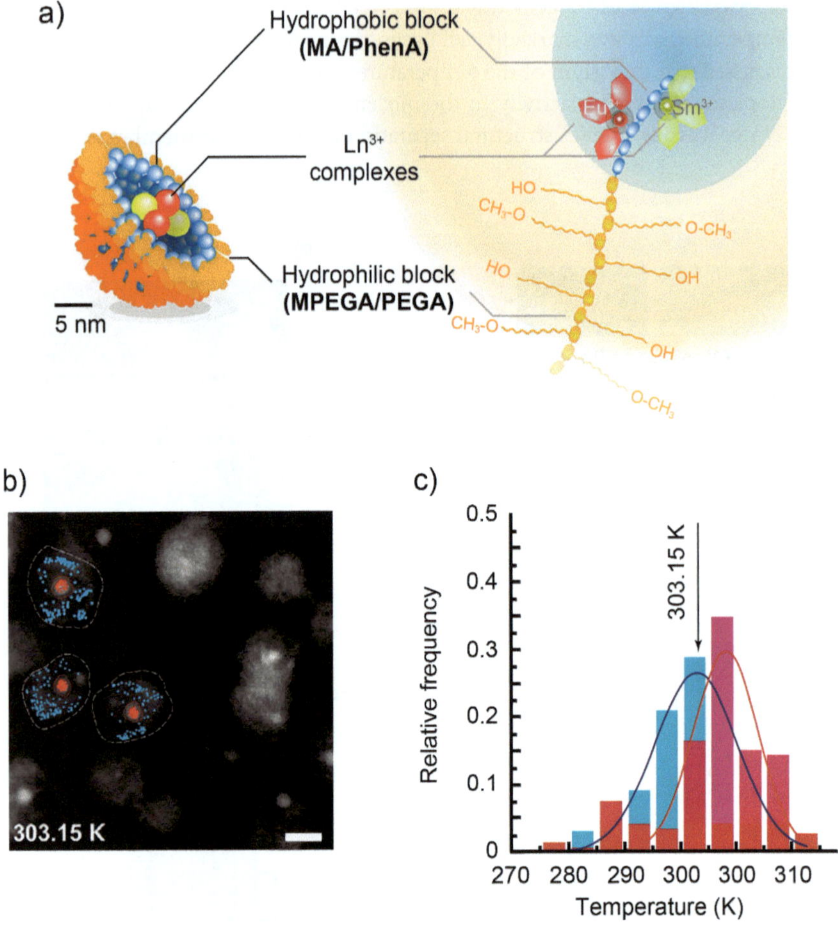

Figure 3.13 Intracellular temperature imaging with lanthanide-bearing polymeric micellar probes. (a) Schematic diagram of lanthanide-bearing polymeric micellar probes. MA: methyl acrylate, PhenA: 4-(acryloyloxymethyl)-1,10-phenanthroline, MPEGA: poly(ethylene glycol) methyl ether acrylate, PEGA: poly(ethylene glycol) acrylate. (b) Fluorescence image of MDA-MB-468 cells incubated with the polymeric micellar probes. Scale bar: 10 µm. Blue and red dots indicate the positions applied for temperature measurements. These points correspond to the nucleolus and the cytoplasm, respectively. The stated temperature is of the culture medium. (c) Temperature histograms at the blue and red points indicated in panel (b). Adapted from Piñol et al. (2020) *Nano Lett.*, **20**, 6466–6472 / American Chemical Society.

probes is shown in Figure 3.13b. The temperature can be calculated from the emission intensity ratio of Sm^{3+} to Eu^{3+} with the uncertainty of 0.2 °C. The histogram in Figure 3.13c indicates that the temperature of the nucleolus (at the red points in Figure 3.13b) is higher than that of other areas (at the blue points) by approximately 5 °C at the maximum.

Fluorescent nanodiamonds (see Section 2.16.4) also supported a temperature ascent in the nucleus compared with the cytoplasm (Wu et al., 2022). After the introduction of fluorescent nanodiamonds (diameter: 100 ± 30 nm) by endocytosis of HMEC-1 (human brain microvascular endothelial) cells, aggregation of fluorescent nanodiamonds to microspheres in a micrometer scale was provoked by optical forces. With the assistance of a scanning optical tweezing system, the position of fluorescent nanodiamond microspheres can be precisely controlled inside cells. As demonstrated in Figure 3.14, fluorescent nanodiamond microspheres at different locations emit at different peak wavelengths, which depend on environmental temperature. Under the temperature resolution of 0.5 °C, the temperature at the nuclear membrane was 2.5 °C higher than that at the plasma membrane.

Intracellular thermometry with genetically encoded fluorescent thermometers resulted in controversy about the temperature difference between the nucleus and the cytoplasm, while both conclusions were raised from the same research group. A genetically encoded ratiometric fluorescent thermometer gTEMP comprises two fluorescent proteins (mTsapphire and Sirius) with different temperature sensitivities and maximum emission wavelengths to produce a temperature-dependent fluorescence intensity ratio. As shown in Figure 3.15a, the fluorescence intensity ratio of gTEMP in the nucleus of HeLa cells was larger than that in the cytoplasm, indicating that the temperature of the nucleus is 2.9 ± 0.3 °C higher than that of the cytoplasm

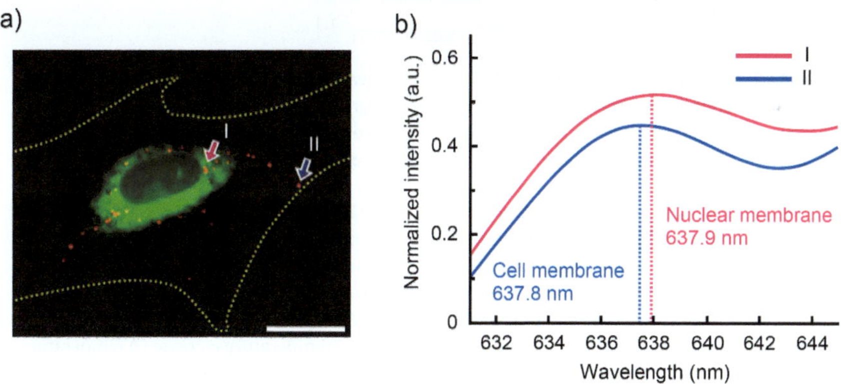

Figure 3.14 Position-specific intracellular thermometry with fluorescent nanodiamonds. (a) Merged fluorescence image of the fluorescent nanodiamond microspheres (red) and Mito-Tracker Green FM (green) in an HMEC-1 cell. Dotted lines represent the boundary of the cell. Pink and blue arrows (I and II) point to the microspheres near the nuclear and plasma membranes, respectively. Scale bar: 10 µm. (b) Fluorescence spectra of the fluorescent nanodiamond microspheres (I and II indicated in panel (a)) close to the nuclear membrane and the plasma membrane of the HMEC-1 cell. Adapted from Wu et al. (2022) *Adv. Sci.*, **9**, 2103354 / JOHN WILEY & SONS, INC. / CC BY 4.0.

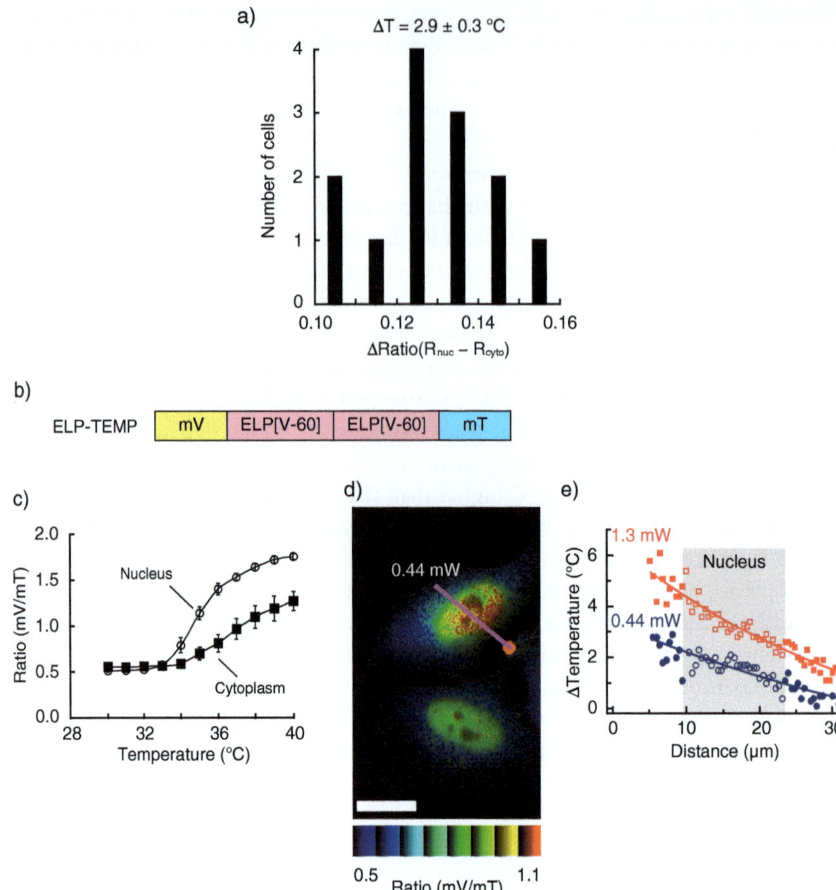

Figure 3.15 Evaluation of temperature difference between the nucleus and the cytoplasm by genetically encoded fluorescent thermometers. (a) Histogram of the difference in the fluorescence intensity ratio of gTEMP between the nucleus and the cytoplasm (ΔRatio(R_{nuc}−R_{cyto})) in a HeLa cell. ΔT represents the average corresponding temperature differences between the nucleus and the cytoplasm. The fluorescence image of gTEMP expressed in HeLa cells is shown in Figure 3.11a. Adapted from Nakano et al. (2017) *PLoS ONE*, **12**, e01712344 / CCBY 4.0 / Public domain. (b) Gene design of ELP-TEMP. ELP[V-60]: elastin-like polypeptide (VPGVG)$_{60}$, mV: mVenus, mT: mTurquoise2. (c) Temperature-dependent fluorescence intensity ratio of mV to mT in the nucleus and the cytoplasm in HeLa cells. (d) Pseudo-colored ratio image of HeLa cells expressing ELP-TEMP under local heating with the laser power of 0.44 mW. A pink line indicates the ROI (region of interest). Scale bar: 20 μm. (e) Relationships between the temperature increase (ΔTemperature) on the line in the panel (d) and the distance from the heat spot under heating with the laser power of 0.44 and 1.3 mW. Closed and open symbols represent the temperature increase in the cytoplasm and the nucleus, respectively. The medium temperature was 34 °C. Panels (b–e) are adapted from Vu et al. (2021) *Sci. Rep.*, **11**, 16519 / Springer Nature / CCBY 4.0 / Public domain.

(Nakano et al., 2017). However, ELP-TEMP, a new version of genetically encoded fluorescent thermometers developed in the same group, denied this temperature gap between the nucleus and the cytoplasm (Vu et al., 2021). ELP-TEMP comprises a temperature-responsive elastin-like peptide (ELP) fused with a yellow fluorescent protein, mVenus (mV), with a cyan fluorescent protein, mTurquoise2 (mT) (Figure

3.15b). The distance between mV and mT changes with a temperature variation, and accordingly, the fluorescence intensity ratio controlled by the efficiency of Förster resonance energy transfer (FRET) between the fluorescent proteins is influenced by temperature. Unfortunately, the fluorescence response curves of ELP-TEMP to a temperature variation are considerably different between the nucleus and the cytoplasm (Figure 3.15c) because it also depends on the concentration of ELP-TEMP (cf. estimated concentration of ELP-TEMP was 1.3 and 0.4 μmol L^{-1} in the nucleus and the cytoplasm, respectively). When the rises in temperature inside a HeLa cell upon laser irradiation were evaluated using the proper calibration curve for the nucleus and the cytoplasm they were found to be smoothly related to the distance from the heat spot irrelevant of the positions (Figure 3.15d, 3.15e). Therefore, it was concluded that the temperatures in the nucleus and the cytoplasm in the absence of a heat source should be almost the same within a temperature resolution.

3.4.1.2 Centrosome

A remarkable feature of intracellular temperature mapping with the anionic fluorescent thermometer (Figure 3.12a, 3.12b) was a single spot in the perinuclear region (Okabe et al., 2012). Co-localization with γ-tubulin identified the spot as a centrosome (Figure 3.16a). Analysis of many COS7 cell samples (n = 35) revealed that the temperature of the centrosome was 0.75 °C higher than the cytoplasm with certain cell–cell variations (Figure 3.16b). The thermogenesis in the centrosome might be associated with its diverse function, such as mitosis and the organization of microtubules (Doxsey, 2001; Conduit et al., 2015), as implied in the temperature map of a dividing HeLa cell (Figure 3.16c). The hydrolysis of tubulin GTP, ATP-driven motion of motor proteins (e.g., dynein and kinesin), and phosphorylation/dephosphorylation of centrosomal proteins by kinase/phosphatase are all possible heat sources at the centrosome (Andersen et al., 2003).

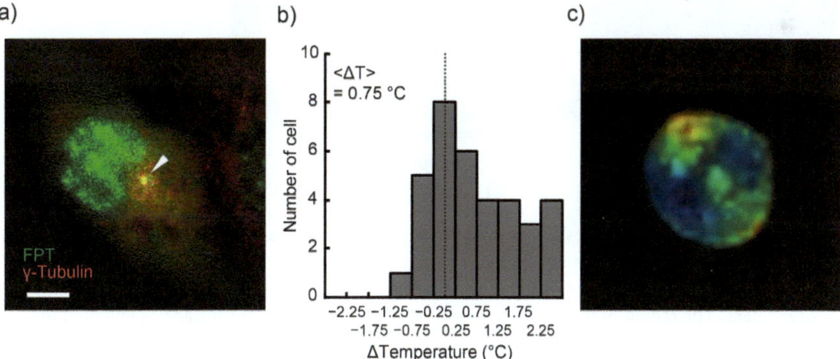

Figure 3.16 Thermogenic centrosome. (a) Merged epifluorescence image of an anionic fluorescent polymeric thermometer (green) and the centrosome visualized with γ-tubulin antibody (red) in a COS7 cell. An arrowhead indicates the centrosome. Scale bar: 10 μm. (b) Histogram of temperature differences between the centrosome and the cytoplasm (ΔTemperature) (cell number *n* = 35). <ΔT> represents an average of the histogram. Panels (a,b) are adapted from Okabe et al. (2012) *Nat. Commun.*, **3**, 705 / Springer Nature / CCBY 3.0 / Public domain. (c) Fluorescence lifetime image of the fluorescent polymeric thermometer in a dividing HeLa cell. Panel (c) was recorded by Uchiyama's research group.

3.4.1.3 **Mitochondria**

It is well-known that mitochondria release heat through respiration (Lowell and Spiegelman, 2000). In intracellular temperature mapping with the anionic fluorescent polymeric thermometer (see Figure 3.12a), the temperature distribution and locations of the mitochondria in COS7 cells were co-visualized by the fluorescent polymeric thermometer and a mitochondria indicator MitoTracker Deep Red (Okabe et al., 2012). An enlarged fluorescence lifetime image showed localized thermogenesis near the mitochondria with inhomogeneity (Figure 3.17).

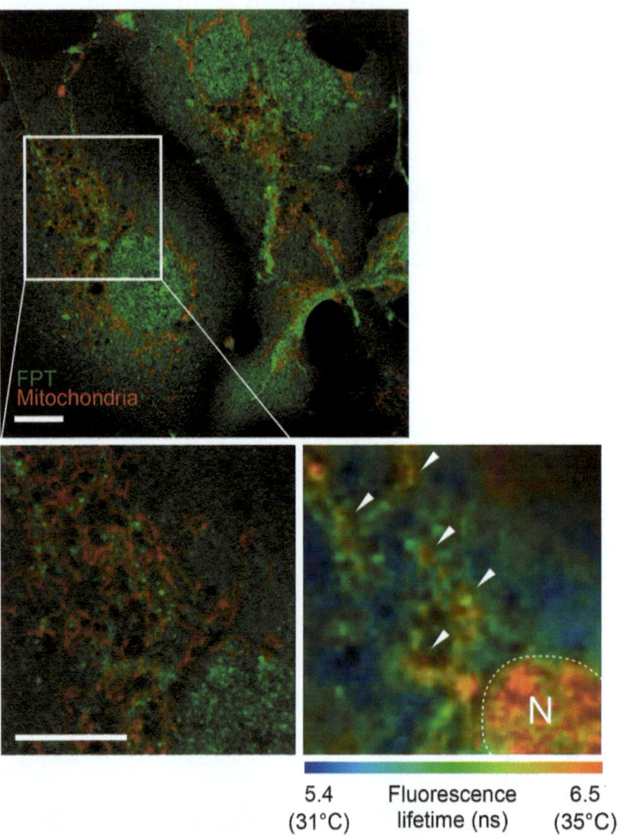

Figure 3.17 Local thermogenesis near mitochondria. Confocal fluorescence images of a fluorescent polymeric thermometer (green) and MitoTracker Deep Red FM (red) (upper and left lower) and fluorescence lifetime image of the fluorescent polymeric thermometer (right lower) in COS7 cells. The region of interest shown in a white square in the upper panel is enlarged in the lower panels. Arrowheads point to local heat production. N represents the nucleus. The temperature of the culture medium was maintained at 30 °C. Scale bars: 10 µm. Adapted from Okabe et al. (2012) *Nat. Commun.*, **3**, 705 / Springer Nature.

3.4.1.4 Neurite in a Neuronal Cell

Temperature distribution of neuronal cells was reported in intracellular thermometry with quantum dots (Tanimoto et al., 2016). Quantum dots were incorporated into differentiated SH-SY5Y cells (human-derived neuronal cell line) using a commercial labeling kit (QTracker™ 655). Calculated from a temperature-dependent emission intensity ratio of quantum dots, the temperature in the neurite was 1.6 °C lower than that in the cell body (Figure 3.18). The reason for this temperature heterogeneity was unclear, but it was assumed that the neurite having a larger surface area could dissipate more heat than the cell body and that the cell body contains warm organelles, such as a nucleus.

3.4.2
Cell-cycle Dependent Intracellular Thermogenesis

Although the number is not large, there are some reports on the effects of the cell cycle on the intracellular temperature. In intracellular thermometry with the fluorescent polymeric thermometer (Figure 3.12a) and FLIM conducted in my laboratory, it was observed that the temperature gap between the nucleus and cytoplasm, described in Section 3.4.1.1, was dependent on the cell cycle (Okabe et al., 2012). That is, the nucleus of COS7 cells synchronized to the G1 phase was 0.70 °C warmer than the cytoplasm (Figure 3.19a), whereas there was little temperature gap (−0.03 °C) in S/G2-phased COS7 cells (Figure 3.19b). Analyzing the data suggested that the temperature homogeneity at the S/G2 phase was due to a temperature increase in the cytoplasm rather than a temperature decrease in the nucleus. This intracellular heat production accelerated at the S/G2 phase did not contradict

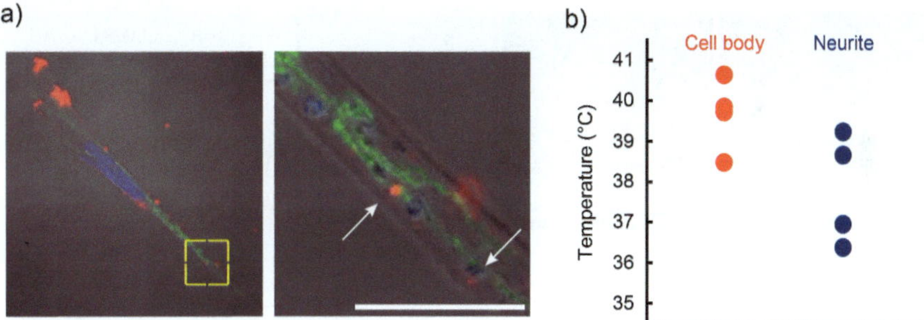

Figure 3.18 Temperature difference between the cell body and the neurite in differentiated neuronal SH-SY5Y cells. (a) Merge of differential interference contrast and fluorescence images of SH-SY5Y cells (blue: nuclei, green: mitochondria, red: quantum dots). The area in the yellow box in the left image is zoomed in the right image. The white arrows point to single quantum dots. Scale bar: 10 μm. (b) The temperatures of the cell body and the neurite measured by the quantum dots. The temperature of a stage-top incubator was set to 37 °C. Adapted from Tanimoto et al. (2016) *Sci. Rep.*, **6**, 22071 / Springer Nature / CCBY 4.0 / Public domain.

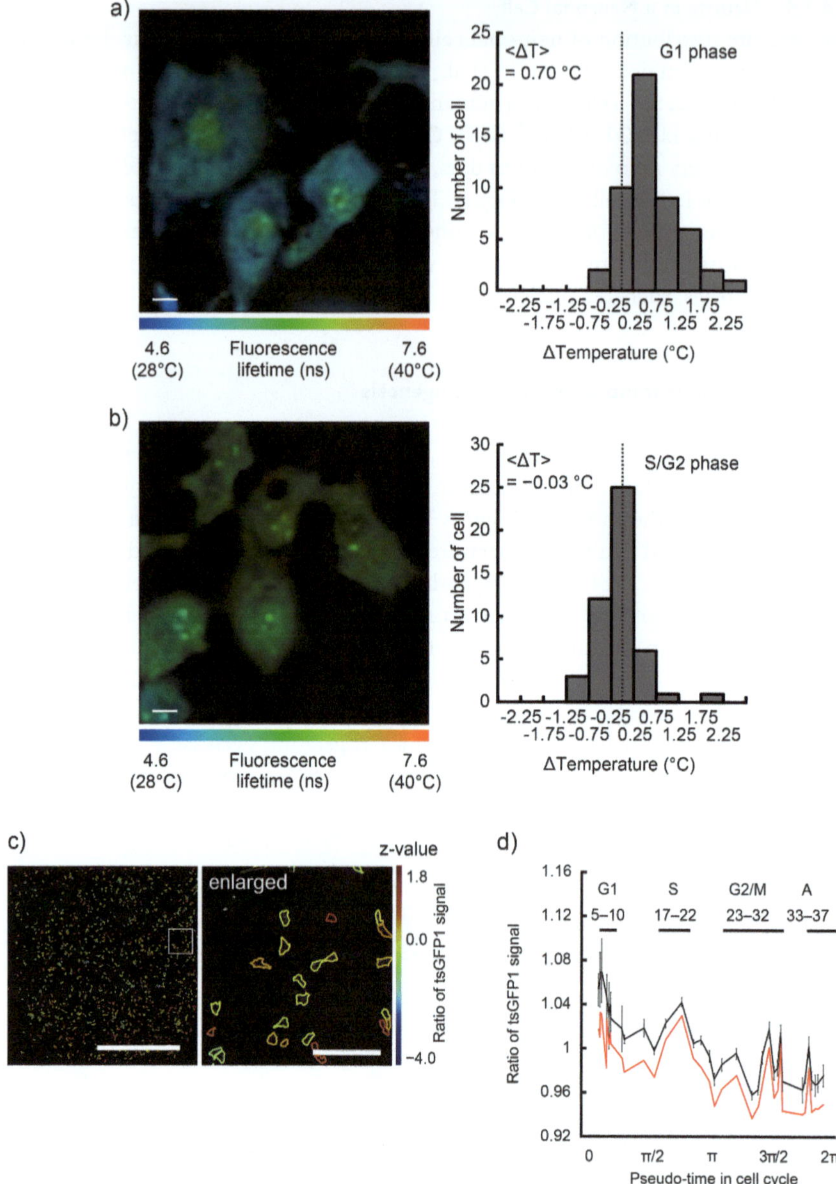

Figure 3.19 Cell cycle-dependent thermogenesis. (a,b) Fluorescence lifetime images of the fluorescent polymeric thermometer in COS7 cells (left) and histograms of the temperature difference (ΔTemperature) between the nucleus and the cytoplasm (right) in COS7 cells synchronized to (a) G1 phase and (b) S/G2 phase. ΔTemperature was calculated by subtracting the average temperature of the cytoplasm from that of the nucleus. <ΔT> represents an average of the histogram. Scale bars: 10 μm. Panels (a,b) are adapted from Okabe et al. (2012) *Nat. Commun.*, **3**, 705 / Springer Nature. (c) Images of automatically identified HeLa cells expressing tsGFP1. The cells are colored by a z-value of the fluorescence intensity ratio of tsGFP1 when excited at 473 nm and 405 nm. Scale bars: 3 mm (left) and 300 μm (right). (d) Intracellular temperature dynamics during cell cycle progression of HeLa cells. The intracellular temperature was evaluated by the fluorescence intensity ratio of tsGFP1 when excited at 473 nm and 405 nm. Black and red lines represent the mean and the medium of data for 25 163 cells, respectively. Panels (c,d) are adapted from Yamanaka et al. (2020) *Biochem. Biophys. Res. Commun.*, **533**, 70–76 / with permission ELSEVIER.

a previous observation for a mass of cells using a microcalorimeter (Yamamura et al., 1986). The opposite result was recorded in a study on cell cycle-dependent intracellular temperature in the cytoplasm of HeLa cells, in which genetically encoded tsGFP1 was used as a fluorescent molecular thermometer (Yamanaka et al., 2020). Numerous data at single-cell resolution were acquired from a wide-field image (Figure 3.19c), and the intracellular temperature was estimated from a fluorescence signal of tsGFP1 expressed in cells. The cell cycle phase of each cell was judged from fluorescence signals of DAPI (4',6-diamidino-2-phenylindole) and geminin after fixing the same cell population. A comparison of intracellular temperature with a cell cycle phase indicated that the temperature in the cytoplasm of HeLa cells was the highest at the G1 phase and gradually decreased with progression through the cell cycle (Figure 3.19d). In the same work, it was also reported that cells were efficiently heated up and hardly cooled down, especially at the G1/S phase. Still, much is unclear on the biological mechanisms in the cell cycle-dependent intracellular temperature fluctuation.

3.4.3
Difference in Intracellular Temperature Between Leader Cells and Follower Cells

Live cells can be categorized into "leader cells" and "follower cells" in the activity. The leader cells develop at wound edges and prepare for a reorganization of the structure for migration and deformation. The the follower cells are next to the leader cells. So, both leader cells and follower cells live in a single culturing environment. With the assumption that the cellular activity is reflected by temperature, the intracellular temperatures of leader and follower cells were measured with two different fluorescent thermometers (Nakamura et al., 2022). Figure 3.20 shows a temperature map of NIH-3T3 (mouse embryonic fibroblast) leader and follower cells, which was obtained with a cationic fluorescent polymeric thermometer (for chemical structure, see later Figure 3.21b). The average temperature of leader cells (33.7 $\pm$ 0.58 °C) was approximately 1 °C higher than that of follower cells (32.7 $\pm$ 0.70 °C). While the temperature resolution is relatively lower than the intracellular thermometry with the fluorescent polymeric thermometer, the intracellular thermometry with quantum dots supported the temperature difference between leader and follower cells. The same group also confirmed that cells started moving only when artificially warmed with an infrared laser. The warmth of leader cells would result from their active movement in addition to a trigger for cell migration and deformation.

3.4.4
Intracellular Temperature Variation Accompanying Chemical Stimulus

One of the standard procedures in developing intracellular thermometry systems is monitoring intracellular temperature variations by treatment with chemical compounds, including a drug. Biological mechanisms in intracellular temperature variation, or concretely saying intracellular thermogenesis, are pursued when a chemical compound, whose effects on intracellular temperature are unknown, is used as a chemical stimulus. In addition, functional assessment of a new fluorescent molecular thermometer in intracellular thermometry, i.e., whether the fluorescent molecular thermometer functions even in biological cells or not, is also conducted when a

a)

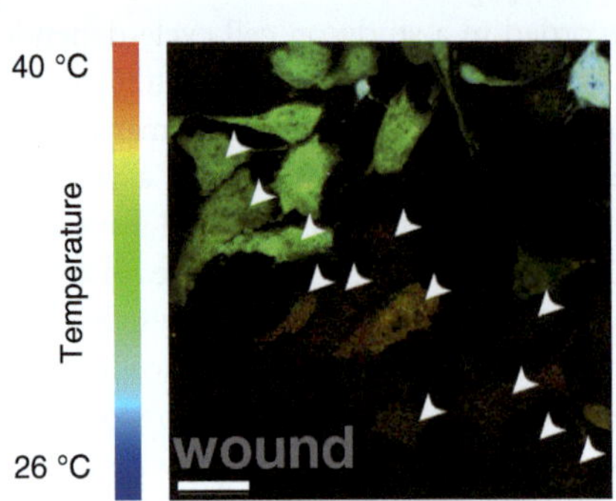

b)

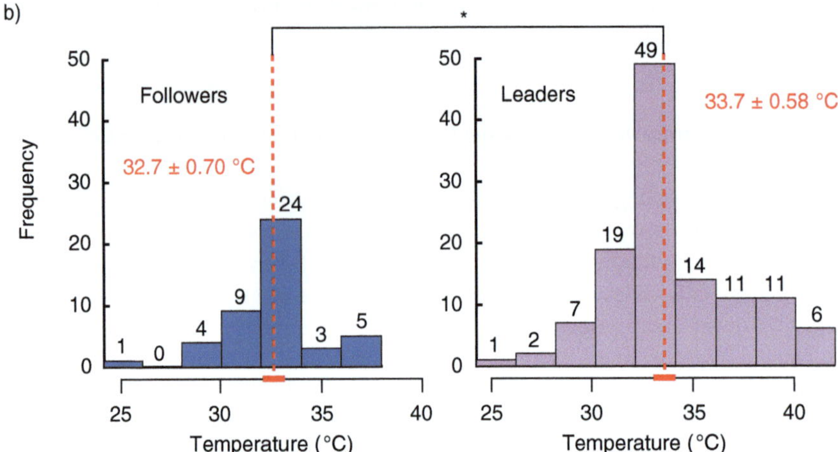

Figure 3.20 Temperature difference between leader cells and follower cells. (a) Temperature image of NIH-3T3 cells. The intracellular temperature was evaluated from the fluorescence lifetime of the fluorescent polymeric thermometer. Leader cells were prepared by scraping off cells at the wound area (lower left corner) and are indicated with white arrowheads. Scale bar: 10 μm. (b) Histograms of intracellular temperature in the leader NIH-3T3 cells and the follower NIH-3T3 cells. *$P < 0.05$. Adapted from Nakamura et al. (2022) *Opt. Contin.*, **1**, 1085–1097 / Optica Publishing Group.

chemical compound, whose effects on cellular temperature have been elucidated (like uncouplers), is used as a chemical stimulus. The following sections give the particulars of chemical compounds influential on intracellular temperature. Examples of intracellular thermometry using organelle-targeted fluorescent molecular thermometers will be described later (Section 3.5).

3.4.4.1 **Uncouplers**

Uncouplers stall oxidative phosphorylation in ATP synthesis in mitochondria. CCCP (carbonyl cyanide 3-chlorophenylhydrazone) and FCCP (carbonyl cyanide

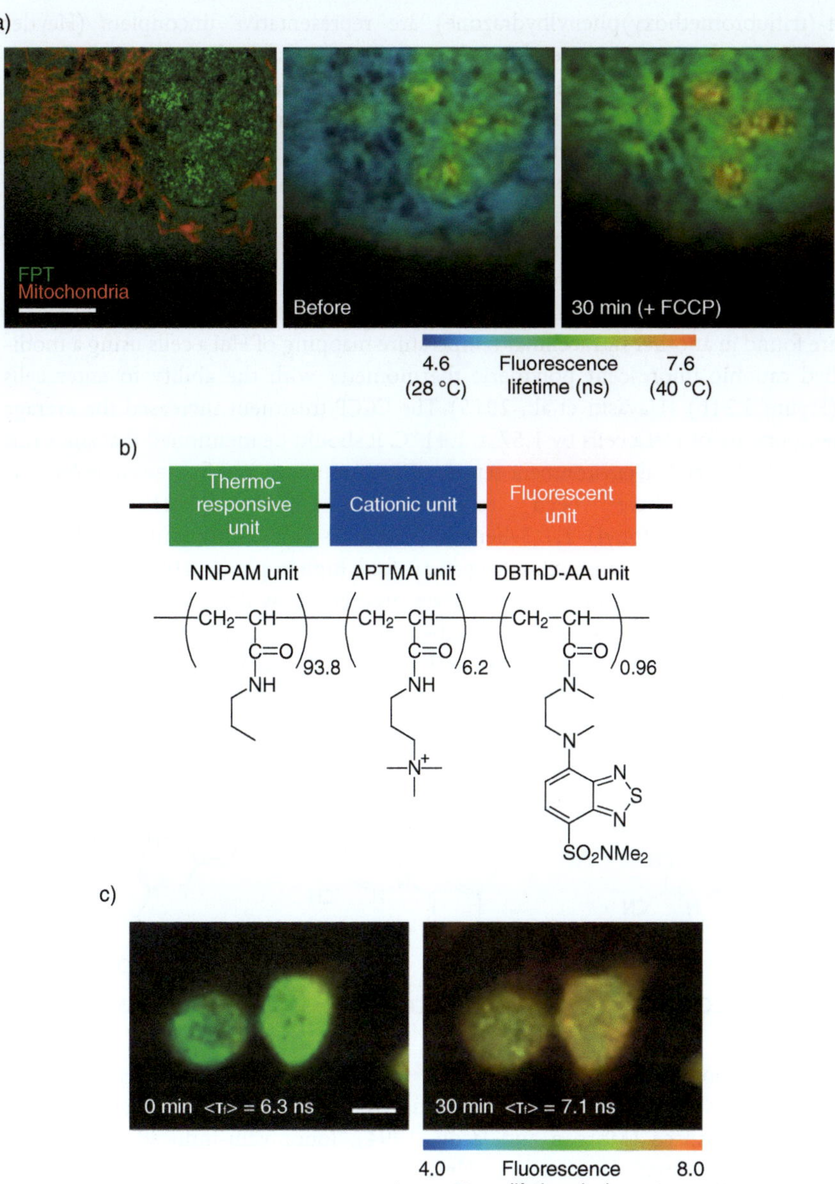

Figure 3.21 Intracellular thermogenesis induced by uncouplers. (a) Confocal fluorescence image of the fluorescent polymeric thermometer (green) and MitoTracker Deep Red FM (red) in living COS7 cells (left) and fluorescence lifetime images of the fluorescent polymeric thermometer before and after the uncoupling by 100 µmol L^{-1} FCCP (middle and right). Scale bar: 10 µm. Adapted from Okabe et al. (2012) *Nat. Commun.*, **3**, 705 / Springer Nature. (b) Chemical structure of a modified cationic fluorescent polymeric thermometer. (c) Fluorescence lifetime images of the cationic fluorescent polymeric thermometer in living HeLa cells before and after the uncoupling by 10 µmol L^{-1} CCCP. <Δτf> represents an average fluorescence lifetime in the HeLa cells. Scale bar: 10 µm. Adapted from Hayashi et al. (2015) *PLoS ONE.*, **10**, e0117677 / PLOS / CC BY 4.0.

4-(trifluoromethoxy)phenylhydrazone) are representative uncouplers (Heytler and Prichard, 1962). In an early work using the masses of cells and microcalorimetry, FCCP remarkably caused heat production by respiration (Nakamura and Matsuoka, 1978). In our early intracellular temperature mapping with a fluorescence polymeric thermometer microinjected into COS7 cells, a temperature increase due to local heat production from the mitochondria following the treatment of FCCP could be imaged at a single-cell level (Figure 3.21a) (Okabe et al., 2012). As evaluated from fluorescence lifetime images of whole COS7 cells, this heat production by FCCP resulted in an average temperature increase of $1.02 \pm 0.17\,°C$ inside COS7 cells 30 min after the chemical stimuli. Reproducible results are found in another intracellular temperature mapping of HeLa cells using a modified cationic fluorescent polymeric thermometer with the ability to enter cells (Figure 3.21b) (Hayashi et al., 2015). The CCCP treatment increased the average temperature of HeLa cells by $1.57 \pm 1.41\,°C$. It should be mentioned that since this research, the original environment-sensitive DBD-AA units in our fluorescent polymeric thermometers have been replaced by DBThD-AA (N-(2-{[7-(N,N-dimethylaminosulfonyl)-2,1,3-benzothiadiazol-4-yl]-(methyl)amino}ethyl)-N-methylacrylamide) units exhibiting a 10-fold higher photostability (Uchiyama et al., 2012). In 2014, an international research group, which was led by Professor Hoehn and which I was also a member, discovered a novel uncoupler BAM15 ((2-fluorophenyl){6-[(2-fluorophenyl)amino](1,2,5-oxadiazolo[3,4-e]pyrazin-5-yl)}amine) having a different action mechanism (Kenwood et al., 2014). The effects of BAM15 on intracellular thermogenesis are currently under investigation in my research group.

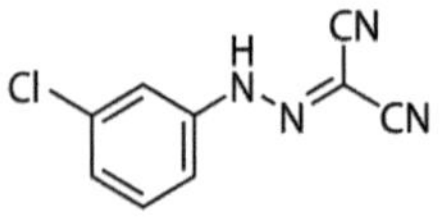

CCCP 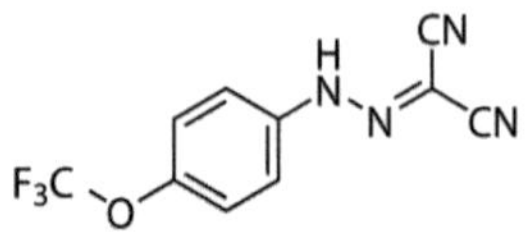   FCCP 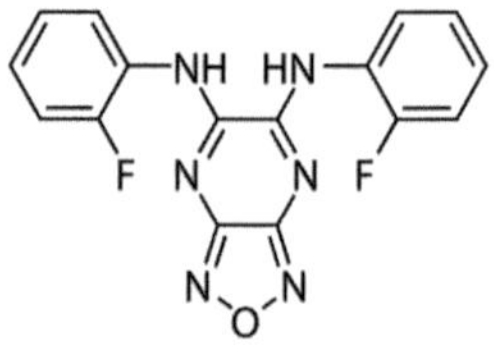   BAM15

3.4.4.2 Ionomycin

Ionomycin is an ionophore that can strongly bind to Ca^{2+} ion and distribute it to intracellular space (Morgan and Jacob, 1994). Ionomycin-induced temperature variation was initially observed in a HeLa cell attached to a glass micropipette filled with a dimethyl sulfoxide (DMSO) solution of a fluorescent molecular thermometer EuTTA (Suzuki et al., 2007). After an increase in Ca^{2+} concentration inside a HeLa cell by the treatment with ionomycin, remarkable heat production occurred with some time delay (28–126 s). The increase in temperature was suppressed by a Ca^{2+}-ATPases blocker, thapsigargin, suggesting that the heat production in a HeLa cell was led by the enzymatic activity of Ca^{2+}-ATPases in the endoplasmic reticulum. As a related work, Ca^{2+} dynamics under microscopic heat pulses in HeLa cells was observed with cellular thermometry by the glass micropipette filled with the EuTTA solution and one of the fluorescent polymeric thermometers (Tseeb et al., 2009).

A rise in intracellular temperature by ionomycin was detected by quantum dots for the first time (Yang et al., 2011). Quantum dots were delivered into NIH/3T3 (mouse embryonic fibroblast) cells using a commercially available QTracker cell labeling kit. Figure 3.22a shows the intracellular temperature variation of NIH/3T3 cells after the treatment with ionomycin. Position-dependent local heat production (~8 °C at the maximum) could be seen. Another illustration used ratiometric fluorescent nanoparticles that contained a fluorescent molecular thermometer EuTTA and a reference fluorophore rhodamine 101 in a hydrophobic poly(methyl methacrylate) (PMMA) core and were enveloped by a hydrophilic poly(allylamine)

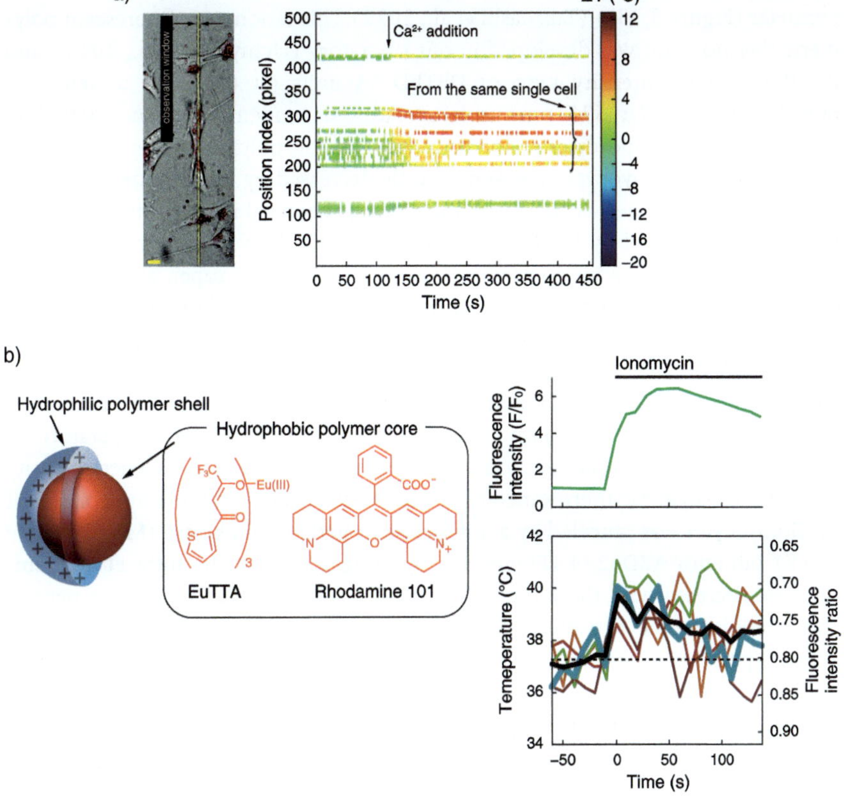

Figure 3.22 Intracellular heat production induced by ionomycin. (a) Bright field image of NIH/3T3 cells overlaid with a fluorescence image of quantum dots (left) and position-dependent intracellular temperature variation (ΔT) by ionomycin calcium complex in NIH/3T3 cells (right). Scale bar: 20 μm. Adapted from Yang et al. (2011) *ACS Nano*, **5**, 5067–5071 / American Chemical Society. (b) Structure of a ratiometric fluorescent nanothermometer consisting of EuTTA and rhodamine 101 (left) and effects of ionomycin (4 μmol L^{-1}) on the Ca^{2+} concentration (evaluated by a variation in the fluorescence intensity F/F$_0$ of Fluo-4, upper right) and intracellular temperature (evaluated from the fluorescence intensity ratio of the nanothermometer (colored: individual cells, black: average), lower right) in HeLa cells. Adapted from Takei et al. (2014) *ACS Nano*, **8**, 198–206.

hydrochloride (PAH) shell (Figure 3.22b) (Takei et al., 2014). Due to the cationic PAH shells (see also Figure 3.6a), the ratiometric fluorescent sensor can spontaneously enter HeLa cells by incubation at 37 °C for 2 hours and exist mainly at the endosomes. Ionomycin-induced concurrent increases in intracellular Ca^{2+} concentration and temperature in HeLa cells were monitored.

3.4.4.3 3-Iodothyronamine (T1AM)

3-Iodothyronamine (T1AM, Figure 3.23a) is an active thyroid hormone metabolite. The T1AM level in serum increases in heart failure subjects with cardiac cachexia. Temperature variation induced by T1AM was monitored in neonatal rat cardiomyocytes (Figure 3.23b) with the ratiometric fluorescent polymeric thermometer (Figure 3.23c) (Takahashi et al., 2022). This ratiometric fluorescent polymeric thermometer was developed in our laboratory (Uchiyama et al., 2015), and the fluorescence intensity ratio of DBThD-AA units (at 605 nm) to reference BODIPY-AA units (at 525 nm) is temperature-dependent. As demonstrated in Figure 3.23d, the fluorescence intensity ratio of the polymeric thermometer in neonatal rat cardiomyocytes was significantly decreased by the treatment of T1AM (50 μmol L^{-1}). In other words, the temperature of neonatal rat cardiomyocytes was remarkably decreased by T1AM (50 μmol L^{-1}). Only the variation in the fluorescence intensity ratio was reported in the original research paper, and it was not converted to a value of temperature variation. The treatment of neonatal rat cardiomyocytes with T1AM (50 μmol L^{-1}) induced the expression of B-type natriuretic peptide mRNA. An A-type natriuretic peptide (carperitide) and a β-adrenergic receptor agonist (CL316,243) did not affect the temperature of neonatal rat cardiomyocytes, whereas these compounds significantly increase the temperature of brown adipocytes (see later Section 3.6). The decrease in the fluorescence intensity ratio of the polymeric thermometer (i.e., intracellular temperature) in neonatal rat cardiomyocytes was canceled by a mitogen-activated protein kinase (MAPK) kinase (MEK) inhibitor AZD6244 (Figure 3.23d), suggesting that the intracellular temperature decrease with the increase in mRNA levels of B-type natriuretic peptide was caused by the activation of the MEK/ERK (extracellular signal-regulated kinases) pathway.

From the viewpoint of intracellular thermometry, this report is exceptional because a stimulus-induced temperature decrease inside a cell was unusual.

3.4.5
Intracellular Temperature Variation Accompanying Virus Infection

The recent pandemic of COVID-19 (Coronavirus disease 2019) caused by SARS-CoV-2 (severe acute respiratory syndrome coronavirus 2) was a major global issue with serious impacts on our daily lives. Initial screening of patients with symptoms is performed by monitoring the body temperature. To understand the biological mechanisms of such infectious diseases at the cellular level, intracellular thermometry during the infection is indispensable.

To date, there have been two reports on temperature measurements of live cells infected by the influenza virus. In the first report, polystyrene microspheres containing temperature-sensitive rhodamine B (Section 2.4.1) were utilized as fluorescent thermometers (Maruyama et al., 2018). The polystyrene microspheres were placed on the outer plasma membrane of H292 (human lung

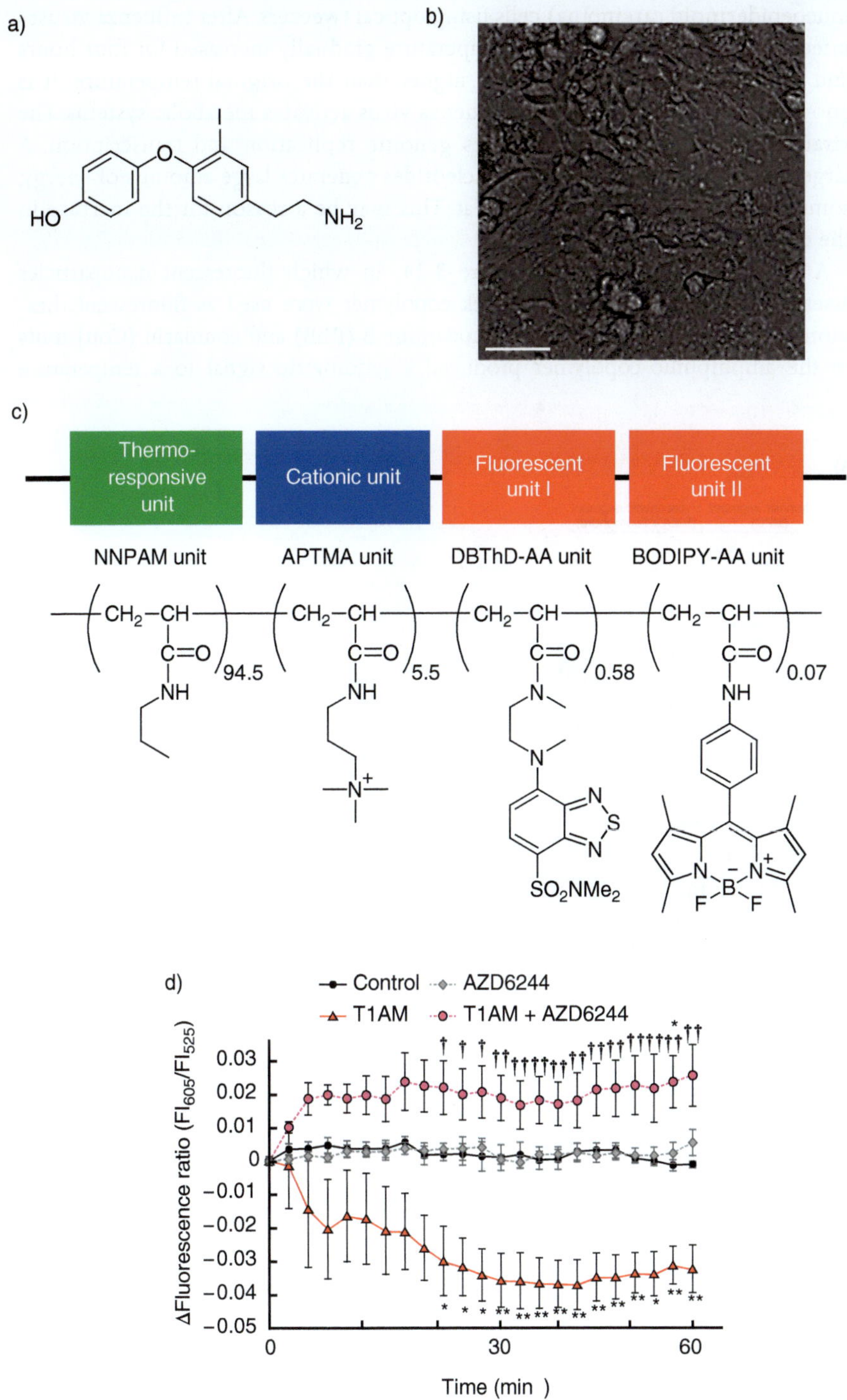

Figure 3.23 Effects of 3-iodothyronamine (T1AM) on the intracellular temperature of neonatal rat cardiomyocytes. (a) Chemical structure of T1AM (b) Differential interference contrast image of neonatal rat cardiomyocytes. Scale bar: 40 μm. (c) Chemical structure of the ratiometric fluorescent polymeric thermometer. (d) Effects of T1AM and a mitogen-activated protein kinase (MAPK)-kinase (MEK) 1/2 inhibitor, AZD6244, on the change in the fluorescence intensity ratio of the ratiometric fluorescent polymeric thermometer at 605 nm and 525 nm in neonatal rat cardiomyocytes. $*P < 0.05$ and $**P < 0.01$ versus control; $†P < 0.05$ and $††P < 0.01$ versus T1AM at each time point. Adapted from Takahashi et al. (2022) *Sci. Rep.*, **12**, 12740 / Springer Nature / CCBY 4.0 / Public domain.

mucoepidermoid carcinoma) cells using optical tweezers. After influenza viruses infected H292 cells, the cellular temperature gradually increased for four hours and reached a maximum of 4–5 °C higher than the original temperature. It is known that the infection of the influenza virus activates metabolic systems. The viral multiplication process involves genome replication and transcription. A large quantity of consumption of nucleotides generates large amounts of energy, some of which is converted into heat. This may be a reason for the increase in the cellular temperature.

A newer result is shown in Figure 3.24, in which fluorescent nanoparticles assembled with an amphiphilic block copolymer were used as fluorescent thermometers (Zhao et al., 2021). The rhodamine B (RhB) and coumarin (Cou) units in the amphiphilic copolymer produced a ratiometric signal to a temperature

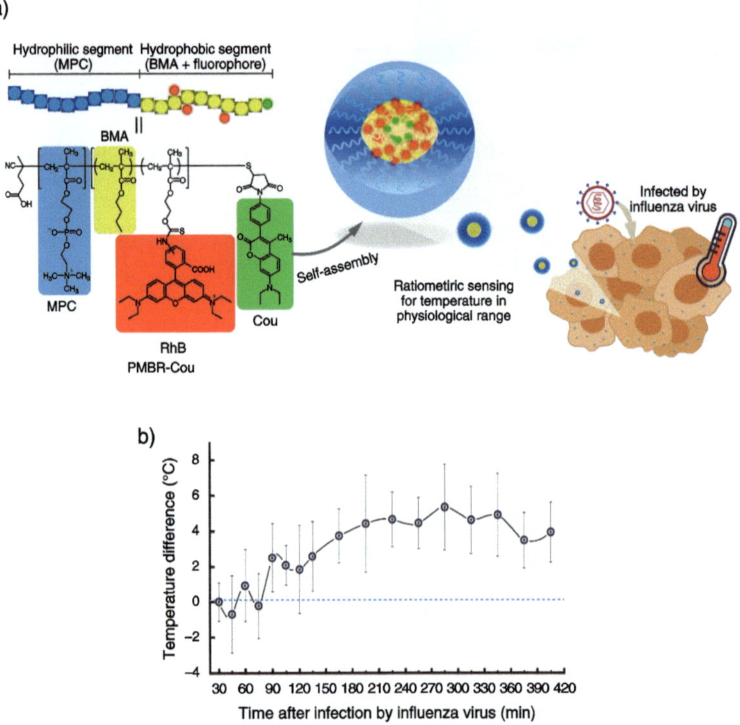

Figure 3.24 Effects of the infection of influenza virus on the temperature of H292 cells. (a) Schematic diagram of an amphiphilic block copolymer-based fluorescent thermometer (PMBR-Cou) for intracellular thermometry. MPC: 2-methacryloyloxyethyl phosphorylcholine; BMA: *n*-butyl methacrylate; RhB: methacryloxyethyl thiocarbamoyl rhodamine B; Cou: 7-diethylamino-3-(4-maleimidophenyl)-4-methylcoumarin. (b) Variation in intracellular temperature after the influenza virus infection. Adapted from Zhao et al. (2021) *J. Colloid Interface Sci.*, **601**, 825–832.

variation. Cationic 2-methacryloyloxyethyl phosphorylcholine (MPC) units in the copolymer supported spontaneous entry of the fluorescence nanoparticles to living cells under gentle incubation (i.e., 34 °C for 30 min). Figure 3.24b shows a time-course of the intracellular temperature of H292 cells after infection by the influenza virus A/PR/8/34 at a moi (multiplicity of infection) of 1. A lasting increase in intracellular temperature (approximately 4–5 °C) was observed for 3–6 hours after the infection. Because the sensitivity of rhodamine B to a temperature variation is not very high for live cell experiments, intracellular thermometry with more sensitive fluorescent molecular thermometers would be preferred. A detailed molecular mechanism of virus infection at a single-cell level is still an exciting theme of intracellular thermometry.

3.4.6
Intracellular Temperature Variation Accompanying Local Heating

Cellular response to physical stimulation is another matter of interest. Above all, local heating by an external source is the most often attempted theme ever in intracellular thermometry. Temperature monitoring of living cells upon local heating implies three independent purposes. First, the function of fluorescent molecular thermometers inside live cells can be assessed by local heating, the same as the whole heating of a dish to culture cells (the latter case can be seen in Figure 3.3 in the previous section). Second, the relationships between local temperature and intracellular heat shock responses (Lindquist, 1986) can be investigated. Third, efficient treatment in thermal therapy can be expected at the single-cell level by accurately controlling local temperature. The last category will be discussed in detail in Chapter 6, which describes advanced research fields of intracellular thermometry. Here, methods for local heating and early representative results of intracellular temperature measurement upon local heating will be summarized (see Table 3.2 for comprehensive information, including recent examples).

Figure 3.25a shows the intracellular thermometry of a HeLa cell upon local heating by irradiation of a near-infrared (NIR) laser beam at 920 nm, which is absorbed by water molecules to release heat (Vetrone et al., 2010). The temperature inside a HeLa cell was evaluated by the temperature-dependent fluorescence intensity ratio of $NaYF_4$:Er^{3+},Yb^{3+} nanoparticles (at 525 and 545 nm) internalized by incubation for 1.5 hours. A relationship between an applied voltage of the laser beam and intracellular temperature was obtained, and a tiny membrane fragment due to cell death was observed when the intracellular temperature reached 45 °C.

Figure 3.25b is the relationship between an increase in intracellular temperature of HeLa cells and heating time with Pd nanosheets that have large absorption (molar extinction coefficient: 4.1×10^9 M^{-1} cm^{-1} at 1045 nm) in the NIR region (Ke et al., 2012). The Pd nanosheets were added into Dulbecco's modified eagle medium (DMEM) and were irradiated by laser at 808 nm to make heat. The intracellular temperature of HeLa cells was measured by the molecular beacon (see Figure 2.31). As shown in Figure 3.25b, the temperature of HeLa cells rose gradually from 20.0 to 34.3 °C after 10 min laser irradiation to Pd nanosheets at 808 nm.

Table 3.2 Intracellular thermometry by fluorescent thermometers upon local heating.

Case	Method for heating	Fluorescent thermometer	Subject cell[a]	Detection range (°C)	Ref.
1	Radio-frequency magnetic-field heating with $MnFe_2O_4$ nanoparticles	DyLight549	HEK293		[1]
2	NIR[b] laser at 920 nm	$NaYF_4$:Er^{3+},Yb^{3+} nanoparticles	HeLa	25–45	[2]
3	Laser irradiation at 808 nm to Pd nanosheets in a culture medium	L-DNA molecular beacon	HeLa	20–37	[3]
4	Laser irradiation at 532 nm to gold nanoparticles in cells	Fluorescent nanodiamonds	WS1		[4]
5	A specially designed sample stage equipped with hot and cool aluminum blocks with a 100 μm gap	Lipoic acid-protected gold nanoclusters	Fixed HeLa	15–42.5	[5]
6	IR[c] laser	CDSeS/ZnS quantum dots in the matrix of poly(methyl methacrylate-co-methacrylic acid)	HepG2	25–39	[6]
7	Microwave irradiation (20–40 W)	Rhodamine B and rhodamine 110-incorporated F127-melamine-formaldehyde polymer nanoparticles	HeLa	20–50	[7]

a) HEK293: human embryonic kidney cell, HeLa: human epithelial carcinoma cell, WS1: human skin fibroblast cell, HepG2: human hepatocellular carcinoma cell.
b) Near-infrared.
c) Infrared.

1 Huang, H., Delikanli, S., Zeng, H., Ferkey, D. M., and Pralle, A. (2010) *Nat. Nanotechnol.*, **5**, 602–606.
2 Vetrone, F., Naccache, R., Zamarrón, A., de la Fuente, A. J., Sanz-Rodríguez, F., Maestro, L. M., Rodriguez, E. M., Jaque, D., Solé, J. G., and Capobianco, J. A. (2010) *ACS Nano*, **4**, 3254–3258.
3 Ke, G., Wang, C., Ge, Y., Zheng, N., Zhu, Z., and Yang, C. J. (2012) *J. Am. Chem. Soc.*, **134**, 18908–18911.
4 Kucsko, G., Maurer, P. C., Yao, N. Y., Kubo, M., Noh, H. J., Lo, P. K., Park, H., and Lukin, M. D. (2013) *Nature*, **500**, 54–58.
5 Shang, L., Stockmar, F., Azadfar, N., and Nienhaus, G. U. (2013) *Angew. Chem. Int. Ed.*, **52**, 11154–11157.
6 Liu, H., Fan, Y., Wang, J., Song, Z., Shi, H., Han, R., Sha, Y., and Jiang, Y. (2015) *Sci. Rep.*, **5**, 14879.
7 Wu, Y., Liu, J., Ma., J., Liu, Y., Wang, Y., and Wu, D. (2016) *ACS Appl. Mater. Interfaces*, **8**, 14396–14405.

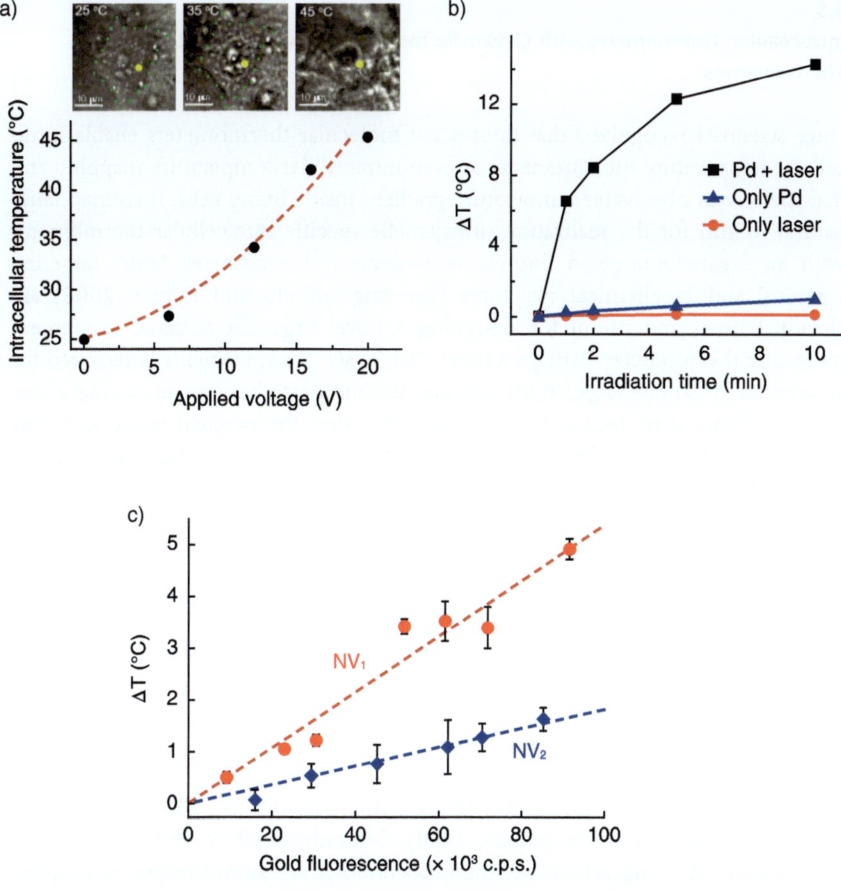

Figure 3.25 Monitoring of intracellular temperature during heating by external sources. (a) Optical transmission images of a HeLa cell containing $NaYF_4$:Er^{3+},Yb^{3+} nanoparticles (top) and a laser beam-induced increase in the intracellular temperature determined by $NaYF_4$:Er^{3+},Yb^{3+} nanoparticles (bottom). The wavelength of the laser beam (focused into the yellow points in the transmission images) was 920 nm. Adapted from Vetrone et al. (2010) *ACS Nano*, **4**, 3254–3258 / American Chemical Society. (b) Intracellular temperature variation (ΔT) in HeLa cells by heating with a Pd nanosheet upon laser irradiation at 808 nm. An L-DNA molecular beacon monitored the intracellular temperature. Adapted from Ke et al. (2012) *J. Am. Chem. Soc.*, **134**, 18909. (c) Intracellular temperature variation (ΔT) at the positions of NV_1 and NV_2 (see Figure 3.4d) under laser irradiation with different power on a gold nanoparticle. Adapted from Kucsko et al. (2013) *Nature*, **500**, 54–58.

In the first report on intracellular thermometry using fluorescent nanodiamonds, a gold nanoparticle introduced into a WS1 cell was used as a heater under laser illumination at 532 nm (Kucsko et al., 2013). As indicated in Figure 3.25c, the position at a fluorescent nanodiamond NV_1 located nearer to the gold nanoparticle as a heat source was efficiently heated with a function of laser power compared to the position at a fluorescent nanodiamond NV_2 (see Figure 3.4d for the fluorescence image).

3.5
Intracellular Thermometry with Organelle-targeted Fluorescent Molecular Thermometers

Once scientists recognized that fluorescent molecular thermometers enable intracellular temperature measurements or even intracellular temperature mapping and that there is an observable temperature gradient inside living cells, it seemed fairly natural to aim for the realization of organelle-specific intracellular thermometry with an organelle-targeted fluorescent molecular thermometer. Many targeting chemical and biochemical structures (van Engelenburg and Palmer, 2008) are already known and useful for designing a novel organelle-targeted fluorescent molecular thermometer. Perhaps a more careful set of experiments is required for an accurate organelle-targeted intracellular thermometry because an organelle-targeted fluorescent molecular thermometer may alter the original function of the organelle by the contact between them. Nevertheless, until now, various organelle-targeted fluorescent molecular thermometers have been developed to bring interesting findings on organelle-specific temperature, as follows. A recent paper on a series of BODIPY-based organelle-targeted fluorescent molecular viscometers may be interesting to readers because these BODIPY derivatives can be used as fluorescent molecular thermometers, once localized at organelles, to sense temperature-dependent viscosity changes of their membranes (Liu et al., 2022).

3.5.1
Mitochondria-targeted Intracellular Thermometry

It has been believed for a long time that the mitochondria are heat sources inside living cells (Lowell and Spiegelman, 2000). Certainly, much more heat is released from a mass of living cells when the mitochondria are stimulated by uncouplers such as CCCP and FCCP (Nakamura and Matsuoka, 1978). Furthermore, intracellular temperature mapping also proved that local spaces near the mitochondria are warmer than other spaces (see Figure 3.17). These facts have accelerated the development of mitochondria-targeted fluorescent molecular thermometers.

The first illustration of mitochondria-targeted fluorescent molecular thermometers is based on a genetically encoded fluorescent protein tsGFP1 (Kiyonaka et al., 2013). The original tsGFP1 consists of GFP and a thermo-responsive coiled-coil protein TlpA, an autoregulatory repressor protein in *Salmonella* (Figure 3.26a). The temperature-dependent conformational change of tsGFP1 results in a measurable change in the fluorescence properties of a GFP moiety. Although tsGFP1 is expressed uniformly throughout the cytoplasm, tsGFP1-mito, i.e., the fused tsGFP1 to a mitochondria-targeting sequence, is expressed across the entire mitochondrial population (Figure 3.26b). The heat production at the mitochondria upon the treatment with CCCP could be qualitatively monitored in a HeLa cell by tsGFP1-mito. The highly flexible design of genetically encoded fluorescent sensors allows us to easily create other organelle-targeted tsGFP1: plasma membrane and endoplasmic reticulum (ER). Another mitochondria-targeted genetically encoded fluorescent protein, emGFP-Mito, has also been reported (Savchuk et al., 2019). The maximum emission wavelength of emGFP-Mito shifted bathochromically with increasing temperature, which was utilized for mitochondrial thermometry of HeLa cells.

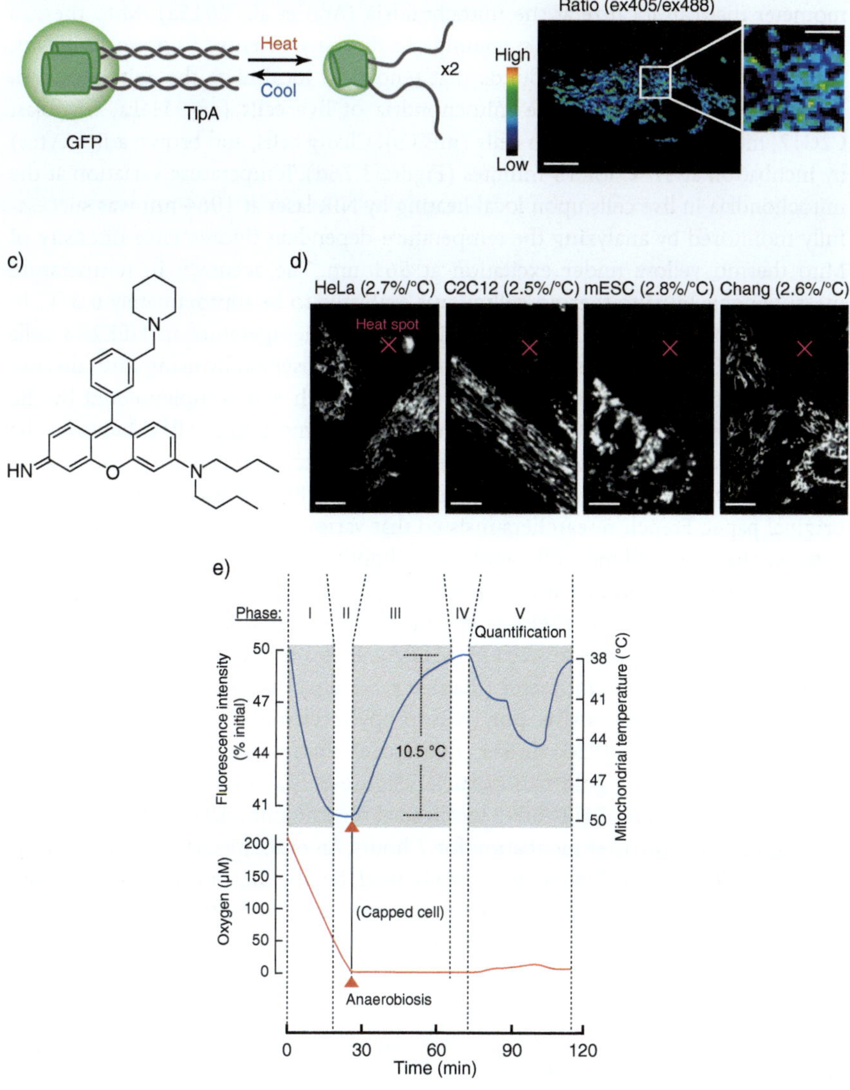

Figure 3.26 Mitochondria-targeted fluorescent molecular thermometers. (a,b) tsGFP1-mito. (a) Functional mechanism of tsGFP1. (b) Pseudo-color confocal ratio images of mitochondria-targeted tsGFP1-mito expressed in a HeLa cell. The region of interest, as shown in the square in the left panel, is enlarged in the right panel. Scale bars: 10 μm and 3 μm, respectively. Panels (a,b) are Kiyonaka et al. (2013) *Nat. Methods*, **10**, 1232–1238 / Springer Nature. (c–e) Mito thermo yellow. (c) Chemical structure. (d) Confocal fluorescence images of Mito thermo yellow in HeLa, C2C12, mESC, and Chang cells. Indicated sensitivities in the fluorescence intensity of Mito thermo yellow (%/°C) were evaluated by local heating at the spot. Scale bars: 10 μm. Panels (c,d) are adapted from Arai et al. (2015a) *Chem. Commun.*, **51**, 8044–8047 / Royal Society of Chemistry. (e) Temperature increase of the mitochondria by respiration in HEK293 cells. Temperature was estimated from the fluorescence intensity of Mito thermo yellow. Phase I: cell respiration after cells are exposed to aerobic condition in PBS. Phase II: cell respiration under aerobic condition (maximal warming of mitochondria). Phase III: cell respiration arrested due to oxygen exhaustion. Phase IV: respiration remains stalled due to anaerobiosis. Phase V: Temperature of the cell suspension medium is changed for confirming the function of Mito thermo yellow. Panel (e) is adapted from Chréiten et al. (2018) *PLoS Biol.*, **16**, e2003992 / PLOS / CCBY 4.0 / Public domain.

Mito thermo yellow (Figure 3.26c) is a small organic fluorescent molecular thermometer that can localize at the mitochondria (Arai et al., 2015a). Mito thermo yellow was discovered through screening of a diversity-oriented fluorescent library, especially on rosamine compounds that tended to localize at the mitochondria. Mito thermo yellow stains the mitochondria of live cells (e.g., HeLa, myoblast C2C12, mouse embryonic stem cells (mESCs), Chang cells, and brown adipocytes) by incubation at 37 °C for 15 minutes (Figure 3.26d). Temperature variation at the mitochondria in live cells upon local heating by NIR laser at 1064 nm was successfully monitored by analyzing the temperature-dependent fluorescence intensity of Mito thermo yellow under excitation at 561 nm. The accuracy in temperature measurements by Mito thermo yellow was evaluated to be approximately 0.3 °C. In 2018, the surprising result that the mitochondrial temperature in HEK293 cells reached 50 °C at the highest during respiration was observed by using Mito thermo yellow (Figure 3.26e) (Chrétien et al., 2018), which was complemented by the work of Chrétien et al. (2020). This unusual gap of more than 10 °C between the mitochondrial and physiological temperatures caused arguments for and against the reliability of intracellular thermometry (Lane, 2018) (see also Chapter 5). In the original paper, French researchers insisted that various enzymes related to the respiratory chain worked most efficiently at or slightly above 50 °C as an indirect proof of reliability in their intracellular thermometry.

Triphenylphosphonium (TPP) is the most well-known targeting structure for mitochondria (Zielonka et al., 2017). MTT-AuNP is a mitochondria-targeted fluorescent nanoparticle thermometer (diameter: ~5 nm) that is composed of the TPP moiety, a thermo-responsive poly(N-isopropylacrylamide) (PNIPAM) chain, a poly(ethylene glycol) (PEG) linker, fluorescent fluorescein (FITC) and cyanine (Cy5) moieties, and a gold nanoparticle (Figure 3.27a) (Wang et al., 2019). As shown in Figure 3.27b, MTT-AuNP localized at the mitochondria of MB49 (murine bladder cancer) cells after incubation for 2 hours. An exact mechanism of temperature-dependency of MTT-AuNP was not clarified, but the thermo-responsive behavior of the PNIPAM chain with a temperature variation may affect the fluorescence properties of Cy5 units. The fluorescence intensity ratio of Cy5 to FITC (reference) units increased with increasing temperature. According to the fluorescence intensity ratio of Cy5 to FITC units, the intracellular temperature of MB49 cells, HUVEC (human umbilical vein endothelial) cells, and murine RAW 264.7 macrophages were compared. As displayed in Figure 3.27c, MB49 cells were warmer than the others. Another example of utilizing a TPP structure to develop a mitochondria-targeted fluorescent thermometer is seen in a modified gold nanocluster with a diameter of approximately 2 nm (Wang et al., 2021). The mitochondria-targeting TPP moieties were attached to gold nanoclusters via a condensation reaction of 4-(carboxybutyl) triphenylphosphonium bromide and an amino group of glutathione chemically adsorbed to gold nanoclusters (Figure 3.27d). This modified gold nanocluster stained the mitochondria of L929 (mouse fibroblast) cells by incubation at 37 °C for 4 hours and illustrated the non-uniform temperature distribution of the mitochondria via its temperature-dependent fluorescence lifetime (i.e., area "a" vs. area "b" in Figure 3.27e). Furthermore, Er^{3+}-based upconversion nanoparticles (UCNPs) were also delivered to mitochondria after the surface modification with TPP (Di et al., 2021). Er^{3+}-based UCNPs stained with HeLa cells after

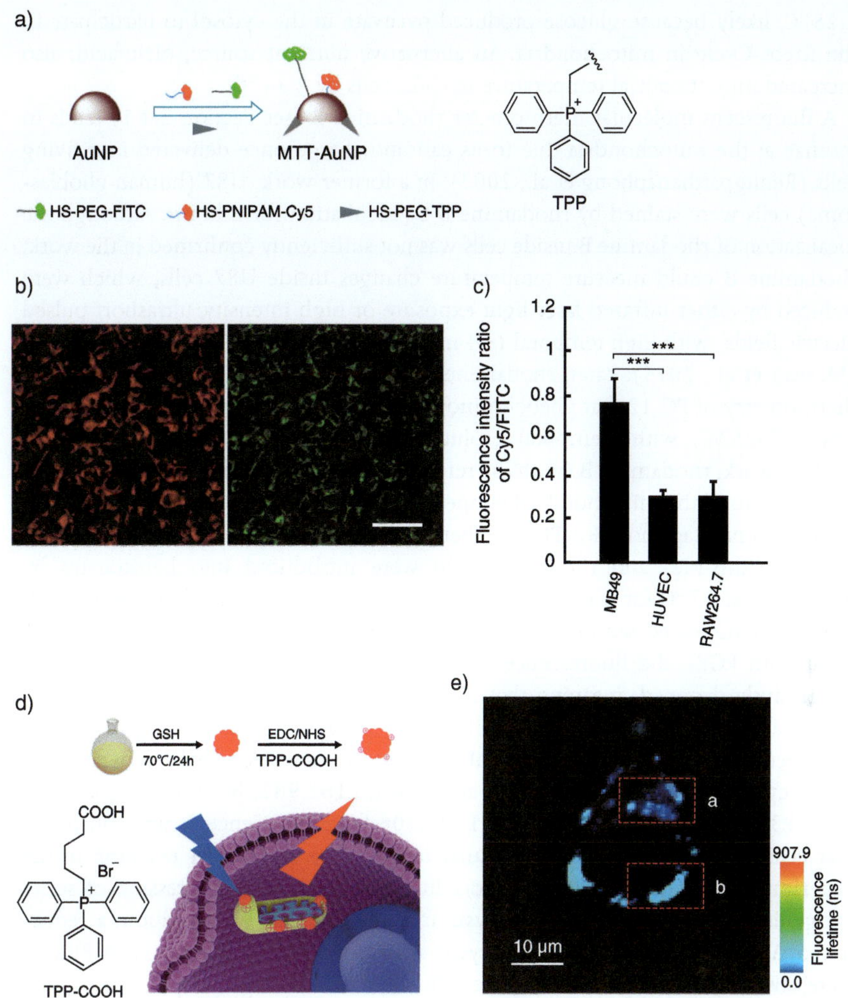

Figure 3.27 Mitochondria-targeted fluorescent thermometers based on gold nanoparticles. (a–c) Thermo-responsive MTT-AuNP. (a) Preparation. (b) Fluorescence images of Cy5 moieties in MTT-AuNP (left) and Mito Tracker (right) in MB49 cells. Scale bar: 50 µm. (c) Comparison of mitochondrial temperature among MB49, HUVEC, and RAW264.7 cells by the fluorescence intensity ratio of Cy5 units to FITC units in MTT-AuNPs. Data are mean ± SD, *n* = 4. ***, P < 0.001. Panels (a–c) are adapted from Wang et al. (2019) *ACS Appl. Bio Mater.*, **2**, 3178–3182 / American Chemical Society. (d,e) TPP-modified gold nanoparticles. (d) Schematic diagram of gold nanoclusters modified by 4-(carboxybutyl) triphenylphosphonium bromide (TPP-COOH). GSH: glutathione, EDC: 1-(3-dimethylaminopropyl)-3-ethylcarbodiimide, NHS: *N*-hydroxysulfosuccinimide sodium salt. (e) Fluorescence lifetime image of the gold nanoclusters in L929 cells. The excitation and emission wavelengths were 405 nm and 545–620 nm, respectively. Panels (d,e) are adapted from Wang et al. (2021) *Anal. Chem.*, **93**, 15072–15079 / American Chemical Society.

the incubation for 12 hours, and the fluorescence intensity ratio at 525 nm (due to $^{2}H_{11/2} \rightarrow {}^{4}I_{15/2}$ of Er^{3+}) and 545 nm (due to $^{4}S_{3/2} \rightarrow {}^{4}I_{15/2}$ of Er^{3+}) under the excitation at 980 nm was temperature-dependent. It was observed that a high glucose medium (5 mg mL^{-1}) increased the mitochondrial temperature of HeLa cells by

2.25 °C, likely because glucose produced pyruvate in the cytosol to participate in the Krebs Cycle in mitochondria. An alternative nutrient source, oleic acid, also increased mitochondrial temperature in HeLa cells by 2.74 °C.

A fluorescent molecular thermometer rhodamine B (see Section 2.4.1) tends to localize at the mitochondria due to its cationic charge once delivered into living cells (Reungpatthanaphong et al., 2003). In a former work, U87 (human glioblastoma) cells were stained by rhodamine B by incubation for 2 hours. Although the localization of rhodamine B inside cells was not sufficiently confirmed in the work, rhodamine B could measure temperature changes inside U87 cells, which were induced by either infrared laser light exposure or high intensity, ultrashort pulsed electric fields, with high temporal (~4 ms) and temperature (~0.1 °C) resolutions (Moreau et al., 2015). Later, rhodamine B was utilized for real-time intracellular thermometry of PC-12 (rat pheochromocytoma) cells that were exposed to microwaves (8–25 W), with a temporal resolution of 1.5 seconds (Dong et al., 2021). In another work, rhodamine B (with the reference temperature-insensitive rhodamine 800) measured the mitochondrial temperature of primary hepatocytes under treatment with prostaglandin E_2 (PGE$_2$) (Shen et al., 2018). The methyl ester of rhodamine B (RhB-Me) and rhodamine 800 were introduced into hepatocytes by incubation at 37 °C for 1 hour. The fluorescence intensity ratio of rhodamine 800 to RhB-Me increased with increasing the mitochondrial temperature. By the treatment with PGE$_2$, the fluorescence intensity ratio of rhodamine 800 to RhB-Me significantly dropped, meaning that the temperature of mitochondria in hepatocytes was decreased (Figure 3.28). PGE$_2$ exerts biological effects through one of its four receptors EP1–EP4. The effects of PGE$_2$ on the mitochondrial temperature in hepatocytes were blocked by an EP4 antagonist, L161,982, but not antagonists of EP1–EP3 (AH6809 for EP1 and EP2; L798,106 for EP3). Hence, it was concluded that PGE$_2$ decreased hepatic intracellular temperature via an EP4 receptor. In the experiments following biological procedures, an increase in expression of some lipogenic genes, hampering of lipolysis and mitochondrial β-oxidation, a reduction of ATP level, and an elevation of cyclic adenosine monophosphate (cAMP) level were observed in connection with the PGE$_2$-induced cooling of hepatocytes.

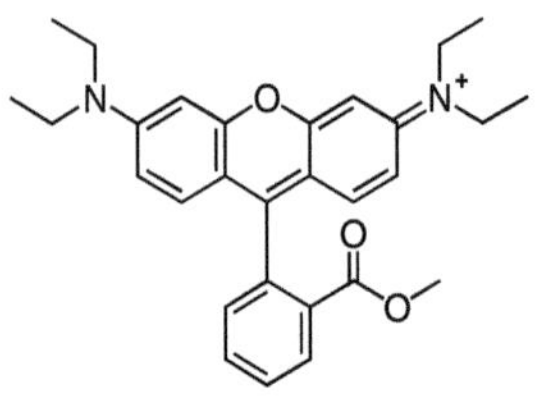

RhB-ME

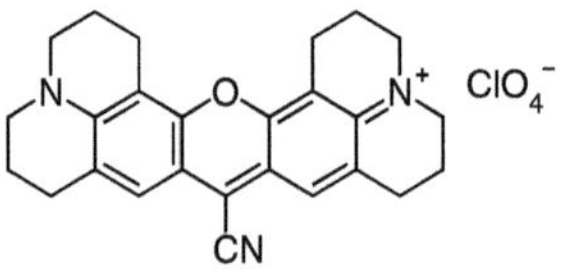

Rhodamine 800

It is well-known that mitochondrial potentials are changeable so that cationic mitochondria-targeted fluorescent probes may inevitably diffuse out from the mitochondria during intracellular thermometry. Mito-TEM is the first fixable, mitochondria-targeted fluorescent molecular thermometer (Huang et al., 2018). The benzaldehyde moiety in Mito-TEM can react with amino groups of mitochondrial

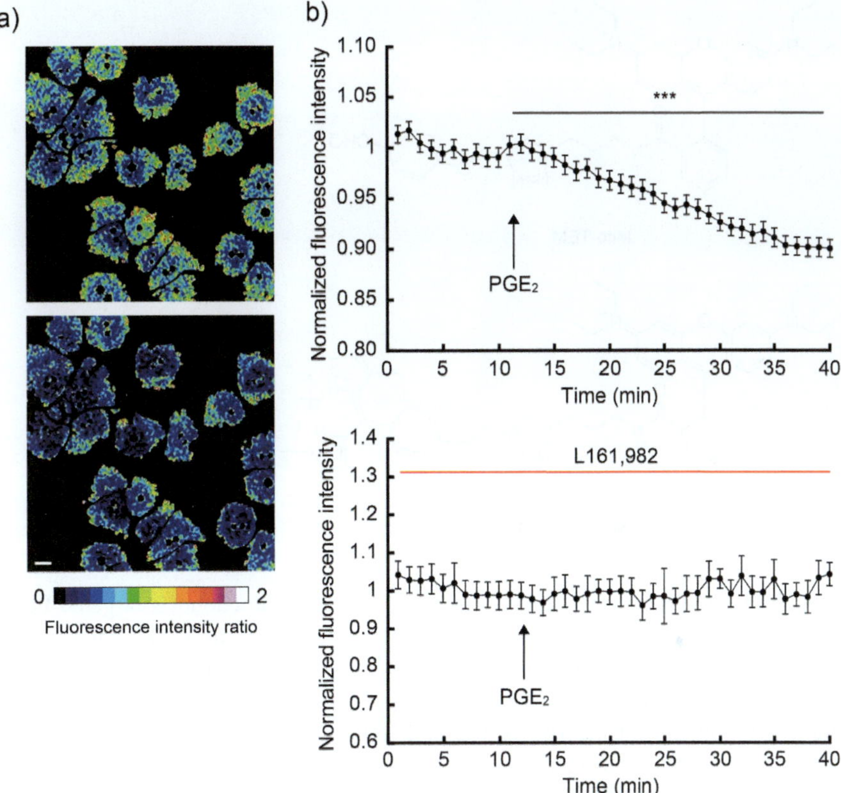

Figure 3.28 Effects of prostaglandin E_2 (PGE_2) on the mitochondrial temperature of hepatocytes. (a) Pseudo-colored images of the fluorescence intensity ratio of rhodamine 800 and RhB-Me before (top) and after (bottom) the treatment with 10 nmol L^{-1} PGE_2. Scale bar: 20 µm. (b) Variation in the fluorescence intensity ratio of rhodamine 800 (excited at 635 nm and collected at 655–755 nm) and RhB-Me (excited at 559 nm and collected at 575–620 nm) with the treatment by 10 nmol L^{-1} PGE_2 without and with the preincubation with an EP4 antagonist 1 µmol L^{-1} L161,982. The data are shown as mean ± SEM with *** for $p < 0.001$. Adapted from Shen et al. (2018) *Sci. Rep.*, **8**, 13065 / The Authors / CCBY 4.0 / Public domain.

proteins for immobilization at the mitochondria. Additional care was required to check whether the immobilization of Mito-TEM alters the biological status of the mitochondria or not. Mito-TEM labeled the mitochondria of MCF-7 (human breast carcinoma) cells and monitored the mitochondrial temperature during the treatment with phorbol 12-myristate 13-acetate as an activator of the protein kinase C system. The same research group developed a ratiometric version, Mito-TEM 2.0, which added a BODIPY structure as a second fluorophore to Mito-TEM compositions (Huang et al., 2021). Mitochondrial temperature variation in HeLa cells in inflammation states induced by the stimulation with *Staphylococcus aureus* or lipopolysaccharide was recorded with Mito-TEM 2.0.

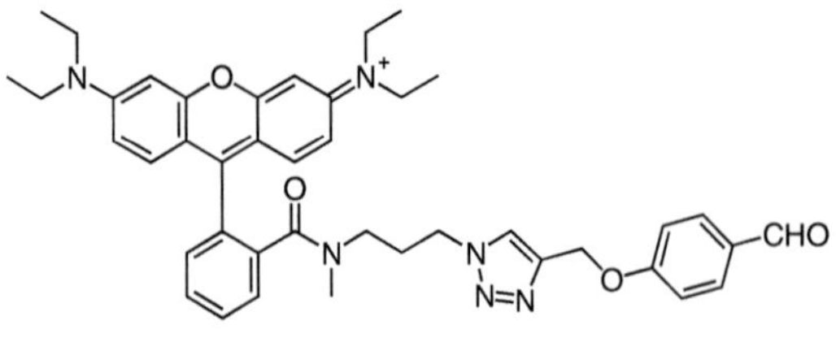

Mito-TEM

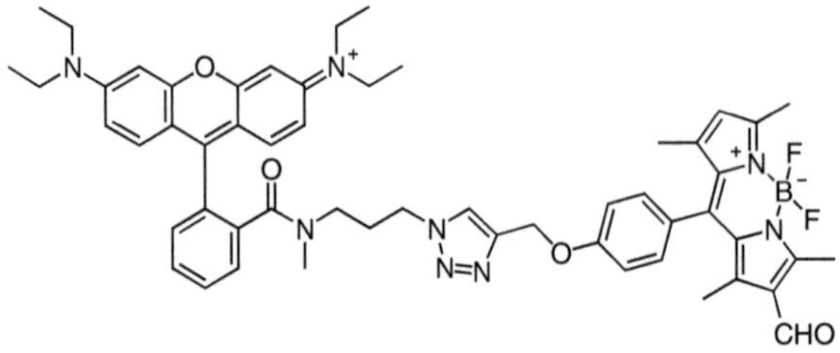

Mito-TEM 2.0

A rhodamine derivative RhBIV was developed as a mitochondria-targeted fluorescent molecular thermometer more sensitive than rhodamine B (Shen et al., 2021). Under the excitation at 520 nm, the fluorescence intensity of RhBIV at 592 nm at 42 °C was 45% of that at 20 °C in phosphate-buffered saline (PBS). Although the photophysical mechanism of the high sensitivity of RhBIV to a temperature variation remains unclear, mitochondrial temperature measurements of HepG2 cells and living mice (see Chapter 6) were conducted.

Si-rhodamine, in which a silicon atom replaces the bridging oxygen atom, shows a drastic red-shift in maximum emission wavelength (i.e., 630–650 nm) compared to original rhodamine, with retaining high fluorescence efficiency (Fu et al., 2008). This bathochromic shift of Si-rhodamine, which was ascribed to the σ^*-π^* conjugation between the σ^* orbital of silicon atom and the π^* system of the xanthene moiety (Yamaguchi and Tamao, 1998), has been desired especially in fluorescence imaging of organs and individuals as tissue transparency is high in the long wavelength region (Ikeno et al., 2017). In the same way as rhodamine B, SiR–CH$_2$Cl was localized at mitochondria due to its hydrophobic, cationic nature once it spontaneously entered into live cells by a short incubation (e.g., 30 minutes) (Tang et al., 2020). In HepG2 cells, SiR–CH$_2$Cl emitted temperature-dependent fluorescence at around 670 nm (i.e., relative fluorescence intensity at 40 °C was 0.5 compared to that at 30 °C). The residence of SiR–CH$_2$Cl in the mitochondria might be supported by covalent bonding between the chloromethyl group of SiR–CH$_2$Cl and nucleophiles in the mitochondria. A remarkable temperature increase of the mitochondria in HepG2 cells was observed when a chemical reagent, phorbol 12-myristate-13-acetate, activated the protein kinase C system.

RhBIV

SiR-CH$_2$Cl

3.5.2
Endoplasmic Reticulum-targeted Intracellular Thermometry

The endoplasmic reticulum (ER) is considered to be a heat source in skeletal muscle in birds and animals (de Meis, 2001; de Meis et al., 2005). In a proposed mechanism, thermogenesis in skeletal muscle is likely related to sarco-endoplasmic reticulum Ca^{2+}-ATPase (SERCA) via uncoupling of ATP hydrolysis from pumping of Ca^{2+} from the cytosol into the sarcoplasmic reticulum lumen. The flexibility in function of genetically encoded fluorescent molecular thermometers straightforwardly enabled the preparation of ER-targeted fluorescent molecular thermometers (named tsGFP1-ER) by fusion of tsGFP1 (see also Section 3.5.1) and an ER-targeting signal (Kiyonaka et al., 2013). Figure 3.29a indicates the temperature map of ER in differentiated C2C12 myotubes, which was drawn by tsGFP1-ER having temperature-dependent fluorescence properties (i.e., a fluorescence intensity ratio in this case). The temperature at ER in differentiated C2C12 myotubes remarkably dropped by the treatment with a SERCA inhibitor, cyclopiazonic acid.

ER thermo yellow (Figure 3.29b) is an ER-targeted fluorescent molecular thermometer developed by combinatorial synthesis by modifying side chains of different fluorophore backbones based on a diversity oriented fluorescent library approach (Arai et al., 2014). In this screening approach, the chemical structure of a fluorescent molecular thermometer was optimized for acquiring the ability to target ER and high sensitivity to a temperature variation. Staining of ER in living cells was conducted by incubation with ER thermo yellow at 37 °C for 30 minutes. Later, ER thermo yellow was used for intracellular thermometry of HeLa cells in the establishment of a precise heat-induced drug delivery system, in which the release of a model compound calcein from liposomes was thermally well-controlled inside cells (Arai et al., 2015b). Moreover, intracellular temperature mapping of C2C12 myotubes was performed using ER thermo yellow and FLIM (Figure 3.29c) (Itoh et al., 2016). When a ryanodine receptor (a Ca^{2+} channel located at the sarcoplasmic reticulum) was stimulated by caffeine, the intracellular temperature increased by 1.6 ± 0.6 °C with a Ca^{2+} burst inside cells (Figure 3.29d). In contrast, an inhibitor of SERCA, thapsigargin, also induced a Ca^{2+} burst but not thermogenesis in C2C12 myotubes. It should be noted that a fluorescent protein-based thermometer mCherry that existed in the cytoplasm could not detect the caffeine-induced temperature increase of ER. Thus, organelle-targeted fluorescent thermometers can exclusively detect a temperature variation occurring at specific organelles.

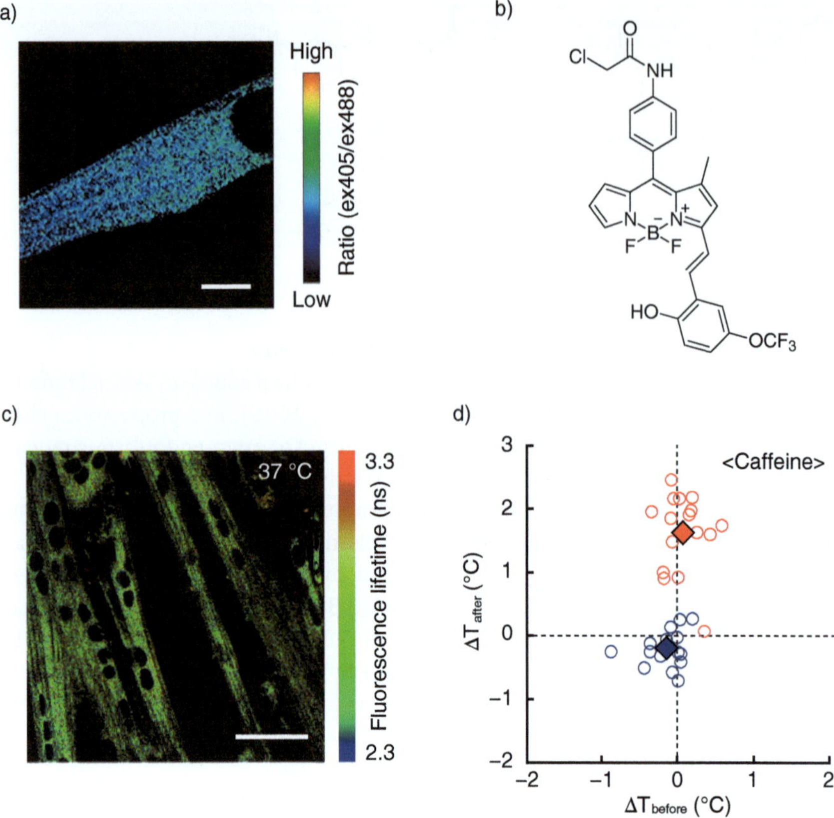

Figure 3.29 Endoplasmic reticulum (ER)-targeted fluorescent molecular thermometers. (a) Ratiometric pseudo-color image of the fluorescence intensity ratio of tsGFP1-ER with the excitation at 405 and 488 nm in a myotube differentiated from myogenic C2C12 cells. Scale bar: 10 μm. Adapted from Kiyonaka et al. (2013) *Nat. Methods*, **10**, 1232–1238 / Springer Nature. (b–d) ER thermo yellow. (b) Chemical structure. (c) Fluorescence lifetime image of ER thermo yellow in C2C12 myotubes at 37 °C. Scale bar: 50 μm. (d) Effects of caffeine on the temperature of the sarcoplasmic reticulum. ΔT_{before} and ΔT_{after} are the temperature differences of the sarcoplasmic reticulum between sample groups before and after the stimulation with caffeine (red) and dimethyl sulfoxide (DMSO, control, blue). The red and blue diamonds are the mean values. Panels (c,d) are adapted from Itoh et al. (2016) *Chem. Commun.*, **52**, 4458–4461 / Royal Society of Chemistry / CC BY 4.0.

3.5.3
Plasma Membrane-targeted Intracellular Thermometry

There are some examples of fluorescent thermometers that are located at the plasma membrane. As the plasma membrane of living cells is generally considered a border between biological species and non-biological matters, plasma membrane-targeted fluorescent thermometers would be utilized to monitor the surface boundary temperature of living cells. First, as a physical approach, optical tweezers were effective for attaching rhodamine B-containing polystyrene microspheres onto the plasma membrane of H292 cells (Figure 3.30). Then, cellular thermometry was performed during influenza virus multiplication inside the H292 cells (Maruyama et al., 2018). In addition to the genetically encoded plasma membrane-targeted fluorescent molecular thermometer (tsGFP1-F) (Kiyonaka et al., 2013), the following have been reported as plasma membrane-targeted fluorescent thermometers.

The first plasma membrane-specific temperature measurement was performed using a fluorescent molecular thermometer DyLight549 (DyLight® is a trademark of Thermo Fisher Scientific Inc., USA), of which the chemical structure remains unclear, unfortunately (Huang et al., 2010). Streptavidin-DyLight549-coated superparamagnetic MnF_2O_4 nanoparticles were located on the cellular membrane of HEK293 cells, where an enzymatically biotinylated acceptor peptide was present. The fluorescence intensity of DyLight549 in the vicinity of MnF_2O_4 nanoparticles decreased upon heating the nanoparticles by application of a radiofrequency magnetic field (B = 13 G), providing an estimate of the temperature rise by approximately 17 °C (Figure 3.31). Interestingly, the fluorescence intensity of Golgi-targeted

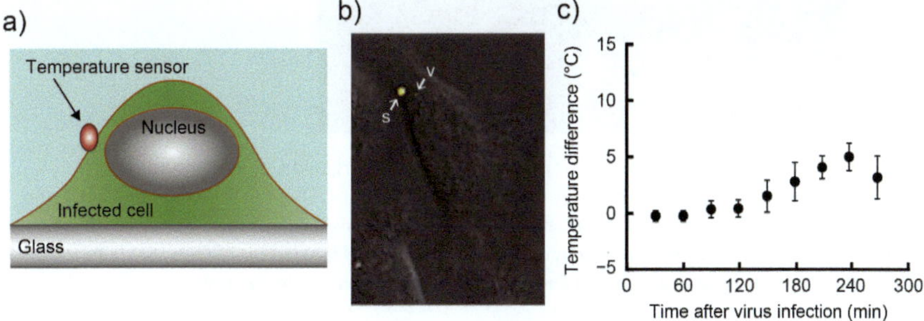

Figure 3.30 Cellular thermometry by rhodamine B-containing polystyrene microspheres attached to the plasma membrane of H292 cells during influenza virus replication. (a) Schematic diagram of the thermometry. (b) Microscopic image. The labels "s" and "v" represent the sensor particle and the Syto21-labeled influenza virus, respectively. (c) Temperature variation of an H292 cell after the influenza virus infection. Adapted from Maruyama et al. (2018) *Virus Res.*, **257**, 94–101 / with permission from Elsevier.

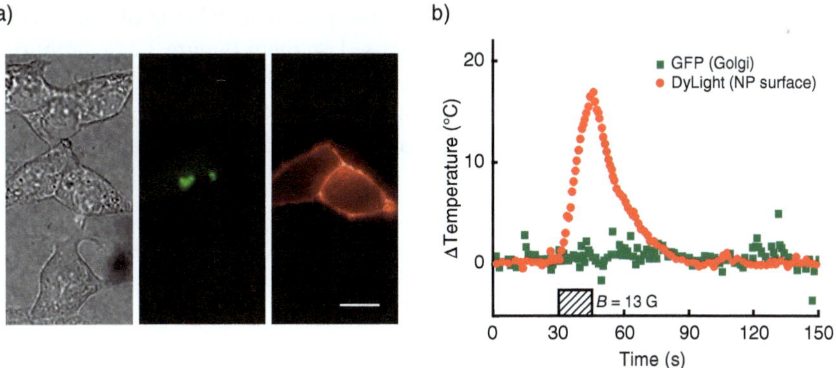

Figure 3.31 Temperature monitoring of HEK293 cells under plasma membrane-specific magnetic-field heating. (a) Differential interference contrast image (left) and fluorescence images of the Golgi body labeled with a green fluorescent protein (middle) and the plasma membrane stained with DyLight549 conjugated to the streptavidin-coated cell membrane-targeted $MnFe_2O_4$ nanoparticle (right). Scale bar: 20 μm. (b) Temperature variation (ΔTemperature) of the plasma membrane (red) and the Golgi apparatus (green) in HEK293 cells during the application of the radio-frequency magnetic field to the superparamagnetic ferrite nanoparticles targeted to the plasma membrane of HEK293 cells. The temperatures of the plasma membrane and the Golgi apparatus were evaluated from the fluorescence intensities of DyLight549 and green fluorescent protein, respectively. Adapted from Huang et al. (2010) *Nat. Nanotehnol.*, **5**, 602–606 / Springer Nature.

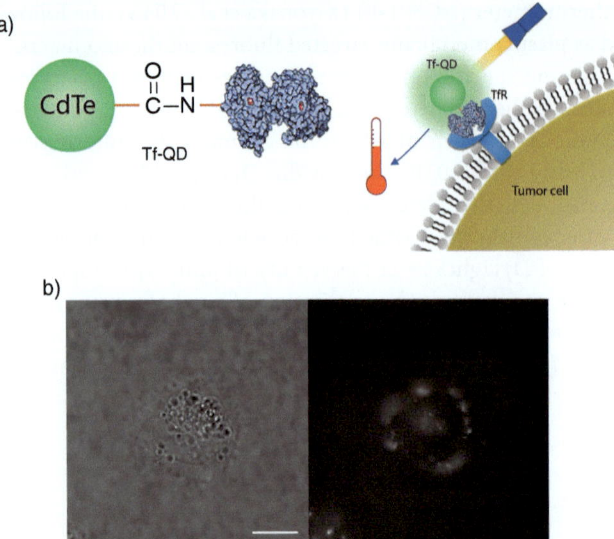

Figure 3.32 Specific labeling of tumor cells by membrane-targeted CdTe quantum dots conjugated with a membrane transport protein transferrin (Tf-QDs). (a) Structure of Tf-QD and schematic diagram of the interaction between Tf-QD and transferrin receptor (TfR). (b) Bright-field image (left) and fluorescence image (right) of a HOS cell labeled by Tf-QDs. Scale bar: 20 μm. Adapted from Yang et al. (2021) *Small*, **17**, 2102807 / John Wiley & Sons.

GFP was unchanged upon applying the radiofrequency magnetic field. This means that limited local heating at the plasma membrane can be carried out with strepta-vidin-DyLight549-coated superparamagnetic MnF_2O_4 nanoparticles while monitoring the membrane's temperature.

Quantum dots conjugated with the membrane-bound transport protein transferrin (Tf-QDs) can be specifically localized at the plasma membrane of tumor cells because the transferrin receptor is overexpressed on them (Figure 3.32). From the temperature-dependent emission peak wavelength of CdTe quantum dots in Tf-QDs which were localized at the plasma membrane of a HOS (human osteosarcoma) cell, an observable temperature distribution (standard deviation: 1.1 °C) was confirmed there (Yang et al., 2021).

3.6
Intracellular Thermometry of Brown Adipocytes

Brown adipocytes (also called "brown adipose cells" or "brown fat cells") were originally found in rodents in the middle of the 16th century and have been scientifically discussed from the 1960s (Smith and Hock, 1963; Cannon and Nedergaard, 2004). Brown adipocytes play a significant role both in maintaining a stable body temperature through central efferent pathways mediating non-shivering thermogenesis (Nakamura and Morrison, 2007) and in regulating whole-body energy expenditure (Rosen and Spiegelman, 2006): Brown adipose tissue can generate heat at up to 500 W kg^{-1}, while most other tissues produce heat at only 2–3 W kg^{-1} (Power, 1989). In 2009, four research groups independently discovered brown adipocytes in adult humans (Cypess et al., 2009; Lichtenbelt et al., 2009; Saito et al., 2009; Virtanen et al., 2009). Then, the biological study started to focus on strategies to increase the activity of brown adipocytes and inducible brown-like adipocytes having similar morphology and functions (called "beige adipocytes") as a potential approach for treating obesity and related metabolic disorders, which are becoming severe global public health concerns (Harms and Seale, 2013; Betz and Enerbäck, 2015; Sidossis and Kajimura, 2015). Also, in a related research area, e.g., food science studies have shown that food ingredients, such as capsaicin (Yoneshiro et al., 2012), fish oil (Kim et al., 2015), Thai black ginger extract (Matsushita et al., 2015), and hop extract (Morimoto-Kobayashi et al., 2015), can activate brown adipocytes and increase their energy expenditure by stimulating the sympathetic nervous system (Saito et al., 2020). Therefore, brown adipocytes are the most appealing subject in intracellular thermometry at the time of writing. Until now, around ten papers have been published on the intracellular thermometry of brown adipocytes upon chemical stimulation, biological defects of proteins, and environmental temperature variation. Mainly, the ratiometric fluorescent polymeric thermometers (Figure 3.23c) developed in my laboratory were utilized for monitoring the intracellular temperature of brown adipocytes in the experiments in the literature. Published research is summarized in Table 3.3 and will be outlined below.

Table 3.3 Intracellular thermometry of brown adipocytes by fluorescent molecular thermometers

Case	Examined factor	Used fluorescent thermometer	Finding	Ref.
1	ASK1[a]	FNT[b]	CL316,243-induced thermogenesis is reduced in ASK1-deficient adipocytes.	[1]
2	Cold exposure	RFPT[c]	When external temperature declines from 34 to 29 °C, intracellular temperature rises.	[2]
3	Isoproterenol and forskolin	ERthermAC[d]	β-Adrenergic receptor agonist (isoproterenol) and adenylate cyclase activator (forskolin) increase intracellular temperature.	[3]
4	Norepinephrine	Mixture of rhodamine B methyl ester and rhodamine 800[e]	Norepinephrine increases intracellular temperature.	[4]
5	Norepinephrine and CL316,243	RFPT[c]	β-Adrenergic receptor agonists (norepinephrine and CL316,243) increase intracellular temperature by more than 1 °C.	[5]
6	Maturation	RFPT[c]	β-Adrenergic receptor agonists-induced thermogenesis is abolished in precursor brown adipocytes.	[5]

(Continued)

Table 3.3 (*Continued*)

Case	Examined factor	Used fluorescent thermometer	Finding	Ref.
7	Isoproterenol, CL316,243, and carperitide	RFPT[c]	β-Adrenergic receptor agonists (isoproterenol and CL316,243) and an A-type natriuretic peptide (carperitide) increase intracellular temperature.	[6]
8	Environmental temperature	RFPT[c]	The temperature increases induced by the chemical stimuli with isoproterenol, CL316,243, and carperitide are observed only at 35 °C but not 37 °C.	[6]
9	PKR[f]-like ER kinase (PERK)	Modified FNT[g]	An ER (endoplasmic reticulum)-resident sensor, PERK is required for CL316,243-induced intracellular thermogenesis.	[7]
10	Pannexin-1 channel	RFPT[c]	Pretreatment of brown adipocytes with pannexin-1 inhibitors (carbenoxolone or spironolactone) reduces β3-adrenergic receptor agonists-induced intracellular temperature increase.	[8]
11	Selenoprotein P	RFPT[c]	Selenoprotein P impairs intracellular thermogenesis via reduction of noradrenaline-induced mitochondrial reactive oxygen species, which is confirmed by the treatment with selenoprotein P and the production of knockout mice.	[9]

a) Apoptosis signal-regulating kinase 1.
b) Fluorescent nanogel thermometer (Figure 3.3a).
c) Ratiometric fluorescent polymeric thermometer (Figure 3.23c or its different lots).
d) See Section 3.5.2.
e) See Section 3.5.1.
f) Protein kinase R.
g) Composed of DBThD-AA units instead of DBD-AA units. See text.

1 Hattori, K., Naguro, I., Okabe, K., Funatsu, T., Furutani, S., Takeda, K., and Ichijo, H. (2016) *Nat. Commun.*, **7**, 11158.
2 Rexius-Hall, M. L., Uchiyama, S., Eddington, D., and Rehman, J. (2017) *FASEB J.*, **31**, Suppl. 886.14.
3 Kriszt, R., Arai, S., Itoh, H., Lee, M. H., Goralczyk, A. G., Ang, X. M., Cypess, A. M., White, A. P., Shamsi, F., Xue, R., Lee, J. Y., Lee, S.-C., Hou, Y., Kitaguchi, T., Sudhaharan, T., Ishiwata, S., Lane, E. B., Chang, Y.-T., Tseng, Y.-H., Suzuki, M., and Raghunath, M. (2017) *Sci. Rep.*, **7**, 1383.
4 Xie, T.-R., Liu, C.-F., and Kang, J.-S. (2017) *Biophys. Rep.*, **3**, 85–91.
5 Tsuji, T., Ikado, K., Koizumi, H., Uchiyama, S., and Kajimoto, K. (2017) *Sci. Rep.*, **7**, 12889.
6 Kimura, H., Nagoshi, T., Yoshii, A., Kashiwagi, Y., Tanaka, Y., Ito, K., Yoshino, T., Tanaka, T. D., and Yoshimura, M. (2017) *Sci. Rep.*, **7**, 12978.
7 Kato, H., Okabe, K., Miyake, M., Hattori, K., Fukaya, T., Tanimoto, K., Beini, S., Mizuguchi, M., Torii, S., Arakawa, S., Ono, M., Saito, Y., Sugiyama, T., Funatsu, T., Sato, K., Shimizu, S., Oyadomari, S., Ichijo, H., Kadowaki, H., and Nishitoh, H. (2020) *Life Sci. Alliance*, **3**, e201900576.
8 Senthivinayagam, S., Serbulea, V., Upchurch, C. M., Polanowska-Grabowska, R., Mendu, S. K., Sahu, S., Jayaguru, P., Aylor, K. W., Chordia, M. D., Steinberg, L., Oberholtzer, N., Uchiyama, S., Inada, N., Lorenz, U. M., Harris, T. E., Keller, S. R., Meher, A. K., Kadl, A., Desai, B. N., Kundu, B. K., and Leitinger, N. (2021) *Mol. Metab.*, **44**, 10130.
9 Oo, S. M., Oo, H. K., Takayama, H., Ishii, K.-a., Takeshita, Y., Goto, H., Nakano, Y., Kohno, S., Takahashi, C., Nakamura, H., Saito, Y., Matsushita, M., Okamatsu-Ogura, Y., Saito, M., and Takamura, T. (2022) *Cell Rep.*, **38**, 110566.

In my research project, brown adipocytes were prepared by browning primary adipocytes collected from stromal vascular cells of interscapular brown adipose tissue in three-week-old male Wistar rats (Tsuji et al., 2017). The maturation of brown adipocytes was confirmed by measuring mRNA expression of uncoupling protein 1 (UCP1) and other gene markers. Differentiated brown adipocytes are morphologically characterized by multi-locular lipid droplets (Figure 3.33), whereas primary undifferentiated pre-brown adipocytes show a fibroblast-like structure.

Figure 3.34a displays the intracellular temperature variation by the chemical stimulation with a β-adrenergic receptor (β-AR) agonist, norepinephrine (NE), and a β3-AR-specific agonist, CL316,243 (Tsuji et al., 2017). The temperature was evaluated from the fluorescence intensity ratio of the ratiometric fluorescent polymeric thermometers (Figure 3.23c) at 585 $\pm$ 25 nm and 510 $\pm$ 10 nm. The mean temperature increases ($\pm$ SE) at 31 min after the stimulation were 1.25 $\pm$ 0.25 °C for NE and 1.39 $\pm$ 0.38 °C for CL316,243. The intracellular temperature increase of brown adipocytes by NE was remarkably suppressed by pretreatment with a β3-AR-specific antagonist, SR59230A, for 4 hours, which supported that thermogenesis by NE (and CL316,243) was induced through the activation of β3-AR.

Figure 3.34b shows the results of a different experimental setup using an A-type natriuretic peptide (ANP, carperitide), CL316,243, and an activator of β-AR-cAMP pathway, isoproterenol (Kimura et al., 2017). The temperature changes evaluated from the fluorescence intensity ratio of the ratiometric fluorescent polymeric thermometers (Figure 3.23c) at 605 and 525 nm were +1.55 °C for 10^{-9} mol L^{-1} ANP, +2.18 °C for 10^{-7} mol L^{-1} ANP, +1.98 °C for 0.5 µmol L^{-1} CL316,243, and +1.94 °C for 10^{-7} mol L^{-1} isoproterenol. The ANP treatment increased mRNA levels of uncoupling protein-1 (UCP1), while the increase in intracellular temperature and the UCP1 expression were canceled by inhibiting p38MAPK-UCP1 pathway by SB203580.

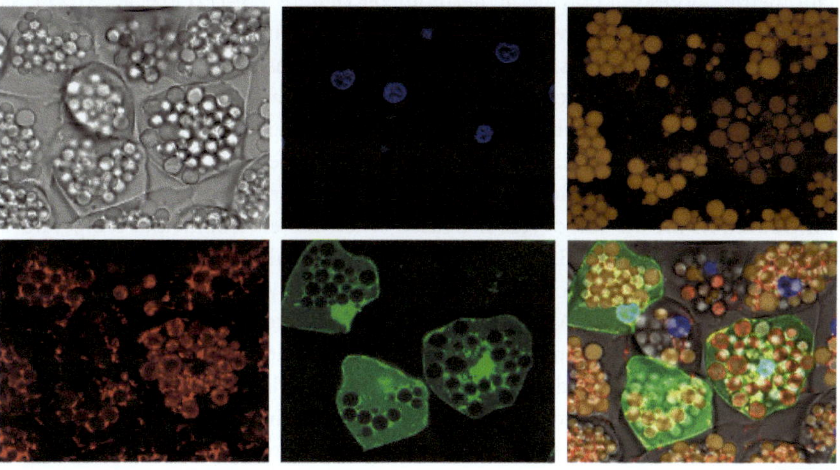

Figure 3.33 Brown adipocytes. Differential interference contrast image (upper left), confocal fluorescence images of Hoechst33342 (nuclei, upper middle), Nile Red (lipid, upper right), MitoTracker Deep Red (mitochondria, lower left), the ratiometric fluorescent polymeric thermometer (lower middle, chemical structure is shown in Figure 3.23c) and the merged image (lower right). Scale bar: 20 µm. Adapted from Tsuji et al. (2017) *Sci. Rep.*, **7**, 12889 / Springer Nature / CC BY 4.0.

Effects of apoptosis signal-regulating kinase 1 (ASK1) on the CL316,243-induced thermogenesis of brown adipocytes are indicated in Figure 3.34c, where temperature variations were evaluated from the changes in fluorescence intensity of the fluorescent nanogel thermometer (shown in Figure 3.3) (Hattori et al., 2016). CL316,243-induced thermogenesis was reduced in ASK1-deficient adipocytes, likely due to the ablation of UCP1.

Figure 3.34d demonstrates the effects of PKR (protein kinase R)-like ER kinase (PERK) on the CL316,243-induced thermogenesis of brown adipocytes (Kato et al., 2020). The intracellular temperature was evaluated by the temperature-dependent fluorescence intensity of a fluorescent nanogel thermometer with fluorescent DBThD-AA units (Uchiyama et al., 2012). The temperature increase induced by CL316,243 was significantly reduced in PERK-deficient brown adipocytes. Nevertheless, the thermogenic defect in PERK-deficient brown adipocytes was recovered by the infection with retroviruses especially expressing PERK-ΔLD. Thus, PERK activation may be indispensable for thermogenesis in brown adipocytes.

Figure 3.34e shows the effects of pannexin-1 (Panx1) channels on CL316,243-induced thermogenesis of brown adipocytes (Senthivinayagam et al., 2021). Pretreatment of brown adipocytes with Panx1 inhibitors, carbenoxolone (CBX) and spironolactone (Spiro) decreased the CL316,243-induced change in the fluorescence intensity ratio of the ratiometric fluorescent polymeric thermometer (Figure 3.23c) at 575 and 513 nm, meaning that the Panx1 inhibitors significantly reduced the β3-AR-specific agonist-induced intracellular temperature increase of brown adipocytes.

Selenoprotein P (SeP) is a hepatokine whose primary function is the antioxidation of reactive oxygen species (ROS). Interestingly, human SeP (hSeP) treatment remarkably suppressed the noradrenaline(NA)-induced intracellular thermogenesis of brown adipocytes, which was evaluated from the fluorescence intensity ratio of the ratiometric fluorescent polymeric thermometer (Figure 3.34f) (Oo et al., 2022). As treating brown adipocytes with SeP eliminated the NA-induced mitochondrial ROS required for thermogenesis, SeP is considered an intrinsic factor that impairs thermogenesis in brown adipocytes and, thus, a potential therapeutic target.

Interestingly, environmental temperature also influences the thermogenesis of brown adipocytes. As indicated in Figure 3.35a, the normalized fluorescence intensity of the ratiometric fluorescent polymeric thermometer (Figure 3.23c) introduced in brown adipocytes increased when the environmental temperature gradually decreased from 34 to 25 °C (Rexius-Hall et al., 2017). That is, the intracellular temperature of brown adipocytes increased by cold exposure. The intracellular thermogenesis by ANP, CL316,243, and isoproterenol in brown adipocytes shown in Figure 3.34b happened when the environmental temperature (i.e., the temperature of culture media) was 35 °C, but wholly dismissed at a slightly higher temperature, 37 °C. Furthermore, such cold exposure-induced thermogenesis was not observed in pre-matured brown adipocytes (Figure 3.35a) as neither a β-AR agonist NE nor a β3-AR-specific agonist CL316,243 had effects on the intracellular temperature of pre-matured brown adipocytes (Figure 3.35b) (Tsuji et al.,

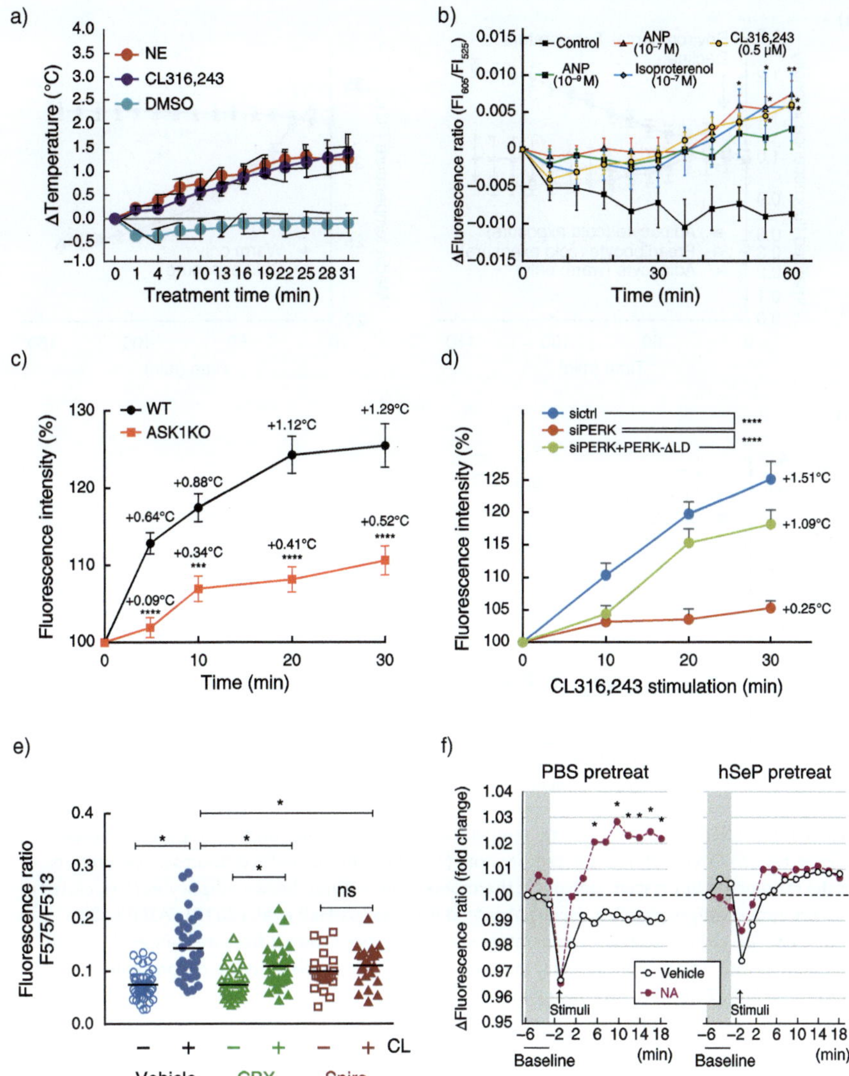

Figure 3.34 Effects of various chemical factors on thermogenesis of brown adipocytes. (a) Norepinephrine (NE) and CL316,243. Adapted from Tsuji et al. (2017) *Sci. Rep.*, **7**, 12889 / Springer Nature / CCBY 4.0 / Public domain. (b) A-type natriuretic peptide (ANP, carperitide), CL316,243, and isoproterenol. Adapted from Kimura et al. (2017) *Sci. Rep.*, **7**, 12978 / Springer Nature / CCBY 4.0. (c) Apoptosis signal-regulating kinase 1 (ASK1). WT: wildtype; ASK1KO: ASK1 knockout cell. ***P < 0.001, ****P < 0.0001 compared with WT. Adapted from Hattori et al. (2016) *Nat. Commun.*, **7**, 11158 / Springer Nature / CCBY 4.0. (d) Protein kinase R (PKR)-like endoplasmic reticulum kinase (PERK). ****P < 0.0001. (adapted from Kato et al. (2020) *Life Sci. Alliance*, **3**, e201900576 / CCBY 4.0. (e) Pannexin-1 (Panx1) inhibitors, carbenoxolone (CBX) and spironolactone (Spiro). CL: CL316,243. *P < 0.05. Adapted from Senthivinayagam et al. (2021) *Mol. Metab.*, **44**, 101130 / Elsevier / CCBY 4.0. (f) Human selenoprotein P (hSeP). *P < 0.05. Adapted from Oo et al. (2022) *Cell Rep.*, **38**, 110566. Fluorescent molecular thermometers utilized in the intracellular thermometry are summarized in Table 3.3.

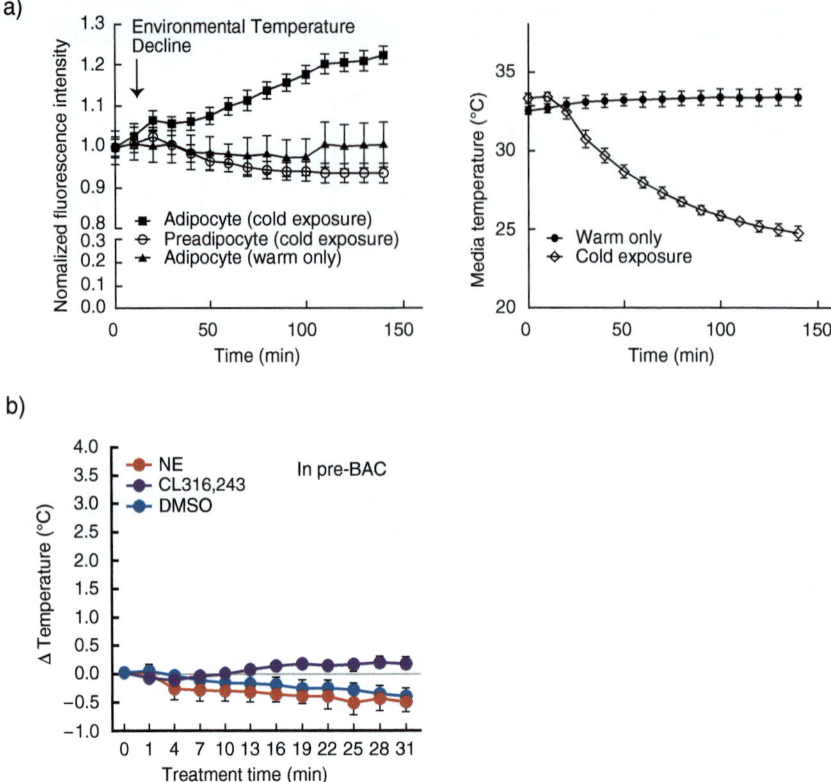

Figure 3.35 Effects of environmental temperature and maturation on thermogenesis of brown adipocytes. (a) Effects of medium temperature (right) on the normalized fluorescence intensity of the ratiometric fluorescent polymeric thermometer in matured brown adipocytes (left, closed) and pre-brown adipocytes (left, open). Adapted from Rexius-Hall et al. (2017) *FASEB J.*, **31**, Suppl. 886.14. (b) Intracellular temperature variation in pre-brown adipocytes under chemical stimulation with norepinephrine (NE, red) and CL316,243 (purple). Adapted from Tsuji et al. (2017) *Sci. Rep.*, **7**, 12889 / Springer Nature / CCBY 4.0.

2017), concluding that the maturation is necessary for thermogenesis of brown adipocytes. The molecular mechanisms in brown adipocyte thermogenesis are gradually unveiled by intracellular thermometry with fluorescent molecular thermometers, as illustrated in Figure 3.36. Remarkably, the intracellular temperature changes in brown adipocytes mentioned above have been correlated with the phenotype of temperature maintenance and respiratory activity in individuals. Moreover, the results of intracellular thermometry in brown adipocytes are under discussion for their association with body temperature (Kang et al., 2021) and efficient hypothermia therapy (Kashiwagi et al., 2020) in humans.

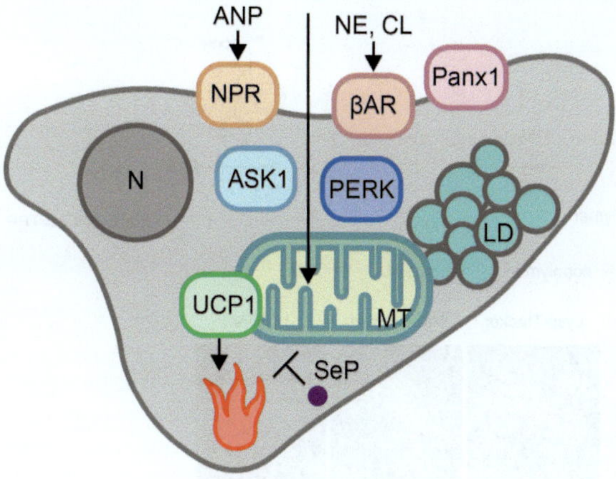

Figure 3.36 Schematic drawing of the molecular mechanism of heat production in brown adipocytes. Physiologically active molecules such as A-type natriuretic peptide (ANP) and β adrenergic agonists (norepinephrine (NE) and CL316,243 (CL)) are detected by receptors such as a natriuretic peptide receptor (NPR), β-adrenoreceptor (βAR), and Pannexin1 (Panx1) at the cell membrane, followed by cell signaling, including apoptosis signal-regulating kinase 1 (ASK1) and protein kinase R-like ER kinase (PERK), to the mitochondria (MT), where the uncoupling reaction by uncoupling protein 1 (UCP1) is provoked to produce heat that can be inhibited by selenoprotein P (SeP). N and LD indicate nucleus and lipid droplets, respectively. Adapted from Okabe and Uchiyama (2021) *Commun. Biol.*, **4**, 1377 / Springer Nature / CCBY 4.0.

3.7
Application of Intracellular Thermometry in Various Biological Fields

3.7.1
Simultaneous Temperature Monitoring at the Lysosome and the Mitochondria by Organelle-targeted Fluorescent Thermometers

The newest progress in organelle-targeted intracellular thermometry is simultaneous temperature measurements at the lysosome and the mitochondria with upconversion nanoparticles (UCNPs)-based fluorescent thermometers (Di et al., 2022). As shown in Figure 3.37a, both lysosome-targeted UCNPs and mitochondria-targeted UCNPs were prepared from $NaYF_4$: 20%Yb^{3+}, 2%Er^{3+} nanoparticles modified with oleic acid (UCNP@OA). To make lysosome-targeted UCNPs (UCNP@copolymer), UCNP@OA were coated by a hydrophilic PEG-MEMA$_{80}$-b-EGMP$_3$ di-block copolymer. Then, further functionalization with poly-L-lysine (PLL) and (3-carboxypropyl)triphenylphosphonium bromide (TPP) provided mitochondria-targeted UCNPs (UCNP@PLL@TPP). After incubation at 37 °C for 12 hours, these organelle-targeted UCNPs are localized at lysosome and mitochondria, respectively, in HeLa cells (Figure 3.37b) to show temperature-dependent fluorescence intensity ratio at 525 nm (due to $^2H_{11/2} \rightarrow {}^4I_{15/2}$ of Er^{3+}) and 545 nm (due to $^4S_{3/2} \rightarrow {}^4I_{15/2}$ of Er^{3+}) that corresponded to the temperature uncertainty of 0.8 °C. As indicated in

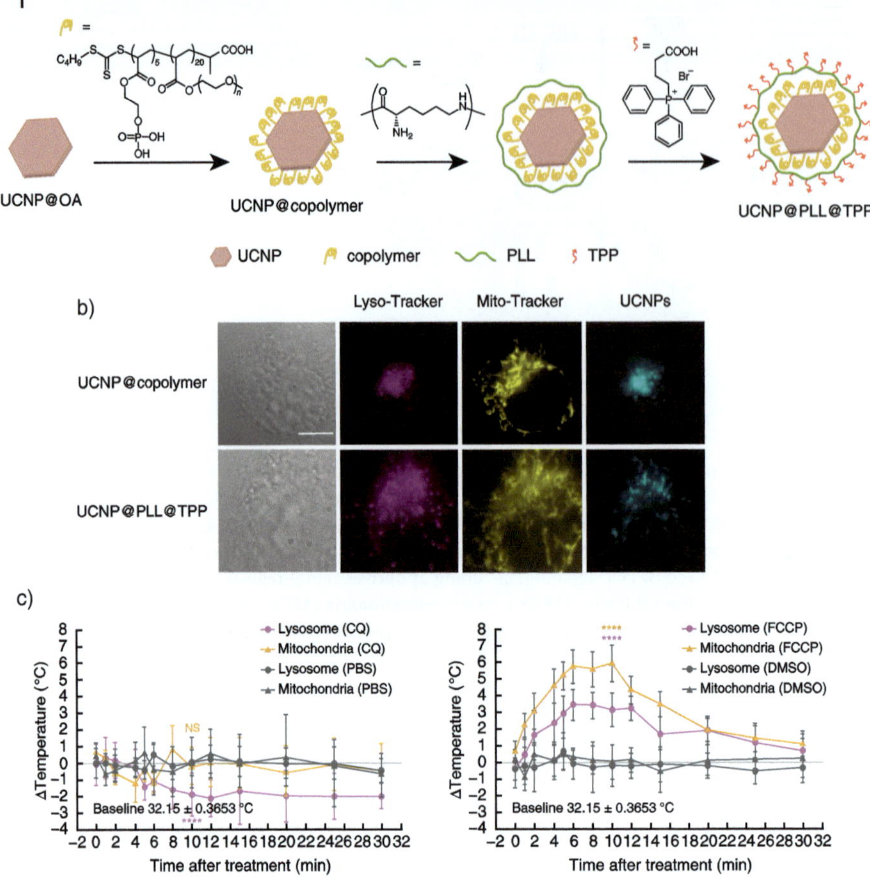

Figure 3.37 Simultaneous temperature measurements of lysosome and mitochondria by lysosome-targeted UCNP@copolymer and mitochondria-targeted UCNP@PLL@TPP. (a) Preparation of UCNP@copolymer and UCNP@PLL@TPP. (b) Bright-field images of a HeLa cell stained by LysoTracker Deep Red FM, MitoTracker Deep Red FM, and UCNPs (UCNP@copolymer or UCNP@PLL@TPP) (leftmost) and the corresponding fluorescence images of LysoTracker Deep Red FM (second left), MitoTracker Deep Red FM (second right), and UCNPs (rightmost). Scale bar: 10 μm. (c) Effects of chloroquine (CQ) (left) and FCCP (right) on lysosomal and mitochondrial temperatures. PBS and DMSO were used as controls. NS: no significance; ****P < 0.0001. Di et al. (2022) *Proc. Natl. Acad. Sci. USA*, **119**, e2207402119.

Figure 3.37c, a lysosomotropic reagent, chloroquine (CQ), decreased temperature by ~2 °C selectively at the lysosome (but not at mitochondria). When treated with an uncoupler FCCP, the temperature increased at the mitochondria by 6 °C at maximum, and a thermal transition from the mitochondria to the lysosomes was observed. These results strongly suggested the presence of a remarkable transient temperature gradient inside living cells. Therefore, high spatial resolution is important for intracellular thermometry correlating intracellular temperature with biological events inside cells.

3.7.2
Intracellular Temperature Regulation by Unsaturation of Phospholipids that Constitute Mitochondrial Membranes

One of the answers to the question "how to control the intracellular temperature in living cells?" was recently found from a biochemical viewpoint with intracellular thermometry using the cationic fluorescent polymeric thermometer (see Figure 3.21b) and a genetically encoded fluorescent thermometer (Murakami et al., 2022). In *Drosophila* S2 cells, $\Delta 9$-fatty acid desaturase DESAT1, which introduces a double bond at the $\Delta 9$ position of the acyl moiety of acyl-CoA, has an important role in mitochondrial thermogenesis. In *Desat1*-deficient cells or cells treated by a DESAT1 inhibitor (stearoyl-CoA desaturase 1 (SCD1) inhibitor), the intracellular temperature evaluated by the cationic fluorescent polymeric thermometer was significantly lower than that of the original S2 cells but recovered by the supplement of monounsaturated fatty acids such as palmitoleic acid (C16:1) and oleic acid (C18:1), or overexpression of FLAG-DESAT1 (a calpain-mediated degradation-resistant DESAT1 protein) (Figure 3.38a, 3.38b). Similar results were obtained in intracellular thermometry with a genetically encoded fluorescent thermometer tsGFP1-LP (based on tsGFP1 (see Section 3.5.1) but having a lower functional temperature range) and mitochondria-targeted tsGFP1-LP-mito which was prepared by fusing a target signal to tsGFP1-LP (Figure 3.38c, 3.38d). These DESAT1-mediated increases in intracellular temperature were caused by the enhancement of F_1F_o-ATPase-dependent mitochondrial respiration. Also, the F_1F_o-ATPase-dependent mitochondrial respiration was potentiated by cold exposure, where mitochondrial cristae structures were remodeled with DESAT1-mediated unsaturation of mitochondrial phospholipids.

3.7.3
Intracellular Thermogenesis by Aβ42 Aggregation

Amyloid-β42 (Aβ42) is a 42-residue variant of Aβ and a hallmark of Alzheimer's disease. The deposition of insoluble plaques of Aβ42 fibrils is a characteristic of the disease. The most applied model to produce Aβ42 fibrils is a nucleation-elongation with a high energy barrier (Figure 3.39a) (Young et al., 2017). Once Aβ42 is aggregated to fibrils with sufficient energy to overcome the activation barrier, the temperature of a local area increases due to the exothermic process. Then, initial nucleation of Aβ42 is spurred, and heat release will accelerate with further Aβ42 aggregation. Considering that mitochondria are hot spots in living cells, the effects of Aβ42 on the intracellular temperature of HEK293T (human embryonic kidney) cells were investigated in detail (Chung et al., 2022). The incubation of HEK293T cells with 500 nM recombinant Aβ42 increased the intracellular temperature that was monitored via the fluorescence lifetime of the cationic fluorescent polymeric thermometer (see Figure 3.21b) with fluorescent lifetime imaging microscopy (Figure 3.39b). The significant rise in intracellular temperature by Aβ42 ($+2.8\,°C$) with the formation of fibrillar structures in the cytoplasm (Figure 3.39c) was

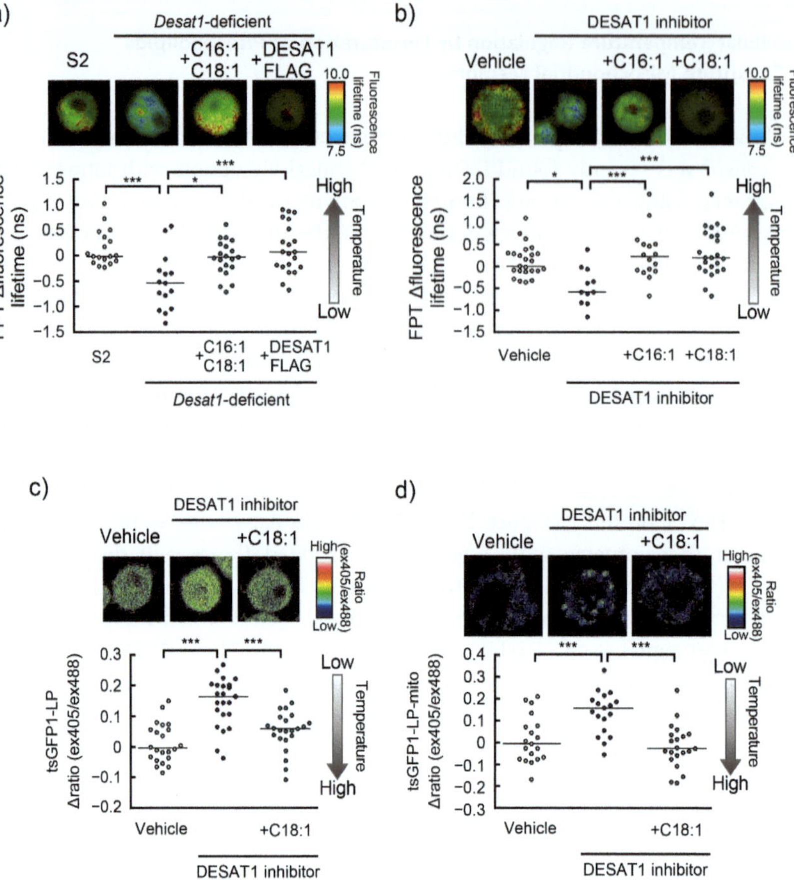

Figure 3.38 Effects of unsaturation of phospholipid acyl chains constituting mitochondrial inner membranes on thermogenesis in *Drosophila* S2 cells. (a) Effects of the deletion of Δ9-fatty acid desaturase (*Desat1*) and the addition of palmitoleic acid (C16:1), oleic acid (C18:1), and FLAG-DESAT1 (a calpain-mediated degradation-resistant DESAT1 protein) on the intracellular temperature evaluated by the fluorescence lifetime of the cationic fluorescent polymeric thermometer. (b) Effects of a DESAT1 inhibitor (stearoyl-CoA desaturase 1 (SCD1) inhibitor), C16:1, and C18:1 on the intracellular temperature evaluated by the fluorescence lifetime of the cationic fluorescent polymeric thermometer. (c) Effects of a DESAT1 inhibitor and C18:1 on the intracellular temperature evaluated by the fluorescence intensity ratio of tsGFP1-LP with the excitation at 405 and 488 nm. (d) Effects of a DESAT1 inhibitor and C18:1 on the intracellular temperature evaluated by the fluorescence intensity ratio of tsGFP1-LP-mito with the excitation at 405 and 488 nm. *$p < 0.05$, ***$p < 0.001$. Adapted from Murakami et al. (2022) *Cell Rep.*, **38**, 110487 / Elsevier / CC BY 4.0.

alleviated upon the treatment with MJ040X, which is the esterified derivative of a small molecular drug preventing the aggregation of Aβ42 by binding to the C-terminus. Comparing the intracellular temperature, ATP level in the cytoplasm, extracellular acidification rate, and oxygen consumption rate upon the treatment by Aβ42 with those by an uncoupler FCCP as a positive control, effects of Aβ42 were

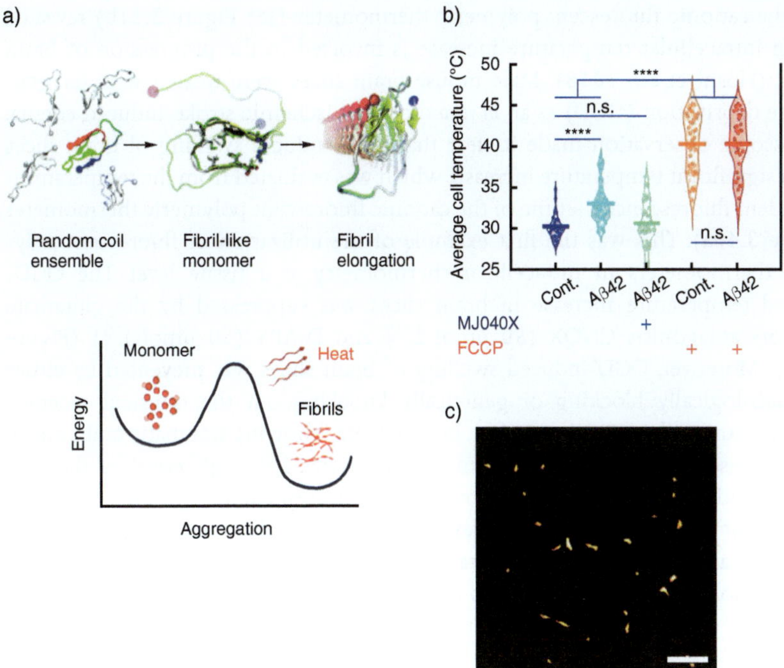

Figure 3.39 Intracellular thermogenesis by aggregation of Aβ42 inside cells. (a) Schematic image (top) and energy diagram (bottom) of Aβ42 between monomer and aggregated fibril states. (b) Effects of intracellular Aβ42 on the average cellular temperature of HEK293T cells. The intracellular temperature was evaluated from the temperature-dependent fluorescence lifetime of the cationic fluorescence polymeric thermometer in HEK293T cells. For the fluorescence lifetime imaging, the temperature of a stage top heater was set to 30 °C. MJ040X inhibits intracellular aggregation of Aβ42. **** is p <0.0001, and n.s. is not significant. (c) Super-resolution direct stochastic optical reconstruction microscopy (dSTORM) image of AF647 tagged antibody labeled Aβ42. Scale bar: 500 nm. The upper part in panel (a) is adapted from Barz et al. (2021) *Chem. Commun.*, **57**, 947–950. The others are adapted from Chung et al. (2022) *J. Am. Chem. Soc.*, **144**, 10034–10041.

mainly limited to exothermic elongation but not significantly expanded to biological phenomena such as mitochondrial damage. It is profound that a chemical event (i.e., exothermic aggregation of Aβ42) in live cells varies the intracellular temperature and could accelerate disease progression in biological objects.

3.7.4
Intracellular Temperature Variation Involved in Brain Edema

Brain edema (Klatzo et al., 1967; Rosenberg et al., 1999) is characterized by an increase in net brain water content, which increases brain volume. Although brain edema has a high fatality rate, cellular and molecular mechanisms underlying brain edema remain unclear. In 2018, intracellular thermometry of mouse brain tissues with the cationic fluorescent polymeric thermometer (see Figure 3.21b) revealed that an intracellular temperature increase is involved in the progression of brain edema (Hoshi et al., 2018). Male mouse brain slices were treated with oxygen-glucose deprivation (OGD) as an *in vitro* model of ischemic stroke-induced edema. Microscopic observation made it clear that OGD induces swelling of brain slices with a significant temperature increase, which was evaluated from the temperature-dependent fluorescence lifetime of the cationic fluorescent polymeric thermometer (Figure 3.40a). This was the first example of the utilization of fluorescent polymeric thermometers in intracellular thermometry at a tissue level. The OGD-induced temperature increase in brain slices was suppressed by the glutamate receptors antagonists CNQX (50 μmol L^{-1}) and D-AP5 (50 μmol L^{-1}) (Figure 3.40b). Moreover, OGD-induced swelling of brain slices was prevented by either pharmacologically blocking or genetically knocking out the transient receptor potential vanilloid 4 (TRPV4). In fact, brain edema following traumatic brain injury was suppressed in TRPV4-deficient male mice *in vivo*. I strongly consider that this result is quite crucial for future progress in intracellular thermometry because intracellular temperature variation was undoubtedly involved in a complex cascade of biological events. That is, temperature fluctuations inside cells are not only a result of biological reactions but also a reason for them.

3.7.5
Intracellular Temperature Variation by Ca^{2+} through Ryanodine Receptor 1 and its Dysfunction in Malignant Hyperthermia

The commitment of intracellular temperature variation in a biologically complex cascade has also been found in malignant hyperthermia. Type 1 ryanodine receptor (RYR1) is a Ca^{2+} release channel in the sarcoplasmic reticulum (SR) of the skeletal

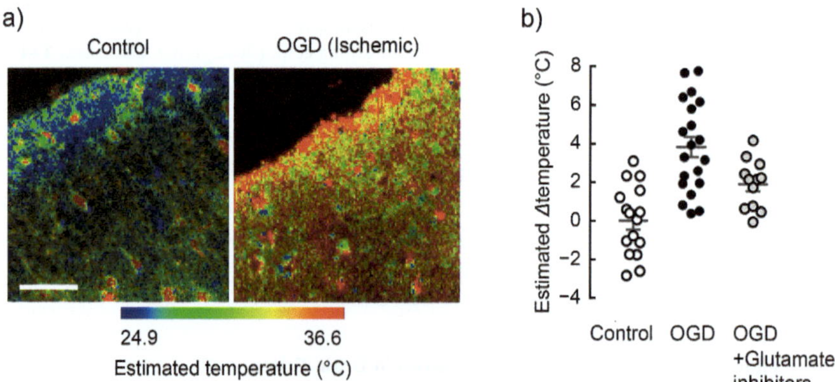

Figure 3.40 Thermal signaling in mouse brain during ischemia. (a) Temperature maps of non-treated (control) and oxygen-glucose deprivation (OGD)-treated brain slices. The temperature was measured by the cationic fluorescent polymeric thermometer. Scale bar: 50 μm. (b) Variations in the brain temperature. Glutamate inhibitors: 6-cyano-7-nitroquinoxaline-2,3-dione (CNQX) and D-2-amino-5-phosphonopentanoic acid (D-AP5). Hoshi et al. (2018) *J. Neurosci.*, **38**, 5700–5709 / The Authors / CC BY 4.0.

muscle, and mutations in RYR1 cause severe muscle diseases such as malignant hyperthermia with a disorder of Ca^{2+}-induced Ca^{2+} release through RYR1 from the SR (Fujii et al., 1991). A heterozygous R2509C-RYR1 mouse is a model animal with malignant hyperthermia-like symptoms when anesthetized by isoflurane (Yamazawa et al., 2021). In experiments, heterozygous myocytes were prepared from R2509C-RYR1 mice, and their intracellular Ca^{2+} concentration and temperature under treatment with isoflurane were spontaneously monitored by a fluorescent Ca^{2+} sensor Cal-520 and an ER-targeted fluorescent molecular thermometer ERthermAC (in which the chlorine atom of ER thermo yellow (Figure 3.29b) is replaced by a hydrogen atom), respectively (Tsuboi et al., 2022). As indicated in Figure 3.41, under the treatment of isoflurane, neither variation was observed in Ca^{2+} concentration nor intracellular temperature in flexor digitorum brevis fibers isolated from wild-type mice, while remarkable increases in both Ca^{2+} concentration (represented by fluorescence increase of Cal-520) and intracellular temperature (reflected by fluorescence decrease of ERthermAC) were confirmed in those from heterozygous R2509C-RYR1 mice. Dysfunctional Ca^{2+} dynamics-induced temperature increase in malignant hyperthermia occurs at a single-cell level.

In independent research, the relationship between Ca^{2+} leak through RYR1 and thermogenesis in resting skeletal muscle was investigated in detail (Meizoso-Huesca et al., 2022). In accordance with an experimental scheme shown in Figure 3.42a, the contribution of RYR1 to SERCA-dependent temperature variation at SR could be evaluated using a fluorescent molecular thermometer ER thermo yellow, a Ca^{2+} indicator Fluo-5N acetoxymethyl ester, an RYR1 inhibitor ryanodine, and an SR Ca^{2+} pump inhibitor cyclopiazonic acid. The Ca^{2+} leak through RYR1 raised cytosolic Ca^{2+} concentration in the local vicinity of Ca^{2+} pumps at the SR, resulting in amplified thermogenesis. Furthermore, gene-dose-dependent increases in RYR1 Ca^{2+} leak in RYR1 mutant mice resulted in SERCA-dependent temperature increases (Figure 3.42b), which was consistent with raised Ca^{2+} concentration at the SR Ca^{2+} pump via RYR1 Ca^{2+} leak. These results can elucidate the biological mechanism that regulates non-shivering thermogenesis from skeletal muscle in mammals.

3.7.6
Intracellular Switch between ATP Synthesis and Thermogenesis

A biologically active switch between ATP synthesis and thermogenesis has been disclosed inside THP-1 (human monocytic leukemia) cells with the aid of intracellular thermometry with the cationic fluorescent polymeric thermometer (see Figure 3.21b) (Li et al., 2020). As a proposed mechanism is illustrated in Figure 3.43a, when a heme level is low, the major facilitator superfamily domain containing 7C (MFSD7C) promotes ATP synthesis by interacting with the mitochondrial components (the electron transport chain (ETC) complexes III, IV, and V (F_0 and F_1)) and destabilizing sarcoendoplasmic reticulum Ca^{2+}-ATPase 2b (SERCA2b). When a heme is bound to the N-terminal domain, MFSD7C dissociates from the electron transport chain components and SERCA2b, resulting in thermogenesis due to the stabilization of SERCA2b. An intracellular temperature increase was detected

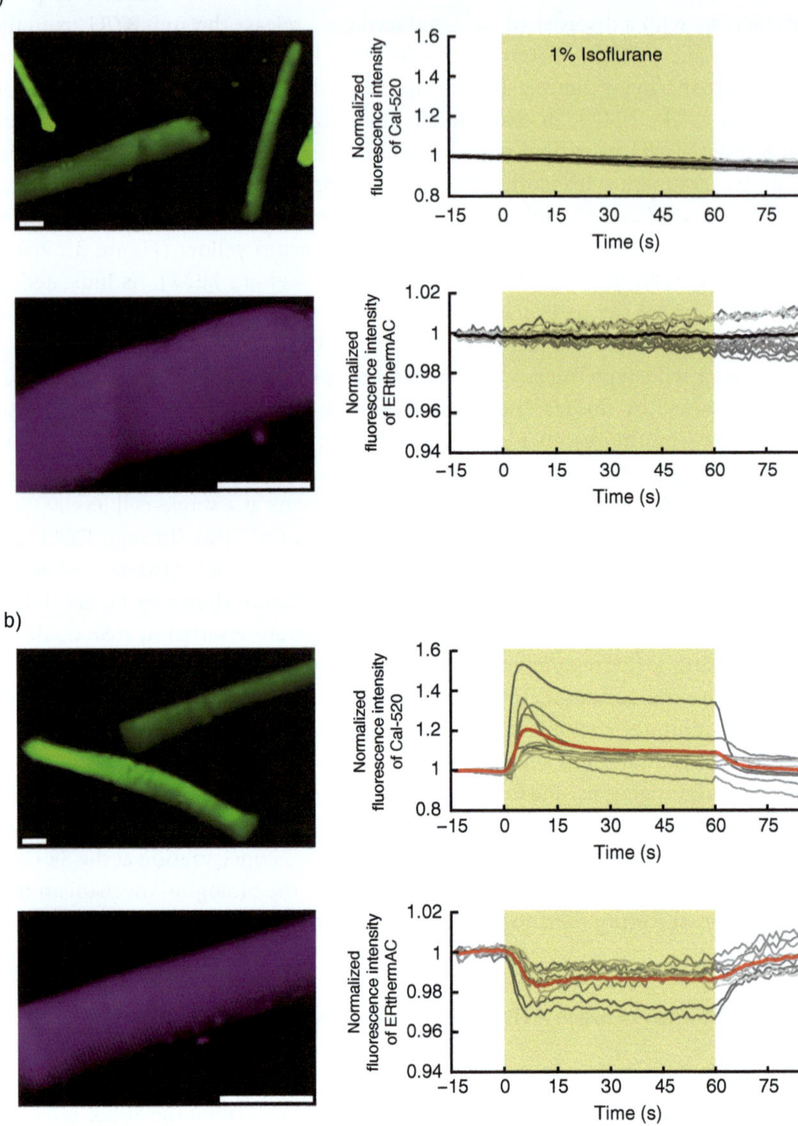

Figure 3.41 Effects of isoflurane on the Ca^{2+} concentration in the cytoplasm and the temperature of sarcoplasmic reticulum in flexor digitorum brevis from (a) a wild-type mouse and (b) an R2509C-RYR1 mouse. Fluorescence images of a fluorescent Ca^{2+} sensor Cal-520 (upper left) and an ER-targeted fluorescent molecular thermometer ERthermAC (lower left) in flexor digitorum brevis and changes in the fluorescence intensity (gray: individual cell, black and red: average) of Cal-520 (upper right) and ERthermAC (lower right) under the isoflurane application. Scale bars: 50 μm. Tsuboi et al. (2022) *J. Gen. Physiol.*, **154**, e202213136 / Rockefeller University Press.

a)

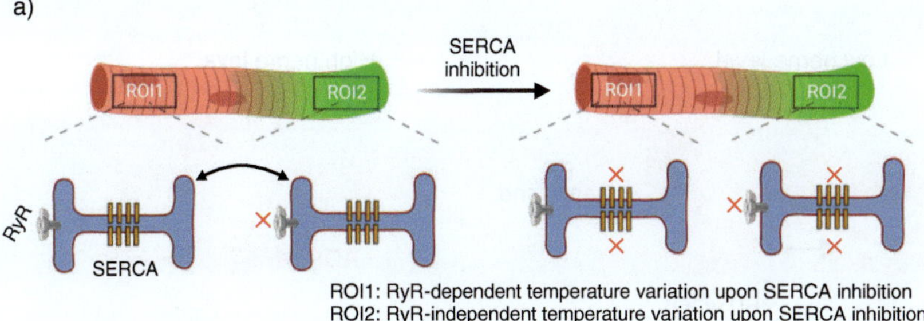

b)

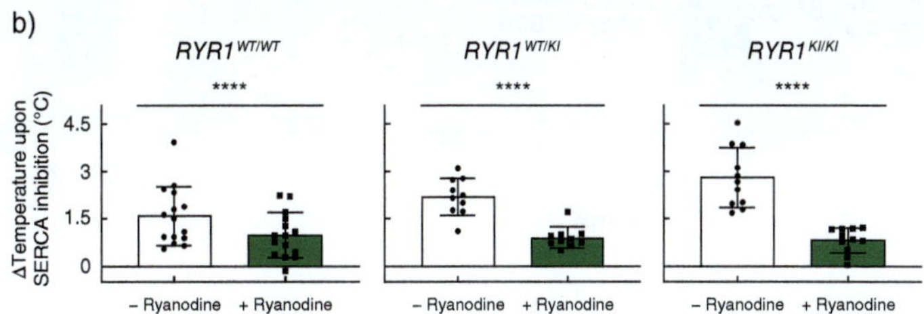

Figure 3.42 Role of Ca^{2+} leak through ryanodine receptor 1 (RYR1) in the thermogenesis of skeletal muscle. (a) Schematic diagram of the experiment to define RYR1 contribution to Ca^{2+} pump-dependent thermogenesis in the sarcoplasmic reticulum (SR). (b) Temperature variations of SR upon sarco(endo)plasmic reticulum Ca^{2+}-ATPase (SERCA) inhibition with cyclopiazonic acid in mouse $RYR1^{WT/WT}$, $RYR1^{WT/KI}$, and $RYR1^{KI/KI}$ fibers. SR temperature was monitored by ER thermo yellow. Meizoso-Huesca et al. (2022) *Proc. Natl. Acad. Sci. USA*, **119**, e2119203119.

by the temperature-dependent fluorescence intensity of the cationic fluorescent polymeric thermometer (Figure 3.43b). Intracellular temperature significantly increased by 0.12 °C by the treatment with heme for 1 hour in THP-1 cells. On the other hand, *Mfsd7c* knockout clones (A11, B11, 3D12, and 4B8) showed a higher temperature (0.15–0.3 °C) than non-treated THP-1 cells even without heme treatment (i.e., thermogenesis is always on). In $SERCA2b^{-/-}$ cells, thermogenesis by the heme treatment was significantly diminished. This intracellular switch between ATP synthesis and thermogenesis represents a cellular intrinsic mechanism to regulate mitochondrial energy metabolism, likely including strict temperature control inside cells.

3.7.7
Embryonic Development by Controlled Local Heating

Understanding the coordination of cell-division timing is a question in developmental biology. Temperature is a possible influential parameter in determining cell-cycle duration, but to precisely control intracellular temperature is challenging.

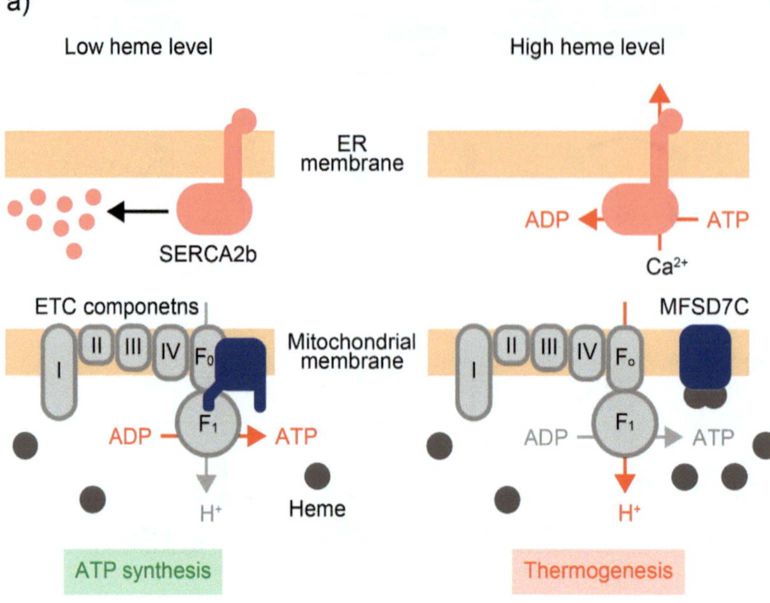

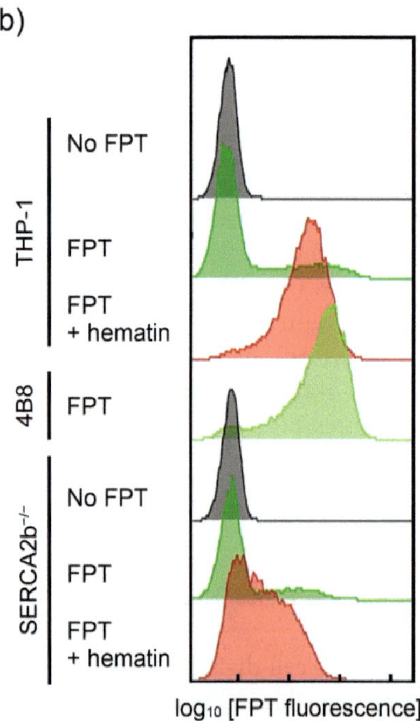

Figure 3.43 Endogenous switching between ATP synthesis and thermogenesis inside THP-1 cells. (a) Proposed model of MFSD7C functions according to heme level. (b) Thermogenesis by hematin in parental THP-1 cells, *Mfsd7c* knockout clone 4B8, and *Serca2b*⁻/⁻ cells. The higher fluorescence intensity of a cationic fluorescent polymeric thermometer (FPT) indicates a higher intracellular temperature. Adapted from Li et al. (2020) *Nat. Commun.*, **11**, 4837. Springer Nature / CC BY 4.0.

Using a local laser heating at 1480 nm and a fluorescent nanodiamonds-based intracellular thermometry, a method for controlling the cell-division timing in *Caenorhabditis elegans* embryos (Figure 3.44a) has been established (Choi et al., 2020). First, intracellular temperature distribution during local IR illumination to an embryo of *C. elegans* was recorded using nitrogen-vacancy centers of fluorescent nanodiamonds with optically detected magnetic resonance (ODMR) techniques (Figure 3.44b). Then, selective and controlled acceleration of cell divisions of *C. elegans* embryos, even an inversion of division order at a two-cell (AB and P1) stage, was enabled by local IR irradiation with appropriate power (Figure 3.44c). These data suggested that the asynchronous cell cycle of early embryonic development in *C. elegans* is determined independently by each temperature of individual cells rather than via cell-to-cell communication. Evidently, intracellular temperature determines a cell's destiny.

3.7.8
Regulation of Mitochondrial Temperature in Oocytes by Mito-Q

Mito-Q is a commercially available mitochondria-targeted antioxidant (www. mitoq.com). Mito-Q consists of cationic triphenylphosphine and the ubiquinone portion of coenzyme Q and removes excess reactive oxygen species in

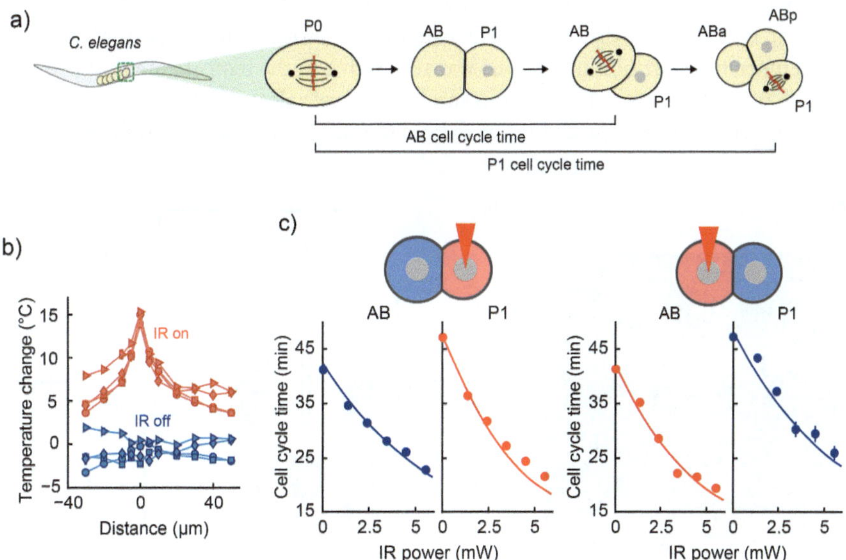

Figure 3.44 Effects of local laser heating on cell division timing of *C. elegans* embryos. (a) Early embryonic development of *C. elegans*. At the end of the one-cell stage, the chromosomes in P0 split and are pulled apart by spindle fibers, resulting in two daughter cells, AB and P1. The two daughter cells exhibit asynchronous division, in which AB undergoes mitosis earlier than P1. (b) Position-dependent intracellular temperature change by IR laser heating (Power: 5.7 mW). The temperature was measured by the nitrogen-vacancy center of fluorescent nanodiamonds. The different symbols correspond to different iterative runs. (c) Selective acceleration of cell-cycle times by P1 nucleus heating (left) and AB nucleus heating (right). Choi et al. (2020) *Proc. Natl. Acad. Sci. USA*, **117**, 14636–14641.

mitochondria. In oocytes, Mito-Q diminishes oxidative stresses caused by *in vitro* maturation. In order to clarify the effects of Mito-Q on mitochondrial thermogenesis as well as oocyte development and ATP production, the temperature of porcine oocytes during *in vitro* maturation was measured with a mitochondria-targeted fluorescent molecular thermometer, Mito thermo yellow (see Figure 3.26c) under treatment with Mito-Q (Zhou et al., 2022). The results shown in Figure 3.45 indicated that Mito-Q increased the temperature of mitochondria in oocytes at 24 and 44 hours during *in vitro* maturation and decreased that at 36 hours. Compared with control experiments, the production of ATP and the expression of uncoupling protein 2 (UCP2) were significantly suppressed by Mito-Q. From the study, it was concluded that Mito-Q promoted oocyte maturation *in vitro* and maintained the stability of thermogenesis in mitochondria by inhibiting UCP2 expression.

3.7.9
Temperature Variation Associated to Hippocampal Neuron Firing

In the late stages of preparing this book, the relationship between intracellular temperature and neuronal activity was reported for the first time (Petrini et al., 2022). The ODMR technique with nitrogen-vacancy centers in fluorescent nanodiamonds revealed that the intracellular temperature of hippocampal neurons, obtained from C57BL/6 mouse 16-day embryos, was significantly increased by 1.02 ± 0.24 °C under treatment with picrotoxin ($100\ \mu\text{mol L}^{-1}$), a selective GABA$_A$ (γ-aminobutyric acid type A) receptor inhibitor that drastically potentiated the neuronal firing activity. Moreover, this temperature variation in neurons was reversed (i.e., the temperature dropped by 0.50 ± 0.17 °C compared with the control) when tetrodotoxin ($0.3\ \mu\text{mol L}^{-1}$) and cadmium chloride ($500\ \mu\text{mol L}^{-1}$) were used to inhibit the spontaneous firing.

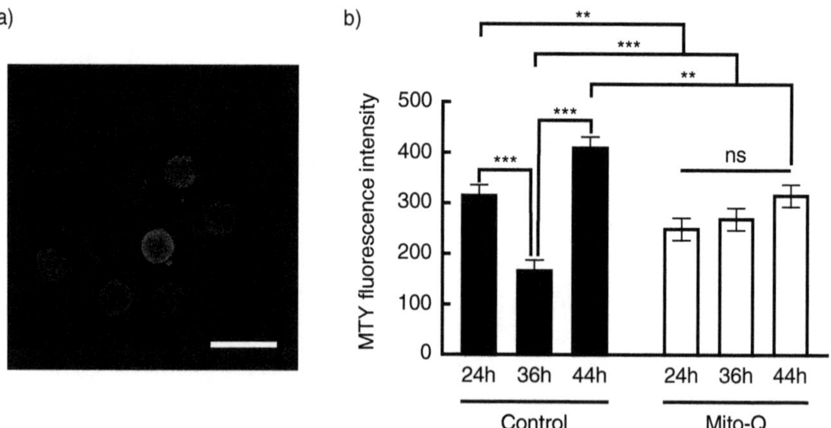

Figure 3.45 Effects of Mito-Q on mitochondrial thermogenesis during *in vitro* maturation of porcine oocytes. (a) Fluorescence image of Mito thermo yellow in porcine oocytes. Scale bar: 200 µm. (b) Effects of Mito-Q on the fluorescence intensity of Mito thermo yellow (MTY) in porcine oocytes. **P < 0.01; ***P < 0.001; ns: non-significant (P > 0.05). Error bars indicate the standard error of the mean. Zhou et al. (2022) *Theriogenology*, **187**, 205–214 / Elsevier.

References

Albers, A. E., Chan, E. M., McBride, P. M., Ajo-Franklin, C. M., Cohen, B. E., and Helms, B. A. (2012) Dual-emitting quantum dot/quantum rod-based nanothermometers with enhanced response and sensitivity in live cells. *J. Am. Chem. Soc.*, **134**, 9565–9568.

Andersen, J. S., Wilkinson, C. J., Mayor, T., Mortensen, P., Nigg, E. A., and Mann, M. (2003) Proteomic characterization of the human centrosome by protein correlation profiling. *Nature*, **426**, 570–574.

Arai, S., Lee, S.-C., Zhai, D., Suzuki, M., and Chang, Y. T. (2014) A molecular fluorescent probe for targeted visualization of temperature at the endoplasmic reticulum. *Sci. Rep.*, **4**, 6701.

Arai, S., Suzuki, M., Park, S.-J., Yoo, J. S., Wang, L., Kang, N.-Y., Ha, H.-H., and Chang, Y.-T. (2015a) Mitochondria-targeted fluorescent thermometer monitors intracellular temperature gradient. *Chem. Commun.*, **51**, 8044–8047.

Arai, S., Lee, C.-L. K., Chang, Y.-T., Sato, H., and Sou, K. (2015b) Thermosensitive nanoplatforms for photothermal release of cargo from liposomes under intracellular temperature monitoring. *RSC Adv.*, **5**, 93530–93538.

Behr, J.-P. (1993) Synthetic gene-transfer vectors. *Acc. Chem. Res.*, **26**, 274–278.

Betz, M. J. and Enerbäck, S. (2015) Human brown adipose tissue: what we have learned so far. *Diabetes*, **64**, 2352–2360.

Cannon, B. and Nedergaard, J. (2004) Brown adipose tissue: function and physiological significance. *Physiol. Rev.*, **84**, 277–359.

Chapman, C. F., Liu, Y., Sonek, G. J., and Tromberg, B. J. (1995) The use of exogenous fluorescent probes for temperature measurements in single living cells. *Photochem. Photobiol.*, **62**, 416–425.

Choi, J., Zhou, H., Landig, R., Wu, H.-Y., Yu, X., von Stetina, S. E., Kucsko, G., Mango, S. E., Needleman, D. J., Samuel, A. D. T., Maurer, P. C., Park, H., and Lukin, M. D. (2020) Probing and manipulating embryogenesis via nanoscale thermometry and temperature control. *Proc. Natl. Acad. Sci. USA*, **117**, 14636–14641.

Chrétien, D., Bénit, P., Ha, H.-H., Keipert, S., El-Khoury, R., Chang, Y.-T., Jastroch, M., Jacobs, H. T., Rustin, P., and Rak, M. (2018) Mitochondria are physiologically maintained at close to 50 °C. *PLoS Biol.*, **16**, e2003992.

Chrétien, D., Bénit, P., Leroy, C., El-Khoury, R., Park, S., Lee, J. Y., Chang, Y.-T., Lenaers, G., Rustin, P., and Rak, M. (2020) Pitfalls in monitoring mitochondrial temperature using charged thermosensitive fluorophores. *Chemosensors*, **8**, 124.

Chung, C. W., Stephens, A. D., Konno, T., Ward, E., Avezov, E., Kaminski, C. F., Hassanali, A. A., and Schierle, G. S. K. (2022) Intracellular Aβ42 aggregation leads to cellular thermogenesis. *J. Am. Chem. Soc.*, **144**, 10034–10041.

Conduit, P. T., Wainman, A., and Raff, J. W. (2015) Centrosome function and assembly in animal cells. *Nat. Rev. Mol. Cell Biol.*, **16**, 611–624.

Copolovici, D. M., Langel, K., Eriste, E., and Langel, Ü (2014) Cell-penetrating peptides: design, synthesis, and applications. *ACS Nano.*, **8**, 1972–1994.

Craggs, T. D. (2009) Green fluorescent protein: structure, folding and chromophore maturation. *Chem. Soc. Rev.*, **38**, 2865–2875.

Crouch, S. P. M., Kozlowski, R., Slater, K. J., and Fletcher, J. (1993) The use of ATP bioluminescence as a measure of cell proliferation and cytotoxicity. *J. Immunol. Methods*, **160**, 81–88.

Cypess, A. M., Lehman, S., Williams, G., Tal, I., Rodman, D., Goldfine, A. B., Kuo, F. C., Palmer, E. L., Tseng, Y.-H., Doria, A., Kolodny, G. M., and Kahn, C. R. (2009) Identification and importance of brown adipose tissue in adult humans. *N. Eng. J. Med.*, **360**, 1509–1517.

de Meis, L. (2001) Uncoupled ATPase activity and heat production by the sarcoplasmic reticulum Ca^{2+}-ATPase. *J. Biol. Chem.*, **276**, 25078–25087.

de Meis, L., Arruda, A. P., and Carvalho, D. P. (2005) Role of sarco/endoplasmic reticulum Ca^{2+}-ATPase in thermogenesis. *Biosci. Rep.*, **25**, 181–190.

Di, X., Wang, D., Zhou, J., Zhang, L., Stenzel, M. H., Su, Q. P., and Jin, D. (2021) Quantitatively monitoring in situ mitochondrial thermal dynamics by upconversion nanoparticles. *Nano Lett.*, **21**, 1651–1658.

Di, X., Wang, D., Su, Q. P., Liu, Y., Liao, J., Maddahfar, M., Zhou, J., and Jin, D. (2022) Spatiotemporally mapping temperature dynamics of lysosomes and mitochondria using cascade organelle-targeting upconversion nanoparticles. *Proc. Natl. Acad. Sci. USA*, **119**, e2207402119.

Dong, G., Lu, W., Zhao, X., Guan, X., Ji, Y., Zhang, X., Pan, J., Ning, J., Zhou, H., and Wang, C. (2021) Real-time temperature measurement of living cells exposed to microwaves using a temperature-dependent fluorescent dye. *Chem. Phys.*, **547**, 111190.

Donner, J. S., Thompson, S. A., Kreuzer, M. P., Baffou, G., and Quidant, R. (2012) Mapping intracellular temperature using green fluorescent protein. *Nano Lett.*, **12**, 2107–2111.

Doxsey, S. (2001) Re-evaluating centrosome function. *Nat. Rev. Mol. Cell Biol.*, **2**, 688–698.

European Brewery Convention (1977) E.B.C. analytica microbiologica. *J. Inst. Brew.*, **83**, 109–118.

Frankel, A. D. and Pabo, C. O. (1988) Cellular uptake of the Tat protein from human immunodeficiency virus. *Cell*, **55**, 1189–1193.

Fu, M., Xiao, Y., Qian, X., Zhao, D., and Xu, Y. (2008) A design concept of long-wavelength fluorescent analogs of rhodamine dyes: replacement of oxygen with silicon atom. *Chem. Commun.*, 1780–1782.

Fujii, J., Otsu, K., Zorzato, F., deLeon, S., Khanna, V. K., Weiler, J. E., O'Brien, P. J., and MacLennan, D. H. (1991) Identification of a mutation in porcine ryanodine receptor associated with malignant hyperthermia. *Science*, **253**, 448–451.

Godbey, W. T., Wu, K. K., Hirasaki, G. J., and Mikos, A. G. (1999) Improved packing of poly(ethylenimine)/DNA complexes increases transfection efficiency. *Gene Ther.*, **6**, 1380–1388.

Gota, C., Okabe, K., Funatsu, T., Harada, Y., and Uchiyama, S. (2009) Hydrophilic fluorescent nanogel thermometer for intracellular thermometry. *J. Am. Chem. Soc.*, **131**, 2766–2767.

Green, M. and Loewenstein, P. M. (1988) Autonomous functional domains of chemically synthesized human immunodeficiency virus Tat *trans*-activator protein. *Cell*, **55**, 1179–1188.

Harms, M. and Seale, P. (2013) Brown and beige fat: development, function and therapeutic potential. *Nat. Med.*, **19**, 1252–1263.

Hattori, K., Naguro, I., Okabe, K., Funatsu, T., Furutani, S., Takeda, K., and Ichijo, H. (2016) ASK1 signalling regulates brown and beige adipocyte function. *Nat. Commun.*, **7**, 11158.

Hayashi, T., Fukuda, N., Uchiyama, S., and Inada, N. (2015) A cell-permeable fluorescent polymeric thermometer for intracellular temperature mapping in mammalian cell lines. *PLoS ONE*, **10**, e0117677.

Heytler, P. G. and Prichard, W. W. (1962) A new class of uncoupling agents—carbonyl cyanide phenylhydrazones. *Biochem. Biophys. Res. Commun.*, **7**, 272–275.

Homma, M., Takei, Y., Murata, A., Inoue, T., and Takeoka, S. (2015) A ratiometric fluorescent molecular probe for visualization of mitochondrial temperature in living cells. *Chem. Commun.*, **51**, 6194–6197.

Hoshi, Y., Okabe, K., Shibasaki, K., Funatsu, T., Matsuki, N., Ikegaya, Y., and Koyama, R. (2018) Ischemic brain injury leads to brain edema via hyperthermia-induced TRPV4 activation. *J. Neurosci.*, **38**, 5700–5709.

Huang, H., Delikanli, S., Zeng, H., Ferkey, D. M., and Pralle, A. (2010) Remote control of ion channels and neurons through magnetic-field heating of nanoparticles. *Nat. Nanotechnol.*, **5**, 602–606.

Huang, Z., Li, N., Zhang, X., Wang, C., and Xiao, Y. (2018) Fixable molecular thermometer for real-time visualization and quantification of mitochondrial temperature. *Anal. Chem.*, **90**, 13953–13959.

Huang, Z., Li, N., Zhang, X., and Xiao, Y. (2021) Mitochondria-anchored molecular thermometer quantitatively monitoring cellular inflammations. *Anal. Chem.*, **93**, 5081–5088.

Ikeno, T., Nagano, T., and Hanaoka, K. (2017) Silicon-substituted xanthene dyes and their unique photophysical properties for fluorescent probes. *Chem. Asian J.*, **12**, 1435–1446.

Ishiyama, M., Shiga, M., Sasamoto, K., Mizoguchi, M., and He, P.-g. (1993) A new sulfonated tetrazolium salt that produces a highly water-soluble formazan dye. *Chem. Pharm. Bull.*, **41**, 1118–1122.

Ishiyama, M., Miyazono, Y., Sasamoto, K., Ohkura, Y., and Ueno, K. (1997) A highly water-soluble disulfonated tetrazolium salt as a chromogenic indicator for NADH as well as cell viability. *Talanta*, **44**, 1299–1305.

Itoh, H., Arai, S., Sudhaharan, T., Lee, S.-C., Chang, Y.-T., Ishiwata, S., Suzuki, M., and Lane, E. B. (2016) Direct organelle thermometry with fluorescence lifetime imaging microscopy in single myotubes. *Chem. Commun.*, **52**, 4458–4461.

Kang, R., Nagoshi, T., Kimura, H., Tanaka, T. D., Yoshii, A., Inoue, Y., Morimoto, S., Ogawa, K., Minai, K., Ogawa, T., Kawai, M., and Yoshimura, M. (2021) Possible association between body temperature and B-type natriuretic peptide in patients with cardiovascular diseases. *J. Card. Fail.*, **27**, 75–82.

Kashiwagi, Y., Komukai, K., Kimura, H., Okuyama, T., Maehara, T., Fukushima, K., Kamba, T., Oki, Y., Shirasaki, K., Kubota, T., Miyanaga, S., Nagoshi, T., and Yoshimura, M. (2020) Therapeutic hypothermia after cardiac arrest increases the plasma level of B-type natriuretic peptide. *Sci. Rep.*, **10**, 15545.

Kato, H., Okabe, K., Miyake, M., Hattori, K., Fukaya, T., Tanimoto, K., Beini, S., Mizuguchi, M., Torii, S., Arakawa, S., Ono, M., Saito, Y., Sugiyama, T., Funatsu, T., Sato, K., Shimizu, S., Oyadomari, S., Ichijo, H., Kadowaki, H., and Nishitoh, H. (2020) ER-resident sensor PERK is essential for mitochondrial thermogenesis in brown adipose tissue. *Life Sci. Alliance*, **3**, e201900576.

Ke, G., Wang, C., Ge, Y., Zheng, N., Zhu, Z., and Yang, C. J. (2012) L-DNA molecular beacon: a safe, stable, and accurate intracellular nano-thermometer for temperature sensing in living cells. *J. Am. Chem. Soc.*, **134**, 18908–18911.

Kenwood, B. M., Weaver, J. L., Bajwa, A., Poon, I. K., Byrne, F. L., Murrow, B. A., Calderone, J. A., Huang, L., Divakaruni, A. S., Tomsig, J. L., Okabe, K., Lo, R. H., Coleman, G. C., Columbus, L., Yan, Z., Saucerman, J. J., Smith, J. S., Holmes, J. W., Lynch, K. R., Ravichandran, K. S., Uchiyama, S., Santos, W. L., Rogers, G. W., Okusa, M. D., Bayliss, D. A., and Hoehn, K. L. (2014) Identification of a novel mitochondrial uncoupler that does not depolarize the plasma membrane. *Mol. Metab.*, **3**, 114–123.

Kim, M., Goto, T., Yu, R., Uchida, K., Tominaga, M., Kano, Y., Takahashi, N., and Kawada, T. (2015) Fish oil intake induces UCP1 upregulation in brown and white adipose tissue via the sympathetic nervous system. *Sci. Rep.*, **5**, 18013.

Kimura, H., Nagoshi, T., Yoshii, A., Kashiwagi, Y., Tanaka, Y., Ito, K., Yoshino, T., Tanaka, T. D., and Yoshimura, M. (2017) The thermogenic actions of natriuretic peptide in brown adipocytes: the direct measurement of the intracellular temperature using a fluorescent thermoprobe. *Sci. Rep.*, **7**, 12978.

Kiyonaka, S., Kajimoto, T., Sakaguchi, R., Shinmi, D., Omatsu-Kanbe, M., Matsuura, H., Imamura, H., Yoshizaki, T., Hamachi, I., Morii, T., and Mori, Y. (2013) Genetically encoded fluorescent thermosensors visualize subcellular thermoregulation in living cells. *Nat. Methods*, **10**, 1232–1238.

Klatzo, I. (1967) Neuropathological aspects of brain edema. *J. Neuropathol. Exp. Neurol.*, **26**, 1–14.

Krishan, A. (1975) Rapid flow cytofluorometric analysis of mammalian cell cycle by propidium iodide staining. *J. Cell Biol.*, **66**, 188–193.

Kucsko, G., Maurer, P. C., Yao, N. Y., Kubo, M., Noh, H. J., Lo, P. K., Park, H., and Lukin, M. D. (2013) Nanometre-scale thermometry in a living cell. *Nature*, **500**, 54–58.

Lane, N. (2018) Hot mitochondria? *PLoS Biol.*, **16**, e2005113.

Lewinski, N., Colvin, V., and Drezek, R. (2008) Cytotoxicity of nanoparticles. *Small*, **4**, 26–49.

Li, Y., Ivica, N. A., Dong, T., Papageorgiou, D. P., He, Y., Brown, D. R., Kleyman, M., Hu, G., Chen, W. W., Sullivan, L. B., Rosario, A. D., Hammond, P. T., Heiden, M. G. V., and Chen, J. (2020) MFSD7C switches mitochondrial ATP synthesis to thermogenesis in response to heme. *Nat. Commun.*, **11**, 4837.

Lichtenbelt, W. D. van M., Vanhommerig, J. W., Smulders, N. M., Drossaerts, J. M. A. F. L., Kemerink, G. T., Bouvy, N. D., Schrauwen, P., and Teule, G. J. J. (2009) Cold-activated brown adipose tissue in healthy men. *N. Eng. J. Med.*, **360**, 1500–1508.

Lindquist, S. (1986) The heat-shock response. *Ann. Rev. Biochem.*, **55**, 1151–1191.

Liu, Y., Cheng, D. K., Sonek, G. J., Berns, M. W., Chapman, C. F., and Tromberg, B. J. (1995) Evidence for localized cell heating induced by infrared optical tweezers. *Biophys. J.*, **68**, 2137–2144.

Liu, X., Yamazaki, T., Kwon, H.-Y., Arai, S., and Chang, Y.-T. (2022) A palette of site-specific organelle fluorescent thermometers. *Mater. Today Bio*, **16**, 100405.

Lowell, B. B. and Spiegelman, B. M. (2000) Towards a molecular understanding of adaptive thermogenesis. *Nature*, **404**, 652–660.

Maruyama, H., Kimura, T., Liu, H., Ohtsuki, S., Miyake, Y., Isogai, M., Arai, F., and Honda, A. (2018) Influenza virus replication raises the temperature of cells. *Virus Res.*, **257**, 94–101.

Maruyama, H., Hashim, H., Yanagawa, R., and Arai, F. (2020) Injection of a fluorescent microsensor into a specific cell by laser manipulation and heating with multiple wavelengths of light. *IEEE Int. Conf. Robot. Autom.*, 3437–3442.

Matsushita, M., Yoneshiro, T., Aita, S., Kamiya, T., Kusaba, N., Yamaguchi, K., Takagaki, K., Kameya, T., Sugie, H., and Saito, M. (2015)

Kaempferia parviflora extract increases whole-body energy expenditure in humans: roles of brown adipose tissue. *J. Nutr. Sci. Vitaminol.*, **61**, 79–83.

Meizoso-Huesca, A., Pearce, L., Barclay, C. J., and Launikonis, B. S. (2022) Ca^{2+} leak through ryanodine receptor 1 regulates thermogenesis in resting skeletal muscle. *Proc. Natl. Acad. Sci. USA*, **119**, e2119203119.

Moreau, D., Lefort, C., Burke, R., Leveque, P., and O'Connor, R. P. (2015) Rhodamine B as an optical thermometer in cells focally exposed to infrared laser light or nanosecond pulsed electric fields. *Biomed. Opt. Express*, **6**, 4105–4117.

Mosmann, T. (1983) Rapid colorimetric assay for cellular growth and survival: application to proliferation and cytotoxicity assays. *J. Immunol. Methods*, **65**, 55–63.

Murakami, A., Nagao, K., Sakaguchi, R., Kida, K., Hara, Y., Mori, Y., Okabe, K., Harada, Y., and Umeda, M. (2022) Cell-autonomous control of intracellular temperature by unsaturation of phospholipid acyl chains. *Cell Rep.*, **38**, 110487.

Morgan, A. J. and Jacob, R. (1994) Ionomycin enhances Ca^{2+} influx by stimulating store-regulated cation entry and not by a direct action at the plasma membrane. *Biochem. J.*, **300**, 665–672.

Morimoto-Kobayashi, Y., Ohara, K., Takahashi, C., Kitao, S., Wang, G., Taniguchi, Y., Katayama, M., and Nagai, K. (2015) Matured hop bittering components induce thermogenesis in brown adipose tissue *via* sympathetic nerve activity. *PLoS ONE*, **10**, e0131042.

Nakamura, T. and Matsuoka, I. (1978) Calorimetric studies of heat of respiration of mitochondria. *J. Biochem.*, **84**, 39–46.

Nakamura, K. and Morrison, S. F. (2007) Central efferent pathways mediating skin cooling-evoked sympathetic thermogenesis in brown adipose tissue. *Am. J. Physiol. Regul. Integr. Comp. Physiol.*, **292**, R127–R136.

Nakamura, T., Sakamoto, J., Okabe, K., Taniguchi, A., Yamada, T. G., Nonaka, S., Kamei, Y., Funahashi, A., Tominaga, M., and Hiroi, N. F. (2022) Temperature elevation detection in migrating cells. *Opt. Contin.*, **1**, 1085–1097.

Nakano, M., Arai, Y., Kotera, I., Okabe, K., Kamei, Y., and Nagai, T. (2017) Genetically encoded ratiometric fluorescent thermometer with wide range and rapid response. *PLoS ONE*, **12**, e0172344.

Okabe, K., Inada, N., Gota, C., Harada, Y., Funatsu, T., and Uchiyama, S. (2012) Intracellular temperature mapping with a fluorescent polymeric thermometer and fluorescence lifetime imaging microscopy. *Nat. Commun.*, **3**, 705.

Oo, S. M., Oo, H. K., Takayama, H., Ishii, K.-a., Takeshita, Y., Goto, H., Nakano, Y., Kohno, S., Takahashi, C., Nakamura, H., Saito, Y., Matsushita, M., Okamatsu-Ogura, Y., Saito, M., and Takamura, T. (2022) Selenoprotein P-mediated reductive stress impairs cold-induced thermogenesis in brown fat. *Cell Rep.*, **38**, 110566.

Oparka, M., Walczak, J., Malinska, D., van Oppen, L. M. P. E., Szczepanowska, J., Koopman, W. J. H., and Wieckowski, M. R. (2016) Quantifying ROS levels using $CM-H_2DCFDA$ and HyPer. *Methods*, **109**, 3–11.

Oyama, K., Takabayashi, M., Takei, Y., Arai, S., Takeoka, S., Ishiwata, S., and Suzuki, M. (2012) Walking nanothermometers: spatiotemporal temperature measurement of transported acidic organelles in single living cells. *Lab Chip*, **12**, 1591–1593.

Painting, K. and Kirsop, B. (1990) A quick method for estimating the percentage of viable cells in a yeast population, using methylene blue staining. *World J. Microbiol. Biotechnol.*, **6**, 346–347.

Petrini, G., Tomagra, G., Bernardi, E., Moreva, E., Traina, P., Marcantoni, A., Picollo, F., Kvaková, K., Cígler, P., Degiovanni, I. P., Carabelli, V., and Genovese, M. (2022) Nanodiamond-quantum sensors reveal temperature variation associated to hippocampal neurons firing. *Adv. Sci.*, **9**, 2202014.

Piñol, R., Zeler, J., Brites, C. D. S., Gu, Y., Téllez, P., Neto, A. N. C., da Silva, T. E., Moreno-Loshuertos, R., Fernandez-Silva, P., Gallego, A. I., Martinez-Lostao, L., Martínez, A., Carlos, L. D., and Millán, A. (2020) Real-time intracellular temperature imaging using lanthanide-bearing polymeric micelles. *Nano Lett.*, **20**, 6466–6472.

Power, G. G. (1989) Biology of temperature: the mammalian fetus. *J. Dev. Physiol.*, **12**, 295–304.

Reungpatthanaphong, P., Dechsupa, S., Meesungnoen, J., Loetchutinat, C., and Mankhetkorn, S. (2003) Rhodamine B as a mitochondrial probe for measurement and monitoring of mitochondrial membrane potential in drug-sensitive and -resistant cells. *J. Biochem. Biophys. Methods*, **57**, 1–16.

Rexius-Hall, M. L., Uchiyama, S., Eddington, D., and Rehman, J. (2017) Glycolysis is required for rapid adipocyte thermogenesis induced by cold stress. *FASEB J.*, **31**, Suppl. 886.14.

Riveline, D. and Nurse, P. (2009) 'Injecting' yeast. *Nat. Methods*, **6**, 513–514.

Rosen, E. D. and Spiegelman, B. M. (2006) Adipocytes as regulators of energy balance and glucose homeostasis. *Nature*, **444**, 847–853.

Rosenberg, G. A. (1999) Ischemic brain edema. *Prog. Cardiovasc. Dis.*, **42**, 209–216.

Saito, M., Okamatsu-Ogura, Y., Matsushita, M., Watanabe, K., Yoneshiro, T., Nio-Kobayashi, J., Iwanaga, T., Miyagawa, M., Kameya, T., Nakada, K., Kawai, Y., and Tsujisaki, M. (2009) High incidence of metabolically active brown adipose tissue in healthy adult humans. *Diabetes*, **58**, 1526–1531.

Saito, M., Matsushita, M., Yoneshiro, T., and Okamatsu-Ogura, Y. (2020) Brown adipose tissue, diet-induced thermogenesis, and thermogenic food ingredients: from mice to men. *Front. Endocrinol.*, **11**, 222.

Samanta, A., Paul, B. K., and Guchhait, N. (2012) Photophysics of DNA staining dye propidium iodide encapsulated in bio-mimetic micelle and genomic fish sperm DNA. *J. Photochem. Photobiol. B: Biol.*, **109**, 58–67.

Savchuk, O. A., Silvestre, O. F., Adão, R. M. R., and Nieder, J. B. (2019) GFP fluorescence peak fraction analysis based nanothermometer for the assessment of exothermal mitochondria activity in live cells. *Sci. Rep.*, **9**, 7535.

Senthivinayagam, S., Serbulea, V., Upchurch, C. M., Polanowska-Grabowska, R., Mendu, S. K., Sahu, S., Jayaguru, P., Aylor, K. W., Chordia, M. D., Steinberg, L., Oberholtzer, N., Uchiyama, S., Inada, N., Lorenz, U. M., Harris, T. E., Keller, S. R., Meher, A. K., Kadl, A., Desai, B. N., Kundu, B. K., and Leitinger, N. (2021) Adaptive thermogenesis in brown adipose tissue involves activation of pannexin-1 channels. *Mol. Metab.*, **44**, 101130.

Shalek, A. K., Robinson, J. T., Karp, E. S., Lee, J. S., Ahn, D.-R., Yoon, M.-H., Sutton, A., Jorgolli, M., Gertner, R. S., Gujral, T. S., MacBeath, G., Yang, E. G., and Park, H. (2010) Vertical silicon nanowires as a universal platform for delivering biomolecules into living cells. *Proc. Natl. Acad. Sci. USA*, **107**, 1870–1875.

Shen, L., Xie, T.-R., Yang, R.-Z., Chen, Y., and Kang, J.-S. (2018) Application of a dye-based mitochondrion-thermometry to determine the receptor downstream of prostaglandin E_2 involved in the regulation of hepatocyte metabolism. *Sci. Rep.*, **8**, 13065.

Shen, F., Yang, W., Cui, J., Hou, Y., and Bai, G. (2021) Small-molecule fluorogenic probe for the detection of mitochondrial temperature in vivo. *Anal. Chem.*, **93**, 13417–13420.

Sidossis, L. and Kajimura, S. (2015) Brown and beige fat in humans: thermogenic adipocytes that control energy and glucose homeostasis. *J. Clin. Invest.*, **125**, 478–486.

Slater, T. F., Sawyer, B., and Sträuli, U. (1963) Studies on succinate-tetrazolium reductase systems III. Points of coupling of four different tetrazolium salts. *Biochim. Biophys. Acta*, **77**, 383–393.

Smith, R. E. and Hock, R. J. (1963) Brown fat: thermogenic effector of arousal in hibernators. *Science*, **140**, 199–200.

Soenen, S. J., Rivera-Gil, P., Montenegro, J.-M., Parak, W. J., DeSmedt, S. C., and Brackmans, K. (2011) Cellular toxicity of inorganic nanoparticles: common aspects and guidelines for improved nanotoxicity evaluation. *Nano Today*, **6**, 446–465.

Stewart, M. P., Sharei, A., Ding, X., Sahay, G., Langer, R., and Jensen, K. F. (2016) In vitro and ex vivo strategies for intracellular delivery. *Nature*, **538**, 183–192.

Suzuki, M., Tseeb, V., Oyama, K. and Ishiwata, S. (2007) Microscopic detection of thermogenesis in a single HeLa cell. *Biophys. J.*, **92**, L46–L48.

Takahashi, H., Nagoshi, T., Kimura, H., Tanaka, Y., Yasutake, R., Oi, Y., Yoshii, A., Tanaka, T. D., Kashiwagi, Y., and Yoshimura, M. (2022) Substantial impact of 3-iodothyronamine (T1AM) on the regulations of fluorescent thermoprobe-measured cellular temperature and natriuretic peptide expression in cardiomyocytes. *Sci. Rep.*, **12**, 12740.

Takei, Y., Arai, S., Murata, A., Takabayashi, M., Oyama, K., Ishiwata, S., Takeoka, S., and Suzuki, M. (2014) A nanoparticle-based ratiometric and self-calibrated fluorescent thermometer for single living cells. *ACS Nano*, **8**, 198–206.

Tang, W., Gao, H., Li, J., Wang, X., Zhou, Z., Gai, L., Feng, X. J., Tian, J., Lu, H., and Guo, Z. (2020) A general strategy for the construction of NIR-emitting Si-rhodamines and their application for mitochondrial temperature visualization. *Chem. Asian J.*, **15**, 2724–2730.

Tanimoto, R., Hiraiwa, T., Nakai, Y., Shindo, Y., Oka, K., Hiroi, N., and Funahashi, A. (2016) Detection of temperature difference in neuronal cells. *Sci. Rep.*, **6**, 22071.

Tseeb, V., Suzuki, M., Oyama, K., Iwai, K., and Ishiwata, S. (2009) Highly thermosensitive Ca^{2+} dynamics in a HeLa cell through IP_3 receptors. *HFSP J.*, **3**, 117–123.

Tsien, R. Y. (1998) The green fluorescent protein. *Annu. Rev. Biochem.*, **67**, 509–544.

Tsuboi, Y., Oyama, K., Kobirumaki-Shimozawa, F., Murayama, T., Kurebayashi, N., Tachibana, T., Manome, Y., Kikuchi, E., Noguchi, S., Inoue, T., Inoue, Y. U., Nishino, I., Mori, S., Ishida, R., Kagechika, H., Suzuki, M., Fukuda, N., and Yamazawa, T. (2022) Mice with R2509C-RYR1 mutation exhibit dysfunctional Ca^{2+} dynamics in primary skeletal myocytes. *J. Gen. Physiol.*, **154**, e202213136.

Tsuji, T., Yoshida, S., Yoshida, A., and Uchiyama, S. (2013) Cationic fluorescent polymeric thermometers with the ability to enter yeast and mammalian cells for practical intracellular temperature measurements. *Anal. Chem.*, **85**, 9815–9823.

Tsuji, T., Ikado, K., Koizumi, H., Uchiyama, S., and Kajimoto, K. (2017) Difference in intracellular temperature rise between matured and precursor brown adipocytes in response to uncoupler and β-adrenergic agonist stimuli. *Sci. Rep.*, **7**, 12889.

Uchiyama, S., Kimura, K., Gota, C., Okabe, K., Kawamoto, K., Inada, N., Yoshihara, T., and Tobita, S. (2012) Environment-sensitive fluorophores with benzothiadiazole and benzoselenadiazole structures as candidate components of a fluorescent polymeric thermometer. *Chem. Eur. J.*, **18**, 9552–9563.

Uchiyama, S., Tsuji, T., Ikado, K., Yoshida, A., Kawamoto, K., Hayashi, T., and Inada, N. (2015) A cationic fluorescent polymeric thermometer for the ratiometric sensing of intracellular temperature. *Analyst*, **140**, 4498–4506.

van Engelenburg, S. B. and Palmer, A. E. (2008) Fluorescent biosensors of protein function. *Curr. Opin. Chem. Biol.*, **12**, 60–65.

Vetrone, F., Naccache, R., Zamarrón, A., dela Fuente, A. J., Sanz-Rodríguez, F., Maestro, L. M., Rodriguez, E. M., Jaque, D., Solé, J. G., and Capobianco, J. A. (2010) Temperature sensing using fluorescent nanothermometers. *ACS Nano*, **4**, 3254–3258.

Virtanen, K. A., Lidell, M. E., Orava, J., Heglind, M., Westergren, R., Niemi, T., Taittonen, M., Laine, J., Savisto, N.-J., Enerbäck, S., and Nuutila, P. (2009) Functional brown adipose tissue in healthy adults. *N. Eng. J. Med.*, **360**, 1518–1525.

Vivès, E., Brodin, P., and Lebleu, B. (1997) A truncated HIV-1 Tat protein basic domain rapidly translocates through the plasma membrane and accumulates in the cell nucleus. *J. Biol. Chem.*, **272**, 16010–16017.

Vu, C. Q., Fukushima, S.-i., Wazawa, T., and Nagai, T. (2021) A highly-sensitive genetically encoded temperature indicator exploiting a temperature-responsive elastin-like polypeptide. *Sci. Rep.*, **11**, 16519.

Wang, D., Zhou, M., Huang, H., Ruan, L., Lu, H., Zhang, J., Chen, J., Gao, J., Chai, Z., and Hu, Y. (2019) Gold nanoparticle-based probe for analyzing mitochondrial temperature in living cells. *ACS Appl. Bio Mater.*, **2**, 3178–3182.

Wang, Y., Liang, S., Mei, M., Zhao, Q., She, G., Shi, W., and Mu, L. (2021) Sensitive and stable thermometer based on the long fluorescence lifetime of Au nanoclusters for mitochondria. *Anal. Chem.*, **93**, 15072–15079.

Wang, M., Da, Y., and Tian, Y. (2023) Fluorescent proteins and genetically encoded biosensors. *Chem. Soc. Rev.*, **52**, 1189–1214.

Wei, L., Ma, Y., Shi, X., Wang, Y., Su, X., Yu, C., Xiang, S., Xiao, L., and Chen, B. (2017) Living cell intracellular temperature imaging with biocompatible dye-conjugated carbon dots. *J. Mater. Chem. B*, **5**, 3383–3390.

Wu, T., Chen, X., Gong, Z., Yan, J., Guo, J., Zhang, Y., Li, Y., and Li, B. (2022) Intracellular thermal probing using aggregated fluorescent nanodiamonds. *Adv. Sci.*, **9**, 2103354.

Yamaguchi, S. and Tamao, K. (1998) Silole-containing σ- and π-conjugated compounds. *J. Chem. Soc. Dalton Trans.*, 3693–3702.

Yamamura, M., Hayatsu, H., and Miyamae, T. (1986) Heat production as a cell cycle monitoring parameter. *Biochem. Biophys. Res. Commun.*, **140**, 414–418.

Yamanaka, R., Shindo, Y., Hotta, K., Hiroi, N., and Oka, K. (2020) Cellular thermogenesis compensates environmental temperature fluctuations for maintaining intracellular temperature. *Biochem. Biophys. Res. Commun.*, **533**, 70–76.

Yamazawa, T., Kobayashi, T., Kurebayashi, N., Konishi, M., Noguchi, S., Inoue, T., Inoue, Y. U., Nishino, I., Mori, S., Iinuma, H., Manaka,

N., Kagechika, H., Uryash, A., Adams, J., Lopez, J. R., Liu, X., Diggle, C., Allen, P. D., Kakizawa, S., Ikeda, K., Lin, B., Ikemi, Y., Nunomura, K., Nakagawa, S., Sakurai, T., and Murayama, T. (2021) A novel RyR1-selective inhibitor prevents and rescues sudden death in mouse models of malignant hyperthermia and heat stroke. *Nat. Commun.*, **12**, 4293.

Yang, J.-P. and Huang, L. (1998) Time-dependent maturation of cationic liposome-DNA complex for serum resistance. *Gene Ther.*, **5**, 380–387.

Yang, J.-M., Yang, H., and Lin, L. (2011) Quantum dot nano thermometers reveal heterogeneous local thermogenesis in living cells. *ACS Nano*, **5**, 5067–5071.

Yang, L., Peng, H.-S., Ding, H., You, F.-T., Hou, L.-L., and Teng, F. (2014) Luminescent Ru(bpy)$_3^{2+}$-doped silica nanoparticles for imaging of intracellular temperature. *Microchim. Acta*, **181**, 743–749.

Yang, J., Du, H., Chai, Z., Ling, Z., Li, B. Q., and Mei, X. (2021) Targeted nanoscale 3D thermal imaging of tumor cell surface with functionalized quantum dots. *Small*, **17**, 2102807.

Yoneshiro, T., Aita, S., Kawai, Y., Iwanaga, T., and Saito, M. (2012) Nonpungent capsaicin analogs (capsinoids) increase energy expenditure through the activation of brown adipose tissue in humans. *Am. J. Clin. Nutr.*, **95**, 845–850.

Young, L. J., Schierle, G. S. K., and Kaminski, C. F. (2017) Imaging Aβ(1–42) fibril elongation reveals strongly polarised growth and growth incompetent states. *Phys. Chem. Chem. Phys.*, **19**, 27987–27996.

Zhao, T., Asawa, K., Masuda, T., Honda, A., Kushiro, K., Cabral, H., and Takai, M. (2021) Fluorescent polymeric nanoparticle for ratiometric temperature sensing allows real-time monitoring in influenza virus-infected cells. *J. Colloid Interface Sci.*, **601**, 825–832.

Zhdanov, A. V., Favre, C., O'Flaherty, L., Adam, J., O'Connor, R., Pollard, P. J., and Papkovsky, D. B. (2011) Comparative bioenergetic assessment of transformed cells using a cell energy budget platform. *Integr. Biol.*, **3**, 1135–1142.

Zhou, D., Zhuan, Q., Luo, Y., Liu, H., Meng, L., Du, X., Wu, G., Hou, Y., Li, J., and Fu, X. (2022) Mito-Q promotes porcine oocytes maturation by maintaining mitochondrial thermogenesis via UCP2 downregulation. *Theriogenology*, **187**, 205–214.

Zielonka, J., Joseph, J., Sikora, A., Hardy, M., Ouari, O., Vasquez-Vivar, J., Cheng, G., Lopez, M., and Kalyanaraman, B. (2017) Mitochondria-targeted triphenylphosphonium-based compounds: syntheses, mechanisms of action, and therapeutic and diagnostic applications. *Chem. Rev.*, **117**, 10043–10120.

Zohar, O., Ikeda, M., Shinagawa, H., Inoue, H., Nakamura, H., Elbaum, D., Alkon, D. L., and Yoshioka, T. (1998) Thermal imaging of receptor-activated heat production in single cells. *Biophys. J.*, **74**, 82–89.

4
Cellular Thermometry Based on Non-fluorometric Principles

Before fluorescent molecular thermometers realized intracellular thermometry around the 2010s, rarely we saw attempts to perform cellular temperature measurements using conventional thermometers for tiny-scale thermometry such as thermocouples and thermistors (Lee and Kotov, 2007; Kim et al., 2015). This was because the conventional techniques established for temperature measurements in a small space lacked the sensitivity sufficient for monitoring temperature variations of cells, the spatial resolution for cells on a micro-meter scale, or both. In addition to these reasons from insufficient performance with the conventional thermometers, scientists had unconsciously accepted the assumption that cellular temperature was homogenously equal to the circumstance, which unfortunately spread indifference to cellular thermometry. However, once intracellular thermometry was conducted by fluorescent molecular thermometers to indicate that intracellular temperature is an important physical variable for individuals in biology and medicine, a great deal of effort was made to improve the function of conventional thermometers to fit the cellular thermometry and to establish novel-principle-based techniques enabling measurements of cellular temperature. These non-fluorometric methods can evaluate temperature either inside a cell (i.e., intracellular temperature) or outside a cell (i.e., cell surface temperature). The cellular temperature recorded by non-fluorometric thermometry can be compared with that measured by fluorescent molecular thermometers to confirm the reliability of the methods with each other. Furthermore, new biological insights start to be revealed independently by non-fluorometric thermometry. In this chapter, non-fluorometric cellular thermometry is reviewed with a detailed discussion. The first part (Section 4.1–4.4) discusses the bottom-up techniques based on molecular spectroscopy, whereas the latter (Section 4.5–4.8) is of the top-down manufacturing of electric devices. Table 4.1 is a general summary of the non-fluorometric thermometry introduced in this chapter. The reader is also referred to an excellent review article on non-fluorometric thermometry (Wu et al., 2022b).

Intracellular Thermometry with Fluorescent Molecular Thermometers, First Edition. Seiichi Uchiyama.
© 2024 Wiley-VCH GmbH. Published 2024 by Wiley-VCH GmbH.

Table 4.1 Summary of non-fluorometric cellular thermometry.

Technique	Measurement principle	Temperature resolution	Spatial resolution	Characteristics
Infrared thermometry	Temperature-dependent infrared radiation emitted from an object	0.1–0.5 °C	Cell suspension level	- Monitoring surface temperature of cell suspension
Photoacoustic thermometry	Temperature-dependent pressure rise at an object absorbing light	0.2–0.7 °C	0.23 μm	- Requiring an external photoacoustic imaging contrast agent - Enabling intracellular temperature mapping - Enabling ratiometric imaging by combining fluorescence detection - Temporal resolution: 0.001–0.1 s
Raman thermometry	Temperature-dependent Raman shift of selected molecules	0.2–0.6 °C	1 μm	- Using Raman shifts of water and phenyl isocyanide for intracellular and cell surface temperatures, respectively - Label-free when detecting Raman shift of water molecules - Temporal resolution: 5–90 s
Transmission spectroscopy	Effects of temperature-dependent morphology of thermoresponsive polymer on transmission spectrum of silica microfiber	0.001 °C	Cell cluster level	- Detectable range: 35–42 °C
Thermocouple	Production of electromotive force in a circuit of two dissimilar conductors experiencing thermal gradient (Seebeck effect)	0.054–0.1 °C	Single point in a cell	- Monitoring temperature at a single point inside a cell - Temporal resolution: 32 μs

Technique	Measurement principle	Temperature resolution	Spatial resolution	Characteristics
Resonant thermal sensor	Temperature-dependent changes in resonant frequency of Si	0.001 °C	Single-cell level	- Monitoring surface temperature of a single cell in solution
Bimaterial microcantilever	Different thermal expansion coefficients of two materials	0.02 °C	Single-cell level	- Monitoring surface temperature of a single cell in solution - Microcantilever bends depending on temperature
Thermistor	Temperature-dependent electrical resistance of VO_2	0.0011 °C	Single-cell level	- Monitoring surface temperature of a single cell in solution - High temporal resolution: 0.107 s

4.1
Infrared Thermometry

Infrared thermometry is based on temperature-dependent infrared radiation emitted from all matters and the thermometer in infrared thermometry is called "infrared thermography" or simply "thermography". Infrared thermography has been widely used in the current era. In addition to industrial usages, infrared thermography is especially useful for rapid healthcare screening, such as breast cancer and neonatal disease diagnosis (Figure 4.1a) (Saxena and Willital, 2008; Lahiri et al., 2012). Thermography has recently been used to recognize individuals who are potentially infected with COVID-19, which is caused by severe acute respiratory syndrome coronavirus 2 (SARS-CoV-2), in world airports. Thermography can rapidly monitor the body temperature of travelers (Khaksari et al., 2021). Although infrared thermometry is only capable of measuring the surface temperature of an object, it was used for assessing the thermogenesis of cultured human adipocytes and yeast cells (Paulik et al., 1998). Figure 4.1b shows infrared images of human adipocyte culture solutions 10 minutes after chemical stimulations by rotenone (a mitochondrial electron transport chain complex I inhibitor) and carbonyl cyanide 4-(trifluoromethoxy)phenylhydrazone (FCCP, an uncoupler of mitochondrial respiration). The treatment with rotenone stopped the thermogenesis of human adipocytes, while that with FCCP accelerated their heat production in dose-dependent ways. Comparable temperature variations were observed in yeast cell culture solutions under treatment with rotenone and FCCP. Moreover, the effects of the uncoupling protein-2 (UCP2), a β3-adrenoceptor agonist CL316.243, an anti-diabetic agent troglitazone, and related compounds on the temperature of a cell culture solution were investigated.

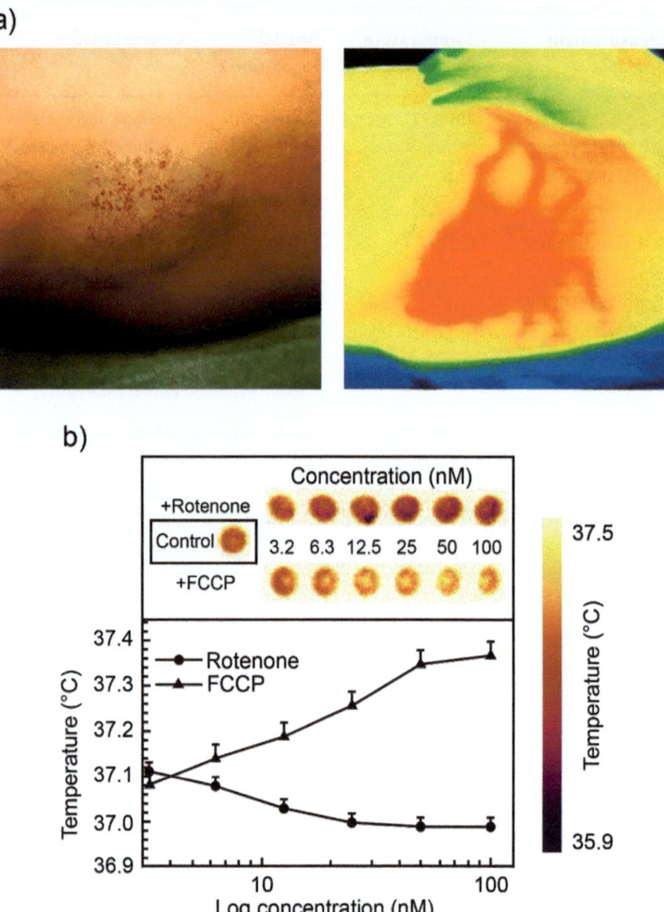

Figure 4.1 Infrared thermometry. (a) Clinical view (left) and infrared image (right) of an 11-year-old female child having angioma on the left buttock. Adapted from Saxena and Willital (2008) *Eur. J. Pediatr.*, **167**, 757–764 / Springer Nature. (b) Infrared images of human adipocyte cultures treated with rotenone and FCCP (top) and dose-dependent effects of rotenone and FCCP on the temperature of human adipocyte cultures (bottom). Paulik et al. (1998) *Pharm. Res.*, **15**, 944–949 / Springer Nature.

4.2
Photoacoustic Thermometry

In 1880, the inventor of the telephone, Alexander Graham Bell, reported the first photoacoustic effects (Bell et al., 1880). Molecules induce pressure waves after absorption of light. Photoacoustic effects are now widely utilized, especially in the medical area, as multi-contrast images of living biological structures from organelles to organs can be created by photoacoustic tomography through the detection of induced pressure waves (Wang and Hu, 2012). The first case of photoacoustic intracellular thermometry took advantage of iron oxide microparticles, which worked as both a photoacoustic imaging contrast agent and a photothermal local heating source (Gao et al., 2013a). The iron oxide microparticles were spontaneously incorporated into HeLa (human epithelial carcinoma) cells during incubation at 37 °C for 24 h, and it was confirmed that a photoacoustic image of the HeLa cell could be acquired with the iron oxide microparticles (Figure 4.2a). After the calibration curve (i.e., the relationship between the photoacoustic signal in live HeLa cells and temperature) was obtained with an external heating pad to increase the temperature of a whole culture media (Figure 4.2b), the intracellular temperature variation during photothermal heating under laser irradiation at 532 nm to iron oxide microparticles was successfully monitored. The temperature resolution of this photoacoustic thermometry of HeLa cells was evaluated to be 0.2 °C.

An improved version of photoacoustic intracellular thermometry was performed with MitoTracker orange (Figure 4.2c) (Gao et al., 2013b). MitoTracker orange stains mitochondria and gives temperature-insensitive fluorescence signals (as a

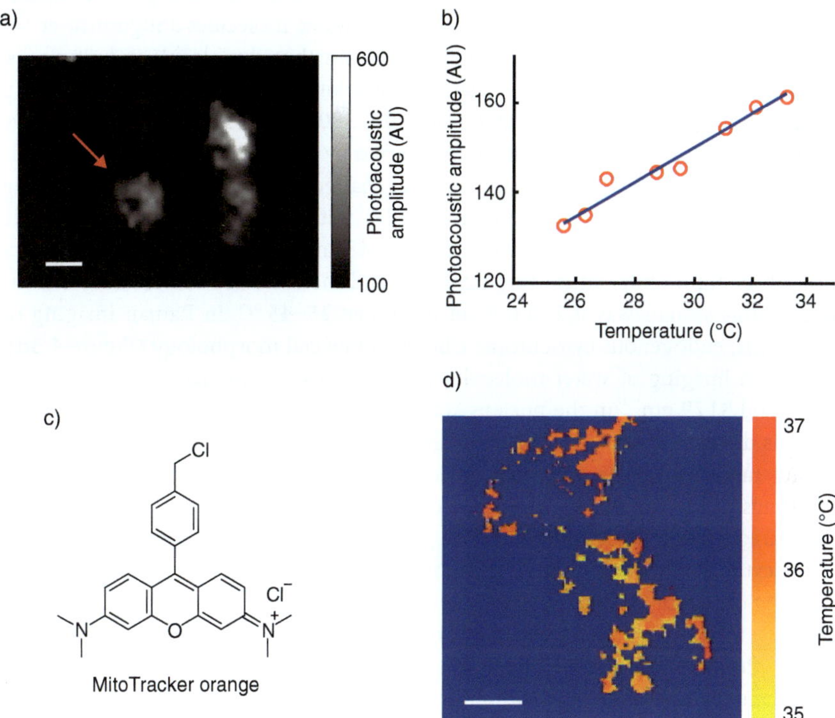

Figure 4.2 Photoacoustic thermometry. (a) Photoacoustic image of HeLa cells loading iron oxide micro-particles. Scale bar: 40 µm. (b) Relationship between photoacoustic signal and temperature of the cell pointed by the red arrow in panel (a). Panels (a,b) are adapted from Gao et al. (2013a) *J. Biomed. Opt.*, **18**, 026003 / SPIE. (c) Chemical structure of MitoTracker Orange. (d) Fluorescence-assisted photoacoustic thermometry for mitochondria of three HeLa cells by MitoTraker orange. (adapted from Gao et al. (2013b) *Appl. Phys. Lett.*, **102**, 193705 / SPIE. The environmental temperature was 36 °C. Scale bar: 10 µm.

reference) and temperature-dependent photoacoustic signals. Thus, ratiometric imaging of mitochondria was realized as a more accurate method for mapping temperature inside cells which were resistant to changes in the concentration of MitoTracker orange and the strength of excitation light. The temperature and spatial resolutions of the established method were 0.7 °C and 0.23 μm, respectively. Figure 4.2d demonstrates mitochondrial temperature mapping of HeLa cells at 36 °C by dividing the photoacoustic signal of MitoTracker orange by its fluorescence signal. It was reported that the standard deviation of mitochondrial temperature distribution of HeLa cells (0.4 °C) at 36 °C was larger than that (0.2 °C) at 27 °C. This may indicate that living cells are more active in heat production and consumption near the body temperature.

4.3
Raman Thermometry

Label-free intracellular temperature mapping was conducted using the temperature-sensitive O–H stretching Raman band of water molecules (Sugimura et al., 2020). Water is the most abundant and widely distributed molecule in living cells. In Raman spectroscopy of Hanks' balanced salt solution (HBSS), the O–H stretching band of water molecules in the region of 3000–3800 cm^{-1} is temperature-sensitive (Figure 4.3a, 4.3b) but not significantly affected by other intracellular environmental factors such as pH. The observed spectral change is likely due to the weakening of hydrogen bonds among water molecules with increasing temperature. The Raman intensity ratio of water molecules at 3548 and 3179 cm^{-1} was a linear function of medium temperature (Figure 4.3c), and the evaluated temperature resolution in this system was 0.2–0.6 °C in the range 25–45 °C. In Raman imaging of HeLa cells, endogenous cytochrome *c* helped draw cell morphology (Figure 4.3d). In Raman imaging of water molecules in HeLa cells, the Raman intensity ratio at 3548 and 3179 cm^{-1} in the nucleus was higher than in the cytoplasm. The reason for this discrepancy could not be exactly determined. However, it might be either strong macromolecular crowding effects that hinder the breaking of hydrogen bondings in the nucleus (Rao et al., 2019) or higher temperature in the nucleus compared to the cytoplasm (see Section 3.4.1.1). Instead, the temperature variation in the cytoplasm with a chemical stimulus could be accurately monitored. In Raman imaging of the O–H stretching bond with a single HeLa cell, the intracellular temperature in the cytoplasm was significantly increased by the treatment with FCCP, while the extracellular temperature was almost unchanged (Figure 4.3e). Simultaneous temperature measurement in both intracellular and extracellular spaces is one of the advantages of Raman cellular thermometry.

The surface temperature measurement was realized by surface-enhanced Raman spectroscopy (SERS) using a temperature-sensitive shift of stretching vibration of a phenyl isocyanide molecule (Hu et al., 2018). Temperature sensing was performed with phenyl isocyanide molecules adsorbed on a gold nanoparticle surface (Figure 4.4a). The increase in temperature induces the orientation change of phenyl isocyanide molecules, leading to the frequency variation of the N≡C stretching vibration of phenyl isocyanide. Figure 4.4b shows temperature-dependent surface-enhanced Raman spectra of phenyl isocyanide in the N≡C stretching vibration region. A linear relationship was found between the peak position and temperature (Figure 4.4c)

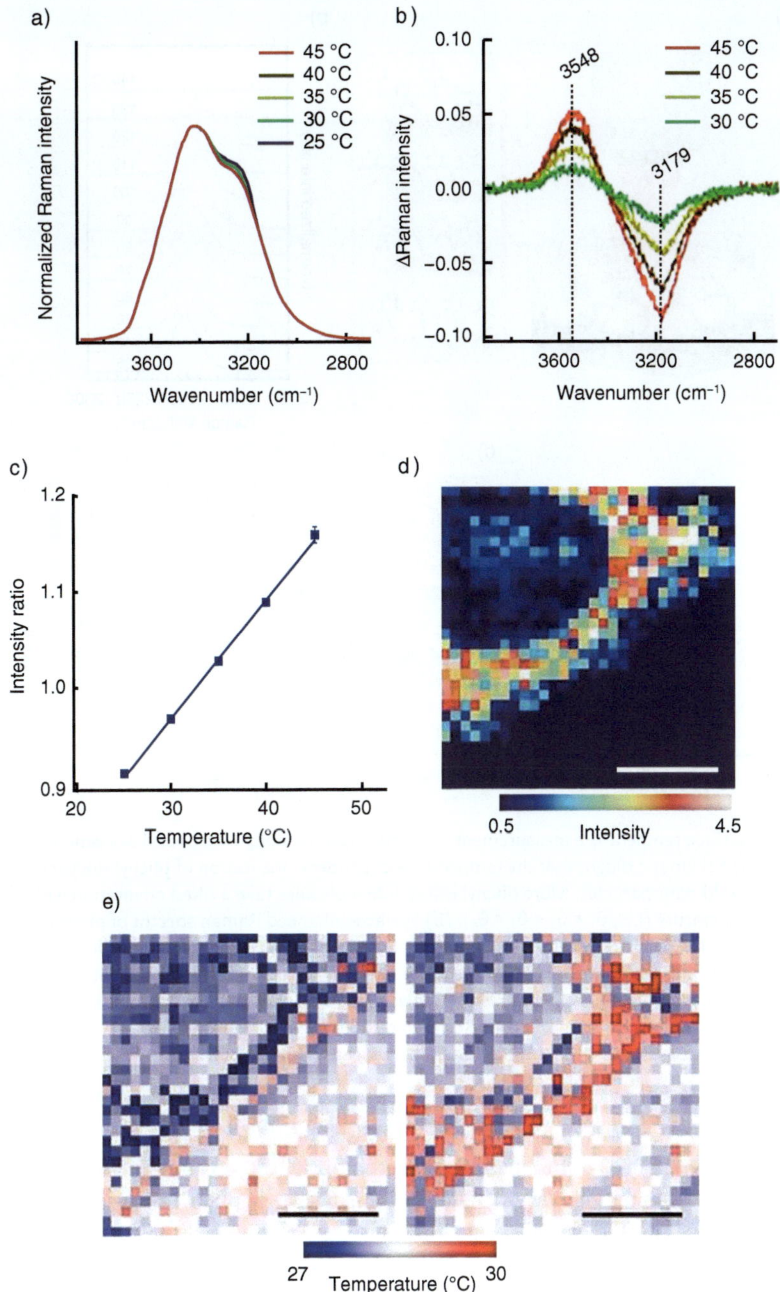

Figure 4.3 Intracellular Raman imaging of the O–H stretching bond of water molecules. (a) Raman spectra of the O–H stretching bond of culture medium (Hanks' balanced salt solution) at 25, 30, 35, 40, and 45 °C. (b) Temperature-dependent variation in Raman spectra compared to 25 °C. (c) Relationship between Raman intensity ratio of the O–H stretching bond at 3548 and 3179 cm^{-1} and temperature. (d) Raman image of a HeLa cell at 750 cm^{-1} (attributed to cytochrome c). Scale bar: 10 µm. (e) Temperature maps of a HeLa cell before and after the treatment with FCCP (left and right, respectively). Temperature of the culture medium: 25 °C. Scale bars: 10 µm. Adapted from Sugimura et al. (2020) *Angew. Chem. Int. Ed.*, **59**, 7755–7760.

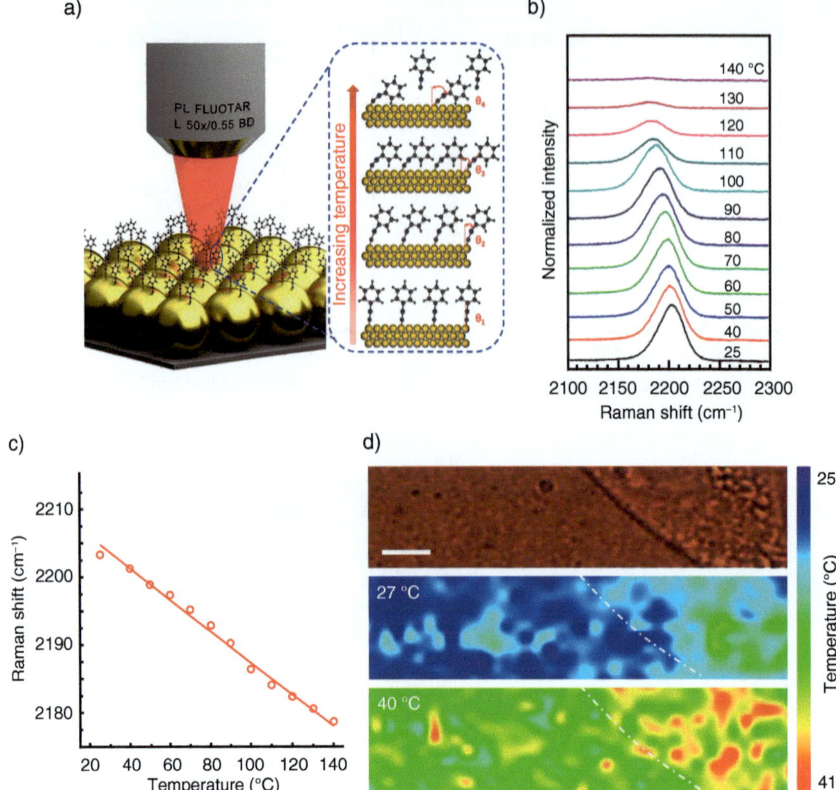

Figure 4.4 Surface temperature measurement by Raman spectroscopy with phenyl isocyanide molecules. (a) Schematic diagram of the temperature-dependent orientation of phenyl isocyanide adsorbed on gold nanoparticles. More phenyl isocyanide molecules take a tilted orientation with increasing temperature (i.e., $\theta_1 < \theta_2 < \theta_3 < \theta_4$). (b) Surface-enhanced Raman spectra of phenyl isocyanide in the N≡C stretching vibration range. (c) Relationship between the Raman shift of phenyl isocyanide and temperature. (d) Temperature mapping of a CaSki cell. White light image (top) and temperature distributions at 27 °C (middle) and 40 °C (bottom). In the panels of temperature distributions, the contour of the cell is indicated by white lines. Scale bar: 10 µm. Hu et al. (2018) *J. Am. Chem. Soc.*, **140**, 13680–13686 /American Chemical Society.

with a sensitivity of 0.232 cm^{-1} per °C (corresponding to 0.43 °C in the temperature resolution). This temperature-dependent Raman shift of phenyl isocyanide was applied for temperature measurements of the surface of living cells with the modification that gold nanoparticles were immobilized with biocompatible PEG-SH (PEGylated thiol) to allow phenyl isocyanide molecules to remain adsorbed on the gold surface in a cell medium. Figure 4.4d demonstrates the cellular thermometry of a living CaSki (human cervix epidermoid carcinoma) cell on the temperature-sensing substrate. The cell contour can be seen in the temperature maps at 27 and 40 °C as the border of two areas with different temperatures, indicating that the cellular temperature is higher than the temperature of the cell medium, probably due to the metabolism. The observed temperature differences between the areas without and with the CaSki cell were 7, 6, 4, and 2 °C at 27, 32, 37, and 40 °C. Hu et al. also tried intracellular thermometry using the gold nanoparticles modified with PEG-SH and phenyl isocyanide. After the introduction of the gold nanoparticles into CaSki cells by incubation and the chemical stimulation with potassium ions, the intracellular temperature rise by 7.6 °C in 10 min was monitored.

Temperature-dependent endogenous production of ergosterol could be utilized for label-free Raman thermometry of fission yeast cells (Figure 4.5a) (Chiu et al., 2013). In contrast to proteins and phospholipids, the fungal sterol ergosterol in yeast cells is significantly diminished when the temperature of the culture medium exceeds 35 °C. Figure 4.5b demonstrates averaged Raman spectra of 30 fission yeast cells cultured at different temperatures for 40 hours. The characteristic Raman bands at 1665, 1440, 1300, and 1003 cm^{-1} were assigned to the C=C stretch of cis-unsaturated lipid chains, the C–H bending, the in-plane CH_2 twisting, and the ring-breathing of phenylalanine residues in proteins, respectively, and were not remarkably temperature sensitive. On the other hand, the Raman band at 1602 cm^{-1} was significantly affected by the environmental temperature during cell culture. The decrease in the Raman band of ergosterol at a higher temperature is due to the repression of enzymes in the initial stage of ergosterol biosynthesis (i.e., the synthesis of mevalonate from acetyl-CoA). Thus, the ratio of the Raman peak area at 1602 cm^{-1} to that at 1440 cm^{-1} is temperature-dependent and dramatically reduced when the medium temperature is higher than 35 °C (Figure 4.5c). Although the method cannot perform real-time cellular thermometry, the environmental temperature at the location where yeast cells grow can be discussed correctly.

4.4
Use of Transmission Spectroscopy

Temperature-dependent transmission spectra of a thermo-responsive poly(N-isopropylacrylamide) (PNIPAM)-coated silica nanofiber in the infrared region were applied for cellular thermometry (Huang et al., 2017). PNIPAM conjugated to a tapered silica nanofiber in water undergoes a morphological change at the lower critical solution temperature (approximately 35 °C in this case) (Figure 4.6a). The abruptly tapered silica microfiber creates an interferometric fringe in the fiber transmission spectrum. The wavelength of the interference fringe is shifted according to the temperature-dependent morphological change of PNIPAM coating (Figure 4.6b). In cellular thermometry, the PNIPAM-coated silica nanofiber was implanted into a cluster of rat breast carcinoma cells (Figure 4.6c). Then, a linear relationship between wavelength shift in the interference fringe and environmental temperature (18.74 nm per °C) was found in the range from 35 to 42 °C (Figure 4.6d). It was reported that the temperature resolution of the system using the PNIPAM-coated tapered silica nanofiber reached 0.001 °C, much better than that of

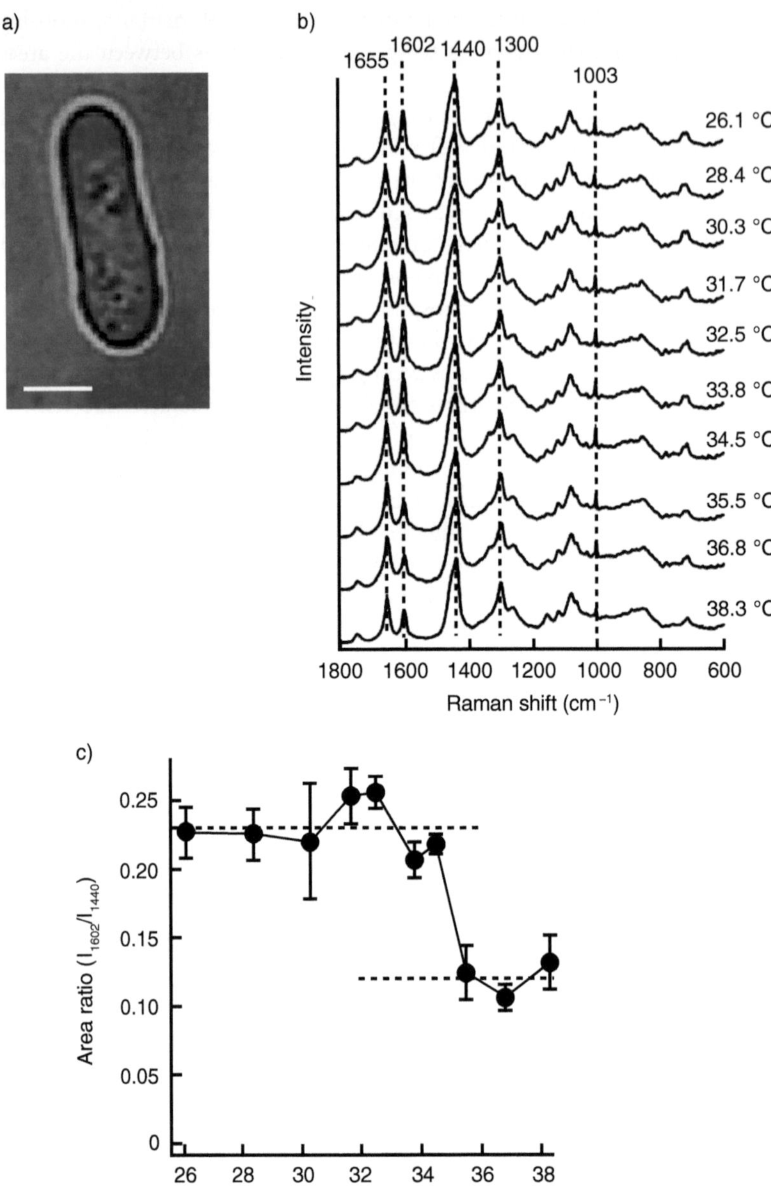

Figure 4.5 Temperature-dependent Raman spectra of fission yeast cells. (a) Bright-field image of a yeast cell. Scale bar: 5 μm. (b) Averaged Raman spectra of yeast cells (cell number n = 30) at different temperatures. (c) Relationship between the Raman peak area ratio (at 1602 and 1440 cm^{-1}) of yeast cells and temperature. The peaks at 1602 and 1440 cm^{-1} are assigned to ergosterol and C–H bending, respectively, in yeast cells. Chiu et al. (2013) *ChemBioChem*, **14**, 1001–1005 / John Wiley & Sons.

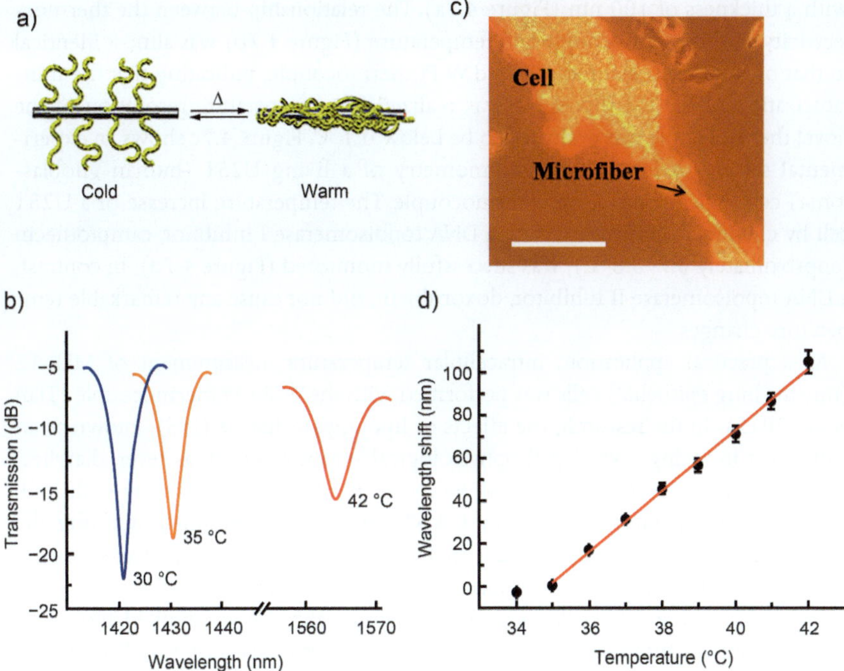

Figure 4.6 Transmission spectra of a silica microfiber coated by thermo-responsive PNIPAM. (a) Heat-induced morphological change of PNIPAM chains attached to the silica microfiber. (b) Transmission spectra of the microfiber at 30 °C (blue), 35 °C (orange), and 42 °C (red). (c) Optical image in the cellular experiment. The microfiber was implanted into a cluster of rat breast carcinoma cells. Scale bar: 100 μm. (d) Temperature-dependent shift in the transmission dip wavelength of the microfiber inside the cells. Adapted from Huang et al. (2017) *ACS Appl. Mater. Interfaces*, **9**, 9024–9028 / American Chemical Society.

fluorescent thermometry utilizing quantum dots with temperature-dependent emission wavelength (see Section 2.16.1).

4.5
Thermocouple

Thermocouples are standard thermoelectric devices in temperature measurements between −270 and 3000 °C. Thermometry by a thermocouple is based on the Seebeck effect: an electromotive force is produced in a circuit consisting of two dissimilar conductors experiencing a thermal gradient (Childs et al., 2000). The first thermocouple to measure intracellular temperature was made from tungsten (W), polyurethane (PU), and platinum (Pt) (Wang et al., 2011). A tungsten substrate with a tip curvature radius smaller than 100 nm was coated by polyurethane except at the tip, followed by sputtering with platinum to make an outermost thin film with a thickness of 100 nm (Figure 4.7a). The relationship between the thermoelectricity of this thermocouple and temperature (Figure 4.7b) was almost identical to that of a conventional macro-sized W-Pt thermocouple, indicating that the miniaturization of the thermocouple was realized. The temperature resolution of the novel thermocouple was evaluated to be below 0.1 °C. Figure 4.7c shows an experimental setting of intracellular thermometry of a living U251 (human glioblastoma) cell with the novel tiny thermocouple. The temperature increase of a U251 cell by chemical stimulation using a DNA topoisomerase I inhibitor, camptothecin (approximately 0.7–0.8 °C), was successfully monitored (Figure 4.7d). In contrast, a DNA topoisomerase II inhibitor, doxorubicin, did not cause any remarkable temperature changes.

As a practical application, intracellular temperature measurement of MLE-12 (murine lung epithelial) cells was performed with the W-Pu-Pt thermocouple (Tian et al., 2015). In the research, the effects of lipopolysaccharide (LPS), known as an endotoxin inducing several pathophysiological symptoms such as fever, diarrhea, and septic shock, on the intracellular temperature of MLE-12 cells were evaluated, and a rapid temperature increase by 0.4 °C was observed within 1 min after the treatment by LPS. In contrast, a heavy metal salt $CoCl_2$ to induce hypoxic stress did not cause significant temperature variation of MLE-12 cells.

In 2022, the tiny W-PU-Pt thermocouple was utilized in an applied biological study on the antitumor effects of attenuated *Salmonella typhimurium* VNP20009 (Wu et al., 2022a). First, a multi-cell thermometry system with a thin-film platinum resistive sensor revealed that VNP20009 increased the temperature of macrophages by 0.2 °C, and phagocytes did not trigger this temperature elevation. Since VNP20009 did not induce a similar temperature variation in B16F10 (mouse melanoma) cells, the temperature increase of macrophages by VNP20009 could be attributed to specific physiological or biochemical responses of the cell. A pyroptosis inducer, nigericin, also induced the warming of macrophages by 0.2 °C, suggesting that VNP20009-mediated cell warming was related to pyroptosis. Complementary single-cell thermometry with the W-Pu-Pt thermocouple confirmed the temperature increase of macrophage by VNP20009 in a single-cell level (Figure 4.7e). In comparing temperature variation by VNP20009 between immortalized bone marrow-derived macrophages (iBMDMs) and $iBMDMp^0$ (iBMDMs that lost mitochondrial biological function), cellular warming was not observed in the latter case. Thus, it can be concluded that mitochondria have a critical role in the temperature increase of a macrophage by the infection with VNP20009. These results are informative for understanding energy metabolism in pyroptosis, the mechanism of temperature variation at the inflammatory site, and immune resistance to microorganisms.

Another microscale thermocouple for intracellular thermometry was fabricated using silicon-based microelectromechanical systems (MEMS) (Rajagopal et al., 2019). Figure 4.8a displays images of a microscale silicon-based Au/Pd thermocouple with a calibration accuracy of 54 mK at 300 ± 10 K. Intracellular temperature measurements with the Au/Pd thermocouple were performed for neurons from the abdominal ganglia in the sea slug *Aplysia californica* (Figure 4.8b), in which the cell health was simultaneously monitored with the real-time membrane potential

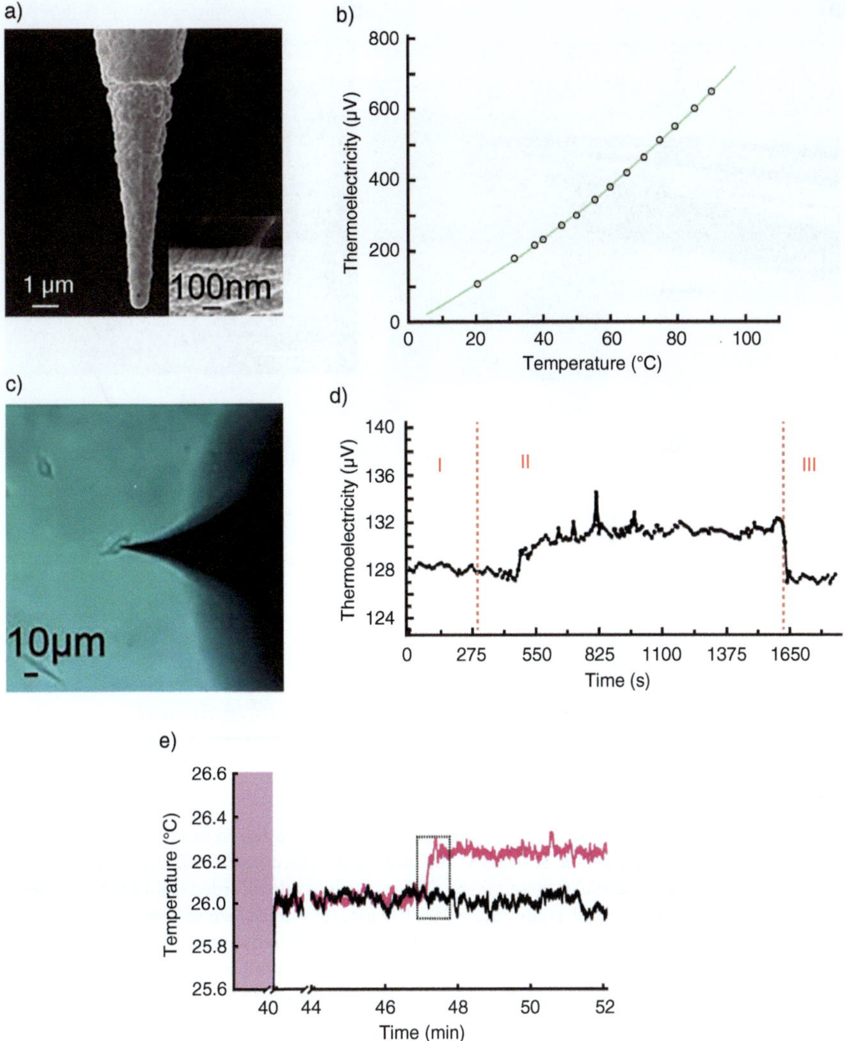

Figure 4.7 Intracellular thermometry by micro-sized W/Pt thermocouple. (a) SEM image of the thermocouple probe composed of tungsten, polyurethane, and platinum. Inset: enlarged SEM image of the probe surface. (b) Relationship between thermoelectric power and temperature. (c) Optical image of a U251 cell with the inserted probe. (d) Temperature variation of a U251 cell by the treatment with camptothecin. The thermocouple was inserted into the cell (I), camptothecin was added (II), and the thermocouple was withdrawn from the cell (III). Panels (a–d) are Adapted from Wang, C. et al. (2011). (e) Temperature variation in a pyroptosis-like macrophage by *Salmonella typhimurium* VNP20009. (adapted from Wu et al. (2022) *J. Am. Chem. Soc.*, **144**, 19396–19409). Intracellular (purple) and extracellular (black) temperature fluctuations with the treatment by VNP20009. VNP20009 was added at 0 s, and the thermocouple probe was inserted into the cell at the dashed square.

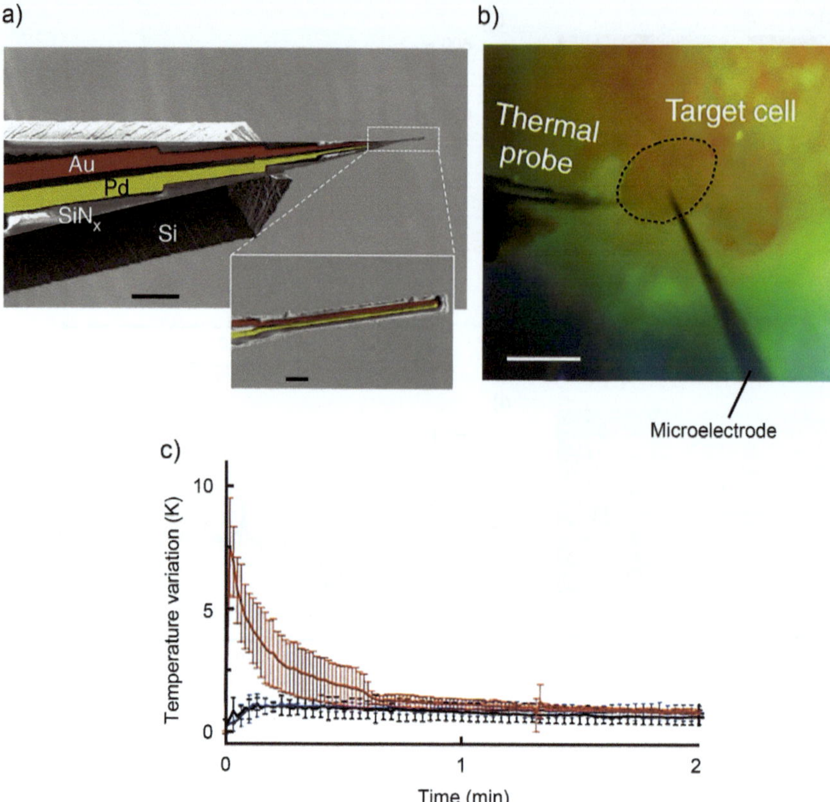

Figure 4.8 Intracellular thermometry by a microscale silicon-based Au/Pd thermocouple. (a) False-colored SEM images of the probe. Scale bars: 100 and 5 μm, respectively. (b) Optical image of *Aplysia* neuron R15 in intracellular thermometry. The microelectrode recorded real-time membrane potential to check cellular health status. Scale bar: 100 μm. (c) Intracellular temperature variation as a response to the mitochondrial proton uncoupling by BAM15 (red) and the corresponding extracellular temperature and the heat of mixing as controls (blue and black, respectively). Adapted from Rajagopal et al. (2019) *Commun. Biol.*, **2**, 279 / Springer Nature / CC BY 4.0.

recorded by a sharp voltage microelectrode. Initially, the off-target responses (including cytotoxicity and rapid depolarization of the plasma membrane) were observed when conventional carbonyl cyanide m-chlorophenylhydrazone (CCCP) was used as an uncoupler to induce thermogenesis in mitochondria. Thus, a newly developed uncoupler, N5,N6-bis(2-fluorophenyl)-[1,2,5]oxadiazolo[3,4-b]pyrazine-5,6-diamine (BAM15) (Kenwood et al., 2014) was adopted to monitor a temperature variation specifically by mitochondrial proton uncoupling. Figure 4.8c represents intracellular responses to BAM15 with a significant temperature spike (7.5 ± 2.0 K) in the first 30 seconds. By fitting the temperature response to BAM15 to a bi-exponential curve, the short-term transient (4.8 ± 3.0 K with a time constant $\tau_1 = 1.0 \pm 0.4$ s) was separated from the long-term one (4.7 ± 0.9 K with a time constant

$\tau_2 = 16.6 \pm 9.2$ s).The time constant of the first component corresponded well to the time scale of proton diffusion during proton transport across a mitochondrial inner membrane. On the other hand, the longer component could be ascribed to a combination of glucose catabolism, the heat of mixing, and response to delayed exposure to BAM15.

The newest microthermocouple composed of carbon (C) and platinum (Pt) is also capable of measuring intracellular temperature with the temperature resolution of 0.08–0.24 °C (Huang et al., 2023).This C/Pt microthermocouple revealed that a human breast cancer MCF7 cell is warmer than a normal human liver LO2 cell by 0.13 °C.

Though not applicable to intracellular temperature measurements, Pd/Cr and Cr/Pt micro-thermocouple arrays are useful in discussing possible temperature variations in living cells (Yang et al., 2017). In a testing device, a Pd/Cr or Cr/ Pt micro-thin-film thermocouple array is fabricated on the glass substrate, and shallow circular wells are confined as "testing zones" by SU-8 photoresist with a lithography process and subsequently polydimethylsiloxane (PDMS) with three-dimensional printing technique (Figure 4.9a). A representative microscopic image of HepG2 (human hepatoblastoma) cells on a Pd/Cr micro-thermocouple array is shown in Figure 4.9b. The stability of the temperature measurement system is 10 mK, which is high enough for monitoring the spontaneous temperature fluctuation of living cells. In most cases, temperature fluctuations of HepG2 cells without additional chemical or mechanical stimulus were less than 60 mK for up to 57 hours. However, in an exceptional case, the maximum cellular temperature variation detected under no stimuli was up to 285 mK (Figure 4.9c).

The last example is a microcapillary coated by a thin line of 60 nm NiCr and 5 nm Ta/60 nm Ni (Type K thermocouple) with a high accuracy of < 0.1 K (Herth et al., 2013). Using this type K thermocouple, temperatures of the *Stenotaphrum variegata* green and white leaves and the *Arabidopsis thaliana* trichome were monitored under light irradiation at 2.4 mmol photons $m^{-2}s^{-1}$. The temperature increase (approximately 1.5 °C) of the *S. variegata* green leaf and the *A. thaliana* trichome was higher than that (0.75 °C) of the *S. variegata* white leaf. The authors insisted that the thermocouple could monitor the temperature at a single-cell level. Although the original paper missed the straightforward evidence that the micro-thermocouple was impaled in a single cell, temperature increment by the same light irradiation in a single cell of an *A. thaliana* trichome was reported to be 0.5 °C.

4.6
Resonant Thermal Sensor

Inomata's group at Tohoku University has developed several unique microfabricated temperature measurement systems by which surface temperature of living cells can be monitored.

The first model is a pico calorimeter based on a heat-detecting Si resonant thermal sensor (termed "resonator") (Inomata et al., 2012, 2016). Figure 4.10a is a

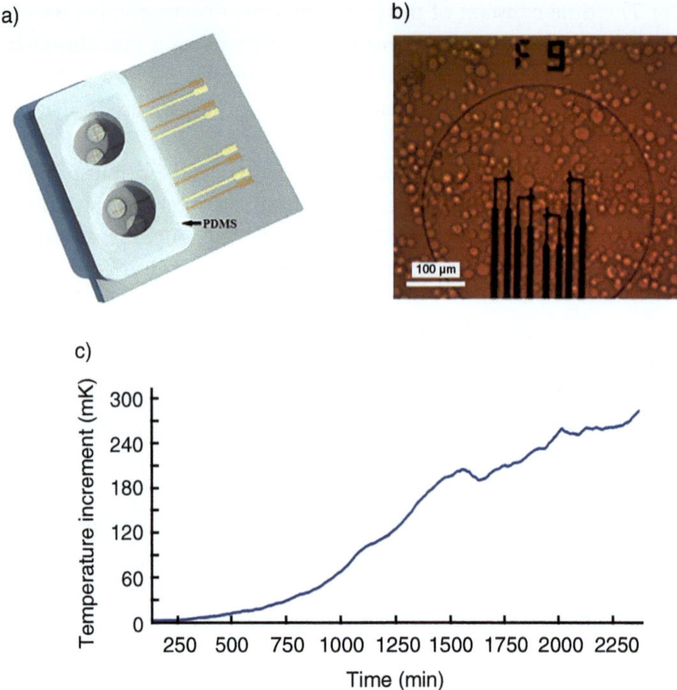

Figure 4.9 Cellular thermometry by a micro-thermocouple array. (a) Schematic diagram of the testing device. Micro-thin-film thermocouple array (yellow and orange) is fabricated on the glass substrate (gray). Polydimethylsiloxane (PDMS) confines cylindroid testing zones. (b) Microscopic image of cultured HepG2 cells on the micro-thermocouple array. (c) Observed spontaneous temperature increment of HepG2 cells. Adapted from Yang et al. (2017) *Sci. Rep.*, **7**, 1721 / Springer Nature / CC BY 4.0.

schematic diagram of the thermal sensing system in which a cantilevered Si sensor is integrated as a "resonator". This resonator enclosed in a vacuum chamber senses heat released from a sample in a microchannel via a Si heat guide. A glass wall separates the vacuum chamber and the microchannel. The change in sample temperature causes a shift in the resonant frequency of the resonator. An actual image of the temperature measurement of a brown fat cell (diameter: approximately 20 µm) is indicated in Figure 4.10b. After introducing cell suspension into the microchannel, the measurement was performed on a single cell occasionally attached to the sample stage at 25 °C. The heat resolution of the fabricated sensor is 5.2 pJ (signal-to-noise = 1), corresponding to 1.3 mK. Interestingly, a non-stimulated brown fat cell spontaneously generated a heat pulse with short intervals (Figure 4.10c upper). The temperature increase of the resonator by 0.1–0.2 °C was observed when the brown fat cell was stimulated with norepinephrine (Figure 4.10c lower), reasoning that the temperature of the brown fat cell itself increased by 0.63 °C under the stimulation (with the calculation from the size of a cell and thermal conductivities of materials). A related work using a silicon pn junction diode in a thermal sensor was conducted in the same research group (Yamada et al., 2016).

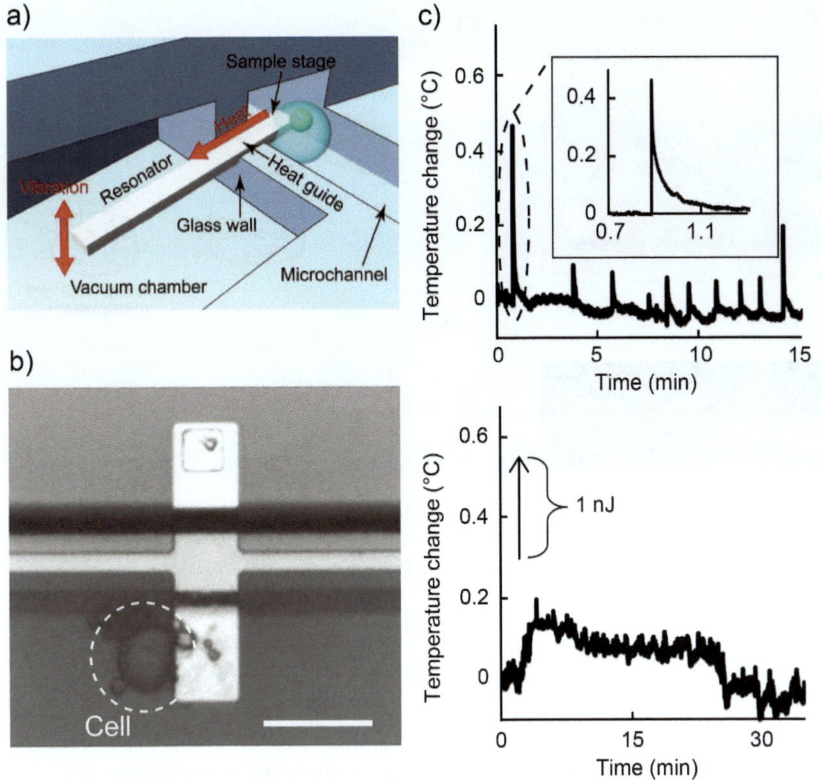

Figure 4.10 Si resonant thermal sensor. (a) Schematic diagram. (b) Image in the measurement of a single brown fat cell. Scale bar: 50 µm. (c) Temperature measurement of a single fat cell without stimulations (top) and with the stimulation by 1 µM norepinephrine solution (bottom). Adapted from Inomata et al. (2012) *Appl. Phys. Lett.*, **100**, 154104 / AIP Publishing.

4.7
Bimaterial Microcantilever

The second microfabricated system for measuring cellular temperature from Inomata's group involves bimaterial microcantilevers (Sato et al., 2014). Figure 4.11a is a scanning electron microscope image of the microcantilevers consisting of a gold layer (0.1 µm thick) and a silicon nitride layer (0.2 µm thick) and the overview of the bimaterial microcantilever and its base. The bimaterial microcantilever bends depending on the environmental temperature, as displayed in Figure 4.11b, and a linear relationship was found between the displacement of the microcantilever and ambient temperature. Figure 4.11c displays a schematic diagram of the experimental setup using a bimaterial microcantilever for detecting cellular temperature variation. The temperature resolution of the microcantilever in cellular thermometry was evaluated to be approximately 20 mK. Temperature variations of murine brown adipocytes under stimulation by norepinephrine were monitored with different sizes of bimaterial microcantilevers, assuming that the temperature

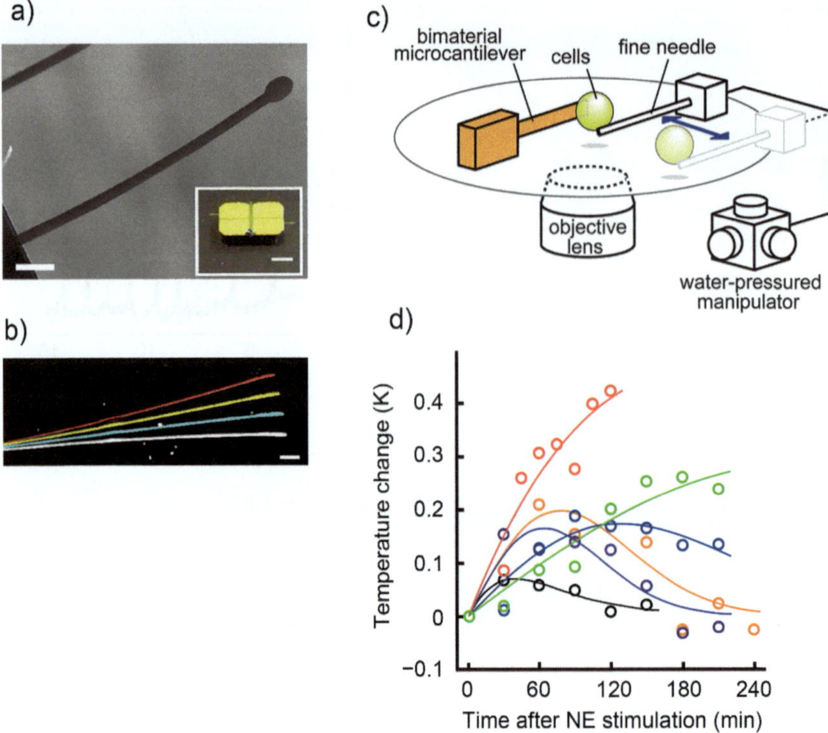

Figure 4.11 Bimaterial microcantilever. (a) SEM image of a bimaterial microcantilever with gold and silicon nitride layers. Scale bar: 100 µm. Inset: an overview of the microcantilever and its base. Scale bar: 1 mm. (b) Merged image of a biomaterial microcantilever at 40 °C (red), 35 °C (yellow), 30 °C (blue), and 25 °C (white). Scale bar: 50 µm. (c) Experimental setup. (d) Temperature measurement of brown adipocytes after the stimulation by norepinephrine (NE). The data were recorded with six microcantilevers of different sizes (in each color). Adapted from Sato et al. (2014) *Biophys. J.*, **106**, 2458–2464 / with permission ELSEVIER.

distribution was uniform inside a cell (Figure 4.11d). The maximum temperature increase of a brown adipocyte by norepinephrine was 0.217 ± 0.120 K, and the temperature rise continued for a few hours.

4.8
Thermistor

The latest microfabricated temperature measurement system developed by the Inomata's group utilized a thermistor (Inomata et al., 2020). "Thermistor" is a portmanteau word from thermally-sensitive resistor, and its functional mechanism is based on the temperature-dependent electrical resistance of a material (Hyde, 1971). The device for thermometry was composed of vanadium dioxide (VO_2) microthermistors on a suspended silicon dioxide (SiO_2) membrane with a PDMS chamber. Figure 4.12a shows a schematic diagram of the thermal sensing device. At the bottom of the PDMS chamber, ten pairs of main and reference sensors were

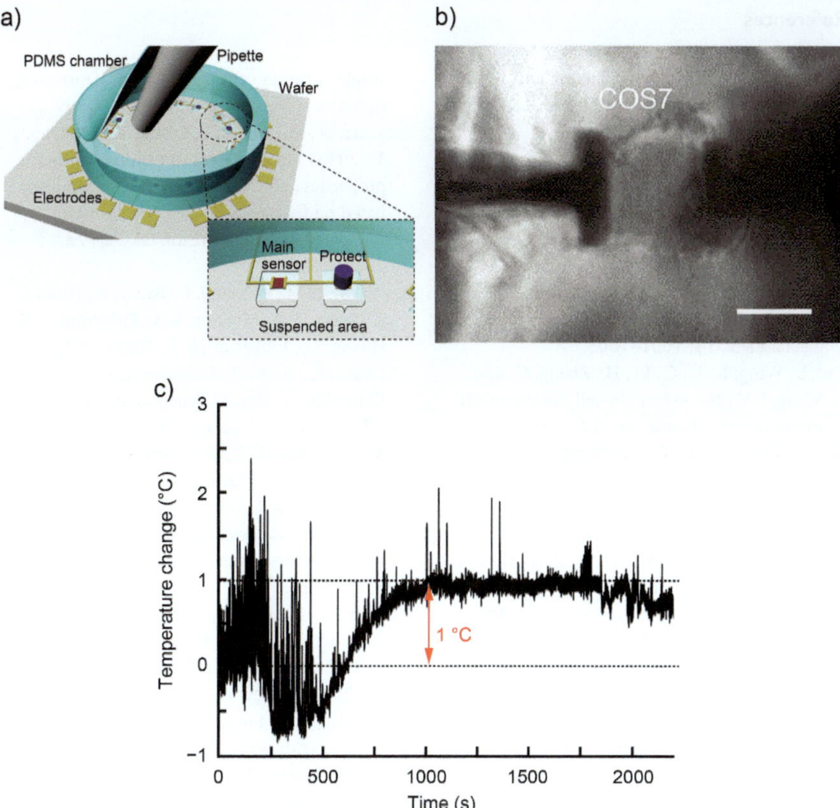

Figure 4.12 Microfabricated thermistor. (a) Schematic diagram of the device. Vanadium dioxide microthermistors (denoted as main sensors) are located on a suspended silicon dioxide membrane in a PDMS chamber. (b) Optical image of a COS7 cell on the sensor. Scale bar: 50 µm. (c) Temperature measurement of COS7 cells with stimulation by FCCP. FCCP solution was added to the chamber at 100 s. Adapted from Inomata et al. (2020) *Sens. Bio-sens. Res.*, **27**, 100309 / with permission from ELSEVIER.

formed on the suspended SiO_2 membrane. Cells were cultured on the bottom surface of the chamber, and a single cell occasionally rested on the main sensor. To prevent contact by cells, the reference sensors were covered with a photosensitive polymer SU-83050. The temperature resolution of this temperature measurement device was evaluated to be 1.1 mK. The microfabricated thermistor monitored the temperature variation of a single COS7 (African green monkey kidney fibroblast) cell under treatment with FCCP (Figure 4.12b). As demonstrated in Figure 4.12c, the temperature of the COS7 cell increased by approximately 1 °C ten minutes after adding FCCP solution to the chamber. Spontaneous temperature fluctuation of COS7 cells was also monitored using the microfabricated thermistor at 25 and 37 °C under non-stimulated conditions. A relatively large amplitude (0.8 °C) was observed at a specific frequency (2 Hz) at 37 °C, while the amplitude was diminished to 0.1–0.2 °C at 25 °C with a frequency of 0–3 Hz. Research progress on intracellular thermometry with the thermistor device can be seen in the latest work (Inomata et al., 2023).

References

Bell, A. G. (1880) On the production and reproduction of sound by light. *Am. J. Sci.*, **s3-20**(118), 305–324.

Childs, P. R. N., Greenwood, J. R., and Long, C. A. (2000) Review of temperature measurement. *Rev. Sci. Instrum.*, **71**, 2959–2978.

Chiu, Y.-F., Huang, C.-K., and Shigeto, S. (2013) In vivo probing of the temperature responses of intracellular biomolecules in yeast cells by label-free Raman microspectroscopy. *ChemBioChem*, **14**, 1001–1005.

Gao, L., Wang, L., Li, C., Ke, H., Zhang, C., and Wang, L. V. (2013a) Single-cell photoacoustic thermometry. *J. Biomed. Opt.*, **18**, 026003.

Gao, L., Zhang, C., Li, C., and Wang, L. V. (2013b) Intracellular temperature mapping with fluorescence-assisted photoacoustic-thermometry. *Appl. Phys. Lett.*, **102**, 193705.

Herth, S., Giesguth, M., Wedel, W., Reiss, G., and Dietz, K.-J. (2013) Thermomicrocapillaries as temperature biosensors in single cells. *Appl. Phys. Lett.*, **102**, 103505.

Hyde, F. J. (1971) *Thermistors*, Ilife books, London.

Hu, S., Liu, B.-J., Feng, J.-M., Zong, C., Lin, K.-Q., Wang, X., Wu, D.-Y., and Ren, B. (2018) Quantifying surface temperature of thermoplasmonic nanostructures. *J. Am. Chem. Soc.*, **140**, 13680–13686.

Huang, Y., Guo, T., Tian, Z., Yu, B., Ding, M., Li, X., and Guan, B.-O. (2017) Nonradiation cellular thermometry based on interfacial thermally induced phase transformation in polymer coating of optical microfiber. *ACS Appl. Mater. Interfaces*, **9**, 9024–9028.

Huang, L.-Q., Ding, X.-L., Pan, X.-T., Li, Z.-Q., Wang, K., and Xia, X.-H. (2023) Single-cell thermometry with a nanothermocouple probe. *Chem. Commun.*, **59**, 876–879.

Inomata, N., Toda, M., Sato, M., Ishijima, A., and Ono, T. (2012) Pico calorimeter for detection of heat produced in an individual brown fat cell. *Appl. Phys. Lett.*, **100**, 154104.

Inomata, N., Toda, M., and Ono, T. (2016) Highly sensitive thermometer using a vacuum-packed Si resonator in a microfluidic chip for the thermal measurement of single cells. *Lab Chip*, **16**, 3597–3603.

Inomata, N., Inaoka, R., Okabe, K., Funatsu, T., and Ono, T. (2020) Short-term temperature change detections and frequency signals in single cultured cells using a microfabricated thermistor. *Sens. Bio-sens. Res.*, **27**, 100309.

Inomata, N., Miyamoto, T., Okabe, K., and Ono, T. (2023) Measurement of cellular thermal properties and their temperature dependence based on frequency spectra via an on-chip-integrated microthermistor. *Lab Chip*, **23**, 2411–2420.

Kenwood, B. M., Weaver, J. L., Bajwa, A., Poon, I. K., Byrne, F. L., Murrow, B. A., Calderone, J. A., Huang, L., Divakaruni, A. S., Tomsig, J. L., Okabe, K., Lo, R. H., Coleman, G. C., Columbus, L., Yan, Z., Saucerman, J. J., Smith, J. S., Holmes, J. W., Lynch, K. R., Ravichandran, K. S., Uchiyama, S., Santos, W. L., Rogers, G. W., Okusa, M. D., Bayliss, D. A., and Hoehn, K. L. (2014) Identification of a novel mitochondrial uncoupler that does not depolarize the plasma membrane. *Mol. Metab.*, **3**, 114–123.

Khaksari, K., Nguyen, T., Hill, B., Quang, T., Perreault, J., Gorti, V., Malpani, R., Blick, E., Cano, T. G., Shadgan, B., and Gandjbakhche, A. H. (2021) Review of the efficacy of infrared thermography for screening infectious diseases with applications to COVID-19. *J. Med. Imaging*, **8**, 010901.

Kim, M. M., Giry, A., Mastiani, M., Rodrigues, G. O., Reis, A., and Mandin, P. (2015) Microscale thermometry: a review. *Microelectron. Eng.*, **148**, 129–142.

Lahiri, B. B., Bagavathiappan, S., Jayakumar, T., and Philip, J. (2012) Medical applications of infrared thermography: a review. *Infrared Phys. Technol.*, **55**, 221–235.

Lee, J. and Kotov, N. A. (2007) Thermometer design at the nanoscale. *Nanotoday*, **2**, 48–51.

Paulik, M. A., Buckholz, R. G., Lancaster, M. E., Dallas, W. S., Hull-Ryde, E. A., Weiel, J. E., and Lenhard, J. M. (1998) Development of infrared imaging to measure thermogenesis in cell culture: thermogenic effects of uncoupling protein-2, troglitazone, and β-adrenoceptor agonists. *Pharm. Res.*, **15**, 944–949.

Rajagopal, M. C., Brown, J. W., Gelda, D., Valavala, K. V., Wang, H., Llano, D. A., Gillette, R., and Sinha, S. (2019) Transient heat release during induced mitochondrial proton uncoupling. *Commun. Biol.*, **2**, 279.

Rao, C., Verma, N. C., and Nandi, C. K. (2019) Unveiling the hydrogen bonding network of intracellular water by fluorescence lifetime imaging microscopy. *J. Phys. Chem. C*, **123**, 2673–2677.

Sato, M. K., Toda, M., Inomata, N., Maruyama, H., Okamatsu-Ogura, Y., Arai, F., Ono, T., Ishijima, A., and Inoue, Y. (2014) Temperature changes in brown adipocytes detected with a biomaterial microcantilever. *Biophys. J.*, **106**, 2458–2464.

Saxena, A. K. and Willital, G. H. (2008) Infrared thermography: experience from a decade of pediatric imaging. *Eur. J. Pediatr.*, **167**, 757–764.

Sugimura, T., Kajimoto, S., and Nakabayashi, T. (2020) Label-free imaging of intracellular temperature by using the O–H stretching Raman band of water. *Angew. Chem. Int. Ed.*, **59**, 7755–7760.

Tian, W., Wang, C., Wang, J., Chen, Q., Sun, J., Li, C., Wang, X., and Gu, N. (2015) A high precision apparatus for intracellular thermal response at single-cell level. *Nanotechnology*, **26**, 355501.

Wang, C., Xu, R., Tian, W., Jiang, X., Cui, Z., Wang, M., Sun, H., Fang, K., and Gu, N. (2011) Determining intracellular temperature at single-cell level by a novel thermocouple method. *Cell Res.*, **21**, 1517–1519.

Wang, L. V. and Hu, S. (2012) Photoacoustic tomography: in vivo imaging from organelles to organs. *Science*, **335**, 1458–1462.

Wu, L., Chen, F., Chang, X., Li, L., Yin, X., Li, C., Wang, F., Li, C., Xu, Q., Zhuang, H., Gu, N., and Hua, Z.-C. (2022a) Combined cellular thermometry reveals that *Salmonella typhimurium* warms macrophages by inducing a pyroptosis-like phenotype. *J. Am. Chem. Soc.*, **144**, 19396–19409.

Wu, W., Song, Z., Chu, Q., Lin, W., Li, X., and Li, X. (2022b) Cell temperature sensing based on non luminescent thermometers – Short review. *Sens. Actuator A Phys.*, **348**, 113990.

Yamada, T., Inomata, N., and Ono, T. (2016) Sensitive thermal microsensor with pn junction for heat measurement of a single cell. *Jpn. J. Appl. Phys.*, **55**, 027001.

Yang, F., Li, G., Yang, J., Wang, Z., Han, D., Zheng, F., and Xu, S. (2017) Measurement of local temperature increments induced by cultured HepG2 cells with micro-thermocouples in a thermally stabilized system. *Sci. Rep.*, **7**, 1721.

5
Reliability Issue in Intracellular Thermometry

Reliability based on accuracy and precision is the most essential feature of analytical tools. Unless the methodology is highly reliable, observed phenomena and numeral data are meaningless. Although researchers developing a new method for intracellular thermometry must consider its reliability for promotion and publication in scientific journals, the users sometimes adopt a preferred approach without special remarks with regard to its reliability. Unfortunately, the reliability of current intracellular thermometry is not perfectly established, and the experimental researchers using intracellular thermometry should understand what affects its reliability. In addition to the characteristics of a fluorescent molecular thermometer and a fluorescence microscope, those of sample living cells impact the reliability of intracellular thermometry. In this chapter, reliability issues in intracellular thermometry will be outlined for acquiring worthwhile experimental data.

5.1
Sensitivity and Temperature Resolution

The primary index of the function of a fluorescent molecular thermometer is the relative sensitivity Sr (% $°C^{-1}$) (Brites et al., 2012) defined as:

$$Sr = \left(\frac{\partial FS}{\partial T}\right)\frac{1}{FS} \times 100 \tag{5.1}$$

where $\partial FS/\partial T$ and FS represent the slope in the diagram of a selected temperature-dependent fluorescence signal (e.g., fluorescence intensity, fluorescence lifetime, and fluorescence intensity ratio at two different wavelengths) and the temperature, and the value of the selected temperature-dependent fluorescence signal, respectively. Table 5.1 indicates the relative sensitivity Sr values of representative fluorescent molecular thermometers used in intracellular thermometry.

Intracellular Thermometry with Fluorescent Molecular Thermometers, First Edition. Seiichi Uchiyama.
© 2024 Wiley-VCH GmbH. Published 2024 by Wiley-VCH GmbH.

A more practical parameter to reflect the reliability of intracellular thermometry with a fluorescent molecular thermometer is the temperature resolution (δT), which is the minimum temperature difference significantly distinguished. In some cases, temperature resolution is termed "temperature uncertainty". The temperature resolution is evaluated by Eqn. (5.2) (Baker et al., 2005):

$$\delta T = \left(\frac{\partial T}{\partial FS} \right) \delta FS \tag{5.2}$$

where $\partial T/\partial FS$ and δFS represent the inverse of the slope in the diagram of a selected temperature-dependent fluorescence signal (e.g., fluorescence intensity, fluorescence lifetime, and fluorescence intensity ratio at two different wavelengths) and the temperature, and the standard deviation of the selected temperature-dependent fluorescence signal, respectively. The $\partial T/\partial FS$ value inversely reflects the sensitivity of a fluorescent molecular thermometer to a temperature variation

Table 5.1 Relative sensitivities (Sr) of representative fluorescent molecular thermometers utilized in intracellular thermometry [Ref. 1].

Fluorescent molecular thermometer	Sr (%/°C)	Temperature-dependent parameter	Temperature range (°C)	Cell line tested[a]	Ref.
Small organic molecules					
ER thermo yellow	3.9	Fluorescence intensity		HeLa	[2]
Mito thermo yellow	2.7	Fluorescence intensity		HeLa	[3]
Quantum dots					
Qtracker655	6.4	Emission intensity ratio	30–40	SH-SY5Y	[4]
Eu^{3+} complex					
EuTTA[b]/rhodamine 101	4.4	Fluorescence intensity ratio	34–40	HeLa	[5]
Fluorescent proteins					
GFP	0.50	Fluorescence polarization anisotropy	24–40	HeLa	[6]
tsGFP1	1.4	Fluorescence intensity ratio	25–45	HeLa	[7]
gTEMP	1.5	Fluorescence intensity ratio	32–40	HeLa	[8]
Fluorescent polymeric thermometers					
Fluorescent polymeric thermometer	4.3	Fluorescence lifetime	28–40	COS7	[9]

(Continued)

Table 5.1 (*Continued*)

Fluorescent molecular thermometer	Sr (%/°C)	Temperature-dependent parameter	Temperature range (°C)	Cell line tested[a]	Ref.
Cationic fluorescent polymeric thermometer	3.2	Fluorescence lifetime	25–40	HeLa	[10]
Ratiometric fluorescent polymeric thermometer	4.1	Fluorescence intensity ratio	25–45	MOLT-4	[11]
Fluorescent nanodiamonds (FNDs)					
FNDs containing nitrogen-vacancy centers	0.006	Zero phonon line	30–60	HMEC-1	[12]
	0.002	Zero-field splitting	27–37	HeLa	[13]

a) HeLa, human epithelial carcinoma cell, SH-SY5Y: human neuroblastoma cell, COS7: African green monkey kidney cell, MOLT-4: human acute lymphoblastic leukaemia cell, HMEC-1: human microvascular endothelial cell.

b) Europium(III) thenoyltrifluoroacetonate.

1 The original table was published in Chung, C. W. and Schierle, G. S. K. (2021) *ChemBioChem*, **22**, 1546–1558 and modified.

2 Arai, S., Lee, S.-C., Zhai, D., Suzuki, M., and Chang, Y. T. (2014) *Sci. Rep.*, **4**, 6701.

3 Arai, S., Suzuki, M., Park, S.-J., Yoo, J. S., Wang, L., Kang, N.-Y., Ha, H.-H., and Chang, Y.-T. (2015) *Chem. Commun.*, **51**, 8044–8047.

4 Tanimoto, R., Hiraiwa, T., Nakai, Y., Shindo, Y., Oka, K., Hiroi, N., and Funahashi, A. (2016) *Sci. Rep.*, **6**, 22071.

5 Takei, Y., Arai, S., Murata, A., Takabayashi, M., Oyama, K., Ishiwata, S., Takeoka, S., and Suzuki, M. (2014) *ACS Nano*, **8**, 198–206.

6 Donner, J. S., Thompson, S. A., Kreuzer, M. P., Baffou, G., and Quidant, R. (2012) *Nano Lett.*, **12**, 2107–2111.

7 Kiyonaka, S., Kajimoto, T., Sakaguchi, R., Shinmi, D., Omatsu-Kanbe, M., Matsuura, H., Imamura, H., Yoshizaki, T., Hamachi, I., Morii, T., and Mori, Y. (2013) *Nat. Methods*, **10**, 1232–1238.

8 Nakano, M., Arai, Y., Kotera, I., Okabe, K., Kamei, Y., and Nagai, T. (2017) *PLoS ONE*, **12**, e0172344.

9 Okabe, K., Inada, N., Gota, C., Harada, Y., Funatsu, T., and Uchiyama, S. (2012) *Nat. Commun.*, **3**, 705.

10 Hayashi, T., Fukuda, N., Uchiyama, S., and Inada, N. (2015) *PLoS ONE*, **10**, e0117677.

11 Uchiyama, S., Tsuji, T., Ikado, K., Yoshida, A., Kawamoto, K., Hayashi, T., and Inada, N. (2015) *Analyst*, **140**, 4498–4506.

12 Wu, T., Chen, X., Gong, Z., Yan, J., Guo, J., Zhang, Y., Li, Y., and Li, B. (2022) *Adv. Sci.*, **9**, 2103354.

13 Sekiguchi, T., Sotoma, S., and Harada, Y. (2018) *Biophys. Physicobiol.*, **15**, 229–234.

(equivalent to Sr described above). At the same time, δFS is affected by the stability of an experimental setup, such as a temperature controller and excitation strength, and the hysteresis in the function of a fluorescent molecular thermometer. For intracellular thermometry to pursue biological events, an experimental setup with a temperature resolution of 0.5 °C, possibly 0.1–0.2 °C, is preferable. In the author's experiences, the stability of a temperature controller is the most influential factor for temperature resolution when the sensitivity of a fluorescent molecular thermometer is high enough (e.g., Sr > 2%). As a technical procedure, it has been reported that the optimization of calibration datasets obtained from Er^{3+}, Yb^{3+}, and Nd^{3+} ions-containing rare-earth nanoparticles and Ag_2S semiconductor nanocrystals can remarkably improve their temperature resolution (Ximendes et al., 2022).

In general, fluorometry is a highly quantitative detection method with low experimental errors. For example, by selecting an adequate concentration range, a linear relationship between the fluorescence intensity of a fluorescent molecule and its concentration is seen using a spectrofluorometer. The case of fluorescein is indicated in Figure 5.1a, 5.1b. Relative standard deviations of fluorescence intensity of fluorescein in repeating measurements with a spectrofluorometer are less than 1% at selected concentrations (0, 25, 50, 75, and 100 nmol L^{-1}). In contrast, intracellular thermometry, including intracellular temperature mapping, is usually performed with a fluorescence microscope. Then, experimental errors due to fluorescence microscopy should also be considered to affect the reliability of intracellular thermometry. Notably, a fluorescence microscope involves several factors that deteriorate a signal-to-noise ratio compared to a spectrofluorometer:

(i)   a larger background noise
(ii) a narrower linearity range of a detector
(iii) significant deviations in an imaging field due to the optical nonuniformity

As indicated in Figure 5.1c, the last factor could include 3–10% deviations where the fluorescence image of a fluorescein solution was acquired with a fluorescence microscope. In the intracellular thermometry with a fluorescence microscope, the nonuniformity of temperature control in an imaging field is also added to the deviations. In our early case, optimization of the experimental setup improved the relative standard deviation of fluorescence lifetime values of a fluorescent polymeric thermometer in an imaging field to approximately 2% (Figure 5.1d, 5.1e) (Okabe et al., 2012). Compared with the relative standard deviation found in Figure 5.1c, the calculation process of a fluorescence lifetime from a fluorescence decay curve might reduce experimental errors due to the optical and thermal nonuniformity in imaging fields of fluorescence microscopy. In performing intracellular thermometry, the validation of its reliability should always be considered to understand whether the observation is biologically meaningful or just an artifact due to experimental errors.

5.2
Functional Independency of Fluorescent Molecular Thermometers

Generally, the functional independence of fluorescent probes is crucial for reliable analysis in live-cell imaging. For a fluorescent molecular thermometer in intracellular thermometry, both its high sensitivity to a temperature variation and insensitivity to fluctuations of other environmental factors, e.g., a pH variation, are ideal. If a fluorescent molecular thermometer also changes its fluorescence signal by variations of other environmental factors, observations will include a false positive temperature variation. Thus, all scientists agree with the functional requirement for fluorescent molecular thermometers: the exclusive sensitivity of fluorescent molecular thermometers to a temperature variation. Nevertheless, its fulfillment with sufficient evidence is complicated by scientific reasons, which sometimes exceeds the reviewers' expectations for scientific papers.

Here, the reliability issue in the viewpoint of the functional independence of fluorescent molecular thermometers is outlined with our actual case of intracellular

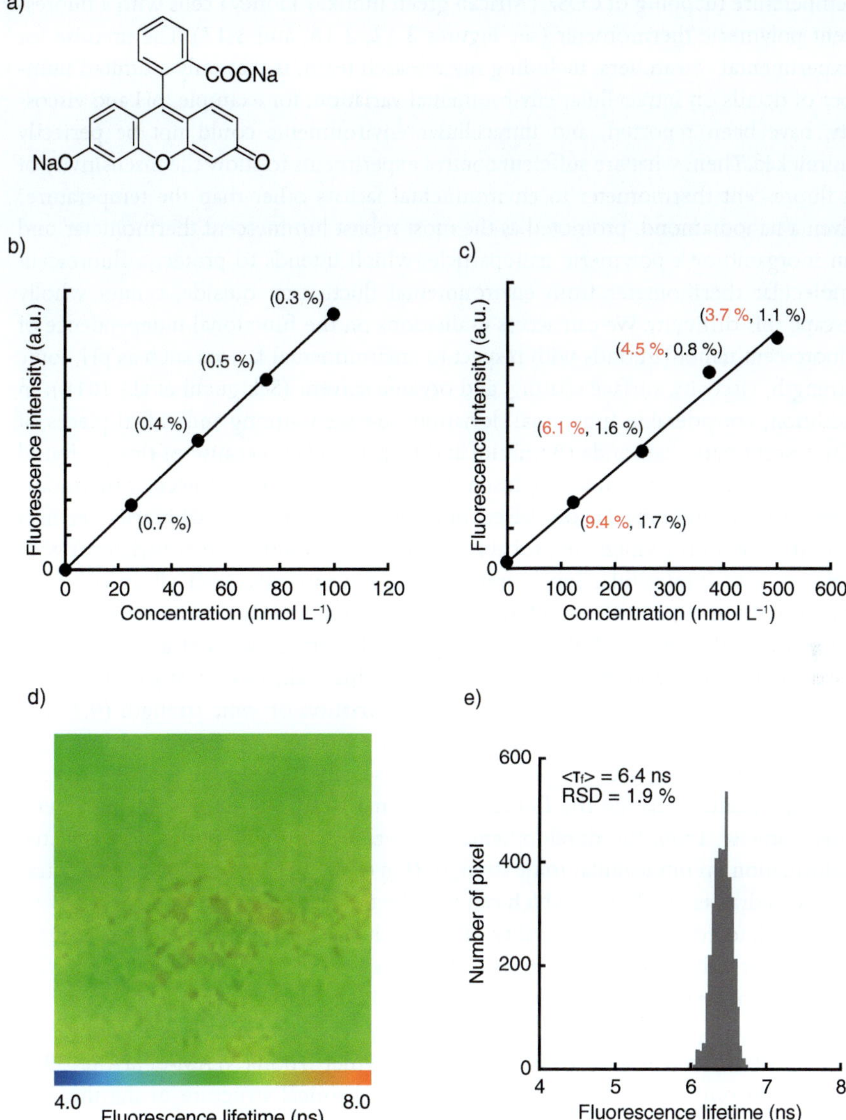

Figure 5.1 Deviations in fluorescence measurements of fluorescein. (a) Chemical structure of fluorescein. (b) Relationship between the fluorescence intensity of fluorescein in a borate buffer (pH 9.18) with the excitation at 490 nm, that was obtained with a spectrofluorometer, and the concentration of fluorescein. The values in parentheses indicate the relative standard deviations in five repeating measurements. (c) Relationship between the fluorescence intensity of fluorescein in a borate buffer (pH 9.18) with excitation at 473 nm, that was obtained with a fluorescence microscope, and the concentration of fluorescein. The red and black values in the parentheses indicate the relative standard deviations of all data in an imaging field and of their average in five repeating measurements, respectively. Panels (b,c) were originally published in Uchiyama et al. (2015, December issue) *Chemistry Today* (in Japanese), 44–47. (d) Fluorescence lifetime image of a fluorescent polymeric thermometer (0.02 w/v%) in a COS7 cell extract at 35 °C. Excitation: 405 nm; emission 500–700 nm. (e) A histogram of fluorescence lifetime in the image in panel (d). <τf> and RSD represent the average and the relative standard deviation of the histogram, respectively. Panels (d,e) Okabe et al. (2012) *Nat. Commun.*, **3**, 705 / Springer Nature.

temperature mapping of COS7 (African green monkey kidney) cells with a fluorescent polymeric thermometer (see Figures 3.12, 3.16, and 3.17). The premise for experimental researchers, including my research team, is that only a limited number of details on intracellular environmental variation, for example, pH and viscosity, have been reported, and intracellular environments could not be perfectly mimicked. Then, what are suffcient control experiments to show the insensitivity of a fluorescent thermometer to environmental factors other than the temperature? Even a nanodiamond, promoted as the most robust luminescent thermometer, and an inorganic or a polymeric nanoparticle, which intends to protect a fluorescent molecular thermometer from environmental fluctuation outside, cannot wholly escape this difficulty. We can access evaluations on the functional independence of fluorescent nanodiamonds with respect to environmental factors such as pH, ionic strength, viscosity, surface coating, and organic solvent (Sekiguchi et al., 2018). In addition, considerable functional deviations are seen among individual pieces of fluorescent nanodiamonds (Bommidi and Pickel, 2021). Because of this profound issue, we could demonstrate the insensitivity of a fluorescent molecular thermometer to environmental factors based only on ranges that we arbitrarily set in a selected medium (which are probably not identical to intracellular variation) with references to intracellular ionic strength, pH, and viscosity. After all, in our early case (Okabe et al., 2012), we referred to the experimental data of intracellular ionic strength (Lodish et al., 2007), pH (Bright et al., 1987; Llopis et al., 1998), and viscosity (Fushimi and Verkman, 1991; Luby-Phelps et al., 1993; Liang et al., 2009), and confirmed that the possible intracellular variation of ionic strength (0.1–0.2, Figure 5.2a), pH (6–10, Figure 5.2b), protein concentration (bovine serum albumin, BSA: 0–3.2 mg mL^{-1}, Figure 5.2c), and viscosity (0.9–1.5 cP, Figure 5.2d) did not significantly change the fluorescence signal from our fluorescent polymeric thermometer. Later, the developments of novel sensing techniques have updated information on intracellular ionic strength (Liu et al., 2017) and protein concentration (Mudrak et al., 2018), which can now be referred to.

Instead, to confirm the reliability of intracellular thermometry, our case using a fluorescent polymeric thermometer collected several indirect supports. (i) Our fluorescent polymeric thermometer lacks pH-sensitive structures (e.g., amino and carboxylic groups), and therefore, pH-insensitivity was expected. (ii) We synthesized a control thermo-insensitive fluorescent polymer (poly(NEAM-*co*-SPA-*co*-DBD-AA)) with only a minor modification on the chemical structure of the thermo-responsive units in order to show that the fluorescence responses of our fluorescent

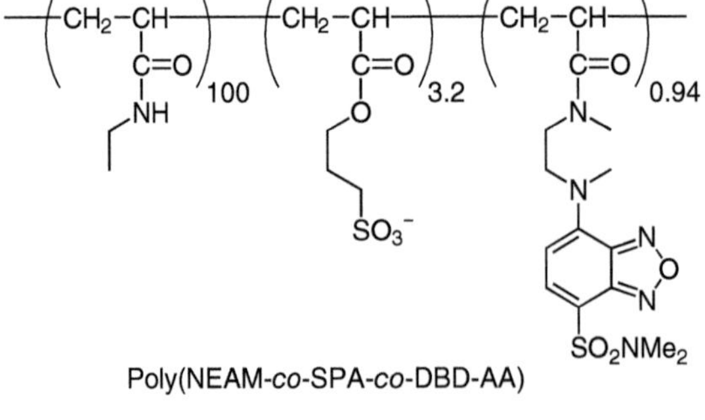

Poly(NEAM-*co*-SPA-*co*-DBD-AA)

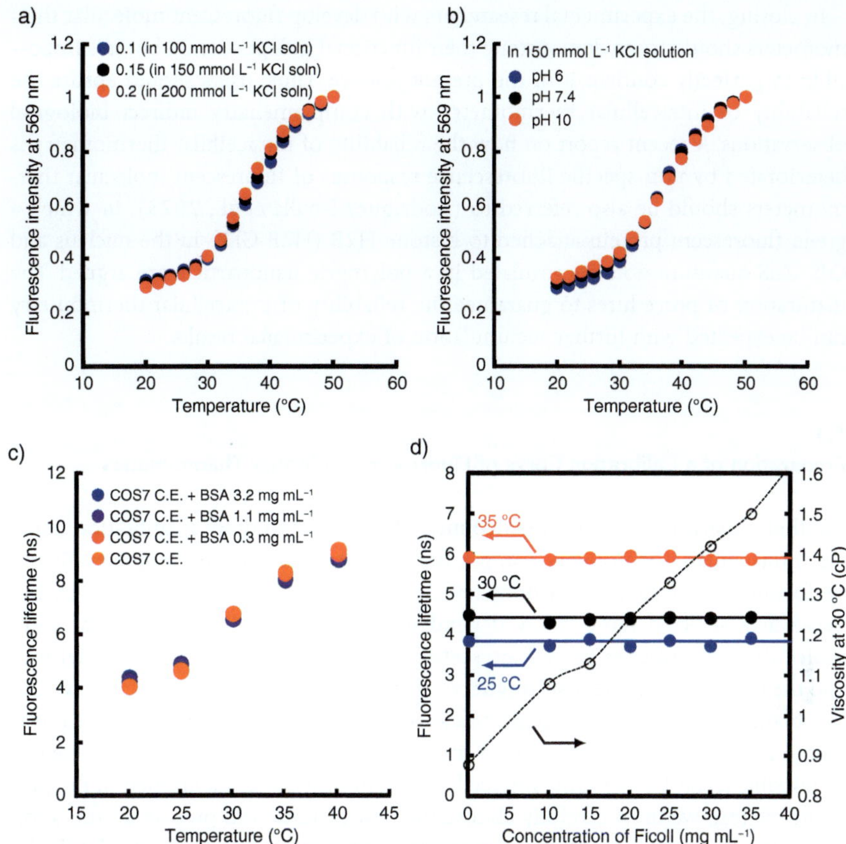

Figure 5.2 Functional independence of a fluorescent polymeric thermometer (FPT). (a,b) Effects of ionic strength (a) and pH (b) on the temperature-dependent fluorescence intensity of FPT (0.001 w/v%). Sample solutions were excited at 456 nm. Fluorescence intensity was normalized at 50 °C. HCl and KOH were used to adjust the pH. (c) Effects of surrounding proteins on the temperature-dependent fluorescence lifetime of FPT (0.001 w/v%, excitation: 405 nm, emission: 560 nm) in a COS7 cell extract (C.E.). Bovine serum albumin (BSA) was used as a model protein. (d) Effects of environmental viscosity on the temperature-dependent fluorescence lifetime of FPT (0.002 w/v%, excitation: 456 nm, emission: 580 nm). Ficoll was added to the KCl solution (150 mmol L^{-1}) to vary the viscosity of the solution. Adapted from Okabe et al. (2012) *Nat. Commun.*, **3**, 705.

polymeric thermometer were not due to the interaction between the fluorophore (i.e., DBD-AA unit) and cellular components. (iii) The observed temperature difference between the nucleus and the cytoplasm (Figure 3.12) was cell cycle-dependent (Figure 3.19a, 3.19b), indicating that the evaluated significant temperature gap between the nucleus and the cytoplasm was not due to the interactions between the fluorescent polymeric thermometer and the constantly abundant intracellular molecules in the nucleus or the cytoplasm. (iv) The pattern of intracellular temperature distribution was identical among different cell lines, i.e., COS7 and HeLa (human epithelial carcinoma) cells. All these facts increased the reliability of the original results on COS7 cells. Finally, it should be mentioned that the most vital support in intracellular thermometry is that comparable phenomena are observed with other thermometers based on different functional mechanisms (see Section 5.4.2).

In closing, the experimental researchers who develop fluorescent molecular thermometers should try to demonstrate their functional independence, but it is impossible to perfectly confirm it in the present science. Thus, users should ensure the reliability of intracellular thermometry with complementary indirect biological observations. A recent report on how the reliability of intracellular thermometry is deteriorated by non-specific fluorescence responses of fluorescent molecular thermometers should be also referred to (Rodríguez-Sevilla et al., 2023), in which a green fluorescent protein attached to histone H2B (H2B-GFP) in the nucleus and CdS/ZnS quantum dots encapsulated in a polymeric nanoparticle are argued. The maturation of procedures to guarantee the reliability of intracellular thermometry can be expected with further accumulation of experimental results.

5.3
Preparation of a Calibration Curve of Fluorescent Molecular Thermometers

Another experimental issue we encountered was how to draw a calibration curve for intracellular thermometry using a fluorescent molecular thermometer. Before performing intracellular thermometry, it is necessary to obtain the relationship between a selected temperature-dependent fluorescence signal of a fluorescent molecular thermometer (e.g., fluorescence intensity ratio or fluorescence lifetime) and the temperature. Scientists might consider that an ideal calibration curve would be obtained under the same conditions as the proposed measurements, i.e., equivalent to intracellular environments for intracellular thermometry. However, since it is impossible to perfectly mimic intracellular environments, as mentioned in the previous section, we must carefully choose the best condition to prepare a calibration curve for accurate intracellular temperature measurements. Again, it should be emphasized that this issue is not exclusive to intracellular thermometry using a fluorescent molecular thermometer but must be considered in all live cell imaging of chemical species (e.g., proton and Ca^{2+} ion) and other physical parameters (e.g., viscosity) by fluorescent sensors. Nevertheless, valuable discussions and suggestions are limited, so we have yet to reach a general conclusion.

Here, two examples are described to show how difficult it is to draw a proper calibration curve using artificial solutions for intracellular environmental measurements. First, is the case using a colorimetric pH sensor 6-carboxyfluorescein (6-CF, Figure 5.3a). Figure 5.3b displays pH-dependent absorption spectra of 6-CF in buffer solutions (Thomas et al., 1979). The calibration curve of 6-CF for pH measurements is obtained with the absorbance ratio at 490 and 465 nm as a pH-dependent detection parameter. As shown in Figure 5.3c, a significant difference can be seen between a series of buffers (50 mmol L^{-1}) containing 110 mmol L^{-1} NaCl, 5 mmol L^{-1} KCl, and 1 mmol L^{-1} $MgCl_2$ and ascites cells, in the relationship between the absorbance ratio at 490 and 465 nm and an environmental pH. The second case was reported as excluded-volume effects of a HeLa cell (Gnutt et al., 2015). In this work, polyethylene glycol (PEG, 10 kDa) was functionalized by end-group labeling using a FRET (Förster resonance energy transfer) pair, Atto488 and Atto565, and a macromolecular crowding effect inside the living cell was evaluated by the FRET efficiency between Atto488 and Atto565. Surprisingly, the FRET efficiencies of PEG

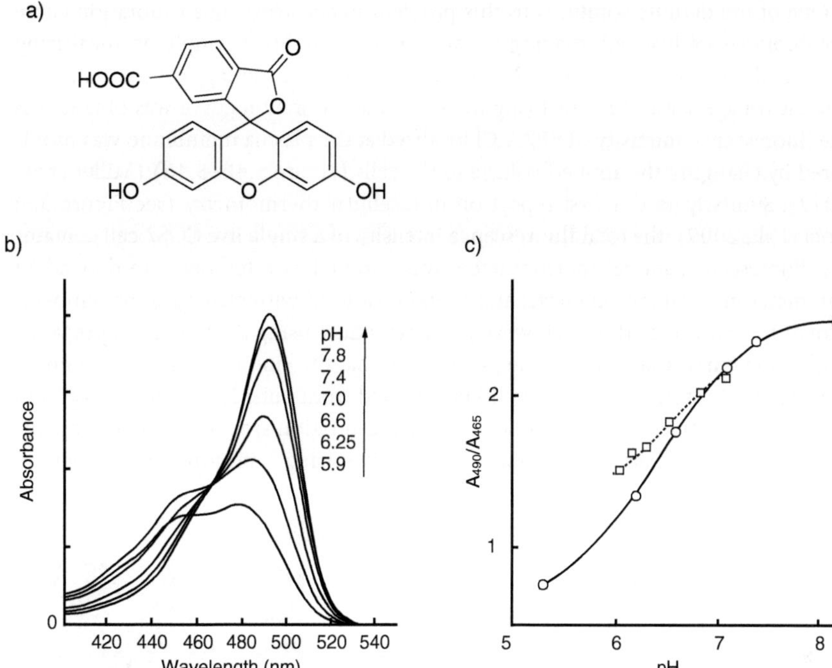

Figure 5.3 Environment-dependent responses of a colorimetric pH sensor 6-carboxyfluorescein (6-CF). (a) Chemical structure of 6-CF. (b) pH-dependent absorption spectra of 6-CF (9 μmol L^{-1}) in 100 mmol L^{-1} Mops (3-morpholinopropane-1-sulfonic acid) buffers. (c) Calibration curves of 6-CF in buffers (circle) and Ehrlich ascites tumor cells (square) at 20 °C. For buffers, tricine (pH 8.2), Hepes (2-[4-(2-hydroxyethyl)piperazin-1-yl]ethanesulfonic acid, pH 7.4), Mops (pH 7.1), Pipes (2,2′-(piperazine-1,4-diyl)di(ethane-1-sulfonic acid), pH 6.6), and Mes (2-(*N*-morpholino)ethanesulfonic acid, pH 6.2 and 5.3) were used. For cells, nigericin (10 μg mL^{-1}) in a Mes buffer (pH 6.54 and below) or a Mops buffer (above pH 6.54) was used. Thomas et al. (1979) *Biochemistry*, **18**, 2210–2218 / American Chemical Society.

labeled with Atto488 and Atto565 in the nucleus of a HeLa cell and in the HeLa cell under the hypertonic shock (with 500 mmol L^{-1} NaCl for 20 seconds) were 36 and ~80%, which were out of the range of the calibration curve using a synthetic polymer of sucrose, Ficoll 70, as a macromolecular crowding agent (c.f., the FRET efficiency in an aqueous solution varies approximately from 41 to 62% with increasing Ficoll 70 from 0 to 400 mg mL^{-1}). These results clearly demonstrate the difficulty in preparing a calibration curve for monitoring intracellular environments.

One of the definite solutions to this problem in constructing a calibration curve for fluorescence live-cell imaging is using living cells. For example, in measuring the membrane potential of HEK293 (human embryonic kidney) cells with a fluorescent voltage sensor VF2.4.Cl (Figure 5.4a), the calibration curve was obtained as the fluorescence intensity of VF2.4.Cl localized at the plasma membrane was monitored by changing the applied voltage to the cells (Figure 5.4b–5.4d) (Miller et al., 2012). Similarly, in our first report on intracellular thermometry (see Figure 3.3; Gota et al., 2009), the total fluorescence intensity of a single live COS7 cell containing fluorescent nanogel thermometers was adopted as a temperature-dependent parameter in a calibration curve, and it was measured with changing the temperature of the culture medium. However, this procedure using living cells to prepare a calibration curve was built on the assumption that the (intra)cellular environment available to fluorescent sensors could be adjusted from outside. Besides voltage and temperature, intracellular pH and Ca^{2+} ion concentration are the only environmental ionic levels that are controllable from outside using a protonophore nigericin

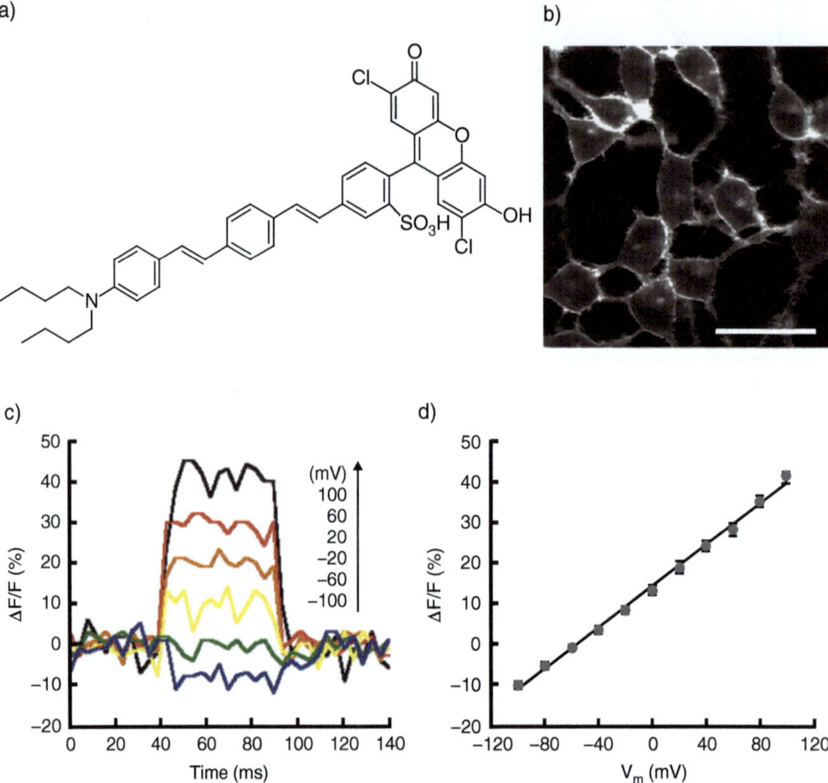

Figure 5.4 Preparation of a calibration curve for measuring cellular membrane potential with a fluorescent voltage sensor VF2.4.Cl. (a) Chemical structure of VF2.4.Cl. (b) Confocal image of VF2.4.Cl localized at the plasma membrane of HEK293 cells. Scale bar: 20 μm. (c) Variation in fluorescence intensity (ΔF/F) of VF2.4.Cl during depolarizing steps to +100 mV or hyperpolarizing steps to −100 mV from a holding potential at −60 mV (black: 100 mV; red: 60 mV; orange: 20 mV; yellow: −20 mV; green: −60 mV; blue: −100 mV). (d) Relationship between the variation in ΔF/F of VF2.4.Cl (from panel (c)) and the membrane potential of HEK293 cells (V_m). Miller et al., (2012) *Proc. Natl. Acad. Sci. USA*, **109**, 2114–2119 / National Academy of Sciences.

(Grillo-Hill et al., 2014) and Ca^{2+} ionophores (e.g., ionomycin and 4-bromo A-23187) (Takahashi et al., 1999), respectively. In a recent report on a phosphorescent probe for oxygen, a calibration curve could be prepared using live AML12 (murine hepatocyte) cells by changing the partial pressure of oxygen in the atmosphere (Mizukami et al., 2022). However, these cases are exceptional.

Furthermore, the situation for fluorescent live-cell imaging of a heterogeneous environment is more complicated. What should be done to create a calibration curve when heterogeneous environments are expected within a live cell, as shown in our first intracellular temperature mapping (Okabe et al., 2012)? Which values among the heterogeneous fluorescence signals should be used? If the average value was reasonable, then how do you calculate one? After all, we must choose a compromise in preparing a calibration curve for fluorescent live-cell imaging to visualize a heterogeneous environment: a fluorescent sensor is dissolved in the solution prepared as close to the intracellular environment as possible. In our case of intracellular temperature mapping (see Figure 3.12; Okabe et al., 2012), the fluorescence signal of a fluorescent polymeric thermometer (i.e., fluorescence lifetime) in a cell extract solution was used in drawing a calibration curve, which enabled a rational analysis of intracellular temperature distribution. In the same way, mixtures of methanol and glycerol with a wide range of viscosity were used as media for a fluorescent molecular rotor in preparing a calibration curve for intracellular viscosity monitoring of SK-OV-3 (human ovarian cancer) cells (Figure 5.5) (Kuimova et al., 2008). Such precedents could suggest the best (scientifically most reliable) calibration curve. At present, we can indicate only a *prima facie* calibration curve made with the environments of live cells or with even simpler solutions. We need to consider other possibilities that are better than the present protocols.

5.4
Objection to Endogenous Thermogenesis

5.4.1
Beginning

In 2014, a French group led by Professor Baffou released an intense critique of the observation of temperature variations in living cells by a sub-degree and more as a *Commentary* in *Nature Methods* (Baffou et al., 2014). This critique was directly triggered by the scientific paper on intracellular thermometry with genetically encoded fluorescent protein thermometers published earlier in the same journal (Kiyonaka et al., 2013). Furthermore, the result of intracellular thermometry with fluorescent polymeric thermometers in our group (Gota et al., 2009; Okabe et al., 2012) was another target in the critique.

Baffou's criticism is based on a viewpoint of the theoretical heat conduction equation established by Fourier in the 1820s (Fourier, 1878), and the summary is as follows. For a liquid medium of thermal conductivity κ and volumetric heat capacity c, the temperature distribution $T(\mathbf{r},t)$ is governed by the heat diffusion equation

$$c\partial_t T(\mathbf{r},t) - \kappa \nabla^2 T(\mathbf{r},t) = p(\mathbf{r},t) \tag{5.3}$$

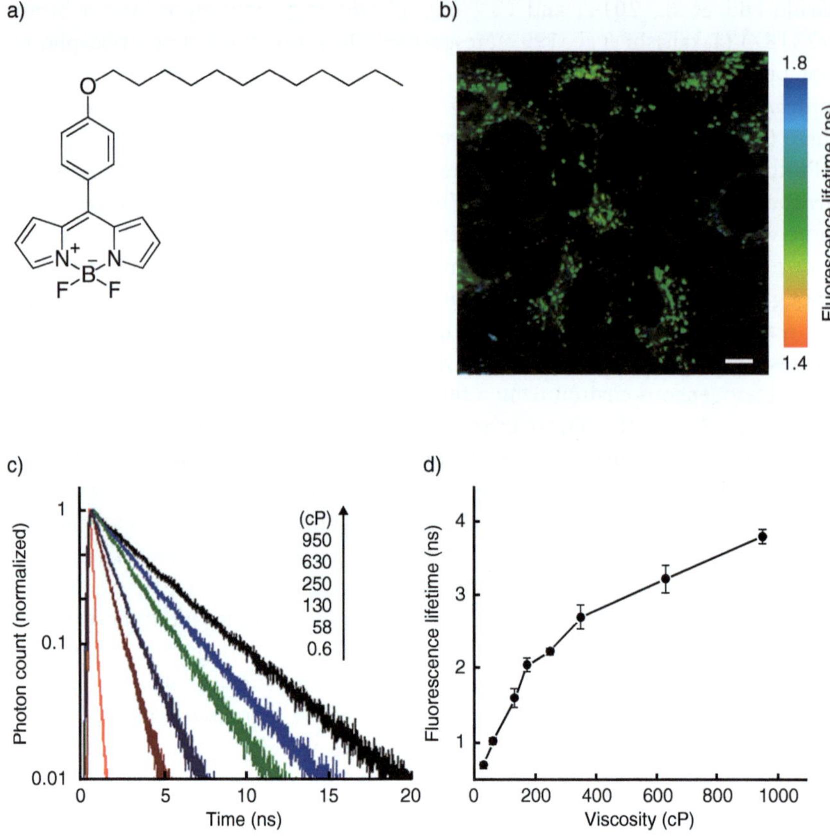

Figure 5.5 Preparation of a calibration curve for measuring intracellular viscosity with a fluorescent molecular rotor (FMR). (a) Chemical structure of FMR. (b) Fluorescence lifetime image of FMR in SK-OV-3 cells. Scale bar: 10 μm. (c) Fluorescence decay curves of FMR in the mixture of methanol and glycerol (black: 950 cP; light blue: 630 cP; green: 250 cP; dark blue: 130 cP; red: 58 cP; orange: 0.6 cP). (d) Relationship between the fluorescence lifetime of FMR and the environmental viscosity. Excitation: 467 nm. Emission: 525 ± 25 nm. Adapted from Kuimova et al. (2008) *J. Am. Chem. Soc.*, **130**, 6672–6673 / American Chemical Society.

where $p(\mathbf{r},t)$ is the heat-source density (power per unit volume). With the simplification of Eqn. (5.3) for discussing intracellular temperature variation, the expected temperature increase (ΔT) can be calculated from the following equation

$$\Delta T = P / (\kappa L) \tag{5.4}$$

where L (m) is the scale of the heat source. Hence, in the case of a typical single cell in the culture medium (L = 10 μm, P = 100 pW, and $\kappa = 1\ \text{W m}^{-1}\ \text{K}^{-1}$ (equivalent to water)), the possible temperature increase ΔT of a single cell should be only 10^{-5} K, which contradicts experimental results, including ours, that the fluorescent nanogel thermometer detected the temperature increase of COS7 cells by 0.45 °C after a chemical stimulus with an uncoupler FCCP (4-(trifluoromethoxy)phenylhydrazone) (Gota et al., 2009).

An intracellular local temperature gradient was also calculated with the assumption that the mitochondria are a heat source (L = 500 nm, P = 100 pW) and the possible temperature increase (ΔT) near the mitochondria should be only 2×10^{-4} K, which also conflicts with our intracellular temperature map obtained by the temperature-dependent fluorescence lifetime of a fluorescent polymeric thermometer (see Figure 3.17; Okabe et al., 2012). Thus, the critique concluded that our observation of intracellular temperature measurements was due to an artifact not an actual temperature variation. Other research groups' results on intracellular thermometry (Kiyonaka et al., 2013; Takei et al., 2014) were also criticized in the critique.

5.4.2
Personal Opinion Concerning Baffou's Critique

To date, I, with my colleagues, expressed personal opinions about Baffou's commentary twice, in a *Feature Article* in *Chemical Communications* (Uchiyama et al., 2017) and a *Perspective* in *Communications Biology* (Okabe and Uchiyama, 2021). Here, the detailed, updated version of my own opinion is described.

Although I respect the introduction of theoretical viewpoints in the discussion of intracellular thermometry to mature the research field, I consider that experimental researchers do not need to be concerned at the gap between experimental results and theoretical calculation by Eqn. (5.4) because the critique seriously lacked scientific merit for the following reasons.

(I)   Although Fourier's law (Eqn. (5.3)) for liquid (and solid) subjects and the conversion of Eqn. (5.3) to Eqn. (5.4) are scientifically reasonable, it is unclear whether Eqn. (5.4) is effective for biological systems. Until experimental proofs of this hypothesis can be obtained, further discussions using Eqn. (5.4) are meaningless. A careful examination of the heat conduction equation used by Baffou et al. shows that it does not accommodate any heat biologically consumed within a living cell. Cells keep entropy low, a phenomenon found exclusively in living organisms (Peterson, 2014). Much energy is expended to regulate the high-dimensional structure of biopolymers such as membranes, proteins, and nucleic acids and to organize controlled one-way irreversible chain reactions (Figure 5.6). In actual biological cells, many endothermic reactions occur (Rolfe and Brown, 1997), and the energy stored in this way may be dissipated through subsequent exothermic reactions. It has been suggested that the heat generated by enzymatic reactions is used for the diffusion

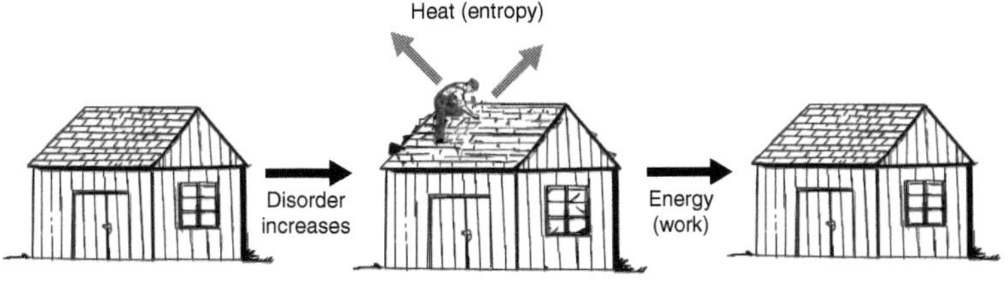

Figure 5.6 Energy consumption to maintain an organized system. Organized systems lose their order and become disarrayed (i.e., entropy increases) (first arrow). Then, energy is needed to repair the disarrayed system (by human work in the illustration) (second arrow). The energy to decrease the entropy of the organized system cannot be evaluated by monitoring the change in its enthalpy. Adapted from Peterson (2014) *Am. Biol. Teach.*, **76**, 88–92 / University of California.

of the enzymes themselves (Riedel et al., 2015). At present, it is impossible to precisely quantify the processes by which heat generated within a cell is dissipated as other types of energy in addition to heat conduction, suggesting that calorimetric measurements do not always reflect the entire possible thermogenesis inside a living cell. Gibbs energy dissipation can now be discussed from the viewpoint of intracellular metabolism (Losa et al., 2022). Thus, treating heat dissipation as if it were solely due to intracellular heat conduction would be an error. The dissipation of energy converted from heat is no longer heat conduction, and the consequence of this conversion on the intracellular temperature change, whether greater or lesser, is an open question.

(II) The critique by Baffou et al. includes several scientific misunderstandings with opportunism. Although 1 (W m^{-1} K^{-1}) was substituted for κ in Eqn. (5.4) with the assumption that a living cell consists of a watery environment in the critique, the actual κ value for water is 0.6096 W m^{-1} K^{-1} (at 300 K) (Ramires et al., 1995). Moreover, the critique adopted the results in temperature measurements by a Si microresonator (see Figure 4.10; Inomata et al., 2012) as a source to support the opinion; in this study, a temperature increase of 0.15 °C or more in a brown adipocyte stimulated with norepinephrine was monitored by the Si sensor attached to the live cell. In the critique, Baffou et al. justified this temperature gap monitored by the microresonator with the reason that the monitoring system was insulated in a vacuum chamber to reduce heat loss (in this case, $\kappa = 10^{-5}$ W m^{-1} K^{-1}). However, this argument is completely inadequate because the monitoring system, including the brown adipocyte and the Si sensor, was located in water in reality. Therefore, in turn, the result by Inomata et al. is a strong support for our conclusion that even living cells located in a culture medium (i.e., water) could increase their average temperature by a sub-degree or more.

(III) In addition, Baffou et al. also rationalized the temperature gap (ΔT, more than a few Kelvins) between our body temperature and the ambient temperature, considering a collective effect (a total heat power and a whole size are PN and $N^{1/3}L$, respectively) with Eqn. (5.4) as

$$\Delta T = PN^{2/3} / (\kappa L) \tag{5.5}$$

where N is the number of cells constructing an organism. However, the cell numbers in endotherms (10^{13}–10^{14}), which Baffou et al. substituted for Eqn. (5.5), did not afford realistic calculation results compared to our body temperature. Moreover, Eqn. (5.5) conflicts with the experimental result that the temperature of the spheroid of HCT116 (human colon cancer) cells (N $\sim 10^4$) incubated at 37 °C is evaluated to be nearly 40 °C by a fluorescent molecular thermometer, dodecyl sulforhodamine (Jenkins et al., 2016). Equation (5.5) can explain neither the temperature gap between our body and ambient temperature nor the similarity of body temperatures of mammalians of different sizes, such as a human, a whale, and a bat (Clarke and Rothery, 2008).

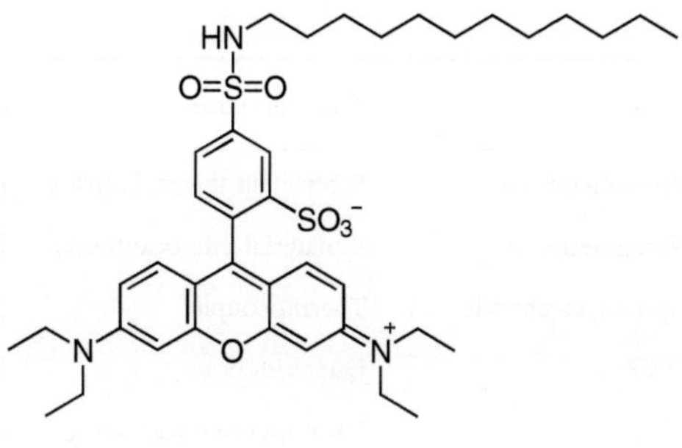

Dodecyl sulforhodamine

(IV) Even after the critique by Baffou et al., many research groups have independently reported cellular temperature variations of 0.1 °C or more with various methods based on different functional mechanisms (Tables 5.2–5.4). A resistance mechanism toward environmental temperature variations is also proposed in RAW (mouse mononuclear macrophages) and CAOV3 (human papillary ovarian adenocarcinoma) cells: intracellular temperatures evaluated by Nd^{3+}, Yb^{3+}, and Er^{3+}-doped UCNPs (upconversion nanoparticles) were significantly higher than that of a cold environment (< 25 °C) but lower than that of a hot environment (> 45 °C) (Wu et al., 2022). As Baffou et al. pointed out in the critique, molecular thermometers may suffer from fluctuations of other environmental factors (e.g., pH or ionic strength) or interactions with cellular components, which is occasionally true and cannot be denied experimentally because the cellular milieu is not fully comprehended and therefore cannot be precisely mimicked; however, probable effects on fluorescence responses depend on thermometers, e.g., the fluorescence signal might be enhanced by acidification in some fluorescent thermometers, but quenched in others. Therefore, the reproducibility of the results by different fluorescent molecular thermometers (Chapter 3) and non-fluorescent thermometers (Chapter 4)

Table 5.2 Reported temperature variation of a whole cell by chemical or physical stimulation.

Cell line[a]	ΔT (°C)	Stimulus[b]	Thermometer	Ref.
COS7	0.45	FCCP	Fluorescent nanogel thermometer	[1]
HeLa	Maximum 1	Ionomycin	EuTTA[c] solution in a pipette	[2]
NIH/3T3	1.84 ± 0.27	Ionomycin	Quantum dot QD655	[3]
U251	0.6–0.8	Camptothecin	Thermocouple	[4]
COS7	1.02 ± 0.17	FCCP	Fluorescent polymeric thermometer	[5]

(Continued)

Table 5.2 (*Continued*)

Cell line[a]	ΔT (°C)	Stimulus[b]	Thermometer	Ref.
Brown adipocyte	0.15	Norepinephrine	Si resonant thermal sensor	[6]
Brown adipocyte	0.217 ± 0.120	Norepinephrine	Bimaterial microcantilever	[7]
MLE-12	0.4	Lipopolysaccharide	Thermocouple	[8]
SH-SY5Y	0.94	CCCP	Quantum dot	[9]
HeLa	1.4	FCCP	Fluorophore-encapsuling UCST[d] polymer	[10]
Mouse primary cortical neuron	0.74	Environmental cooling	Fluorescent nanodiamonds	[11]
H292	5	Influenza virus infection	Rhodamine B-containing polystyrene	[12]
CaSki	7.6	K^+ ion	Surface-enhanced Raman spectroscopy	[13]
COS7	1	FCCP	Microfabricated thermistor	[14]
HeLa	1.8	FCCP	Raman imaging of water molecules	[15]

a) COS7: African green monkey kidney cell, HeLa: human epithelial carcinoma cell, NIH/3T3: mouse embryonic fibroblast cell, U251: human astrocytoma cell, MLE-12: murine lung epithelial cell, SH-SY5Y: human neuroblastoma cell, H292: human lung mucoepidermoid carcinoma cell, CaSKi: human epidermoid cervical cancer cell.

b) FCCP: Carbonyl cyanide 4-(trifluoromethoxy)phenylhydrazone, CCCP: Carbonyl cyanide 3-chlorophenylhydrazone.

c) Europium(III) thenoyltrifluoroacetonate.

d) Upper critical solution temperature.

1 Gota, C., Okabe, K., Funatsu, T., Harada, Y., and Uchiyama, S. (2009) *J. Am. Chem. Soc.*, **131**, 2766–2767.

2 Suzuki, M., Tseeb, V., Oyama, K., and Ishiwata, S. (2007) *Biophys. J.*, **92**, L46–L48.

3 Yang, J.-M., Yang, H., and Lin, L. (2011) *ACS Nano*, **5**, 5067–5071.

4 Wang, C., Xu, R., Tian, W., Jiang, X., Cui, Z., Wang, M., Sun, H., Fang, K., and Gu, N. (2011) *Cell Res.*, **21**, 1517–1519.

5 Okabe, K., Inada, N., Gota, C., Harada, Y., Funatsu, T., and Uchiyama, S. (2012) *Nat. Commun.*, **3**, 705.

6 Inomata, N., Toda, M., Sato, M., Ishijima, A., and Ono, T. (2012) *Appl. Phys. Lett.*, **100**, 154104.

7 Sato, M. K., Toda, M., Inomata, N., Maruyama, H., Okamatsu-Ogura, Y., Arai, F., Ono, T., Ishijima, A., and Inoue, Y. (2014) *Biophys. J.*, **106**, 2458–2464.

8 Tian, W., Wang, C., Wang, J., Chen, Q., Sun, J., Li, C., Wang, X., and Gu, N. (2015) *Nanotechnology*, **26**, 355501.

9 Tanimoto, R., Hiraiwa, T., Nakai, Y., Shindo, Y., Oka, K., Hiroi, N., and Funahashi, A. (2016) *Sci. Rep.*, **6**, 22071.

10 Dong, F., Zheng, T., Zhu, R., Wang, S., and Tian, Y. (2016) *J. Mater. Chem. B*, **4**, 7681–7688.

11 Simpson, D. A., Morrisroe, E., McCoey, J. M., Lombard, A. H., Mendis, D. C., Treussart, F., Hall, L. T., Petrou, S., and Hollenberg, L. C. L. (2017) *ACS Nano*, **11**, 12077–12086.

12 Maruyama, H., Kimura, T., Liu, H., Ohtsuki, S., Miyake, Y., Isogai, M., Arai, F., and Honda, A. (2018) *Virus Res.*, **257**, 94–101.

13 Hu, S., Liu, B.-J., Feng, J.-M., Zong, C., Lin, K.-Q., Wang, X., Wu, D.-Y., and Ren, B. (2018) *J. Am. Chem. Soc.*, **140**, 13680–13686.

14 Inomata, N., Inaoka, R., Okabe, K., Funatsu, T., and Ono, T. (2020) *Sens. Bio-sens. Res.*, **27**, 100309.

15 Sugimura, T., Kajimoto, S., and Nakabayashi, T. (2020) *Angew. Chem. Int. Ed.*, **59**, 7755–7760.

Table 5.3 Reported temperature gradient within a live cell.

Cell line[a]	Area	ΔT (°C)	Thermometer	Ref.
COS7	Nucleus (vs cytoplasm)	0.96	Fluorescent polymeric thermometer	[1]
COS7	Centrosome (vs cytoplasm)	0.75	Fluorescent polymeric thermometer	[1]
SH-SY5Y	Neurite (vs cell body)	−1.6	Quantum dot	[2]
HeLa	Nucleus (vs cytoplasm)	2.9 ± 0.3	Genetically encoded fluorescent protein gTEMP	[3]
MDA-MB468	Nucleolus (vs rest of the cell)	5	Lanthanide-bearing polymeric micelle	[4]
HMEC-1	Nuclear membrane (vs plasma membrane)	2.5	Fluorescent nanodiamond	[5]
A549	Mitochondria (vs cytoplasm)	4.31	Quantum coherence modulation microscopy	[6]

a) COS7: African green monkey kidney cell, SH-SY5Y: human neuroblastoma cell, HeLa: human epithelial carcinoma cell, MDA-MB468: human breast adenocarcinoma cell, HMEC-1: human microvascular endothelial cell, A549: human lung adenocarcinoma cell.

1 Okabe, K., Inada, N., Gota, C., Harada, Y., Funatsu, T., and Uchiyama, S. (2012) *Nat. Commun.*, **3**, 705.
2 Tanimoto, R., Hiraiwa, T., Nakai, Y., Shindo, Y., Oka, K., Hiroi, N., and Funahashi, A. (2016) *Sci. Rep.*, **6**, 22071.
3 Nakano, M., Arai, Y., Kotera, I., Okabe, K., Kamei, Y., and Nagai, T. (2017) *PLoS ONE*, **12**, e0172344.
4 Piñol, R., Zeler, J., Brites, C. D. S., Gu, Y., Téllez, P., Neto, A. N. C., da Silva, T. E., Moreno-Loshuertos, R., Fernandez-Silva, P., Gallego, A. I., Martinez-Lostao, L., Martínez, A., Carlos, L. D., and Millán, A. (2020) *Nano Lett.*, **20**, 6466–6472.
5 Wu, T., Chen, X., Gong, Z., Yan, J., Guo, J., Zhang, Y., Li, Y., and Li, B. (2022) *Adv. Sci.*, **9**, 2103354.
6 Zhou, H., Yao, W., Zhou, X., Dong, S., Wang, R., Guo, Z., Li, W., Qin, C., Xiao, L., Jia, S., Wu, Z., and Li, S. (2023) *ACS Nano*, **17**, 8433–8441.

Table 5.4 Reported local temperature variation within a live cell by chemical stimulation.

Cell line[a]	Area	ΔT (°C)	Stimulus[b]	Thermometer	Ref.
HeLa	Endosome	1.9 ± 0.8	Ionomycin	EuTTA[c] containing ratiometric nanoparticle	[1]
HeLa	Endoplasmic reticulum	1.7 ± 0.4	Ionomycin	ER thermo yellow	[2]
HeLa	Mitochondria	6–9	FCCP	Genetically encoded fluorescent protein, gTEMP	[3]
HEK293	Mitochondria	10.5	Respiration	Mito thermo yellow	[4]
MCF-7	Mitochondria	3	12-Myristate-13-acetate	Mito-TEM	[5]

(Continued)

Table 5.4   (*Continued*)

Cell line[a]	Area	ΔT (°C)	Stimulus[b]	Thermometer	Ref.
HeLa	Mitochondria	3	FCCP	Genetically encoded fluorescent protein, emGFP-Mito	[6]
HeLa	Mitochondria	2.74	Oleic acid	Er^{3+} containing upconversion nanoparticle	[7]
L929	Mitochondria	10	CCCP	Gold nanocluster	[8]
Mouse brain	Mitochondria[d]	4–22	CCCP	Glass microcapillary with a fluorescent diamond crystal	[9]

a) HeLa: human epithelial carcinoma cell, HEK293: human embryonic kidney cell, MCF-7: human breast cancer cell, L929: mouse fibroblast cell.
b) FCCP: carbonyl cyanide 4-(trifluoromethoxy)phenylhydrazone, CCCP: carbonyl cyanide 3-chlorophenylhydrazone.
c) Europium(III) thenoyltrifluoroacetonate.
d) Isolated.

1 Takei, Y., Arai, S., Murata, A., Takabayashi, M., Oyama, K., Ishiwata, S., Takeoka, S., and Suzuki, M. (2014) *ACS Nano*, **8**, 198–206.
2 Arai, S., Lee, S.-C., Zhai, D., Suzuki, M., and Chang, Y. T. (2014) *Sci. Rep.*, **4**, 6701.
3 Nakano, M., Arai, Y., Kotera, I., Okabe, K., Kamei, Y., and Nagai, T. (2017) *PLoS ONE*, **12**, e0172344.
4 Chrétien, D., Bénit, P., Ha, H.-H., Keipert, S., El-Khoury, R., Chang, Y.-T., Jastroch, M., Jacobs, H. T., Rustin, P., and Rak, M. (2018) *PLoS Biol.*, **16**, e2003992.
5 Huang, Z., Li, N., Zhang, X., Wang, C., and Xiao, Y. (2018) *Anal. Chem.*, **90**, 13953–13959.
6 Savchuk, O. A., Silvestre, O. F., Adão, R. M. R., and Nieder, J. B. (2019) *Sci. Rep.*, **9**, 7535.
7 Di, X., Wang, D., Zhou, J., Zhang, L., Stenzel, M. H., Su, Q. P., and Jin, D. (2021) *Nano Lett.*, **21**, 1651–1658.
8 Wang, Y., Liang, S., Mei, M., Zhao, Q., She, G., Shi, W., and Mu, L. (2021) *Anal. Chem.*, **93**, 15072–15079.
9 Romshin, A. M., Osypov, A. A., Popova, I. Y., Zeeb, V. E., Sinogeykin, A. G., and Vlasov, I. I. (2023) *Nanomaterials*, **13**, 98.

strongly supports the concept that the observed intracellular (or cellular) temperature variations by a sub-degree or more were not due to artifacts. For example, microthermocouple arrays revealed that the temperature of HepG2 (human liver carcinoma) cells fluctuates within 0.06 °C and exceptionally 0.285 °C at maximum without any stimulation (Yang et al., 2017). Confirming reproducibility is a common and important process by which the reliability of a novel methodology is experimentally established. In the same issue that carried Baffou's critique in *Nature Methods*, the editorial team described the importance of reproducibility in an *Editorial* (Anonymous, 2014):

> "When possible, comparison with orthogonal methods is an obvious way to test whether a method reports accurately on a biological phenomenon."

(V) The critique included no ideas for further experiments to fill the gap between the experimental results and the theoretical estimation (assuming that Eqn. (5.4) is still effective). Such judgmental attitudes seriously restrain progress in the field of intracellular thermometry. I believe criticisms should be given, with scientific diplomacy and suggestions for improving the original concepts.

As described in (IV) above, I consider that the present techniques for intracellular thermometry (fluorometric and non-fluorometric methods) have already acquired sufficient reliability to evaluate biological events. Nevertheless, we (experimental researchers) must address the issue to bridge the gaps between the experimental results (i.e., intracellular temperature variation by a sub-degree or more) and the theoretical calculation based on the heat conduction equation. Since intracellular temperature is a highly influential physical parameter for thermodynamics, thermochemistry, and heat transfer in a cell, it seems impossible to explain intracellular temperature variation in only one scientific field. Here, I refer to the statement by the physicist Schrödinger in his lecture in 1943 in Dublin (Schrödinger, 1992):

> "What I wish to make clear ...is... we must be prepared to find it [living matter] working in a manner that cannot be reduced to the ordinary laws of physics."

In the future, the careful pursuit of energy income and expense in the thermal, chemical, and mechanical forms will lead to a deeper understanding of the scientific significance of intracellular temperature. Moreover, constructing theories based on entirely new concepts encompassing the abovementioned related fields is also a challenging and promising direction for the research field of intracellular thermometry. To comprehend intracellular temperature variations, it is necessary to experimentally verify and model the three processes involved: thermogenesis, conversion to other energy types, and dissipation. For this purpose, a controllable heat source is helpful because it allows quantifying the energy added to a cell, unlike heat production by chemical stimuli. Intracellular temperature mapping techniques with high temporal resolution will also contribute to kinetic studies. In addition, identifying biological macromolecules (i.e., nucleic acids, proteins, and lipids) involved in intracellular temperature regulation is another important step.

5.4.3
Researchers' Reactions to the Baffou's Critique and Later Discussions

A year after Baffou's critique, two independent Japanese groups that were criticized by it published correspondences in the same journal, *Nature Methods* (Kiyonaka et al., 2015; Suzuki et al., 2015), but they were opposed by Baffou et al. once again (Baffou et al., 2015). Later, Kiyonaka et al. also expressed their detailed opinions in a review article published in *Pflügers Archiv* (Okabe et al., 2018). In both correspondences, the discussion based on Eqn. (5.4) was partly accepted, but the values of the parameters P, κ, and L that Baffou et al. used for Eqn. (5.4) were questioned. For example, Suzuki et al. pointed out that brown adipocytes could produce 1.63 ± 0.73 nW per cell under stimulation with norepinephrine, which was evaluated by a micromachined nanocalorimetric sensor (Johannessen et al., 2002), although Baffou et al. used $P = 100$ pW that was measured by a calorimeter for murine C1300 neuroblastoma cells (Loesberg et al., 1982). Kiyonaka et al. claimed that the actual sizes (L) of the heat sources, i.e., protein complexes and intracellular organelles, are 10–100 nm, while Baffou et al. used $L = 10$ µm as a cellular size. In those days, intracellular thermal conductivity was not experimentally accessed but could be estimated as significantly lower than water. Since then, researchers have tried to determine intracellular thermal conductivity by fluorometric or non-fluorometric methods and confirmed its lowness, which will be described in the following subsection (see Section 5.4.4). The correspondence in *Nature Methods* implied that these re-evaluations of P, κ, and L values could result in a temperature increase (ΔT) in Eqn. (5.4) that is comparable to experimental results.

Although not directly concerning Baffou's critique, the effects of the distance between an intracellular heat source (e.g., the mitochondria) and a fluorescent molecular thermometer were raised for further discussion in the correspondence in *Nature Methods*. This issue will be summarized in another subsection (see Section 5.4.6).

Similar arguments concerning the reliability of intracellular thermometry were later made on the report that the temperature of the mitochondria reached 50 °C (see Figure 3.26e; Chrétien et al., 2018). In addition to individual viewpoints in both in favor of or objecting to the experimental results achieved with the current intracellular thermometry technique (Kang, 2018; Lane, 2018; Yuexuan and Daocheng, 2020; Fahimi and Matta, 2022), in-depth discussions to progress the research field further are now available: modeling of intracellular heat diffusion (Rajagopal and Sinha, 2021) and perspective on the hot mitochondria (Macherel et al., 2021). The literature is conceptual but worth reading to select future research directions carefully.

5.4.4
Determination of Intracellular Thermal Conductivity

Cellular thermal conductivity was initially considered to be an indicator of disease progression, as it could be different between normal and cancer cells (Park et al., 2016; Vishnu et al., 2022). Then, it started to draw much more attention due to the controversy between experimental results and a theoretical calculation in intracellular thermometry, as described in the previous subsections. Until now, intracellular thermal conductivity was measured by non-fluorometric and fluorometric techniques, as summarized in Table 5.5. Interestingly, the intracellular thermal conductivities determined are remarkably different depending on the techniques. At

Table 5.5 Reported thermal conductivity of a living cell.

Case	Method	Cell line[a]	Thermal conductivity (W m^{-1} K^{-1})	Ref.
1	Three-omega (3ω) method with a sensor composed of a micro-well and a metal strip	HeLa	0.604 ± 0.018	[1]
		NIH-3T3 J2	0.579 ± 0.017	[1]
		Rat hepatocyte	0.575 ± 0.017	[1]
		Hs 578Bst	0.585 ± 0.022	[2]
		Hs 578T	0.557 ± 0.014	[2]
		TE 353.Sk	0.593 ± 0.014	[2]
		TE 354.T	0.579 ± 0.013	[2]
		WM-115	0.598	[2]
		WM-266-4	0.590	[2]

(Continued)

Table 5.5 (*Continued*)

Case	Method	Cell line[a]	Thermal conductivity (W m^{-1} K^{-1})	Ref.
2	Fluorometry with gallium nitride (GaN) nanomembranes	HeLa	0.577	[3]
		MCF-7	0.567	[3]
		SK-BR-7	0.564	[3]
		MDA-MB-231	0.547	[3]
3	Micro-pipette thermal sensor technique	Jurkat	0.538 ± 0.006	[4]
4	Fluorometry with fluorescent nanodiamonds	HeLa and MCF-7	0.11 ± 0.04	[5]
5	Transient plasmonic imaging	HeLa	0.31	[6]
		MCF-7	0.32	[6]
		MCF-10A	0.32	[6]
		Chicken	0.31	[6]
		Bullfrog	0.36	[6]
6	On-chip-integrated microthermistor	COS7	0.61–0.66 (25 °C)	[7]
		COS7	0.37–0.55 (37 °C)	[7]
		COS7	0.81–0.82 (45 °C)	[7]
c.f.		Water	0.6096 (300 K)	[8]
c.f.		Air	0.0264 (300 K)	[9]
c.f.	Resistance temperature device	Formalin-fixed tumor tissue	0.309 ± 0.02	[10]
		Adjacent normal tissue	0.563 ± 0.028	[10]

a) HeLa: human epithelial carcinoma cell, NIH-3T3 J2: mouse embryonic fibroblast cell, Hs 578Bst: human mammary gland normal cell, Hs 578T: human mammary gland cancer cell, TE 353.Sk: human skin normal cell, TE 354.T: human skin cancer cell, WM-115: human skin primary cancer cell, WM-266-4: human skin metastatic cancer cell, MCF-7: human breast cancer cell, SK-BR-7: human breast cancer cell, MDA-MB-231: human breast cancer cell, Jurkat: human T-cell acute lymphoblastic leukemia cell, MCF-10A: human breast epithelial cell, COS7: African green monkey kidney cell.

1 Park, B. K., Yi, N., Park, J., and Kim, D. (2013) *Appl. Phys. Lett.*, **102**, 203702.

2 Park, B. K., Woo, Y., Jeong, D., Park, J., Choi, T.-Y., Simmons, D. P., Ha, J., and Kim, D. (2016) *J. Appl. Phys.*, **119**, 224701.

3 El Afandy, R. T., Abu Elela, A. F., Mishra, P., Janjua, B., Oubei, H. M., Büttner, U., Majid, M. A., Ng, T. K., Merzaban, J. S., and Ooi, B. S. (2017) *Small*, **13**, 1603080.

4 Shrestha, R., Atluri, R., Simmons, D. P., Kim, D. S., and Choi, T. Y. (2020) *Int. J. Heat Mass Transf.*, **160**, 120161.

5 Sotoma, S., Zhong, C., Kah, J. C. Y., Yamashita, H., Plakhotnik, T., Harada, Y., and Suzuki, M. (2021) *Sci. Adv.*, **7**, eabd7888.

6 Song, P., Gao, H., Gao, Z., Liu, J., Zhang, R., Kang, B., Xu, J.-J., and Chen, H.-Y. (2021) *Chem*, **7**, 1569–1587.

7 Inomata, N., Miyamoto, T., Okabe, K., and Ono, T. (2023) *Lab Chip*, **23**, 2411–2420.

8 Ramires, M. L. V., de Castro, C. A. N., Nagasaka, Y., Nagashima, A., Assael, M. J., and Wakeham, W. A. (1995) *J. Phys. Chem. Ref. Data*, **24**, 1377–1381.

9 Huber, M. L. and Harvey, A. H. (2012) *CRC Handbook of Chemistry and Physics*, 93rd ed. (eds. W. M. Haynes, D. R. Lide, and T. J. Bruno), CRC Press, Boca Raton, p. 6-240.

10 Vishnu, G. K. A., Gogoi, G., Behera, B., Rila, S., Rangarajan, A., and Pandya, H. J. (2022) *Microsyst. Nanoeng.*, **8**, 1.

present, this difference is not being discussed well but is likely due to the position of sensors, i.e., outside cells (by non-fluorometric sensors) and inside cells (by fluorescent sensors). Some examples will be introduced here.

Figure 5.7 shows the thermal conductivity measurement of a Jurkat (human leukemic T-cell lymphoblast) cell by a transient laser point heating method (Shrestha et al., 2020). The thermal sensor was based on a thermocouple junction with tin alloy filled in a borosilicate glass pipette and nickel-thin film coated outside. A diode-pumped solid-state laser (532 nm) was used as a point heating source, and the temperature increase of the sensor tip during laser irradiation depends on the heat diffused by conduction into the sample. Thus, the thermal conductivity of a Jurkat cell was evaluated to be 0.538 W m^{-1} K^{-1} with an error of $\pm$ 1%.

Before determining the intracellular thermal conductivity fluorometrically, the thermal conductivity of lipid bilayers was evaluated by LiYF$_4$:Er^{3+}/Yb^{3+} upconversion nanoparticles (UCNPs) (Bastos et al., 2019). Figure 5.8a, 5.8b show the schematic diagram and chemical structure of the LiYF$_4$:Er^{3+}/Yb^{3+} UCNP capped by the lipid bilayer, which consists of DOPA (1,2-di-(9Z-octadecenoyl)-sn-glycero-3-phosphate), DOPC (1,2-di-(9Z-octadecenoyl)-sn-glycero-3-phosphocholine), cholesterol, and oleate for the determination of the thermal conductivity. The thermal conductivity of the lipid bilayer was calculated by comparing the temperature of the lipid bilayer-capped LiYF$_4$:Er^{3+}/Yb^{3+} UCNPs with that of uncapped LiYF$_4$:Er^{3+}/Yb^{3+} UCNPs under near-infrared laser irradiation (at 980 nm) in water (Figure 5.8c). It should be noted that the temperatures measured by the lipid bilayer-capped LiYF$_4$:Er^{3+}/Yb^{3+} UCNPs were higher than those of the thermocouple (and uncapped LiYF$_4$:Er^{3+}/Yb^{3+} UCNPs), showing that the lipid bilayer behaves as a thermal barrier between the UCNPs and water. As a result, the thermal conductivity of the lipid bilayer was evaluated to be 0.20 $\pm$ 0.02 W m^{-1} K^{-1} at 300 K.

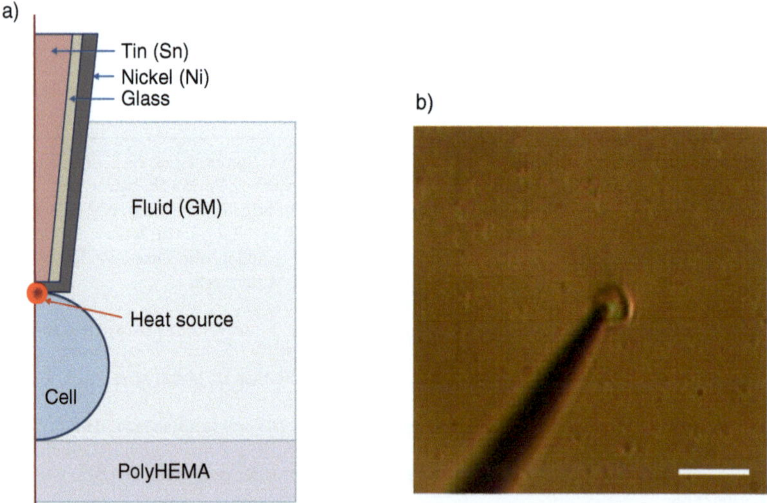

Figure 5.7 Determination of cellular thermal conductivity of a Jurkat cell by a micro-pipette thermal sensor. (a) Schematic diagram of experiments. GM: growth medium, polyHEMA: poly(2-hydroxyethyl methacrylate), heat source: diode-pumped solid-state laser at 532 nm. (b) Thermal conductivity measurement. Scale bar: 25 μm. Shrestha et al. (2020) *Int. J. Heat Mass Transf.*, **160**, 120161 / Elsevier.

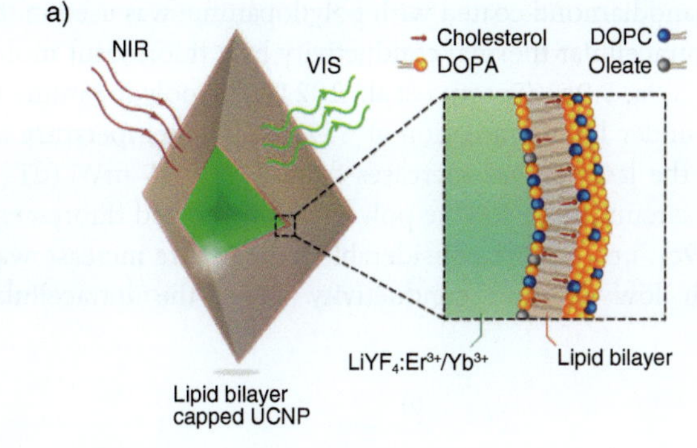

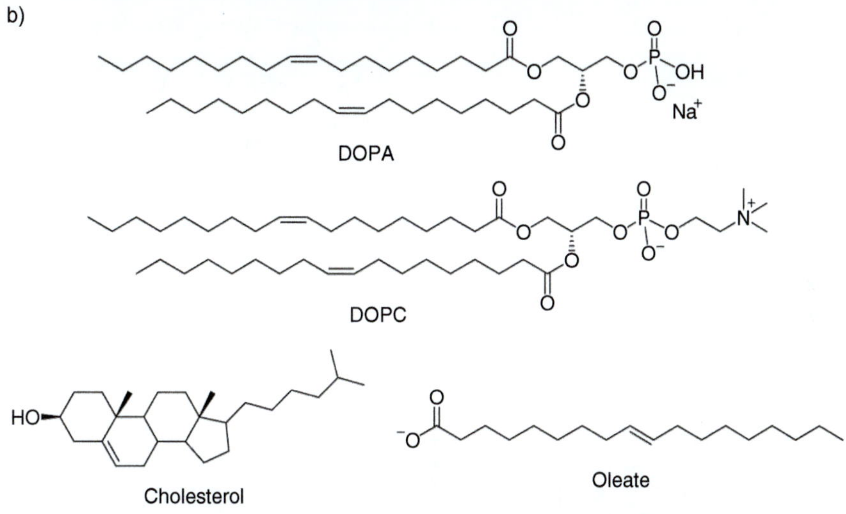

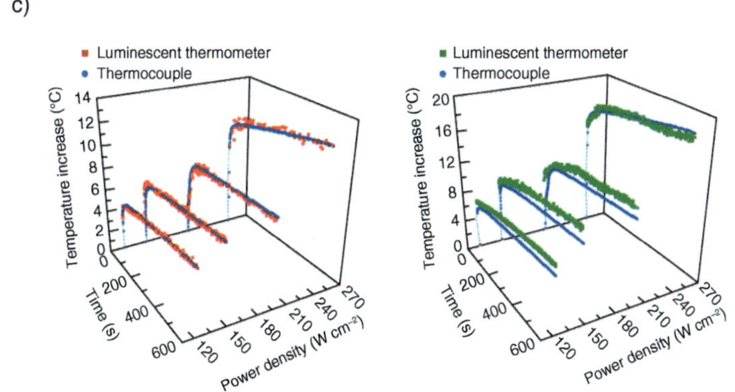

Figure 5.8 Determination of thermal conductivity of a lipid bilayer using $LiYF_4:Er^{3+}/Yb^{3+}$ upconversion nanoparticles (UCNPs). (a) Schematic diagram and structure of the lipid bilayer-capped $LiYF_4:Er^{3+}/Yb^{3+}$ UCNP. (b) Chemical structures of the components, DOPA, DOPC, cholesterol, and oleate. (c) Temperature profiles of the uncapped UCNPs (left) and the lipid bilayer-capped UCNPs (right) dispersed in water under laser irradiation at 980 nm. The temperature was monitored by an immersed thermocouple (blue) and the $LiYF_4:Er^{3+}/Yb^{3+}$ UCNPs (red and green). Bastos et al. (2019) *Adv. Funct. Mater.*, **29**, 1905474 / John Wiley & Sons.

A fluorescent nanodiamond coated with polydopamine was used in the first case of determining intracellular thermal conductivity by a fluorescent molecular thermometer (Figure 5.9a, 5.9b) (Sotoma et al., 2021). The polydopamine layer works as a local heater under laser irradiation at 532 nm. The temperature rises which corresponded to the laser power increases from 7.3 to 25 mW $(dT_{7.3 \to 25}$ (°C)) depended on the circumstance for the polydopamine-coated fluorescent nanodiamond (Figure 5.9c), i.e., a more considerable temperature increase was observed in a space with low thermal conductivity. Thus, the intracellular thermal

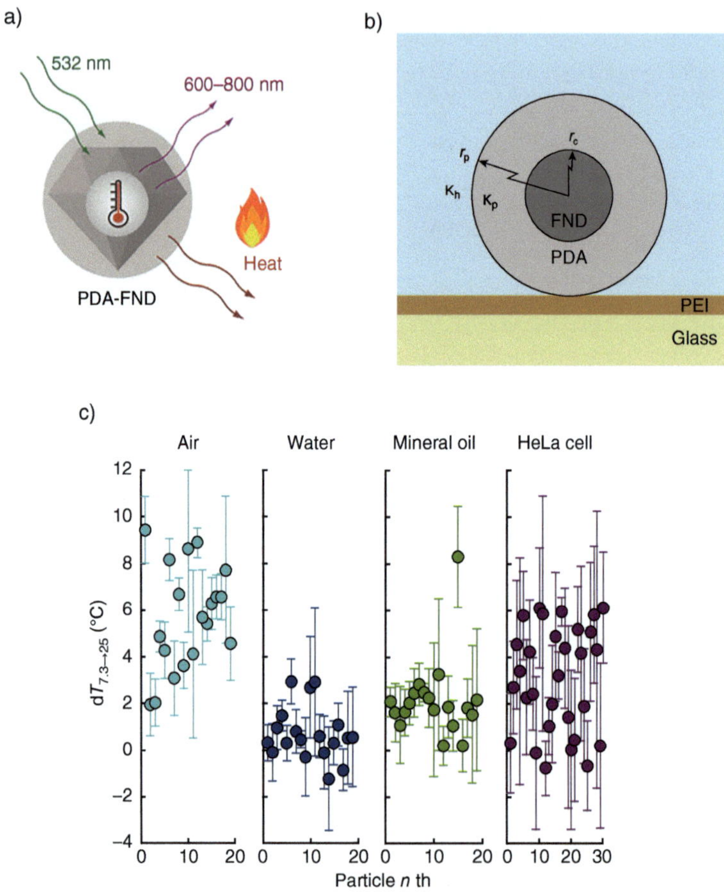

Figure 5.9 Determination of intracellular thermal conductivity using fluorescent nanodiamonds. (a) Schematic illustration of a polydopamine-coated fluorescent nanodiamond (PDA-FND). (b) Model structure of a PDA-FND and its surroundings (r_c: radius of FND, r_p: radius of PDA-FND, κ_p: thermal conductivity of PDA, κ_h: thermal conductivity of outer environment, PEI: polyethylenimine). (c) Temperature increases of PDA-FND when the laser power increases from 7.3 to 25 mW ($dT_{7.3 \to 25}$) in the air (light blue), water (blue), mineral oil (green), and a HeLa cell (purple). Error bars indicate standard deviations. Sotoma et al. (2021) *Sci. Adv.*, **7**, eabd7888 / American Association for the Advancement of Science.

conductivity was determined from particle sizes (r_c and r_p) and thermal conductivities (κ_p and κ_h) to be 0.11 ± 0.04 W m^{-1} K^{-1} for both HeLa and MCF-7 (human breast carcinoma) cells. A relatively large deviation in the experimental data in HeLa cells (Figure 5.9c) was conceivably due to the lack of uniformity among fluorescent nanodiamonds and the difference in the location of fluorescent nanodiamonds. Further investigation using HeLa cells showed that most of the polydopamine-coated fluorescent nanodiamonds (86%) were present in lysosome-associated membrane protein 1 (LAMP1)-green fluorescent protein (GFP)-positive vesicles, and the rest might be located in LAMP1-GFP-negative vesicles or the cytosol.

A transient microscopic method based on the thermal lensing effect can more straightforwardly determine intracellular thermal conductivity (Song et al., 2021). A laser at 532 nm with 5–7 ns pulses was used as a pump beam to heat gold nanoparticles inside a living cell with a temperature difference (ΔT) and a consequently caused refractive index variation (Δn), and a pulsed white light was used as a probe beam to follow the cooling process during Δt (Figure 5.10a). When the heating pulse was over, the photothermal nanolens effects began to evolve with the heat dissipation to a global environment. The photothermal nanolens effects could be profiled by detecting the scattering signal of the probe beam and correlated with heat transfer characteristics (i.e., thermal conductivity) of the surroundings. The measured intracellular thermal conductivity using gold nanoparticles introduced into a HeLa cell was displayed in Figure 5.10b, 5.10c. With a considerable distribution due to an inhomogeneous cellular environment, the mean intracellular thermal conductivity was determined to be 0.31 W m^{-1} K^{-1} at 25 °C. Intracellular thermal conductivities of MCF-7, MCF-10A (human breast epithelial), chicken, and bullfrog cells were also measured to be 0.32, 0.32, 0.31, and 0.36 W m^{-1} K^{-1}, respectively, in a similar way (Figure 5.10d). In addition, the remarkable effects of the temperature on intracellular thermal conductivity could be successfully assessed (Figure 5.10e).

The thermal diffusivity α (m^2 s^{-1}) is the rate of heat transfer from a hot area to a cold area in a material and is defined as

$$\alpha = \frac{\kappa}{\rho C p} \tag{5.6}$$

where κ, ρ, and Cp are thermal conductivity, density (kg m^{-3}), and specific heat capacity (J kg^{-1} K^{-1}). A genetically encoded ratiometric fluorescent molecular thermometer, which is composed of tdTomato and mNeonGreen with emission peaks at 581 and 517 nm, respectively, under the excitation at 495 nm, revealed that the thermal diffusivity of a HeLa cell is $(2.7 \pm 0.4) \times 10^{-8}$ m^2 s^{-1} (Lu et al., 2022). Compared to that of pure water (1.43×10^{-7} m^2 s^{-1}) ($\kappa = 0.60652$ W m^{-1}K^{-1}; ρ = 997.05 kg m^{-3}; Cp = 4.1813 kJ kg^{-1} K^{-1}; Lemmon and Harvey, 2022), the speed of heat transfer inside a living HeLa cell is estimated to be approximately one-fifth of that of a culture medium, which is one of the possible reasons for the remarkable intracellular temperature gradients observed in intracellular thermometry.

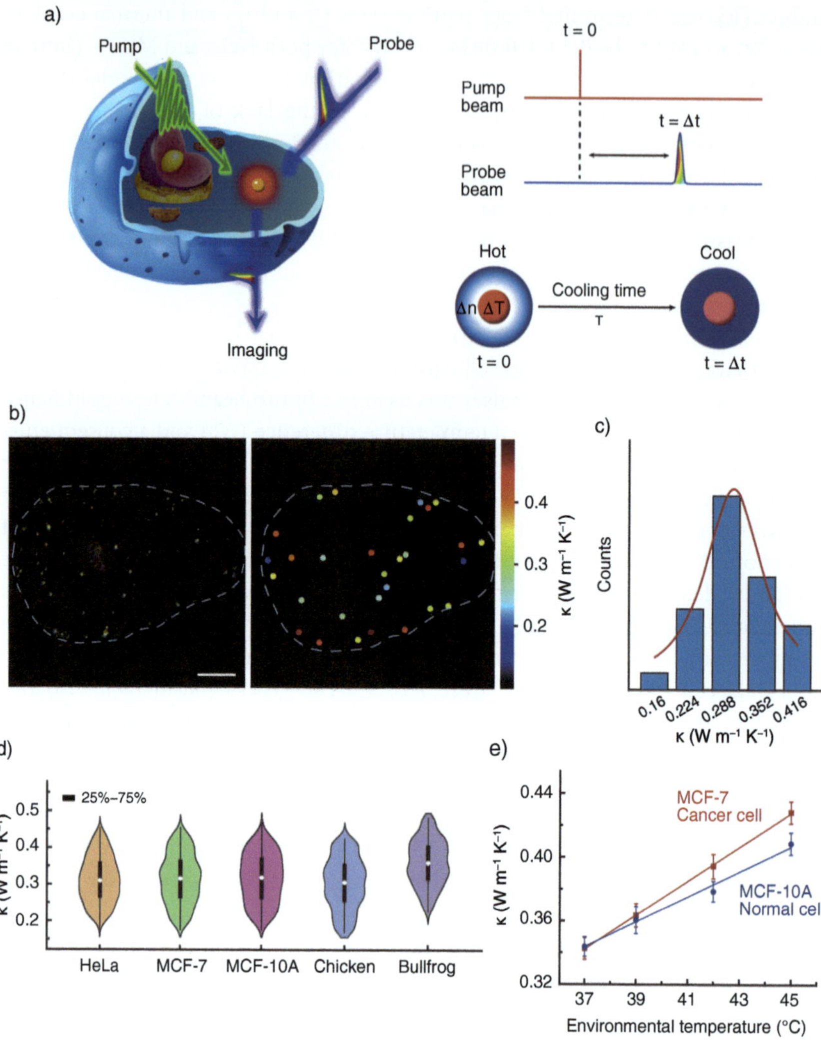

Figure 5.10 Determination of intracellular thermal conductivity based on photothermal lens spectroscopy. (a) Schematic diagram of transient heat dissipation microscopy. See the main text for an experimental procedure. (b) Color dark-field image of a HeLa cell containing gold nanoparticles (left) and its corresponding heat dissipation (i.e., thermal conductivity) map (right). Dotted lines indicate the edges of the HeLa cell. Scale bar: 10 μm. (c) Histogram of the thermal conductivity inside the HeLa cell. (d) Statistical distribution of the intracellular thermal conductivity in HeLa, MCF-7, MCF-10A, chicken, and bullfrog cells. (e) Relationships of the intracellular thermal conductivity and the temperature in MCF-7 (red) and MCF-10A (blue) cells. Song et al. (2021) *Chem*, **7**, 1569–1587 / Elsevier.

5.4.5
Determination of Heat Capacities of Lipid Bilayers

For a better understanding of intracellular thermoregulation and heat transfer, the heat capacity of model lipid bilayers as a component of live cells was also measured by a fluorescent thermometer (Bastos et al., 2020). The heat capacity of lipid bilayers which are composed of oleate, DOPA, DOPC, and cholesterol (21: 51: 5: 24) was evaluated by the lipid bilayer-capped $LiYF_4:Er^{3+}/Yb^{3+}$ UCNPs to be 5.039 ± 0.211 kJ kg^{-1} K^{-1} (cf., heat capacity of water is 4.1813 kJ kg^{-1} K^{-1}).

5.4.6
Effects of the Distance between a Heat Source and a Fluorescent Molecular Thermometer on the Measured Temperature

Although not mentioned in the original critique by Baffou et al., the *Correspondence* to it pointed out that the distance from a heat source (e.g., the mitochondria in living cells) should be considered in evaluating the reliability of intracellular thermometry using theoretical modeling (Kiyonaka et al., 2015).

Fluorescent molecular thermometers can monitor a local temperature variation in a tiny space. When an aqueous suspension of silica and rhodamine B-coated $BaTiO_3$ particles (diameter: ~460 nm) was irradiated with microwaves, a significant difference of approximately 10 °C was observed between the temperature evaluated by rhodamine B located close to the heat source and that measured by a fiber-optic thermometer for the bulk (Ano et al., 2016).

In experiments using HeLa cells under stimulation by ionomycin, the temperature of the endoplasmic reticulum (ER) (that was measured by an ER-targeted fluorescent molecular thermometer ER thermo yellow) increased by approximately 2 °C, while that of the cell surface (measured by a poly(methyl methacrylate) sheet containing a fluorescent molecular thermometer EuTTA and a reference rhodamine 101) remained nearly constant within ±0.2 °C (Oyama et al., 2020). A related *Commentary* on the reality of the intracellular temperature gradient was published in the same issue (Balaban, 2020).

ER thermo yellow

EuTTA

Rhodamine 101

The effects of the nano-scale distance between a heat source and a measured temperature were experimentally clarified using a gold nanorod enveloped by multiple layers of poly(allylamine hydrochloride) (PAH) and poly(sodium 4-styrenesulfonate) (PSS) (Figure 5.11a) (Freddi et al., 2013). A fluorescent molecular thermometer, rhodamine B, was attached to the outer layer of the PAH, PSS-coated gold nanorod. As shown in Figure 5.11b, the measured temperature increase by the irradiation of near-infrared laser (NIR) to the gold nanorod depended on the distance between the gold nanorod and rhodamine B. Four layers of PAH and PSS,

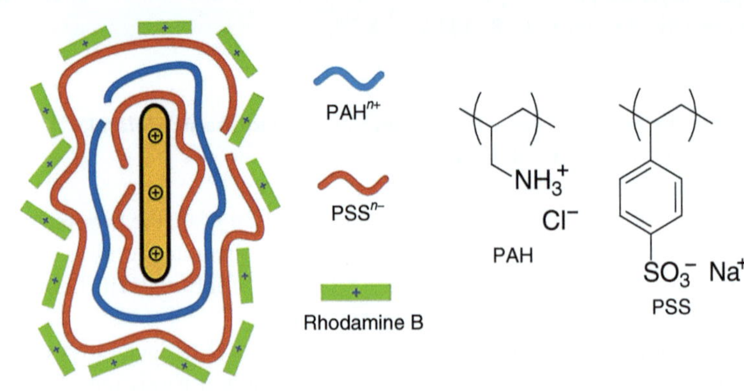

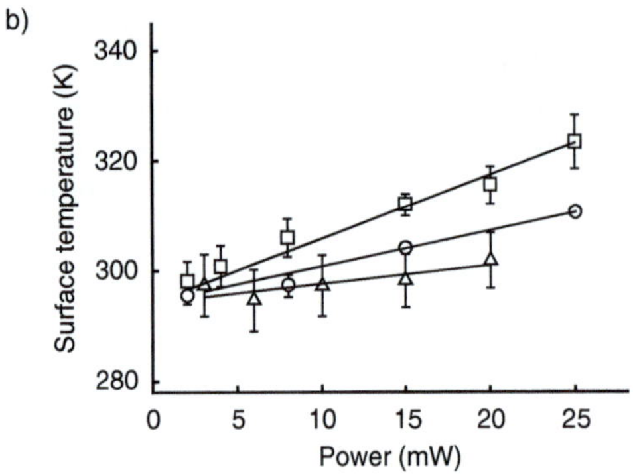

Figure 5.11 Effects of the distance between a heat source and a fluorescent molecular thermometer on the measured temperature. (a) Schematic diagram and chemical structure of a gold nanorod decorated with poly(allylamine hydrochloride) (PAH), poly(sodium 4-styrenesulfonate) (PSS), and a fluorescent molecular thermometer rhodamine B. In the illustration, the number of polymeric layers is three. (b) Temperature evaluated from the fluorescence lifetime of rhodamine B (excitation: 514 nm) on gold nanorods decorated with polymeric layers (3 (square), 5 (circle), and 7 (triangle)) under laser irradiation at 800 nm with different powers. The nanorods were suspended in water. Freddi et al. (2013) *Nano Lett.*, **13**, 2004–2010 / American Chemical Society.

which correspond to approximately 8.5 nm, gave the difference in the measured temperatures by almost 20 °C under the NIR irradiation at 20 mW.

A nano-scale temperature profile under regional heating was also approached using thermally dissociable polymers attached to a gold nanoparticle (Scheme 5.1) (Kabb et al., 2015). The length between the thermally-labile azo group and the gold nanoparticle varied with the different number of polystyrene units (0.51, 0.83, 1.34, and 1.87 nm for repeating number n = 24, 63, 164, and 327, respectively). The fluorescence of an anthracene moiety in the polymer is quenched by a gold nanoparticle but restored after the temperature-dependent dissociation. Thus, the local temperature at the thermally-labile azo linkage could be evaluated from the

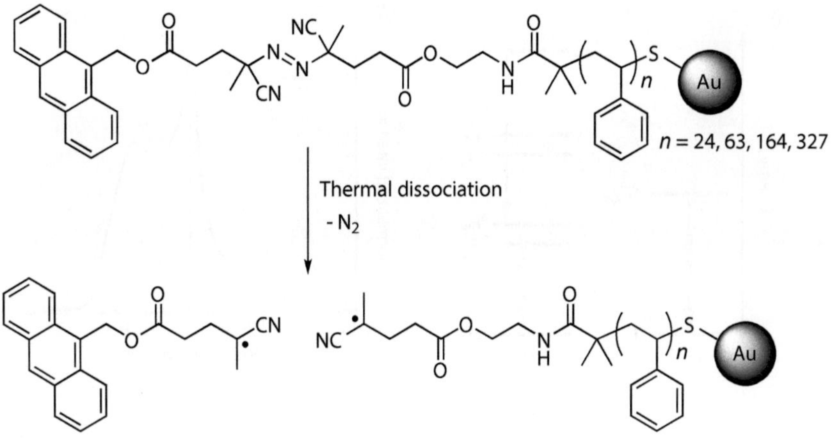

Scheme 5.1

fluorescence intensity of the anthracene moiety. Under the microwave irradiation on the gold nanoparticle in toluene, the temperature of the azo linkage was dependent on the distance from a gold nanoparticle: the temperature near the surface of the gold nanoparticle (0.51 nm away) was nearly 70 °C higher than that of the bulk solution, but no noticeable temperature increase was observed when the azo linkage was 1.87 nm away from the gold nanoparticle.

5.5
Possible Artifacts in Near-infrared Luminescent Thermometry and Proposal for Reliable Thermometry

Since 2018, the research group led by Professor Daniel Jaque, who has published many papers concerning inorganic luminescent thermometers, has clarified several reasons for false-positive signals of rare-earth-doped luminescent thermometers and proposed experimental procedures for more reliable thermometry (Labrador-Páez et al., 2018; Bednarkiewicz et al., 2020). For example, SrF_2:Yb,Tm UCNPs (diameter: 9.5 ± 1.4 nm) dispersed in deuterium oxide shows temperature-dependent fluorescence intensity ratios at 795 nm (due to $^3H_4 \rightarrow {}^3H_6$ transition of Tm^{3+} ion) and 769 nm (due to $^1G_4 \rightarrow {}^3H_5$ transition of Tm^{3+} ion) under excitation at 975 nm (Figure 5.12a). However, the emission spectra of SrF_2:Yb,Tm UCNPs also depend on the excitation power (Figure 5.12b, 5.12c). Thus, the gap in the depth between excitation and collection foci gives different fluorescence intensity ratios at 795 and 769 nm even though the environmental temperature is identical (Figure 5.12d), which causes a significant error (more than 70 °C) in measured temperature (Figure 5.12e). How tissues distorted the emission spectra of Ag_2S, Yb@Nd LaF_3, and Er-Yb@Yb-Tm LaF_3 nanoparticles in the range of 940–1640 nm and caused artifacts during in vivo fluorescence thermometry was also demonstrated in detail in a different report (Shen et al., 2020a). Self-absorption, i.e., partial absorption of the luminescence by the thermometer itself, and absorption of the luminescence signal by the medium's water molecules are other factors that impact the temperature-dependent fluorescence intensity ratio to lower the

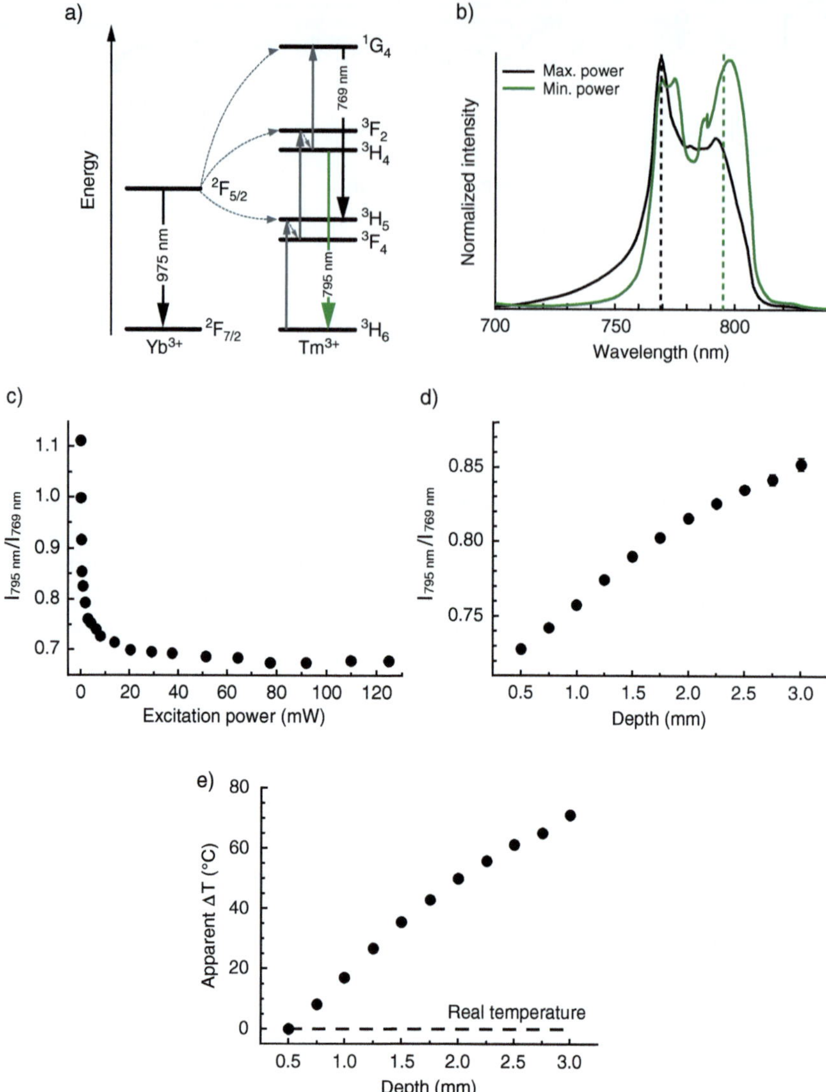

Figure 5.12 Influences of an excitation power-dependent emission band shape in luminescence thermometry. (a) Energy level diagram of Yb^{3+} and Tm^{3+} ions. A grey arrow indicates upconversion by the Yb^{3+} ion. Green and black arrows in the diagram of Tm^{3+} ion display emissions of interest at around 780 nm. (b) Emission spectra of SrF_2:Yb,Tm under excitation at 975 nm at different powers (approximately 2.8×10^6 mW cm^{-2} (green) and 4.0×10^9 mW cm^{-2} (black)). (c) Relationship between the emission intensity ratio of SrF_2:Yb,Tm at 795 and 769 nm (I_{795}/I_{769}) under the excitation at 975 nm and the excitation power at a fixed depth. (d) Relationship between the emission intensity ratio of SrF_2:Yb,Tm (I_{795}/I_{769}) under the excitation with a fixed power at 975 nm and a depth of the excitation and emission focus. (e) Depth-dependent error (ΔT) in temperature measurements by the emission intensity ratio of SrF_2:Yb,Tm (I_{795}/I_{769}) due to the excitation strength-dependent emission band shape. Labrador-Páez et al. (2018) *Nanoscale*, **10**, 22319–22328 / Royal Society of Chemistry.

reliability of thermometry notably. Similar artifacts were also reported in the thermometry using $NaYF_4:Yb^{3+},Er^{3+}$ upconversion microparticles (diameter: 1.4 ± 0.3 μm) from another research group (Pessoa et al., 2022), in which an improved technique for image processing was proposed to avoid misinterpretation.

Then, Jaque et al. also introduced the concept of multiparametric thermal sensing to avoid these artifacts and accomplish precise thermometry (Shen et al., 2020b). Ag_2S nanoparticles dispersed in PBS (phosphate-buffered saline) showed temperature-dependent emission spectra under laser excitation at 808 nm, in which multiparametric thermal sensing was possible by adopting the emission intensity (Figure 5.13a), the maximum emission wavelength (Figure 5.13b), and the emission intensity ratio at 1225 and 1175 nm (Figure 5.13c) as temperature-dependent signals. Independent signal reading led to discrepancies up to 7.5 °C in evaluating temperature during in vivo photothermal treatment using a mouse (Figure 5.13d). Then, the

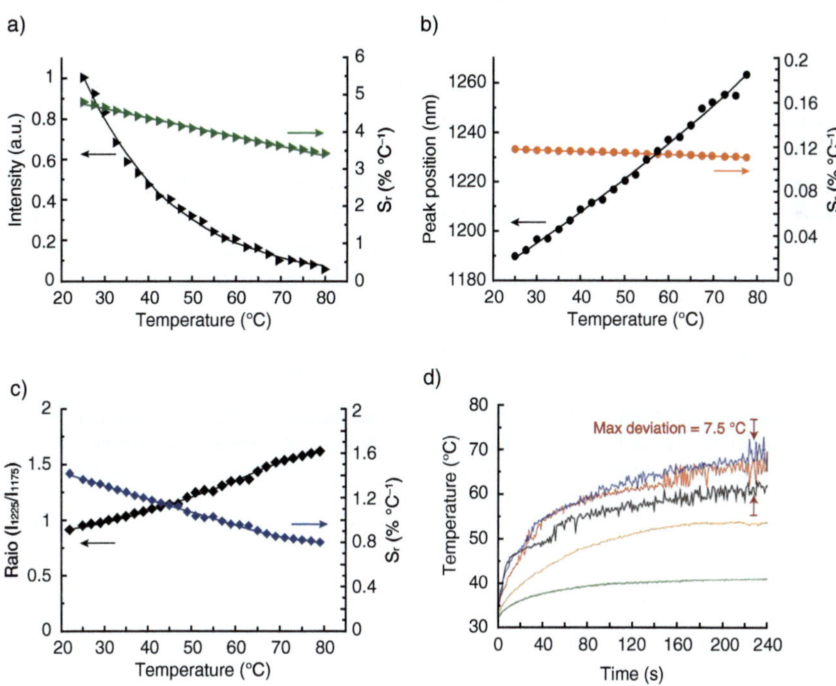

Figure 5.13 Temperature measurements using temperature-dependent multiple parameters of luminescent Ag_2S nanoparticles in PBS under the excitation at 808 nm. (a) Temperature-dependent emission intensity (left) and its sensitivity (right). (b) Temperature-dependent maximum emission wavelength (left) and its sensitivity (right). (c) Temperature-dependent emission intensity ratio at 1175 and 1225 nm (I_{1225}/I_{1175}) and its sensitivity (right). (d) Evaluated temperatures of a tumor in a mouse during photothermal treatment from the emission intensity (red), the maximum emission wavelength (blue), and the emission intensity ratio (I_{1225}/I_{1175}) (black) of Ag_2S nanoparticles. Tumor surface temperatures measured by thermography during photothermal treatment with (orange) and without (green) Ag_2S are also shown. Shen et al. (2020b) *Adv. Funct. Mater.*, **30**, 2002730 / John Wiley & Sons.

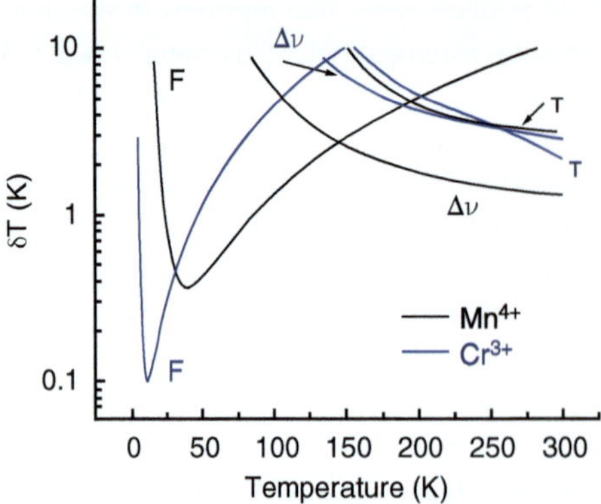

Figure 5.14 Temperature resolutions (δT) of a luminescent thermometer Al_2O_3-Cr,Mn in different sensing modes using luminescence intensity ratio (F), spectral shift ($\Delta \nu$), and luminescence lifetime (τ) of Cr^{3+} ion (blue) and Mn^{4+} ion (black). Mykhaylyk et al. (2021) *Dalton Trans.*, **50**, 14820–14831 / Royal Society of Chemistry.

evaluated temperatures by the different thermal readouts could be converged to the reliable value by statistical analyses with the reasonable assumption that the tissue minimally affected the maximum emission wavelength of Ag_2S nanoparticles. The same concept was found in thermometry using solid Al_2O_3 co-doped with Cr^{3+} and Mn^{4+} as multimodal sensing (Mykhaylyk et al., 2021), in which six different modes, i.e., emission intensity ratio, maximum emission wavelength, and emission lifetime of Cr^{3+} and Mn^{4+} ions, were available for self-validation (Figure 5.14).

References

Ano, T., Kishimoto, F., Sasaki, R., Tsubaki, S., Maitani, M. M., Suzuki, E., and Wada, Y. (2016) In situ temperature measurements of reaction spaces under microwave irradiation using photoluminescent probes. *Phys. Chem. Chem. Phys.*, **18**, 13173–13179.

Anonymous. (2014) The measure of reproducibility. *Nat. Methods*, **11**, 875.

Baffou, G., Rigneault, H., Marguet, D., and Jullien, L. (2014) A critique of methods for temperature imaging in single cells. *Nat. Methods*, **11**, 899–901.

Baffou, G., Rigneault, H., Marguet, D., and Jullien, L. (2015) Reply to: "Validating subcellular thermal changes revealed by fluorescent thermosensors" and "The 10^5 gap issue between calculation and measurement in single-cell thermometry". *Nat. Methods*, **12**, 803.

Baker, S. N., McCleskey, T. M., and Baker, G. A. (2005) An ionic liquid based optical thermometer, in *liquid-based IIIB: Fundamentals, Progress, Challenges and Opportunities: Transformations and Processes* (eds. R. D. Rogers and K. R. Sedden), ACS Symposium Series 902, American Chemical Society, Washington, pp. 171–181.

Balaban, R. S. (2020) How hot are single cells? *J. Gen. Physiol.*, **152**, e202012629.

Bastos, A. R. N., Brites, C. D. S., Rojas-Gutierrez, P. A., DeWolf, C., Ferreira, R. A. S., Capobianco, J. A., and Carlos, L. D. (2019) Thermal properties of lipid bilayers determined using upconversion nanothermometry. *Adv. Funct. Mater.*, **29**, 1905474.

Bastos, A. R. N., Brites, C. D. S., Rojas-Gutierrez, P. A., Ferreira, R. A. S., Longo, R. L., DeWolf, C., Capobianco, J. A., and Carlos, L. D. (2020) Thermal properties of lipid bilayers derived from the transient heating regime of upconverting nanoparticles. *Nanoscale*, **12**, 24169–24176.

Bednarkiewicz, A., Marciniak, L., Carlos, L. D., and Jaque, D. (2020) Standardizing luminescence nanothermometry for biomedical applications. *Nanoscale*, **12**, 14405–14421.

Bommidi, D. K. and Pickel, A. D. (2021) Temperature-dependent excited state lifetimes of nitrogen vacancy centers in individual nanodiamonds. *Appl. Phys. Lett.*, **119**, 254103.

Bright, G. R., Fisher, G. W., Rogowska, J., and Taylor, D. L. (1987) Fluorescence ratio imaging microscopy: temporal and spatial measurements of cytoplasmic pH. *J. Cell Biol.*, **104**, 1019–1033.

Brites, C. D. S., Lima, P. P., Silva, N. J. O., Millán, A., Amaral, V. S., Palacio, F., and Carlos, L. D. (2012) Thermometry at the nanoscale. *Nanoscale*, **4**, 4799–4829.

Chrétien, D., Bénit, P., Ha, H.-H., Keipert, S., El-Khoury, R., Chang, Y.-T., Jastroch, M., Jacobs, H. T., Rustin, P., and Rak, M. (2018) Mitochondria are physiologically maintained at close to 50 °C. *PLoS Biol.*, **16**, e2003992.

Clarke, A. and Rothery, P. (2008) Scaling of body temperature in mammals and birds. *Funct. Ecol.*, **22**, 58–67.

Fahimi, P. and Matta, C. F. (2022) The hot mitochondrion paradox: reconciling theory and experiment. *Trends Chem.*, **4**, 96–110.

Fourier, J. (1878) *The Analytical Theory of Heat*, (Translated by A. Freeman). The University Press, London.

Freddi, S., Sironi, L., D'Antuono, R., Morone, D., Donà, A., Cabrini, E., D'Alfonso, L., Collini, M., Pallavicini, P., Baldi, G., Maggioni, D., and Chirico, G. (2013) A molecular thermometer for nanoparticles for optical hyperthermia. *Nano Lett.*, **13**, 2004–2010.

Fushimi, K. and Verkman, A. S. (1991) Low viscosity in the aqueous domain of cell cytoplasm measured by picosecond polarization microfluorimetry. *J. Cell Biol.*, **112**, 719–725.

Gnutt, D., Gao, M., Brylski, O., Heyden, M., and Ebbinghaus, S. (2015) Excluded-volume effects in living cells. *Angew. Chem. Int. Ed.*, **54**, 2548–2551.

Gota, C., Okabe, K., Funatsu, T., Harada, Y., and Uchiyama, S. (2009) Hydrophilic fluorescent nanogel thermometer for intracellular thermometry. *J. Am. Chem. Soc.*, **131**, 2766–2767.

Grillo-Hill, B. K., Webb, B. A., and Barber, D. L. (2014) Ratiometric imaging of pH probes. *Methods Cell Biol.*, **123**, 429–448.

Inomata, N., Toda, M., Sato, M., Ishijima, A., and Ono, T. (2012) Pico calorimeter for detection of heat produced in an individual brown fat cell. *Appl. Phys. Lett.*, **100**, 154104.

Jenkins, J., Borisov, S. M., Papkovsky, D. B., and Dmitriev, R. I. (2016) Sulforhodamine nanothermometer for multiparametric fluorescence lifetime imaging microscopy. *Anal. Chem.*, **88**, 10566–10572.

Johannessen, E. A., Weaver, J. M. R., Bourova, L., Svoboda, P., Cobbold, P. H., and Cooper, J. M. (2002) Micromachined nanocalorimetric sensor for ultra-low-volume cell-based assays. *Anal. Chem.*, **74**, 2190–2197.

Kabb, C. P., Carmean, R. N., and Sumerlin, B. S. (2015) Probing the surface-localized hyperthermia of gold nanoparticles in a microwave field using polymeric thermometers. *Chem. Sci.*, **6**, 5662–5669.

Kang, J.-S. (2018) Theoretical model and characteristics of mitochondrial thermogenesis. *Biophys. Rep.*, **4**, 63–67.

Kiyonaka, S., Kajimoto, T., Sakaguchi, R., Shinmi, D., Omatsu-Kanbe, M., Matsuura, H., Imamura, H., Yoshizaki, T., Hamachi, I., Morii, T., and Mori, Y. (2013) Genetically encoded fluorescent thermosensors visualize subcellular thermoregulation in living cells. *Nat. Methods*, **10**, 1232–1238.

Kiyonaka, S., Sakaguchi, R., Hamachi, I., Morii, T., Yoshizaki, T., and Mori, Y. (2015) Validating subcellular thermal changes revealed by fluorescent thermosensors. *Nat. Methods*, **12**, 801–802.

Kuimova, M. K., Yahioglu, G., Levitt, J. A., and Suhling, K. (2008) Molecular rotor measures viscosity of live cells via fluorescence lifetime imaging. *J. Am. Chem. Soc.*, **130**, 6672–6673.

Labrador-Páez, L., Pedroni, M., Speghini, A., García-Solé, J., Haro-González, P., and Jaque, D. (2018) Reliability of rare-earth-doped infrared luminescent nanothermometers. *Nanoscale*, **10**, 22319–22328.

Lane, N. (2018) Hot mitochondria? *PLoS Biol.*, **16**, e2005113.

Lemmon, E. W. and Harvey, A. H. (2022) Thermophysical properties of water and steam in *CRC Handbook of Chemistry and Physics*, 103rd ed. (eds. J. R. Rumble, Jr., T. J. Bruno, and M. J. Doa), CRC Press, Boca Raton, pp. 6–1 to 6–4.

Liang, L., Wang, X., Xing, D., and Chen, T. (2009) Noninvasive determination of cell nucleoplasmic viscosity by fluorescence correlation spectroscopy. *J. Biomed. Opt.*, **14**, 024013.

Liu, B., Poolman, B., and Boersma, A. J. (2017) Ionic strength sensing in living cells. *ACS Chem. Biol.*, **12**, 2510–2514.

Llopis, J., McCaffery, J. M., Miyawaki, A., Farquhar, M. G., and Tsien, R. Y. (1998) Measurement of cytosolic, mitochondrial, and Golgi pH in single living cells with green fluorescent proteins. *Proc. Natl. Acad. Sci. USA*, **95**, 6803–6808.

Lodish, H., Berk, A., Kaiser, C. A., Krieger, M., Scott, M. P., Bretscher, A., Ploegh, H., and Matsudaira, P. (2007) *Molecular Cell Biology*, 6th edn, W. H. Freeman, New York, pp. 448–449.

Loesberg, C., van Miltenburg, J. C., and van Wijk, R. (1982) Heat production of mammalian cells at different cell-cycle phases. *J. Therm. Biol.*, **7**, 209–213.

Losa, J., Leupold, S., Alonso-Martinez, D., Vainikka, P., Thallmair, S., Tych, K. M., Marrink, S. J., and Heinemann, M. (2022) Perspective: a stirring role for metabolism in cells. *Mol. Syst. Biol.*, **18**, e10822.

Lu, K., Wazawa, T., Sakamoto, J., Vu, C. Q., Nakano, M., Kamei, Y., and Nagai, T. (2022) Intracellular heat transfer and thermal property revealed by kilohertz temperature imaging with a genetically encoded nanothermometer. *Nano Lett.*, **22**, 5698–5707.

Luby-Phelps, K., Mujumdar, S., Mujumdar, R. B., Ernst, L. A., Galbraith, W., and Waggoner, A. S. (1993) A novel fluorescence ratiometric method confirms the low solvent viscosity of the cytoplasm. *Biophys. J.*, **65**, 236–242.

Macherel, D., Haraux, F., Guillou, H., and Bourgeois, O. (2021) The conundrum of hot mitochondria. *Biochim. Biophys. Acta Bioenerg.*, **1862**, 148348.

Miller, E. W., Lin, J. Y., Frady, E. P., Steinbach, P. A., Kristan, W. B. Jr., and Tsien, R. Y. (2012) Optically monitoring voltage in neurons by photo-induced electron transfer through molecular wires. *Proc. Natl. Acad. Sci. USA*, **109**, 2114–2119.

Mizukami, K., Muraoka, T., Shiozaki, S., Tobita, S., and Yoshihara, T. (2022) Near-infrared emitting Ir(III) complexes bearing a dipyrromethene ligand for oxygen imaging of deeper tissues *in vivo*. *Anal. Chem.*, **94**, 2794–2802.

Mudrak, N. J., Rana, P. S., and Model, M. A. (2018) Calibrated brightfield-based imaging for measuring intracellular protein concentration. *Cytometry A*, **93A**, 297–304.

Mykhaylyk, V. B., Kraus, H., Bulyk, L.-I., Lutsyuk, I., Hreb, V., Vasylechko, L., Zhydachevskyy, Y., Wagner, A., and Suchocki, A. (2021) Al_2O_3 co-doped with Cr^{3+} and Mn^{4+}, a dual-emitter probe for multimodal non-contact luminescence thermometry. *Dalton Trans.*, **50**, 14820–14831.

Okabe, K., Inada, N., Gota, C., Harada, Y., Funatsu, T., and Uchiyama, S. (2012) Intracellular temperature mapping with a fluorescent polymeric thermometer and fluorescence lifetime imaging microscopy. *Nat. Commun.*, **3**, 705.

Okabe, K., Sakaguchi, R., Shi, B., and Kiyonaka, S. (2018) Intracellular thermometry with fluorescent sensors for thermal biology. *Pflügers Arch.*, **470**, 717–731.

Okabe, K. and Uchiyama, S. (2021) Intracellular thermometry uncovers spontaneous thermogenesis and associated thermal signaling. *Commun. Biol.*, **4**, 1377.

Oyama, K., Gotoh, M., Hosaka, Y., Oyama, T. G., Kubonoya, A., Suzuki, Y., Arai, T., Tsukamoto, S., Kawamura, Y., Itoh, H., Shintani, S. A., Yamazawa, T., Taguchi, M., Ishiwata, S., and Fukuda, N. (2020) Single-cell temperature mapping with fluorescent thermometer nanosheets. *J. Gen. Physiol.*, **152**, e201912469.

Park, B. K., Woo, Y., Jeong, D., Park, J., Choi, T.-Y., Simmons, D. P., Ha, J., and Kim, D. (2016) Thermal conductivity of biological cells at cellular level and correlation with disease state. *J. Appl. Phys.*, **119**, 224701.

Pessoa, A. R., Galindo, J. A. O., Serge-Correales, Y. E., Amaral, A. M., Ribeiro, S. J. L., and Menezes, L. de S. (2022) 2D thermal maps using hyperspectral scanning of single upconverting microcrystals: experimental artifact and image processing. *ACS Appl. Mater. Interfaces*, **14**, 38311–38319.

Peterson, J. (2014) Evolution, entropy, & biological information. *Am. Biol. Teach.*, **76**, 88–92.

Rajagopal, M. C. and Sinha, S. (2021) Cellular thermometry considerations for probing biochemical pathways. *Cell Biochem. Biophys.*, **79**, 359–373.

Ramires, M. L. V., de Castro, C. A. N., Nagasaka, Y., Nagashima, A., Assael, M. J., and Wakeham, W. A. (1995) Standard reference data for the thermal conductivity of water. *J. Phys. Chem. Ref. Data*, **24**, 1377–1381.

Riedel, C., Gabizon, R., Wilson, C. A. M., Hamadani, K., Tsekouras, K., Marqusee, S., Pressé, S., and Bustamante, C. (2015) The heat released during catalytic turnover enhances the diffusion of an enzyme. *Nature*, **517**, 227–230.

Rodríguez-Sevilla, P., Spicer, G., Sagrera, A., Adam, A. P., Efeyan, A., Jaque, D., and Thompson, S. A. (2023) Bias in intracellular luminescence thermometry: the case of the green fluorescent protein. *Adv. Optical Mater.*, **11**, 2201664.

Rolfe, D. F. S. and Brown, G. C. (1997) Cellular energy utilization and molecular origin of standard metabolic rate in mammals. *Physiol. Rev.*, **77**, 731–758.

Schrödinger, E. (1992) *What is Life?*, Cambridge University Press, Cambridge.

Sekiguchi, T., Sotoma, S., and Harada, Y. (2018) Fluorescent nanodiamonds as a robust temperature sensor inside a single cell. *Biophys. Physicobiol.*, **15**, 229–234.

Shen, Y., Lifante, J., Fernández, N., Jaque, D., and Ximendes, E. (2020a) In vivo spectral distortions of infrared luminescent nanothermometers compromise their reliability. *ACS Nano.*, **14**, 4122–4133.

Shen, Y., Santos, H. D. A., Ximendes, E. C., Lifante, J., Sanz-Portilla, A., Monge, L., Fernández, N., Chaves-Coira, I., Jacinto, C., Brites, C. D. S., Carlos, L. D., Benayas, A., Iglesias-de la Cruz, M. C., and Jaque, D. (2020b) Ag_2S nanoheaters with multiparameter sensing for reliable thermal feedback during in vivo tumor therapy. *Adv. Funct. Mater.*, **30**, 2002730.

Shrestha, R., Atluri, R., Simmons, D. P., Kim, D. S., and Choi, T. Y. (2020) Thermal conductivity of a Jurkat cell measured by a transient laser point heating method. *Int. J. Heat Mass Transf.*, **160**, 120161.

Song, P., Gao, H., Gao, Z., Liu, J., Zhang, R., Kang, B., Xu, J.-J., and Chen, H.-Y. (2021) Heat transfer and thermoregulation within single cells revealed by transient plasmonic imaging. *Chem*, **7**, 1569–1587.

Sotoma, S., Zhong, C., Kah, J. C. Y., Yamashita, H., Plakhotnik, T., Harada, Y., and Suzuki, M. (2021) In situ measurements of intracellular thermal conductivity using heater-thermometer hybrid diamond nanosensors. *Sci. Adv.*, **7**, eabd7888.

Suzuki, M., Zeeb, V., Arai, K., Oyama, K., and Ishiwata, S. (2015) The 10^5 gap issue between calculation and measurement in single-cell thermometry. *Nat. Methods*, **12**, 802–803.

Takahashi, A., Camacho, P., Lechleiter, J. D., and Herman, B. (1999) Measurement of intracellular calcium. *Physiol. Rev.*, **79**, 1089–1125.

Takei, Y., Arai, S., Murata, A., Takabayashi, M., Oyama, K., Ishiwata, S., Takeoka, S., and Suzuki, M. (2014) A nanoparticle-based ratiometric and self-calibrated fluorescent thermometer for single living cells. *ACS Nano*, **8**, 198–206.

Thomas, J. A., Buchsbaum, R. N., Zimniak, A., and Racker, E. (1979) Intracellular pH measurements in Ehrlich ascites tumor cells utilizing spectroscopic probes generated in situ. *Biochemistry*, **18**, 2210–2218.

Uchiyama, S., Gota, C., Tsuji, T., and Inada, N. (2017) Intracellular temperature measurements with fluorescent polymeric thermometers. *Chem. Commun.*, **53**, 10976–10992.

Vishnu, G. K. A., Gogoi, G., Behera, B., Rila, S., Rangarajan, A., and Pandya, H. J. (2022) RapidET: a MEMS-based platform for label-free and rapid demarcation of tumors from normal breast biopsy tissues. *Microsyst. Nanoeng.*, **8**, 1.

Wu, N., Sun, Y., Kong, M., Lin, X., Cao, C., Li, Z., Feng, W., and Li, F. (2022) Er-based luminescent nanothermometer to explore the real-time temperature of cells under external stimuli. *Small*, **18**, 2107963.

Ximendes, E., Marin, R., Carlos, L. D., and Jaque, D. (2022) Less is more: dimensionality reduction as a general strategy for more precise luminescence thermometry. *Light Sci. Appl.*, **11**, 237.

Yang, F., Li, G., Yang, J., Wang, Z., Han, D., Zheng, F., and Xu, S. (2017) Measurement of local temperature increments induced by cultured HepG2 cells with micro-thermocouples in a thermally stabilized system. *Sci. Rep.*, **7**, 1721.

Yuexuan, Y. and Daocheng, W. (2020) Research shortcomings of fluorescent nanothermometers in biological and medical fields. *Nanomedicine*, **15**, 735–738.

6
Applications of Intracellular Thermometry

The realization of intracellular thermometry has reasonably expanded the realm of scientific research that was found with interest in intracellular temperature. It includes cell biology and cell biotechnology, of course. In vivo temperature measurements and medical applications such as a precisely temperature-controlled drug delivery system are still within our expectations. Even new concepts based on intracellular thermometry have started emerging. This chapter will introduce each application category based on intracellular thermometry. Remember that the application areas are not limited to those described below. You, the readers, have the potential to create new areas of research in intracellular temperature measurement with your unique perspectives and abilities.

6.1
Creation of New Biological Concepts

Based on some observations of the temperature-dependent cellular activity described in Section 3.7.4 (Hoshi et al., 2018) and 3.7.7 (Choi et al., 2020), Dr. Kohki Okabe and I proposed a new biological concept, "thermal signaling", in which a temperature variation within biological cells or bodies acts as an input signal in a cascade of intriguing biological events (Okabe and Uchiyama, 2021). Besides, the reported intracellular temperature gradients between the nucleus and the cytoplasm (Section 3.4.1.1) inspired scientists to propose new concepts. "Intracellular thermophoresis" stands for heat-induced distributions of endogenous biomolecules inside living cells (Talbot et al., 2017). A similar idea has been reported as cytoplasmic convection which could provide a new mechanism for directed, bulk transport within living cells (Howard et al., 2019). A nucleus-involved heat control system has been assumed within living cells as "cell thermoregulation" (Ibraimov, 2017). Theoretical modeling of biological membrane structures exposed to a temperature gradient is also proposed (Atia and Givli,

2014). Intracellular thermometry will also be able to reveal detailed mechanisms of temperature-dependent cellular events such as the thermoresistance after hyperthermia due to the downregulation of glycolytic metabolism (Kanamori et al., 2021) and cooperative regulation of actin-myosin interaction by a regulatory protein drebrin E (Kubota et al., 2021).

Temperature measurements of an intracellular object smaller than a living cell and its organelles have also been proposed (Maciel, 2022). Protein folding and mechanical works can be viewed as thermodynamic processes which are influenced by environmental thermal fluctuations and alter local temperature. A similar viewpoint has also been recognized on RNA (Becskei and Rahaman, 2022); RNA turnover may contribute to intracellular temperature gradients through basal metabolism, and vice versa, RNA may change its function depending on the local environmental temperature. Since fluorescent molecular thermometers are small enough, measuring the temperature of intracellular proteins or RNAs and the temperature gradient near them is possible by a fluorescent labeling method.

6.2
In Vivo Temperature Measurements

One of the practical applications of fluorescent molecular thermometers is in vivo temperature measurements, although in vivo fluorescence imaging does not always make the most of fluorometry's extremely high spatial resolution. Fluorescent probes used for in vivo fluorescence imaging need special characteristics regarding the excitation and emission wavelengths to exclude interference by tissue (Frangioni, 2003; Larson et al., 2003; Hong et al., 2012). Recently, brain thermometry has been proposed as a principal challenge of luminescent thermometers (Rodríguez-Sevilla et al., 2022).

Figure 6.1a indicates temperature maps of a living zebrafish larva at 22 and 28 °C after injection of a phosphorescent polymeric thermometer P3 (Chen et al., 2016). The phosphorescence lifetime, which was calculated from the photons collected at 482 ± 35 nm with the excitation at 405 nm at the time range of 150 to 2000 ns, was temperature-dependent and could be converted to temperature. The time-gated imaging diminished autofluorescence from the yolk sac and belly of the zebrafish remarkably. Figure 6.1b displays real-time thermometry of *Caenorhabditis elegans* (*C. elegans*) with fluorescent nanodiamonds that are coated by polyglycerol to improve dispersibility in aqueous environments and introduced into the gonads by a microinjection technique (Fujiwara et al., 2020). A temperature increase of *C. elegans* by the treatment of an uncoupler FCCP (carbonyl cyanide 4-(trifluoromethoxy) phenylhydrazone) could be monitored by the fluorescent nanodiamonds with a precision of 0.22–0.31 °C.

Figure 6.2 demonstrates in vivo temperature imaging using fluorescent proteins. A fluorescent molecular thermometer GFP (see Section 3.2.9) was tagged to glutamic acid decarboxylase (GAD) and expressed in GABAergic neurons of *C. elegans* (Figure 6.2a) (Donner et al., 2013). With the calibration curve, i.e., the relationship between

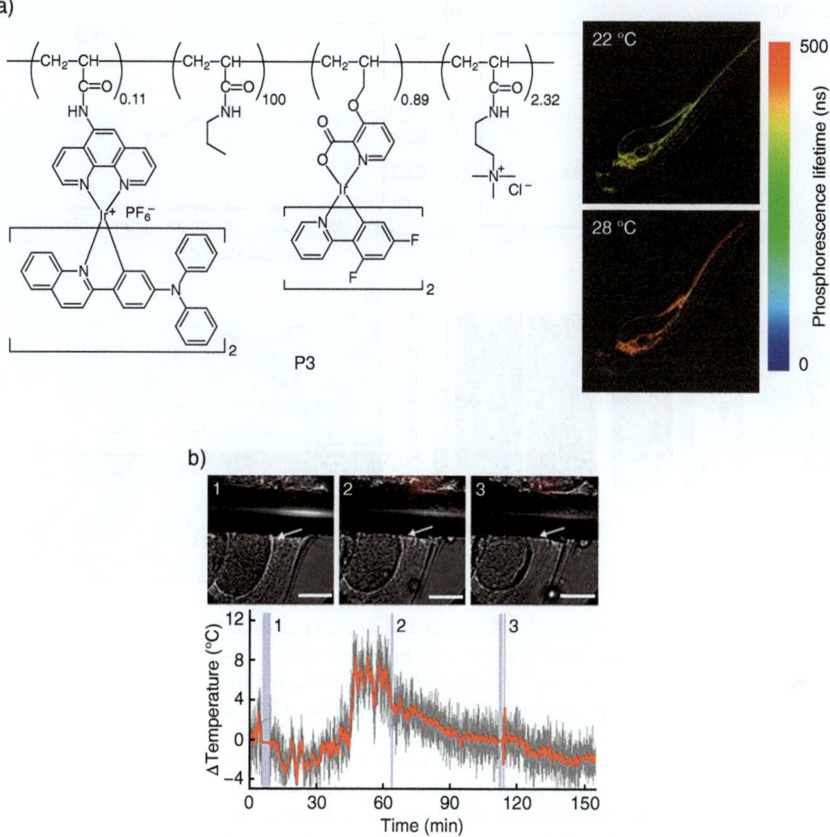

Figure 6.1 In vivo thermometry of a zebrafish larva and *Caenorhabditis elegans* (*C. elegans*).
(a) Temperature mapping of a zebrafish larva by a phosphorescent polymeric thermometer P3.
Chemical structure of P3 (left) and phosphorescence lifetime images of P3 in a zebrafish larva
at 22 and 28 °C (right). The phosphorescence lifetime images were constructed by collecting
photons at 482 ± 35 nm with the excitation at 405 nm in the time range 150–2000 ns. Adapted
from Chen et al. (2016) *Adv. Funct. Mater.*, **26**, 4386–4396 / John Wiley & Sons. (b) Temperature
monitoring inside *C. elegans* by fluorescent nanodiamonds. Merged images of fluorescence
(red) and bright field (gray scale) (top) and a temperature profile evaluated by a fluorescent
nanodiamond (arrows) during the treatment with 60 µmol L^{-1} FCCP (bottom). The numbers
indicated in the merged images are the timestamps corresponding to those in the temperature
profile. Scale bars: 20 µm. Fujiwara et al. (2020) *Sci. Adv.*, **6**, eaba9636 / American Association
for the Advancement of Science - AAAS.

the fluorescence polarization anisotropy (FPA) of GFP and the temperature (Figure
6.2b), obtained with the GFP expressing worms, in vivo temperature imaging of C.
elegans under external and internal heating with the gold nanorods and infrared laser
irradiation at 800 nm was performed (Figure 6.2c). A genetically encoded ratiometric fluorescent thermometer gTEMP, consisting of a temperature-sensitive protein
Sirius and a reference protein mT-Sapphire, also enabled temperature imaging of a
medaka embryo with temperature-dependent fluorescence intensity ratio of Sirius
and mT-Sapphire moieties (Figure 6.2d) (Nakano et al., 2017).

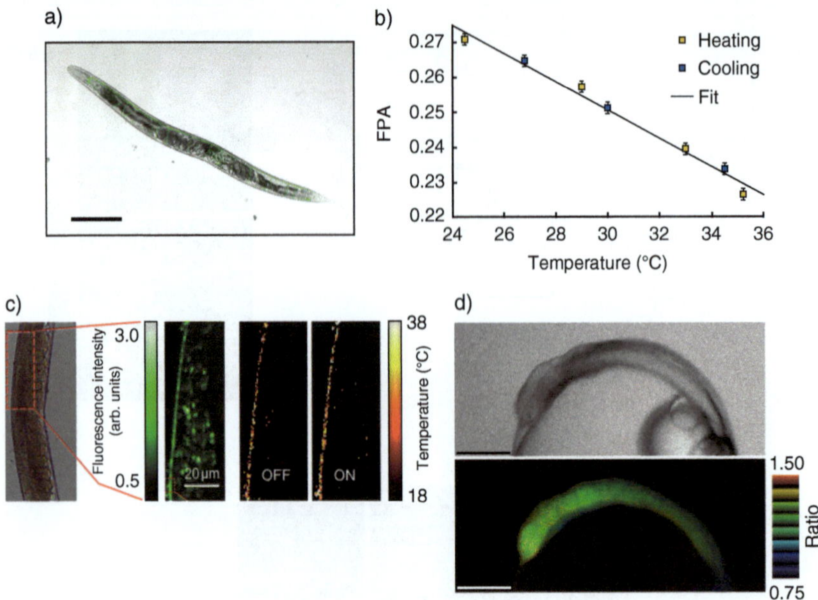

Figure 6.2 In vivo thermometry using fluorescent proteins. (a–c) In vivo thermometry of *Caenorhabditis elegans* (*C. elegans*) by green fluorescent protein (GFP). (a) Bright-field image overlapped with the fluorescence of GFP in GABAergic neurons. Scale bar: 200 μm. (b) Relationship between the fluorescence polarization anisotropy (FPA) and the temperature while heating and cooling. (c) Temperature monitoring of *C. elegans* upon external laser heating. Bright-field image (leftmost), fluorescence image (second left), and temperature images of the neurons before and after the laser heating (second right and rightmost, respectively). Adapted from Donner et al. (2013) *ACS Nano*, **7**, 8666–8672 / American Chemical Society. (d) Temperature mapping of a medaka embryo by a genetically encoded ratiometric fluorescent thermometer gTEMP. Bright-field image (top) and pseudo-colored ratio image (bottom). A higher fluorescence intensity ratio represents a higher temperature. Scale bars: 250 μm. Nakano et al. (2017) *PLoS ONE*, **12**, e0172344 / PUBLIC LIBRARY OF SCIENCE (PLOS) / CC BY 4.0.

In vivo temperature measurements using a europium complex Eu-DT (europium(III)-tris(dinaphthoylmethane)-bis-trioctylphosphine oxide) are summarized in Figure 6.3. Figure 6.3a, 6.3b shows the temperature measurements of fruit fly larvae using poly(styrene-*co*-methacrylic acid) (PS-MA) nanoparticles which contain Eu-DT (as a fluorescent molecular thermometer) and Ir(ppy)$_3$ (tris(2-phenylpyridinato)iridium(III), as a reference) and are coated with polyvinyl alcohol (Arai et al., 2015a). The PS-MA nanoparticles were orally dosed to fruit fly larvae, and their internal temperature was evaluated from the temperature-dependent fluorescence intensity ratio of Eu-DT and Ir(ppy)$_3$ with a moderate temperature resolution (1–1.7 °C). The PS-MA nanoparticles, which contain Eu-DT (as a fluorescent molecular thermometer) and rhodamine 800 (as a reference, Figure 6.3c) and are coated with polyvinyl alcohol, were utilized for temperature imaging of a flight muscle of a living *Dicronorhina derbyana* beetle (Ferdinandus et al., 2016; Miyagawa et al., 2016) after the loading onto the dorsal longitudinal muscle. Through the temperature-dependent fluorescence intensity ratio at 615 nm (due to Eu-DT) and 710 nm (due to rhodamine 800), the transient temperature increase in preflight preparation of *D. derbyana* could be monitored with the temperature resolution of 1–1.4 °C (Figure 6.3d, 6.3e).

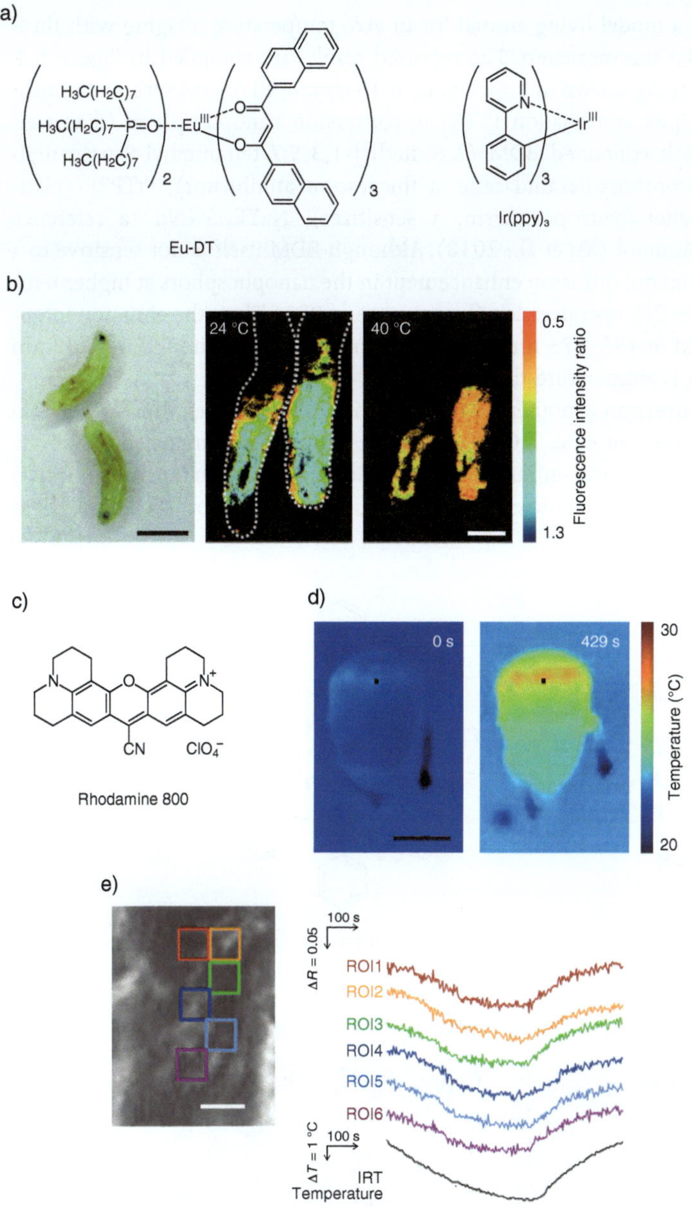

Figure 6.3 In vivo thermometry of fruit fly larvae and a *Dicronorhina derbyana* beetle with a europium complex. (a,b) In vivo thermometry of fruit fly larvae with poly(styrene-*co*-methacrylic acid) nanoparticles containing Eu-DT and Ir(ppy)₃. (a) Chemical structures of Eu-DT and Ir(ppy)₃. (b) Temperature mapping of fruit fly larvae by poly(styrene-*co*-methacrylic acid) nanoparticles containing Eu-DT and Ir(ppy)₃. Bright-field image (left, scale bar: 2 mm) and fluorescence ratio images at 24 and 40 °C (middle and right, respectively, scale bar: 500 μm). Arai et al. (2015a) *Analyst*, **140**, 7534–7539 / Royal Society of Chemistry. (c–e) Temperature monitoring of a *Dicronorhina derbyana* beetle by ratiometric poly(styrene-*co*-methacrylic acid) nanoparticle thermosensor (RNT) containing Eu-DT and rhodamine 800. (c) Chemical structure of rhodamine 800. (d) The increase in muscle temperature at the preflight preparation that was monitored by infrared thermography (IRT). Scale bar: 10 mm. (e) The increase in muscle temperature at the preflight preparation that was monitored by temperature-dependent fluorescence intensity ratio (*R*) of RNT in different ROIs (regions of interest). The black spot in the thermography image in panel (d) was the area indicated in the fluorescence image (left). Scale bar: 100 μm. Ferdinandus et al. (2016) *ACS Sens.*, **1**, 1222–1227 / American Chemical Society.

The mouse is a model living animal for in vivo temperature imaging with fluorescent molecular thermometers. The reported results are compiled in Figure 6.4. Temperature imaging shown in Figure 6.4a, 6.4b employed bovine serum albumin-coated triple–triplet annihilation (TTA) upconversion nanophosphors (diameter: ~160 nm) which contained BDM (2,6-diethyl-1,3,5,7-tetramethyl-8-trimethyl-phenyl-4,4-difluoroboradiazaindacene, a fluorescent annihilator), PtTPBP (platinum tetraphenyltetrabenzoporphyrin, a sensitizer), $NaYF_4$:5%Nd (a reference luminophore) in nujol (Xu et al., 2018). Although BDM itself is not sensitive to a temperature variation, diffusion enhancement in the nanophosphors at higher temperatures induces TTA upconversion fluorescence of BDM. Thus, the emission intensity ratio of BDM in 485–575 nm (λ_{ex} = 635 nm) and Nd^{3+} ion in 980–1300 nm (λ_{ex} = 808 nm) is temperature-dependent.

Another upconversion nanothermometer for in vivo thermometry involves a photochemical reaction (Su et al., 2020). In bovine serum albumin-coated upconversion nanocapsule, a near-infrared photosensitizer $PdPc(OBu)_8$ (palladium(II) 1,4,8,11,15,18,22,25-octabutoxypthalacyanine, Figure 6.4c) absorbs laser light at

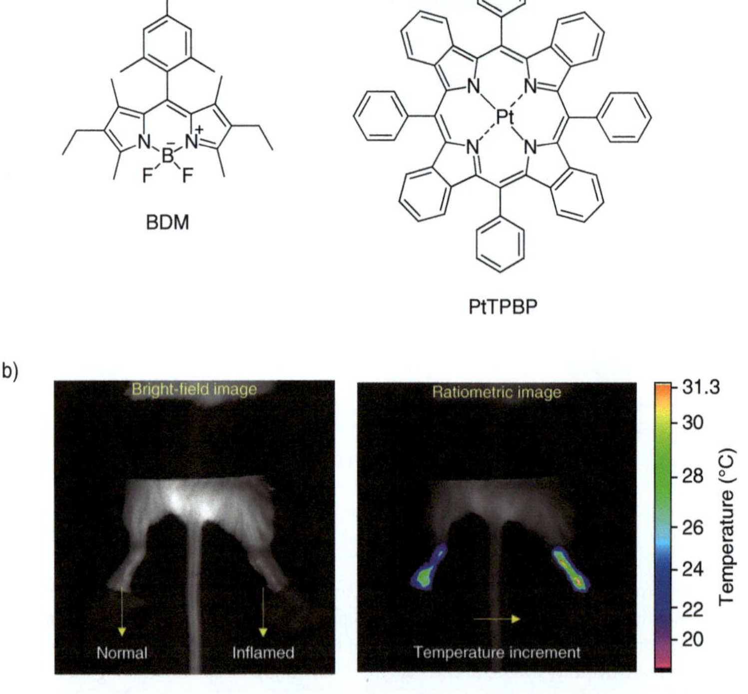

Figure 6.4 In vivo fluorescent thermometry of a mouse. (a,b) Thermometry by triplet–triplet annihilation Nd nanophosphors (TTA-Nd-NPs) composed of BDM (a fluorescent annihilator), PtTPBP (a sensitizer), and $NaYF_4$:5%Nd (a reference luminophore). (a) Chemical structures of BDM and PtTPBP. (b) Bright-field image (left) and temperature image (right) of a Kunming mouse with mild arthritis symptoms in the right leg. TTA-Nd-NPs were used as fluorescent thermometers. The emission intensity ratio of BDM and Nd^{3+} ion was converted to the temperature. Xu et al. (2018) *Nat. Commun.*, **9**, 2698 / Springer Nature / CC BY 4.0.

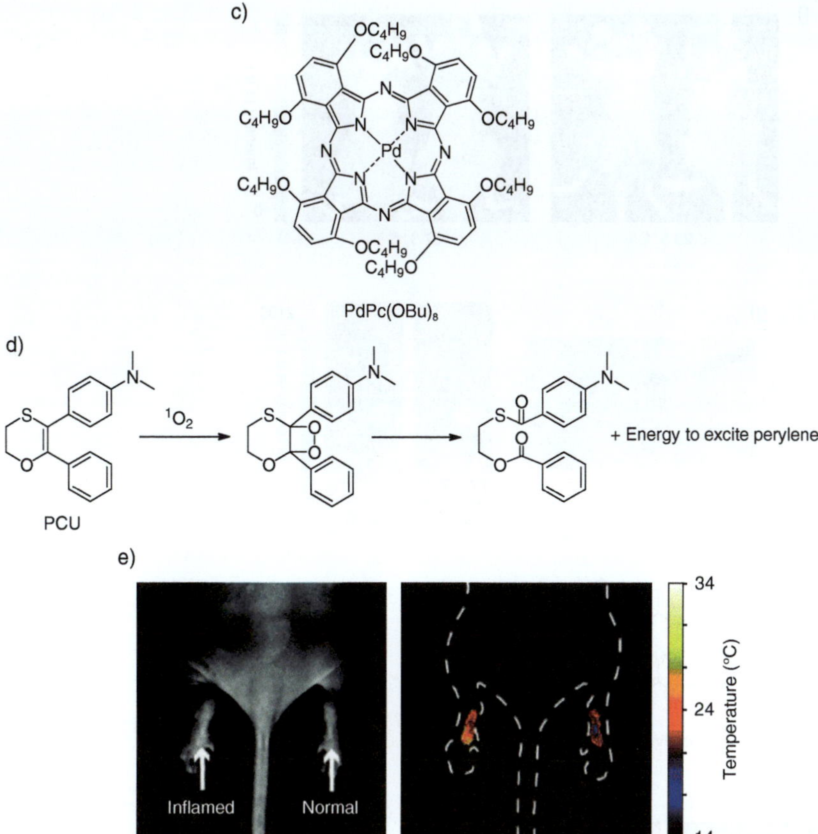

Figure 6.4 (continued) (c–e) Thermometry by ultra-long-lived luminescent nanocapsules (UL-NCs) composed of PdPc(OBu)$_8$, PCU, and perylene. (c) Chemical structure of PdPc(OBu)$_8$. (d) Photochemical reaction of PCU and singlet oxygen. (e) Bright-field image (left) and temperature image (right) of a Kunming mouse with mild arthritis symptoms in the left leg. The fluorescent thermometers UL-NCs were injected into the mouse. Su et al. (2020) *Chem. Commun.*, **56**, 10694–10697.

730 nm and transfers the photoenergy to oxygen to produce singlet oxygen ($^1O_2$). Then, the singlet oxygen reacts with a photoenergy cache unit PCU (4-(5,6-dihydro-2-phenyl-1,4-oxathiin-3-yl)-N,N-dimethylbenzenamine), and the energy is stored in chemical bonds, which is followed by energy transfer to a fluorescent compound perylene to make it emit fluorescence (Figure 6.4d). The photochemical reaction of PCU with singlet oxygen is the rate-determining step in the second scale: a faster fluorescence lifetime of perylene is observed at a higher temperature. Finally, this upconversion system was applied to measure the temperature in the Kunming mouse inflammatory model (Figure 6.4e). The left leg had mild arthritis symptoms, and the right leg was normal. The average temperature, calculated from the fluorescence lifetime of perylene, of the arthritic area in the left leg was 2.6 °C higher than the normal area of the right leg. The thermal resolution of the photochemical reaction-based upconversion nanothermometer was evaluated to be ~0.3 °C at 25 °C.

Figure 6.4f indicates in vivo fluorescence images of tumor-bearing nude mice with polyoxyethylene stearate micelles encapsulating CuInS$_2$/ZnS quantum dots (Zhang

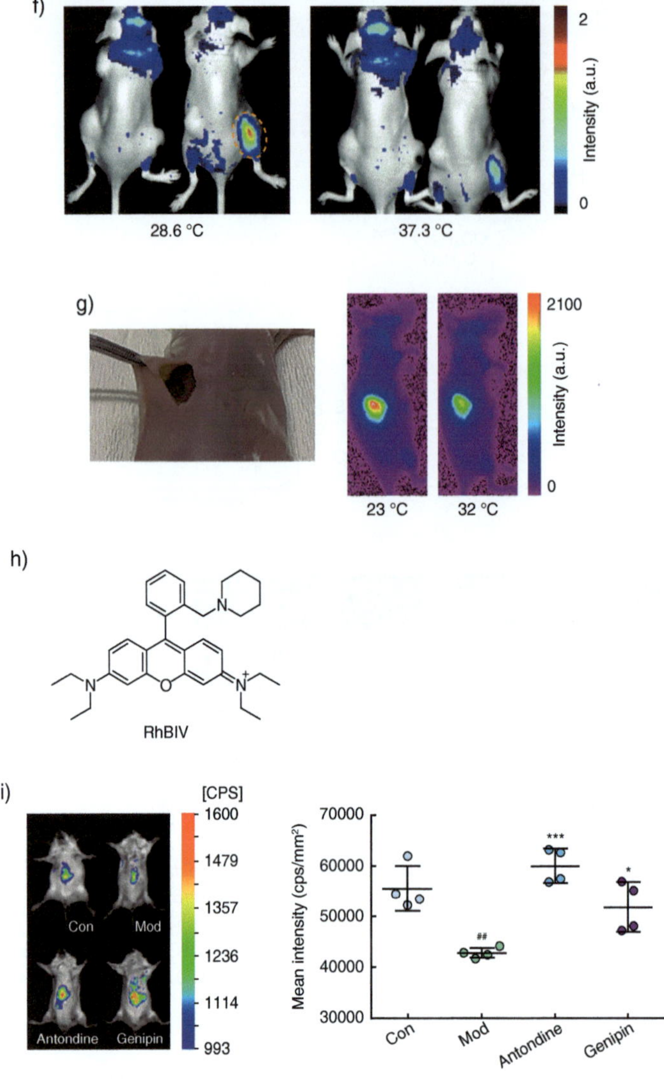

Figure 6.4 (continued) (f) Pseudo-colored fluorescence images of tumor-bearing nude mice with the hypodermic injection of $CuInS_2/ZnS$ containing polyoxyethylene stearate micelles at 28.6 and 37.3 °C (left and right, respectively). The orange dotted circle represents the injected tumor site. Emissions at 650 ± 20 nm with the excitation at 580 ± 10 nm were recorded. Adapted from Zhang et al. (2019) *J. Mater. Chem. B*, **7**, 2835–2844 / Royal Society of Chemistry. (g) Implantation of fluorescent gold nanoparticles with silk film (AuNPs-SF) in the dorsal region of a mouse (left) and pseudo-colored fluorescence images of the implanted mouse at 23 and 32 °C (middle and right, respectively). Adapted from Hua et al. (2020) *Front. Chem.*, **8**, 364 / Frontiers Media S.A. (h,i) Mitochondrial thermometry with a rhodamine B derivative RhBIV. (h) Chemical structure of RhBIV. (i) In vivo temperature imaging (left) and effects of the administration of 2,4-dinitrophenol (28 mg kg^{-1}) to induce fever (Mod) and additional antipyretic drugs (antondine (20 mg kg^{-1}) and genipin (50 mg kg^{-1})) on the body temperature of a mouse (right). Con: control. RhBIV injected into the tail vein was used as a fluorescent molecular thermometer showing temperature dependent fluorescence intensity (i.e., lower intensity at a higher temperature). Significance was determined by Student's t-test: $^{##}p < 0.01$ vs control (Con), $^{***}p < 0.001$, $^{*}p < 0.05$ vs Mod. Adapted from Shen et al. (2021) *Anal. Chem.*, **93**, 13417–13420 / American Chemical Society.

j)

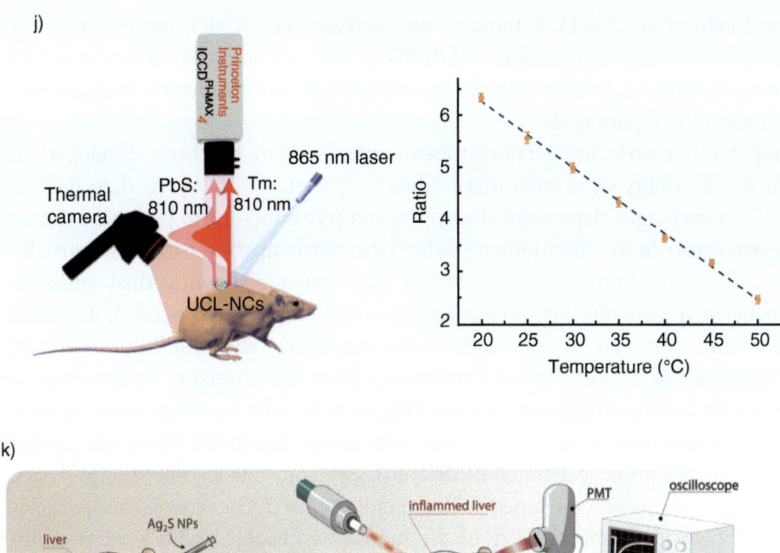

k)

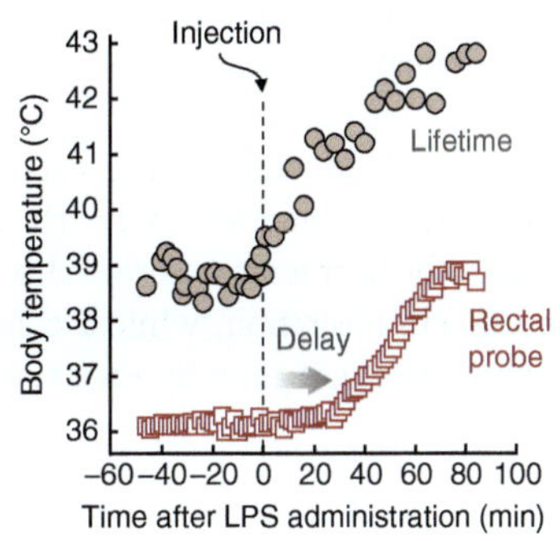

Figure 6.4 (continued) (j) Schematic diagram of in vivo thermometry of a mouse by upconversion nanoclusters (UCL-NCs) containing PbS quantum dots and NaYbF$_4$:0.5%Tm@NaYF$_4$:10%Yb@NaYF$_4$:50%Nd (left), the relationship between fluorescence intensity ratio of UCL-NCs in a mouse and the temperature (right). Qiu et al. (2020) *Nat. Commun.*, **11**, 4 / Springer Nature / CC by 4.0. (k) Schematic representation of in vivo thermometry of a mouse (upper) and time profiles of liver temperature measured by Ag$_2$S nanoparticles (circle) and rectal temperature measured by a rectal probe (square) after the LPS injection (10 mg kg^{-1}) (bottom). Shen et al. (2022) *Adv. Mater.*, **34**, 2107764 / John Wiley & Sons.

et al., 2019). The polyoxyethylene stearate micelle solution in PBS (phosphate-buffered saline) was injected into a mouse tumor, and temperature-dependent emission intensity of the quantum dots at 650 ± 20 nm with excitation at 580 ± 10 nm was recorded. In vivo temperature imaging with glutathione-coated gold nanoparticles in an implantable silk film is demonstrated in Figure 6.4g (Hua et al., 2020). The silk film with gold nanoparticles was implanted into the dorsal region of a female BALB/c mouse, and the temperature-dependent emission intensity of gold nanoparticles was confirmed in a live mouse under anesthesia. The mitochondrial temperature of a mouse could be monitored by a rhodamine B-based fluorescent molecular thermometer RhBIV (Figure 6.4h), which was introduced by a caudal intravenous injection (Shen et al., 2021). A temperature increase (i.e., fever), represented by a decrease in the fluorescence intensity of RhBIV at 600 nm with the excitation at 540 nm, was induced by 2,4-dinitrophenol but canceled by the antipyretic drugs antondine and genipin (Figure 6.4i).

A unique ratiometric temperature measurement system has been developed to improve the reliability of in vivo thermometry (Qiu et al., 2020). As described in Section 5.5, wavelength-dependent absorption and scattering due to biological tissues cause severe errors in the thermometry using ratiometric upconversion nanoparticles (UCNPs). Thus, a ratiometric upconversion thermometer showing dual emission components with different lifetimes at a single wavelength was proposed, in which PbS quantum dots (as a fluorescent thermometer) and NaYbF$_4$:0.5%Tm@NaYF$_4$:10%Yb@NaYF$_4$:50%Nd (as a reference) emits upconversion fluorescence at 810 nm under laser irradiation at 865 nm (Figure 6.4j). The emission intensity ratio could be obtained from a real-time emission spectrum due to the PbS quantum dot and Tm^{3+} ion and a time-gated (20 μs delayed) spectrum due to only the Tm^{3+} ion. Using the calibration curve obtained with the ratiometric upconversion nanoparticles in pork meat, in vivo thermometry of the mouse was enabled with a temperature resolution of 0.5 °C. The last example utilized a poly(ethylene glycol) (PEG)-coated Ag$_2$S nanoparticle with a diameter of 9 ± 1 nm as a luminescent thermometer (Figure 6.4k) (Shen et al., 2022). After intravenous injection, the Ag$_2$S nanoparticles accumulated in the liver with a slow clearance rate (in the order of hours). The administration of lipopolysaccharide (LPS) through intraperitoneal injection (10 mg kg^{-1}) induced a significant liver temperature rise, evaluated from the temperature-dependent emission lifetime of the Ag$_2$S nanoparticles. This inflammatory response was accompanied by the hepatic gene expression of proinflammatory markers such as tumor necrosis factor-alpha (TNF-α), interleukin-6 (IL-6), interleukin-1β (IL-1β), and inducible nitric oxide synthase (iNOS). Notably, the liver temperature evaluated fluorometrically increased immediately after the LPS administration, while a considerable delay was seen in the intracorporeal temperature rise measured by a rectal probe (Figure 6.4k).

6.3
Thermal Medicine at a Single-cell Level

One of the promising applications of intracellular thermometry is "thermal medicine", which explores the manipulation of body, tissue, or cellular temperature for the treatment of disease (Diederich, 2005; Dewhirst et al., 2005). This concept overlaps with the strategy of "theragnostics" (sometimes called "theranostics") that integrates therapeutics with diagnostics to create efficient and safe personalized treatments (Fang and Zhang, 2010). Hyperthermia to kill tumor cells by heating body tissue globally or locally is representative of thermal medicine and occasionally is combined with chemotherapy to improve the efficacy of cancer treatments (Zhang et al., 2013; Jung et al., 2018). Intracellular thermometry enables precise temperature control in thermal medicine at a single-cell level, whereas conventional thermal medicine uses infrared thermography as a thermometer that can measure only skin temperature in a broader area (Tian et al., 2021).

The simplest nanomaterial for thermal medicine at a single-cell level is a hybrid of a molecular heater and a fluorescent molecular thermometer (Oyama et al., 2022).

This concept was first accomplished by the preparation of a fluorescent polymeric thermometer (poly(NIPAM-*co*-DBD-AE))-coated magnetite (i.e., iron oxide) nanoparticles that could be heated under an alternating magnetic field, although not applied to biological systems (Herrera et al., 2008; Polo-Corrales and Rinaldi, 2012). Figure 6.5a indicates the temperature-dependent fluorescence intensity of poly(NIPAM-*co*-DBD-AE)-coated magnetite nanoparticles dispersed in water under an alternating magnetic field (38.4 kA m^{-1} and 233 kHz). Compared to the fiber-optic thermometer measuring the bulk temperature in water, the part of the fluorescent polymeric thermometer, i.e., a poly(NIPAM-*co*-DBD-AE) moiety, sensed an immediate temperature increase of the magnetite nanoparticles by the external

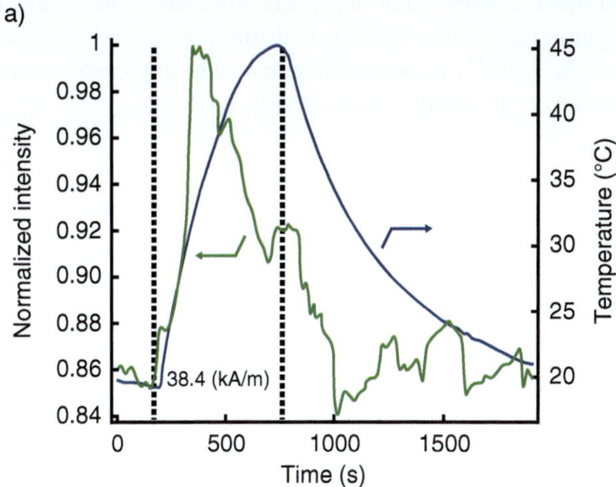

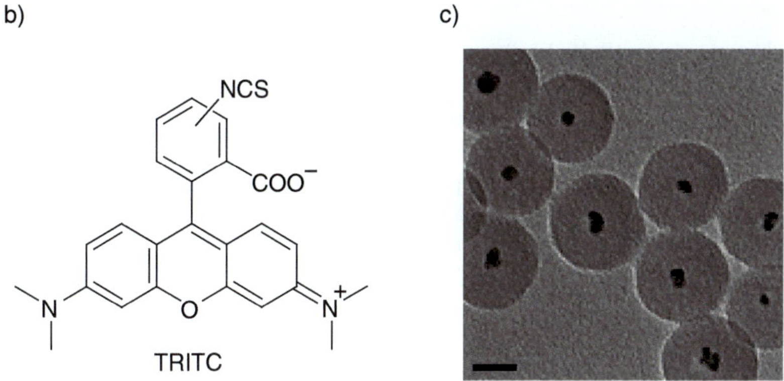

Figure 6.5 Hybrid materials of a magnetite nanoparticle and a fluorescent molecular thermometer. (a) Fluorescence intensity of poly(NIPAM-*co*-DBD-AE) coated iron oxide nanoparticles in water (1% w/v magnetic core) under an alternating magnetic field (38.4 kA m^{-1}, 233 kHz) (green). The fluorescence intensity of the poly(NIPAM-*co*-DBD-AE) moiety increases with a temperature rise. The temperature of the dispersion was monitored with a fiber-optic thermometer (blue). Polo-Corrales and Rinaldi (2012) *J. Appl. Phys.*, **111**, 07B334 / American Institute of Physics. (b,c) Iron-platinum (FePt)-core/SiO$_2$-shell/TRITC-SiO$_2$-shell/ APTES coated magnetic nanoparticles. TRITC: tetramethyl rhodamine isothiocyanate; APTES: (3-aminopropyl)-triethoxysilane. (b) Chemical structure of TRITC. (c) A TEM image of SiO$_2$ coated FePt magnetic nanoparticles. Scale bar: 20 nm. Adapted from Chen et al. (2012) *Chem. Commun.*, **48**, 2501–2503 / Royal Society of Chemistry.

heating. For the same purpose, iron-platinum (FePt)-core/SiO_2-shell/TRITC-SiO_2-shell/APTES-coated magnetic nanoparticles (Figure 6.5b, 6.5c) were also prepared (Chen et al., 2012). Later, a P4VP-*b*-P(MPEGA-*co*-PEGA) nanoparticle, consisting of a poly(4-vinylpyridine) (P4VP) block and a water-soluble and biocompatible block of poly(methoxy poly(ethylene glycol)acrylate) (PMPEGA) and poly(ethylene glycol) acrylate (PEGA) and containing maghemite and Eu^{3+}/Tb^{3+} ion complexes, was prepared and its abilities to monitor intracellular temperature in OK (opossum kidney) cells and to heat up under a magnetic field were confirmed (Piñol et al., 2015). Nanoparticles doping Nd^{3+}, Yb^{3+}, and Er^{3+} ions are capable of heating and measuring local temperatures under single beam excitation at 808 nm, where Nd^{3+} ion releases heat and Er^{3+} ion shows a temperature-dependent emission intensity ratio at 528 and 539 nm. The optimization of nanoparticle structure (i.e., a core/shell pattern containing Yb^{3+} and Er^{3+}/Nd^{3+}, respectively) remarkably improved heating efficiency and thermal sensitivity (Ximendes et al., 2016).

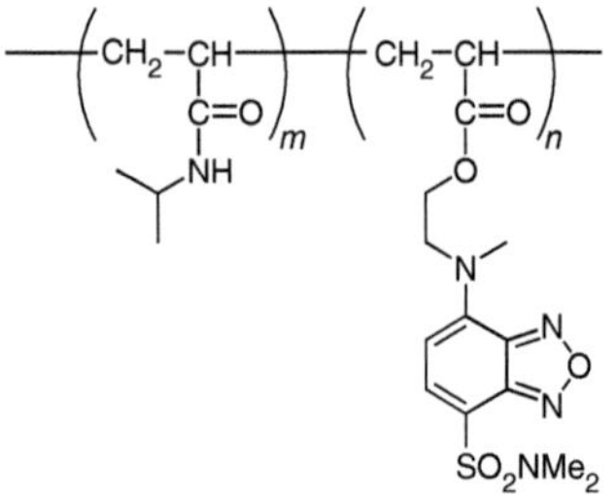

Poly(NIPAM-*co*-DBD-AE)

Simultaneous intracellular heating and temperature monitoring were performed first with multilayered SiO_2 nanoparticles containing gold and CdTe/CdS/ZnS quantum dots (Jiang et al., 2019). After being introduced into HOS (human osteosarcoma) cells, the multilayered SiO_2 nanoparticles informed us of the temperature variation through the temperature-dependent emission peak of the quantum dots during local heating by irradiation of a near-infrared laser at 808 nm to gold in the nanoparticles. When the irradiation (100 mW) was prolonged for five minutes, and the intracellular temperature reached 42 °C, bubbles were generated around the HOS cells, and the cell status worsened. This phenomenon is known as cell damage in cancer therapy (Lapotko, 2009).

A different case of theranostic nanomaterials with the ability to measure intracellular temperature and heat up an intracellular space is summarized in Figure 6.6 (Wu et al., 2021). As shown in Figure 6.6a, fluorescent nanodiamonds were precoated by hyperbranched polyethylenimine ($bPEI_{25k}$) in the presence of polyvinylpyrrolidone (PVP_{10k}), and the PEI structures on the surface of fluorescent nanodiamonds were crosslinked by a four-arm polyethylene glycol N-hydroxysuccinimide (4-arm PEG-NHS_{10k}). Then, an anionic photothermal agent,

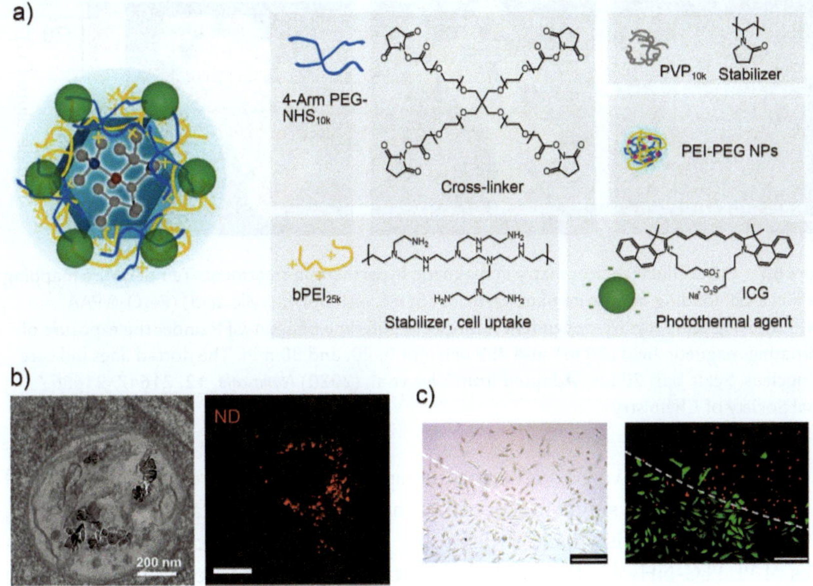

Figure 6.6 Nanodiamond theranostics: intracellular heating and temperature sensing by nanoparticles (NPs) consisting of fluorescence nanodiamonds, nanogels, and indocyanine green (ND-NG-ICG). (a) Chemical structure of ND-NG-ICG. (b) TEM image of an endosomal vesicle of a HeLa cell containing ND-NG-ICG (left, scale bar: 200 nm) and confocal fluorescence image of ND-NG-ICG in a HeLa cell (right, scale bar: 10 μm). (c) Bright-field image (left) and fluorescence image (right) of HeLa cells after the incubation with ND-NG-ICG (100 μg mL^{-1}, 4 h) and the irradiation of near-infrared light (810 nm, 0.35 W cm^{-2}, 20 min) on the upper area (above a dotted line). In the fluorescence image, green and red indicate live and dead cells, respectively. Scale bars: 200 μm. Adapted from Wu et al. (2021) *Nano Lett.*, **21**, 3780–3788 / American Chemical Society / CC BY 4.0.

indocyanine green (ICG), was adsorbed on the nanomaterials by electrostatic interactions to yield ND-NG-ICG with a diameter of 99.6 ± 0.53 nm. After the co-incubation for four hours, ND-NG-ICG (concentration in culture media: 100 μg/ml) entered HeLa (human epithelial carcinoma) cells and was located in the endosomal vesicles (Figure 6.6b). Under the irradiation of a near-infrared LED lamp at 810 nm (0.35 W cm^{-2}), the intracellular temperature of HeLa cells containing ND-NG-ICG increased by ~32 °C that was measured by fluorescent nanodiamonds in ND-NG-ICG. Only the HeLa cells containing ND-NG-ICG under near-infrared irradiation were dead, whereas those outside the irradiation field remained alive (Figure 6.6c).

Figure 6.7 indicates the intracellular temperature mapping of a HeLa cell upon a magnetothermal treatment (Silva et al., 2020). In this experiment, actin-GFP was transfected in HeLa cells as a fluorescent molecular thermometer, and Fe$_3$O$_4$ magnetite nanoparticles coated with poly(acrylic acid) were delivered into HeLa cells by incubation overnight. Although the sensitivity of actin-GFP was relatively low (0.23% °C^{-1}), the heating effects by the Fe$_3$O$_4$ nanoparticles under alternating magnetic field exposure (20 mT and 499 kHz) were successfully visualized.

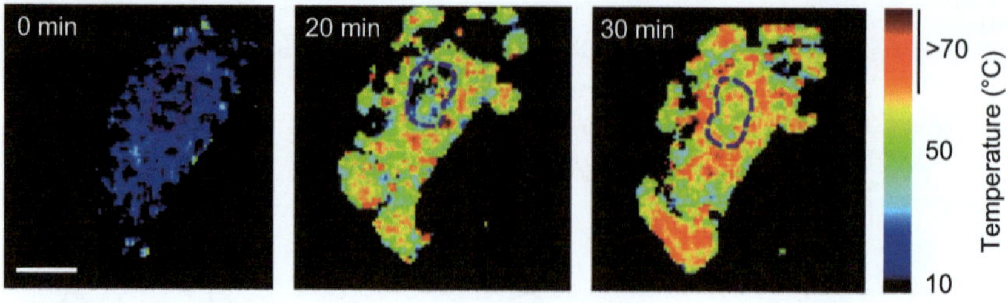

Figure 6.7 Intracellular thermometry in magnetic hyperthermia treatment. Temperature mapping of a HeLa cell loading magnetite nanoparticles coated with poly(acrylic acid) (Fe$_3$O$_4$@PAA MNP) by using temperature-dependent fluorescence lifetime of actin-GFP under the exposure of alternating magnetic field (20 mT and 499 kHz) for 0, 20, and 30 min. The dotted lines indicate the nucleus. Scale bar: 20 μm. Adapted from Silva et al. (2020) *Nanoscale*, **12**, 21647–21656 / Royal Society of Chemistry.

In some cases, in vivo photothermal therapy with temperature monitoring could be demonstrated with single functionalized nanoparticles. For example, NaYF$_4$:Yb^{3+}, Er^{3+}, Nd^{3+} upconversion nanoparticles that were coated with condensation molecules of PL-PEG-NH$_2$ (2-distearoyl-sn-glycero-3-phosphoetha-nolamine-N-[carboxy (polyethyleneglycol)-2000]) and IR-806 and of PL-PEG-NH$_2$ and folic acid, could measure temperature by the emission intensity ratio of Er^{3+} ion at 520 (^{2}H$_{11/2}$ → ^{4}I$_{15/2}$ transition) and 545 nm (^{4}S$_{3/2}$ → ^{4}I$_{15/2}$ transition) and heat a tumor site in a mouse after injected intravenously under the excitation at 808 nm (Wei et al., 2020). In the function of the nanoparticles, IR-806 moieties worked as a molecular antenna to collect the excitation energy and then transfer it to the Yb^{3+} ion or convert it to heat. The efficient hyperthermic effects inducing apoptotic cells were observed through the treatment.

IR-806

A magnetic characteristic of nanoparticles is effective for not only local heating but also guided targeting. Silica nanoparticles that contain magnetic Fe$_3$O$_4$ and are coated by a poly(N-isopropylacrylamide) shell with rhodamine B isothiocyanate and gold showed temperature-dependent fluorescence intensity in HeLa cells (due to the variation in the distance between rhodamine B (a fluorophore) and gold (a quencher)) and guided HeLa cells to a placed magnet (Wang et al., 2015). A noteworthy application of this method is the photothermal muscle contraction using superparamagnetic Fe$_3$O$_4$ nanoparticles-coated *Chlorella pyrenoidosa* (named a "CP@Fe$_3$O$_4$ microswimmer", Figure 6.8a) (Liu et al., 2022). A single CP@Fe$_3$O$_4$ microswimmer could reach the target C2C12 myotube precisely under controlling the direction and frequency of a rotating magnetic field (7 mT), and the laser irradiation at 808 nm on Fe$_3$O$_4$ nanoparticles in the microswimmer (1 W cm^{-2})

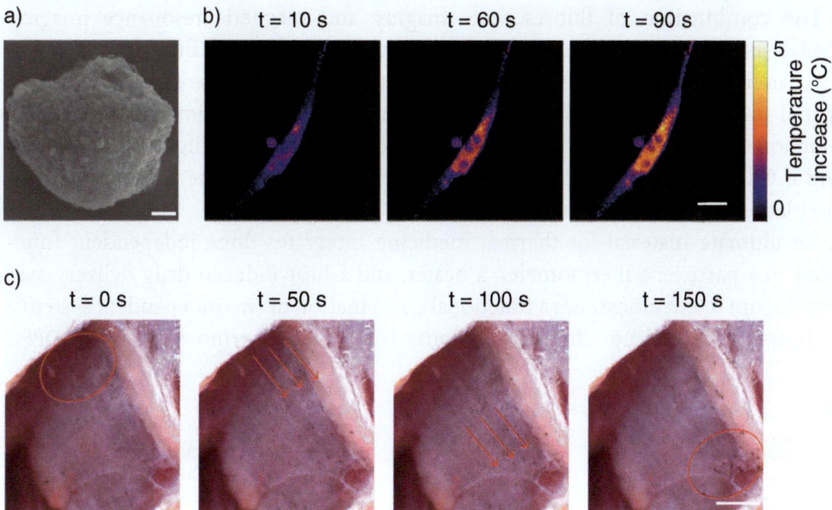

Figure 6.8 Temperature monitoring in photothermal muscle contraction with Fe_3O_4 nanoparticles bonded onto *C. pyrenoidosa* (termed a "CP@Fe_3O_4 microswimmer"). (a) A SEM image of the microswimmer. Scale bar: 1 μm. (b) Time-lapse images of the intracellular temperature of a C2C12 myotube containing Fe_3O_4 microswimmers under near-infrared light irradiation (808 nm, 1 W cm^{-2}). The temperature was evaluated by a ratiometric fluorescent polymeric thermometer. Scale bar: 50 μm. (c) Time-lapse images of the Fe_3O_4 microswimmers moving toward the targeted rat leg muscle fiber under a magnetic field. Scale bar: 500 μm. Adapted from Liu et al. (2022) *ACS Nano*, **16**, 6515–6526 / American Chemical Society.

resulted in a significant increase in intracellular temperature, measured by a ratiometric fluorescent polymeric thermometer, by 5.1 ± 0.4 °C (Figure 6.8b). This temperature increase induced a remarkable contraction of the targeted C2C12 myotube (contraction rate: 7.2 ± 0.8%) due to the actin–myosin interactions. The wireless and precise muscle activation was also demonstrated with a leg sample of an anesthetized rat (Figure 6.8c). The CP@Fe_3O_4 microswimmers moved to the target site within 150 seconds under magnetic manipulation and increased local temperature and muscle contraction frequency under the near-infrared laser irradiation at 808 nm.

The combination of fluorescence imaging and magnetic resonance imaging (MRI) shows complementary advantages in clinical diagnosis: fluorescence imaging has high sensitivity but limited depth of penetration, whereas MRI has high spatial resolution but low sensitivity. A dual-modal polymeric imaging reagent P4 functioned as a fluorescent polymeric thermometer for intracellular thermometry and a Gd-based MRI contrast agent with a high relaxation rate and a prolonged circulation time in vivo (Zhou et al., 2021).

An ultimate material for thermal medicine integrates three independent functions in a particle: a thermometer, a heater, and a heat-induced drug delivery system. Figure 6.9 demonstrates a functional combination of two independent systems: a heat-induced drug delivery system based on thermo-responsive DPPC

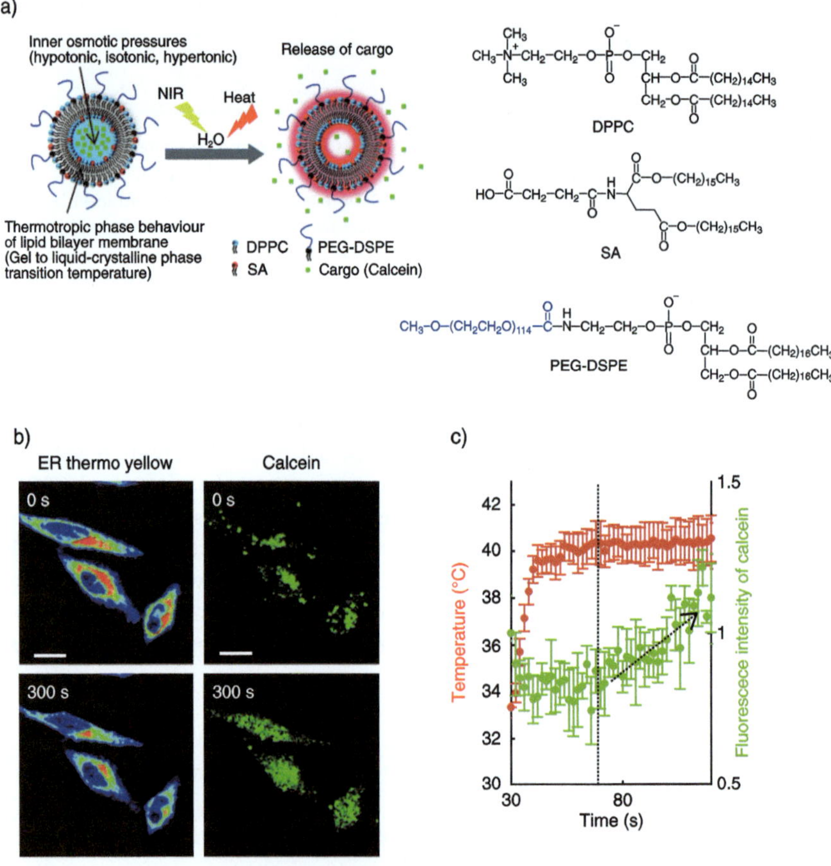

Figure 6.9 Intracellular temperature monitoring in heat-induced drug release inside cells. (a) Schematic diagram of the heat-induced drug delivery system with a 1,2-dipalmitoryl-*sn-glycero*-3-phosphocholine (DPPC)-based liposome containing a model drug, calcein. (b) Fluorescence images of an endoplasmic reticulum (ER)-targeted fluorescent molecular thermometer, ER thermo yellow (left), and calcein (right) in HeLa cells loading the DPPC-based liposomes before (0 s, upper) and after (300 s, lower) laser irradiation at 980 nm. Scale bars: 20 μm. (c) Intracellular temperature rise-induced calcein release from the DPPC-based liposomes in HeLa cells. Adapted from Arai et al. (2015b) *RSC Adv.*, **5**, 93530–93538 / Royal Society of Chemistry.

(1,2-dipalmitoyl-*sn-glycero*-3-phosphocholine) liposomes with an anionic lipid SA (L-glutamic acid, N-(3-carboxy-1-oxopropyl),-1,5-dihexadecyl ester) and a PEG lipid, PEG-DSPE (1,2-distearoyl-*sn-glycero*-3-phosphoethanolamine-N-[monomethoxy poly(ethylene glycol) (5000)]) (Figure 6.9a), and an endoplasmic reticulum (ER)-targeted fluorescent molecular thermometer, ER thermo yellow (see Figure 3.29b for the chemical structure) (Arai et al., 2015b). The DPPC liposomes encapsulating a model drug calcein could be introduced into HeLa cells by incubation at 37 °C for three hours. Then, calcein was released from the DPPC-based liposomes in HeLa cells when a laser at 980 nm was irradiated for heating a water medium (Figure 6.9b). Spontaneous intracellular thermometry with ER thermo yellow indicated a short time lag between the temperature rise to 40 °C and the calcein release (Figure 6.9c).

A prototype that realizes the concept of functional integration of a thermometer, a heater, and a heat-induced drug delivery system is a PEG-type thermo-responsive nanogels composed of 2-(2-methoxyethoxy)ethyl methacrylate (DEGMA) units, oligo(ethylene glycol)methyl ether methacrylate (OEGMA) units, and poly(ethylene glycol) dimethacrylate (PEGDMA) units (Wu et al., 2011). The PEG-type thermo-responsive nanogels contain fluorescent ZnO quantum dots (as a thermometer with excitation and emission at 340 and 530 nm, respectively), gold nanoclusters (as a heater under near-infrared irradiation), and temozolomide (TMZ, as an anti-cancer drug). The release of TMZ from the PEG-type thermo-responsive nanogels was induced by environmental heating and further accelerated by the local heating by near-infrared irradiation to the gold nanoclusters. The effects of combined chemo-photothermal therapy were confirmed by the viability test using B16F10 (mouse melanoma) cells. An effective three-dimensional structure for a heat-induced drug-delivery system was investigated using multilayered capsules combined with fluorescent nanodiamonds (as thermometers) and gold nanoparticles (as heaters) inside B16F10 cells (Gerasimova et al., 2021).

Figure 6.10a displays a newer version of materials combining the three functions (i.e., temperature sensing, heating, and drug releasing), in which a $NaLuF_4$:Yb,Er@ $NaLuF_4$ upconversion nanoparticle (pink and blue, a luminescent thermometer) and a photothermal agent $PdPc(OBu)_8$ (see Figure 6.4c) are encapsulated by a hollow SiO_2 shell (light blue) (Zhu et al., 2018). A model drug doxorubicin (DOX) in DPPC micelles (as thermo-responsive drug release units) is attached to the SiO_2 nanocomposite. In aqueous dispersion, the SiO_2 nanocomposite released DOX efficiently when the local temperature evaluated by temperature-dependent

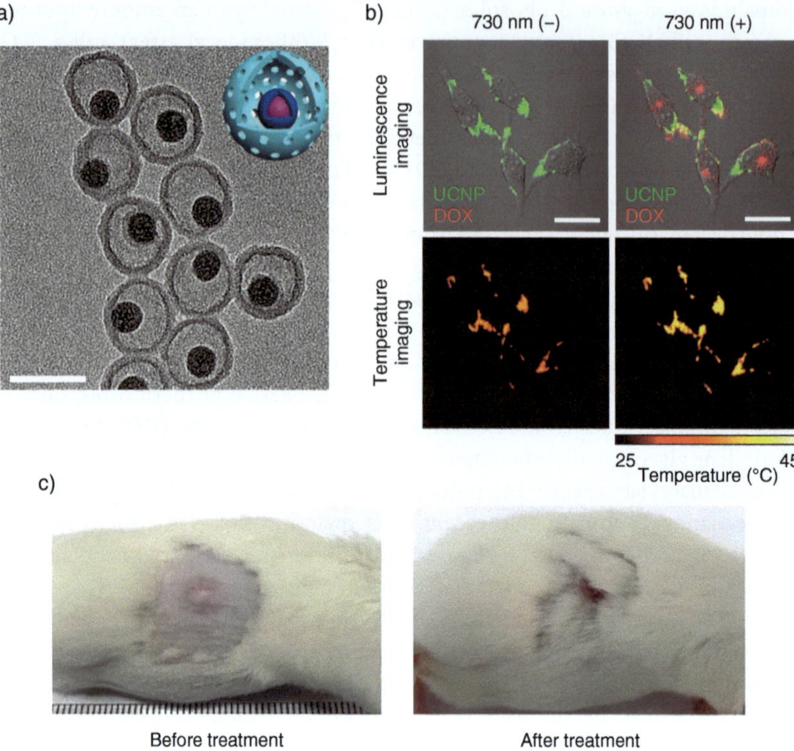

Figure 6.10 Accurate temperature monitoring during heat-induced drug release. (a) Schematic and TEM images of upconversion nanosystems (UCNSs) composed of SiO_2 (light blue), $NaLuF_4$ (blue), temperature-sensing $NaLuF_4$:Yb,Er (pink). Scale bar: 50 nm. (b) Fluorescence images of $NaLuF_4$:Yb,Er (green) and doxorubicin (DOX, red) in MIA PaCa-2 cells loading temperature-responsive UCNSs (TR-UCNSs) modified with DOX, a thermal responsive drug release unit (DPPC micelle), and a photothermal agent $PdPc(OBu)_8$ (upper), and the corresponding temperature maps (lower) without and with laser irradiation at 730 nm (left and right, respectively). Scale bars: 30 μm. (d) Effects of programmed combination therapy using TR-UCNSs on tumor in a mouse. Adapted from Zhu et al. (2018) *Nat. Commun.*, **9**, 2176.

upconversion fluorescence increased to 41.6 °C under laser irradiation at 730 nm. The DOX releases from the SiO_2 nanocomposites under precise temperature control were observed in MIA PaCa-2 (human pancreatic cancer) cells (Figure 6.10b). Changing the laser power of irradiation from 46 to 140 mW cm^{-2} enabled straightforward photothermal therapy in addition to the chemotherapy with DOX. A programmed combination of chemo-/phothermal-therapy resulted in the highly effective in vivo treatment of tumors in mice (Figure 6.10c).

6.4
Utilization of Nanoheater/Fluorescent Thermometer Hybrids in Biotechnology

In an early case involved in biotechnology, intracellular thermometry based on fluorescent proteins was performed to precisely control the local temperature in living cells in which target genes were evoked by infrared laser irradiation (Kamei et al., 2009). Intracellular thermometry could also be effective for precise control of cellular functions such as endocytosis and epidermal growth factor signaling by thermo-responsive elastin-like polypeptides delivered into living cells (Tyrpak et al., 2021).

Nanomaterials, capable of measuring temperature and heating in tiny spaces, are also useful for the practice of intracellular biotechnology. For example, nanohybrids of gold nanorod and fluorescent nanodiamond can determine the rupture temperature of individual membrane nanotubes in HEK293T (human embryonic kidney) cells by generating a temperature gradient on the cell membrane (Tsai et al., 2017). By incubation at 37 °C overnight, gold nanorod/fluorescent nanodiamond hybrids were introduced in the endosome in HEK293T, including membrane nanotubes (Figure 6.11a). Upon laser irradiation at 594 nm for both local heating due to gold nanorods and temperature sensing due to fluorescent nanodiamonds, the membrane-tunneling nanotube was broken when the local temperature reached 28

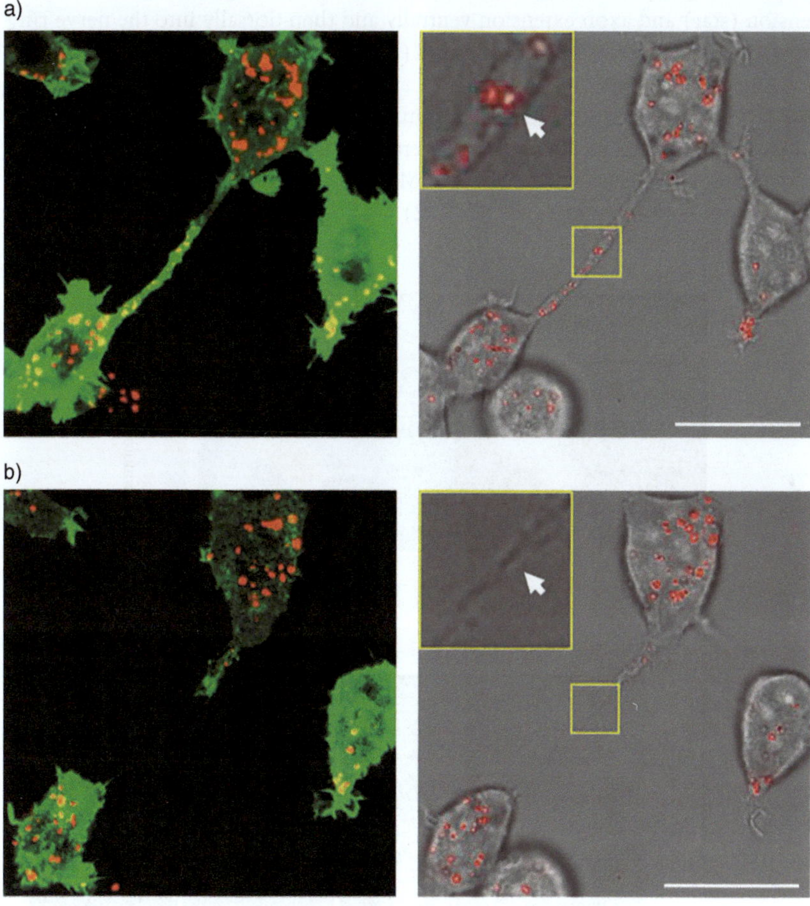

Figure 6.11 Intracellular heating and temperature measurement by gold nanorod-fluorescent nanodiamond (GNR-FND) hybrids for assessing the thermostability of cell membranes. (a,b) Fluorescence images (left) and merged images of bright field and fluorescence (right) of HEK293T cells containing actin-GFP fusion proteins (green) and GNR-FND (red) before (a) and after (b) the laser exposure at 594 nm (330 µW, 6 s) at the point in a membrane-tunneling nanotube. White arrows in the enlarged views (upper left in the merged images) of the membrane-tunneling nanotube (in yellow squares in the main images) indicate the positions of GNR-FNDs irradiated with a 594 nm laser. Scale bars: 25 µm. Adapted from Tsai et al. (2017) *Angew. Chem. Int. Ed.*, **56**, 3025–3030 / John Wiley & Sons.

± 2 °C (Figure 6.11b). As global heating of HEK293T cells with an on-stage incubator resulted in the breaking of membrane-tunneling nanotubes at 38 ± 1 °C, it could be concluded that there exists a significant difference in the damage threshold between local and global heating. A steep temperature gradient created by the gold nanorod/fluorescent nanodiamonds hybrids (~10 °C μm^{-1}) is likely to be the reason for high efficiency in the localized hyperthermia experiments.

Figure 6.12a shows intracellular thermometry of a single *C. elegans* embryonic neuron with a fluorescent protein thermometer *ceh-27*$_{pro}$::mCherry for the process of a reproducible heat-dependent gene expression which used infrared laser irradiation at 1,455 nm (i.e., the absorption peak of water molecules) without radiation damage (Singhal and Shaham, 2017). An example of infrared-laser-induction allowing visualization of morphogenesis under precise temperature control was displayed in Figure 6.12b, in which amphid neurons undergo retrograde dendrite extension (star) and axon extension ventrally and then dorsally into the nerve ring (arrow) and ASI sister-cell death is visible (arrowhead) after infrared laser heating of the ABplaapaa precursor.

The newest case of nanomaterials combining the functions of a heater and a thermometer is illustrated in Figure 6.13a (Ferdinandus et al., 2022). In poly(methyl methacrylate-*co*-methacrylic acid) (PMMA-MA)-based nanoparticles (named nanoHT), EuDT and coumarin102 (C102) are contained as the components of a

a)

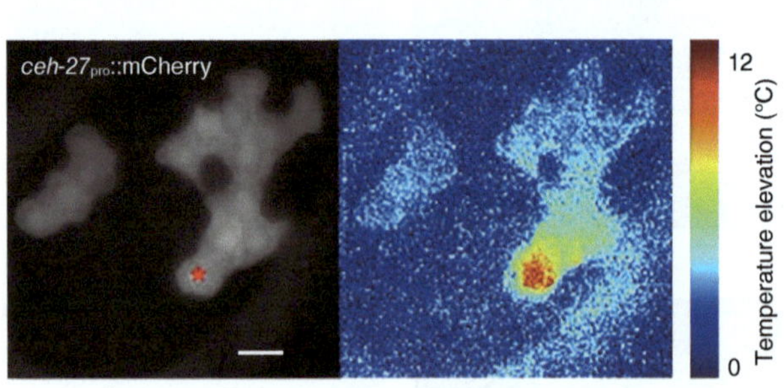

b)

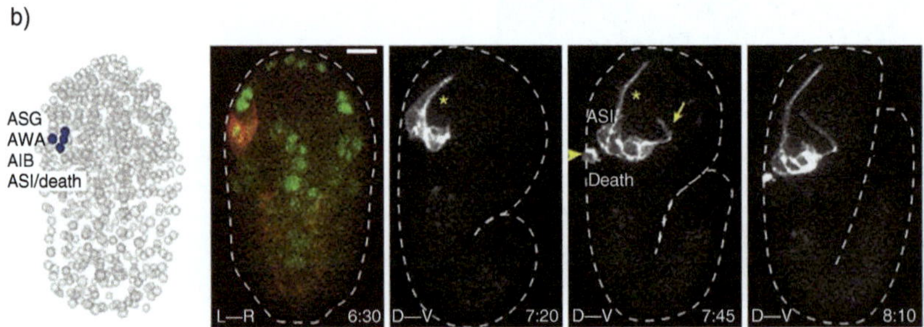

Figure 6.12 Intracellular thermometry by a fluorescent protein during infrared laser-induced gene expression in *Caenorhabditis elegans* embryos. (a) Fluorescence image of *ceh-27*$_{pro}$::mCherry reporter of AB128 of *C. elegans* (left) and temperature map of the single embryo under the laser irradiation (3.5 mV) at the star point (indicated in the fluorescence image) (right). Scale bar: 5 μm. (b) Laser-induced morphogenesis. Retrograde extension of amphid sensory dendrites (star), axon extension ventrally and then dorsally into a nerve ring (arrow), and death of AS1 sister cell (arrowhead). Scale bar: 5 μm. Adapted from Singhal and Shaham (2017) *Nat. Commun.*, **8**, 14100.

ratiometric fluorescent thermometer showing temperature-dependent fluorescence intensity ratio at 612 and 474 nm with the excitation at 405 nm, and vanadyl 2,11,20,29-tetra-*tert*-butyl-2,3-naphthalocyanine (V-Nc) is encapsulated as a local heater under the near-infrared irradiation at 808 nm. Interestingly, the experiment using HeLa cells, including nanoHT in the cytoplasm, revealed that cellular death was induced by near-infrared laser irradiation when the intracellular temperature rose by $13.7 \pm 1.0\,°C$ (Figure 6.13b). In contrast, the cells could be alive with a temperature increase of $10.5 \pm 1.4\,°C$. The correlation between the fluorescence enhancement of a cell death marker (Apopxin Green) and the intracellular temperature variation indicated that the threshold temperature increment required for the death of HeLa cells was $11.4\,°C$ when incubated at $37\,°C$ (Figure 6.13c).

Controllable external thermal stimulation at a measured temperature could be performed on HeLa cells by an artificial submicron-diameter glass capillary bearing

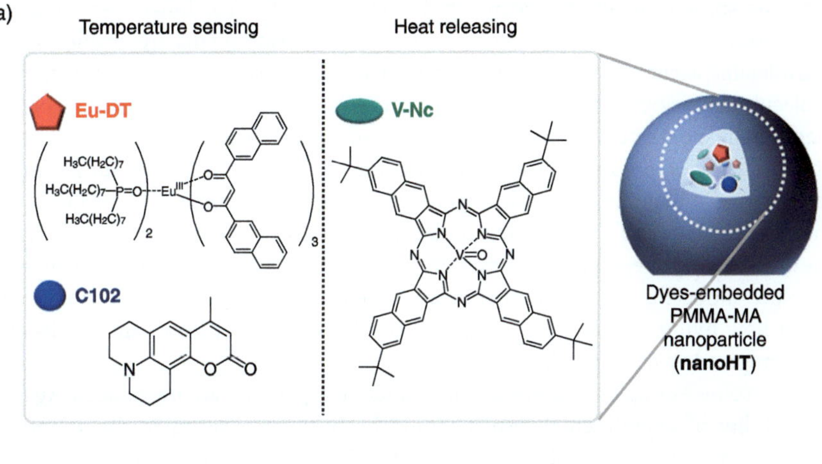

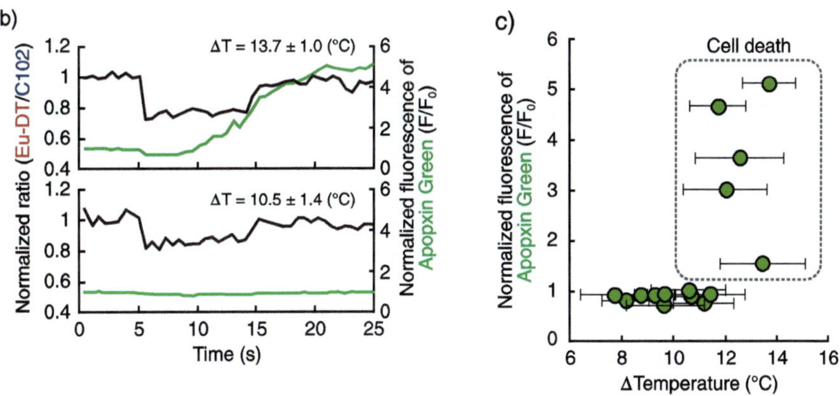

Figure 6.13 Multiplex functions in intracellular thermometry and local heating by dyes-embedded poly(methyl methacrylate-*co*-methacrylic acid) (PMMA-MA) nanoparticles (nanoHT). (a) Chemical structures of the components in nanoHT. (b) Relationships between the temperature variation (ΔT, black, measured from the fluorescence intensity ratio of EuDT and C102 in nanoHT) of HeLa cells and the cell death (green, assessed by fluorescence intensity of Apopxin Green) under near-infrared stimulation at 808 nm (for 10 s). (c) Threshold of the intracellular temperature increase in HeLa cells to provoke cellular death. Ferdinandus et al. (2022) *ACS Nano*, **16**, 9004–9018 / American Chemical Society.

polycrystalline fluorescent nanodiamonds with silicon-vacancy centers (Romshin et al., 2023). Attached to a cellular surface, the glass capillary takes local heating with laser irradiation at 561 nm. The local temperature rise by heating could be evaluated from the temperature-dependent emission maximum of the fluorescent nanodiamonds under excitation at the same wavelength.

6.5
Accurate Measurements of Absolute Intracellular Temperature

I believe absolute intracellular temperature measurements with fluorescent molecular thermometers are much more challenging than relative ones. The latter includes determining temperature differences between the nucleus and the cytoplasm and monitoring a temperature variation upon chemical stimulation, as shown in the experimental data published from our laboratory (Okabe et al., 2012). Although there are some difficulties in assuring the reliability of intracellular thermometry (see Chapter 5), relative intracellular temperature measurements are enabled with considerable accuracy when it can be assumed that the sensitivities of fluorescent molecular thermometers toward a temperature variation (i.e., a differential coefficient of a response curve) are identical between the living cell of a subject and the environment for preparing a calibration curve. In contrast, absolute intracellular temperature measurements with high accuracy require a complete coincidence of the response curve of a fluorescent molecular thermometer inside living cells with the calibration curve, which can never be proved in the present science. Nevertheless, absolute intracellular temperature measurements can answer exciting questions about how living cells are hot compared to culture media or how cancer cells are hot compared to normal cells.

Here, some examples of absolute intracellular temperature are introduced. Again, the validity of experimental results should be further discussed in the scientific community to interpret biological significance. Figure 6.14a demonstrates absolute temperature measurements of the surface of a C127 (mouse breast cancer) cell by a $NaYF_4:Yb^{3+}/Er^{3+}$ UCNPs-decorated spider silk (Gong et al., 2021). With a calibration curve adopting fluorescence intensity ratio at 525 nm (the $^2H_{11/2} \rightarrow {}^4I_{15/2}$ transition) and 548 nm (the $^4S_{3/2} \rightarrow {}^4I_{15/2}$ transition) as a temperature-dependent parameter, a temperature gap of approximately 15 °C was observed between the surface (i.e., R_1) surface and the environment outside the cell (R_2). In another example using the temperature-dependent optically detected magnetic resonance (ODMR) spectrum of fluorescent nanodiamonds, the intracellular temperature in a live HeLa cell when the medium was maintained at 32 °C was determined to be 33.5 °C with the calibration curve prepared in a fixed HeLa cell (Figure 6.14b) (Sekiguchi et al., 2018). This result implied that the temperature inside the cell is 1.5 °C higher than the surrounding medium.

Most non-fluorometric methods for cellular thermometry (see Chapter 4) give absolute temperature values, which is one of the advantages over fluorometric intracellular thermometry. Finally, it is worth mentioning that quantum coherence modulation microscopy might be able to draw intracellular temperature maps with absolute values (Zhou et al., 2023). This work has been published just before the writing of this book was completed and should be noted.

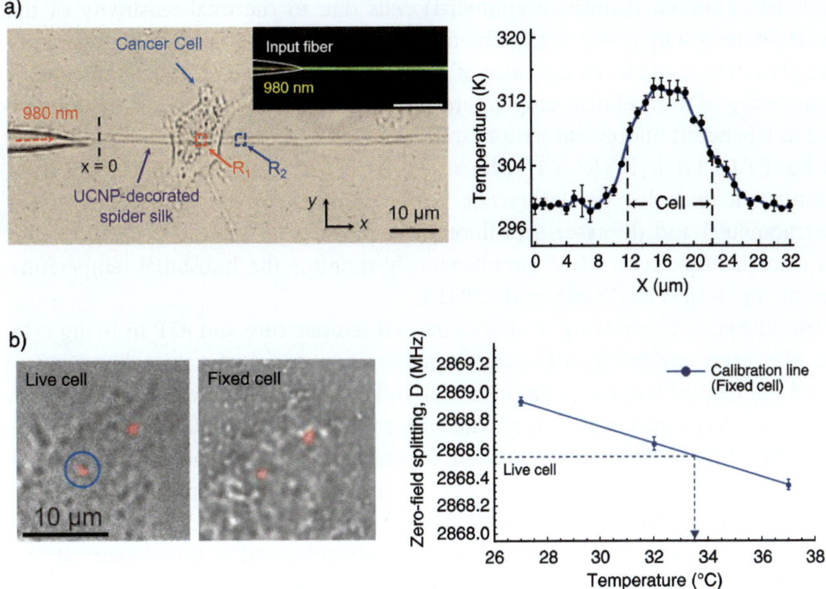

Figure 6.14 Proposed absolute temperature measurements of living cells. (a) Optical microscope image of a breast cancer cell contacted to upconversion nanoparticles (UCNP)-decorated spider silk (inset: a dark-field image of the excited UCNP-decorated spider silk, scale bar: 20 μm) (left) and the temperature distribution along the UCNP-decorated spider silk (right). Adapted from Gong et al., 2021 / American Chemical Society. (b) Merged bright-field and fluorescence images of a HeLa cell containing fluorescent nanodiamonds before and after fixation (left and middle, respectively) and absolute temperature measurement of a living HeLa cell by fluorescent nanodiamonds with a calibration curve obtained using fixed cells (right). Adapted from Sekiguchi et al. (2018) *Biophys. Physicobiol.*, **15**, 229–234 / THE BIOPHYSICAL SOCIETY OF JAPAN.

6.6
Simultaneous Monitoring of Intracellular Temperature and the Concentration of a Chemical Species Related to a Temperature Variation Inside a Living Cell

Understanding the relationship between intracellular temperature and the intracellular level of an endogenous molecule or ion is inevitably important for revealing profound mechanisms of intracellular thermal regulation. Thus, simultaneous monitoring of intracellular temperature and the concentration of a related molecule or ion has been proposed. For example, simultaneous monitoring of temperature and a norepinephrine level inside PC12 (rat pheochromocytoma) cells was proposed using a fluorescent polymeric thermometer PNIPAm-AANBD and a fluorescent sensor for norepinephrine PHE ((E)-1-(4-boronobenzyl)-2-(2-(1,3-dioxo-1H,3H-benzo[*de*]-isochromen-6-yl)vinyl)pyridin-1-ium bromide) (Qiao et al., 2021). It should be noted that highly functionalized molecular sensors that can simultaneously monitor multiple environmental factors are called "molecular logic gates" (see Appendix 5). For example, carbon nanodots prepared from citric acid and Rh6G hydrazide change their fluorescence intensity with both temperature and pH variations in MCF-7 (human breast adenocarcinoma) cells and

MCF-10A (human mammary epithelial) cells due to the dual-sensitivity of the fluorophore (Gadly et al., 2021). Disruptive variations of intracellular temperature and pH values could be an indicator of cancer cells (Swietach, 2019). Simultaneous monitoring of intracellular temperature and pH can also be performed by utilizing two independent fluorescent sensors at the same time. A combination of rhodamine B/Eu-PTTA (N,N,N′,N′-(4′-phenyl-2,2′:6′,2″-terpyridine-6,6″-diyl) bis(methylenenitrilo) tetrakis (acetic acid)) complex (a ratiometric fluorescent thermometer) and fluorescein (a fluorescent pH sensor) immobilized inside and on silica nanoparticles could simultaneously monitor the lysosomal temperature and pH in HeLa cells (Zhang et al., 2018).

Simultaneous monitoring of mitochondrial temperature and ATP in living cells has also been enabled by a fluorescent system composed of a mitochondria-targeted fluorescent polymeric thermometer and a rhodamine B derivative, RhB-ABA (Figure 6.15a) (Qiao et al., 2018). The temperature-dependent fluorescence intensity ratio of the mitochondria-targeted fluorescent polymeric thermometer at 454 and 477 nm under the excitation at 365 nm could be applied for intracellular temperature sensing. The evaluated temperature resolution is $0.98 \pm 0.08\,°C$. In contrast, RhB-ABA with a phenylboronic acid moiety enhanced its fluorescence at 580 nm (with the excitation at 488 nm) due to the cleavage of the ring structure of RhB-ABA by coupling with ATP. After introducing this fluorescent sensing system into HeLa cells by incubation for 20 min, simultaneous monitoring of mitochondrial temperature and an ATP level was performed under the chemical stimulation with FCCP (Figure 6.15b, 6.15c).

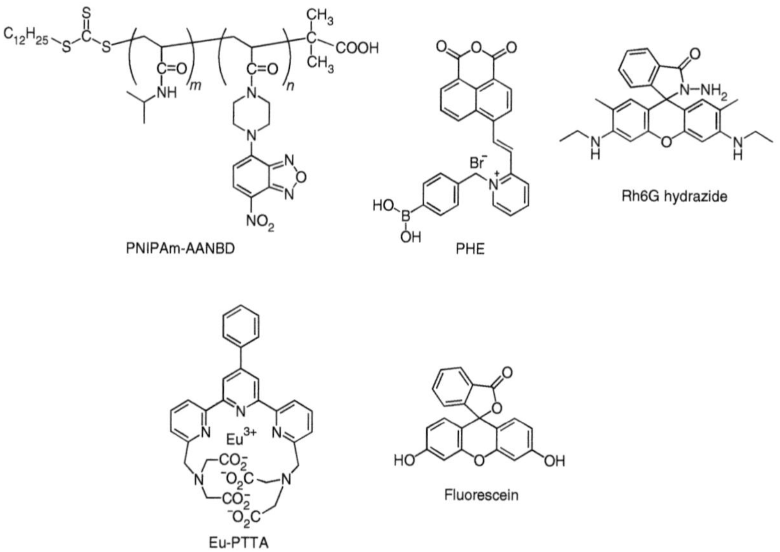

PNIPAm-AANBD

PHE

Rh6G hydrazide

Eu-PTTA

Fluorescein

Later, the same research group developed another simultaneous fluorescent monitoring system PNIPAm-VPBA-C-FAM, that could monitor intracellular temperature and Ca^{2+} concentration (Qiao et al., 2020). In the system, the fluorescent polymeric thermometer PNIPAm-VPBA-C showed temperature-dependent fluorescence intensity at 738 nm under the excitation at 488 nm, and the corresponding temperature resolution in intracellular thermometry of HeLa cells was $0.10–0.63\,°C$ in the range $32.0–39.8\,°C$. The fluorescent Ca^{2+} sensor, Fluo-4 AM (FAM), is highly hydrophobic

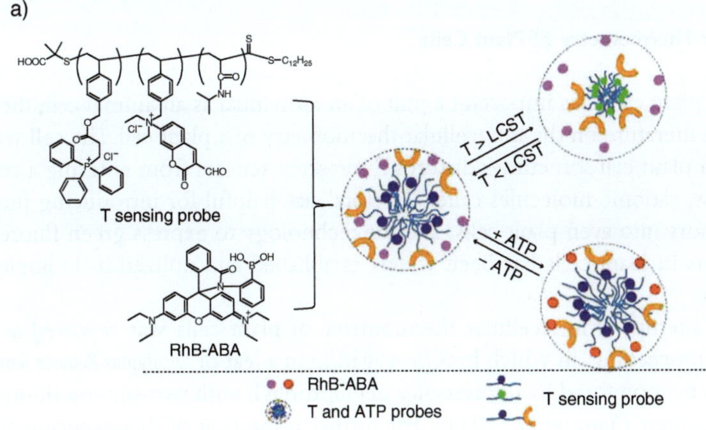

Figure 6.15 Simultaneous monitoring of mitochondrial temperature and ATP level. (a) Chemical structures and function of fluorescent probes used in the study. (b,c) Effects of the FCCP treatment on temperature (b) and an ATP level (c) in mitochondria in HeLa cells. The ATP level was evaluated from the fluorescence intensity of RhB-ABA. HeLa cells were stained with the temperature sensing probe (0.5 mg mL^{-1}) and the ATP sensing probe (5 µmol L^{-1}) for 20 min. Qiao et al. (2018) *Anal. Chem.*, **90**, 12553–12558 / American Chemical Society.

and, therefore, embedded in the self-assembly of PNIPAm-VPBA-C in an aqueous solution. The Ca^{2+} concentration could be calculated from the Ca^{2+}-induced fluorescence enhancement of FAM at 515 nm under the excitation at 488 nm.

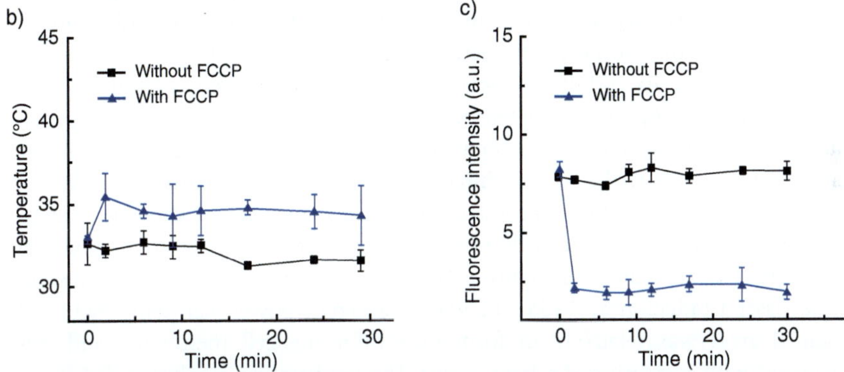

6.7
Intracellular Thermometry of Plant Cells

Although a plant cell is as important a unit of an individual as an animal cell, there is very little literature on the intracellular thermometry of a plant cell. The cell wall protecting a plant cell sometimes disturbs fluorescent sensors from entering a cell. Nevertheless, cationic molecules called "vectors" are helpful for introducing fluorescent sensors into even plant cells, and the technology to express green fluorescent proteins in plant cells has been widely established and utilized in biological laboratories.

An early attempt at intracellular thermometry of plant cells was reported as a conference proceeding, in which heat propagation in a leaf of *Eucalyptus Robusta Smith* was tried to be monitored by fluorescence of chlorophyll with two-photon fluorescence microscopy (Yang et al., 2011). For further utilization of this experimental setup in intracellular thermometry of plant cells, the photophysical study is at least required in terms of temperature-dependent fluorescence properties of chlorophyll. Another case used an onion skin and rhodamine B as an object and a fluorescent molecular thermometer, respectively (Paviolo et al., 2013). In this study, fluorescence intensity and lifetime were compared to be selected as a suitable temperature-dependent parameter in temperature imaging with a microscope. Although fluorescence lifetime was independent of the concentration of rhodamine B, the calibration curves (i.e., the relationship between fluorescence lifetime and temperature) differed between the solution and the onion skins. Thus, fluorescence intensity was adopted as a parameter detected for fluorometric temperature mapping of onion skins by rhodamine B. Figure 6.16 shows temperature maps of onion skins under laser irradiation at 1455 nm with a different power. The temperature of scanned areas was relatively uniform except for the cell membrane and was increased with increasing the laser power. The temperature resolution for single pixel measurements could reach 0.3 °C under laser irradiation with the highest power (96 mV) but was reduced to 1.8 °C in the actual experiments due to the high mobility and instability of rhodamine B inside organic tissues. The authors of the paper suggested that a fluorescent molecular thermometer should be encapsulated and immobilized in biological tissues to achieve intracellular temperature mapping with a high-temperature resolution.

Chlorophyll a
(a type of chlorophyll)

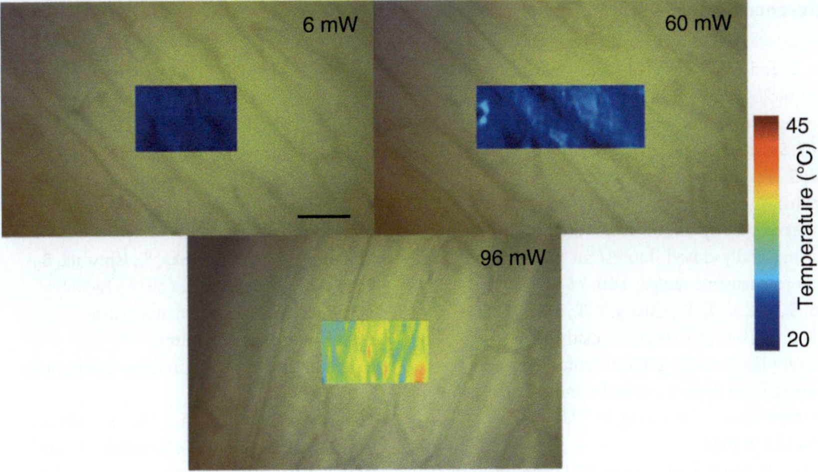

Figure 6.16 Temperature maps of onion skins under heating with different laser powers. The fluorescence of rhodamine B in the colored areas was recorded by a photoluminescence spectrometer with a confocal microscope to obtain the temperature maps. Scale bar: 50 μm. Adapted from Paviolo et al. (2013) *J. Microsc.*, **250**, 179–188 / John Wiley & Sons.

Recently, core/shell $NaGdF_4$:Er^{3+},Yb^{3+}/$NaGdF_4$ upconversion nanoparticles were utilized for intracellular temperature mapping of a moving onion epidermis cell sample. However, this report mainly focused on developing novel high-speed imaging techniques (Liu et al., 2021). Single-shot compressed ultrahigh-speed imaging enabled dynamic thermometry using the temperature-dependent emission lifetime of Er^{3+} ion (400–490 μs) with a temporal resolution of 0.05 s.

In a recent review article, Professor Inada discussed the possibility of intracellular thermometry of plant cells (Inada, 2023). She pointed out two difficulties in the intracellular thermometry of plant cells: (i) the temperature of plant cells is significantly affected by stomatal conductance. Therefore, cultured cells, protoplasts (protoplasms of living plant cells whose cell wall has been removed), and tissues without stomata are suitable to study endogenous temperature variation which is involved in biological events; (ii) most of the experimental settings such as a stage heater for a fluorescence microscope are adapted to mammalian cultured cells. A microscope must be equipped with a temperature control system covering a range below 30 °C for intracellular thermometry of plant cells.

Professor Inada has insisted that one of the unique targets in intracellular thermometry of plant cells is alternative oxidase (AOX), which transfers electrons from ubiquinone to oxygen for the reduction of oxygen into water. AOX is considered to be essential to heat production in an inflorescence of thermogenic plants. Another is a phytohormone, salicylic acid, which can increase the leaf temperature of *Nicotiana tabacum*, likely due to the closure of the stomata (Chaerle et al., 1999). Also, uncoupling effects on mitochondrial respiration are seen by a low level of salicylic acid, suggesting that salicylic acid can induce thermogenesis of plant cells. We are eagerly waiting for experimental research on the intracellular thermometry of plant cells to be conducted.

References

Atia, L. and Givli, S. (2014) A theoretical study of biological membrane response to temperature gradients at the single-cell level. *J. R. Soc. Interface*, **11**, 20131207.

Arai, S., Ferdinandus, Takeoka, S., Ishiwata, S., Sato, H., and Suzuki, M. (2015a) Micro-thermography in millimeter-scale animals by using orally-dosed fluorescent nanoparticle thermosensors. *Analyst*, **140**, 7534–7539.

Arai, S., Lee, C.-L. K., Chang, Y.-T., Sato, H., and Sou, K. (2015b) Thermosensitive nanoplatforms for photothermal release of cargo from liposomes under intracellular temperature monitoring. *RSC Adv.*, **5**, 93530–93538.

Becskei, A. and Rahaman, S. (2022) The life and death of RNA across temperatures. *Comput. Struct. Biotechnol. J.*, **20**, 4325–4336.

Chaerle, L., Van Caeneghem, W., Messens, E., Lambers, H., Van Montagu, M., and Van Der Straeten, D. (1999) Presymptomatic visualization of plant-virus interactions by thermography. *Nat. Biotechnol.*, **17**, 813–816.

Chen, S., Hoskins, C., Wang, L., MacDonald, M. P., and André, P. (2012) A water-soluble temperature nanoprobe based on a multimodal magnetic-luminescent nanocolloid. *Chem. Commun.*, **48**, 2501–2503.

Chen, Z., Zhang, K. Y., Tong, X., Liu, Y., Hu, C., Liu, S., Yu, Q., Zhao, Q., and Huang, W. (2016) Phosphorescent polymeric thermometers for in vitro and in vivo temperature sensing with minimized background interference. *Adv. Funct. Mater.*, **26**, 4386–4396.

Choi, J., Zhou, H., Landig, R., Wu, H.-Y., Yu, X., von Stetina, S. E., Kucsko, G., Mango, S. E., Needleman, D. J., Samuel, A. D. T., Maurer, P. C., Park, H., and Lukin, M. D. (2020) Probing and manipulating embryogenesis via nanoscale thermometry and temperature control. *Proc. Natl. Acad. Sci. USA*, **117**, 14636–14641.

Dewhirst, M. W., Vujaskovic, Z., Jones, E., and Thrall, D. (2005) Re-setting the biologic rationale for thermal therapy. *Int. J. Hyperthermia*, **21**, 779–790.

Diederich, C. J. (2005) Thermal ablation and high-temperature thermal therapy: overview of technology and clinical implementation. *Int. J. Hyperthermia.*, **21**, 745–753.

Donner, J. S., Thompson, S. A., Alonso-Ortega, C., Morales, J., Rico, L. G., Santos, S. I. C. O., and Quidant, R. (2013) Imaging of plasmonic heating in a living organism. *ACS Nano*, **7**, 8666–8672.

Fang, C. and Zhang, M. (2010) Nanoparticle-based theragnostics: integrating diagnostic and therapeutic potentials in nanomedicine. *J. Control. Release*, **146**, 2–5.

Ferdinandus, Arai, S., Takeoka, S., Ishiwata, S., Suzuki, M., and Sato, H. (2016) Facilely fabricated luminescent nanoparticle thermosensor for real-time microthermography in living animals. *ACS Sens.*, **1**, 1222–1227.

Ferdinandus, Suzuki, M., Vu, C. Q., Harada, Y., Sarker, S. R., Ishiwata, S., Kitaguchi, T., and Arai, S. (2022) Modulation of local cellular activities using a photothermal dye-based subcellular-sized heat spot. *ACS Nano*, **16**, 9004–9018.

Frangioni, J. V. (2003) In vivo near-infrared fluorescence imaging. *Curr. Opin. Chem. Biol.*, **7**, 626–634.

Fujiwara, M., Sun, S., Dohms, A., Nishimura, Y., Suto, K., Takezawa, Y., Oshimi, K., Zhao, L., Sadzak, N., Umehara, Y., Teki, Y., Komatsu, N., Benson, O., Shikano, Y., and Kage-Nakadai, E. (2020) Real-time nanodiamond thermometry probing in vivo thermogenic responses. *Sci. Adv.*, **6**, eaba9636.

Gadly, T., Chakraborty, G., Tyagi, M., Patro, B. S., Dutta, B., Potnis, A., Chandwadkar, P., Acharya, C., Suman, S. K., Mukherjee, A., Neogy, S., Wadawale, A., Sahoo, S., Chauhan, N., and Ghosh, S. K. (2021) Carbon nano-dot for cancer studies as dual nano-sensor for imaging intracellular temperature or pH variation. *Sci. Rep.*, **11**, 24341.

Gerasimova, E. N., Yaroshenko, V. V., Talianov, P. M., Peltek, O. O., Baranov, M. A., Kapitanova, P. V., Zuev, D. A., Timin, A. S., and Zyuzin, M. V. (2021) Real-time temperature monitoring of photoinduced cargo release inside living cells using hybrid capsules decorated with gold nanoparticles and fluorescent nanodiamonds. *ACS Appl. Mater. Interfaces*, **13**, 36737–36746.

Gong, Z., Wu, T., Chen, X., Guo, J., Zhang, Y., and Li, Y. (2021) Upconversion nanoparticle decorated spider silks as single-cell thermometers. *Nano Lett.*, **21**, 1469–1476.

Herrera, A. P., Rodríguez, M., Torres-Lugo, M., and Rinaldi, C. (2008) Multifunctional magnetite nanoparticles coated with fluorescent thermo-responsive polymeric shells. *J. Mater. Chem.*, **18**, 855–858.

Hoshi, Y., Okabe, K., Shibasaki, K., Funatsu, T., Matsuki, N., Ikegaya, Y., and Koyama, R. (2018) Ischemic brain injury leads to brain edema via hyperthermia-induced TRPV4 activation. *J. Neurosci.*, **38**, 5700–5709.

Hong, G., Lee, J. C., Robinson, J. T., Raaz, U., Xie, L., Huang, N. F., Cooke, J. P., and Dai, H. (2012) Multifunctional in vivo vascular imaging using near-infrared II fluorescence. *Nat. Med.*, **18**, 1841–1846.

Howard, R., Scheiner, A., Cunningham, J., and Gatenby, R. (2019) Cytoplasmic convection currents and intracellular temperature gradients. *PLoS Comput. Biol.*, **15**, e1007372.

Hua, W., Mao, Y., Zhang, J., Liu, L., Zhang, G., Yang, S., Boyer, D., Zhou, C., Zheng, F., Sun, S., and Lin, S. (2020) Renal clearable gold nanoparticle-functionalized silk film for in vivo fluorescent temperature mapping. *Front. Chem.*, **8**, 364.

Ibraimov, A. I. (2017) Cell thermoregulation: problems, advances and perspectives. *J. Mol. Biol. Res.*, **7**, 58–79.

Inada, N. (2023) A guide to plant intracellular temperature imaging using fluorescent thermometers. *Plant Cell Physiol.*, **64**, 7–18.

Jiang, X., Li, B. Q., Qu, X., Yang, H., Shao, J., and Zhang, H. (2019) Multilayered dual functional $SiO_2@Au@SiO_2@QD$ nanoparticles for simultaneous intracellular heating and temperature measurement. *Langmuir*, **35**, 6367–6378.

Jung, H. S., Verwilst, P., Sharma, A., Shin, J., Sessler, J. L., and Kim, J. S. (2018) Organic molecule-based photothermal agents: an expanding photothermal therapy universe. *Chem. Soc. Rev.*, **47**, 2280–2297.

Kamei, Y., Suzuki, M., Watanabe, K., Fujimori, K., Kawasaki, T., Deguchi, T., Yoneda, Y., Todo, T., Takagi, S., Funatsu, T., and Yuba, S. (2009) Infrared laser-mediated gene induction in targeted single cells in vivo. *Nat. Methods*, **6**, 79–81.

Kanamori, T., Miyazaki, N., Aoki, S., Ito, K., Hisaka, A., and Hatakeyama, H. (2021) Investigation of energy metabolic dynamism in hyperthermia-resistant ovarian and uterine cancer cells under heat stress. *Sci. Rep.*, **11**, 14726.

Kubota, H., Ogawa, H., Miyazaki, M., Ishii, S., Oyama, K., Kawamura, Y., Ishiwata, S., and Suzuki, M. (2021) Microscopic temperature control reveals cooperative regulation of

actin-myosin interaction by drebrin E. *Nano Lett.*, **21**, 9526–9533.

Lapotko, D. (2009) Plasmonic nanoparticle-generated photothermal bubbles and their biomedical applications. *Nanomedicine*, **4**, 813–845.

Larson, D. R., Zipfel, W. R., Williams, R. M., Clark, S. W., Bruchez, M. P., Wise, F. W., and Webb, W. W. (2003) Water-soluble quantum dots for multiphoton fluorescence imaging in vivo. *Science*, **300**, 1434–1436.

Liu, X., Skripka, A., Lai, Y., Jiang, C., Liu, J., Vetrone, F., and Liang, J. (2021) Fast wide-field upconversion luminescence lifetime thermometry enabled by single-shot compressed ultrahigh-speed imaging. *Nat. Commun.*, **12**, 6401.

Liu, L., Wu, J., Chen, B., Gao, J., Li, T., Ye, Y., Tian, H., Wang, S., Wang, F., Jiang, J., Ou, J., Tong, F., Peng, F., and Tu, Y. (2022) Magnetically actuated biohybrid microswimmers for precise photothermal muscle contraction. *ACS Nano*, **16**, 6515–6526.

Maciel, G. S. (2022) Pushing the limits of luminescence thermometry: probing the temperature of proteins in cells. *J. Biol. Phys.*, **48**, 167–175.

Miyagawa, T., Fujie, T., Ferdinandus, Von Doan, T. T., Sato, H., and Takeoka, S. (2016) Glue-free stacked luminescent nanosheets enable high-resolution ratiometric temperature mapping in living small animals. *ACS Appl. Mater. Interfaces*, **8**, 33377–33385.

Nakano, M., Arai, Y., Kotera, I., Okabe, K., Kamei, Y., and Nagai, T. (2017) Genetically encoded ratiometric fluorescent thermometer with wide range and rapid response. *PLoS ONE*, **12**, e0172344.

Okabe, K., Inada, N., Gota, C., Harada, Y., Funatsu, T., and Uchiyama, S. (2012) Intracellular temperature mapping with a fluorescent polymeric thermometer and fluorescence lifetime imaging microscopy. *Nat. Commun.*, **3**, 705.

Okabe, K. and Uchiyama, S. (2021) Intracellular thermometry uncovers spontaneous thermogenesis and associated thermal signaling. *Commun. Biol.*, **4**, 1377.

Oyama, K., Ishii, S., and Suzuki, M. (2022) Opto-thermal technologies for microscopic analysis of cellular temperature-sensing systems. *Biophys. Rev.*, **14**, 41–54.

Paviolo, C., Clayton, A. H. A., Mcarthur, S. L., and Stoddart, P. R. (2013) Temperature measurement in the microscopic regime: a comparison between fluorescence lifetime- and intensity-based methods. *J. Microsc.*, **250**, 179–188.

Piñol, R., Brites, C. D. S., Bustamante, R., Martínez, A., Silva, N. J. O., Murillo, J. L., Cases, R., Carrey, J., Estepa, C., Sosa, C., Palacio, F., Carlos, L. D., and Millán, A. (2015) Joining time-resolved thermometry and magnetic-induced heating in a single nanoparticle unveils intriguing thermal properties. *ACS Nano*, **9**, 3134–3142.

Polo-Corrales, L. and Rinaldi, C. (2012) Monitoring iron oxide nanoparticle surface temperature in an alternating magnetic field using thermoresponsive fluorescent polymers. *J. Appl. Phys.*, **111**, 07B334.

Qiao, J., Chen, C., Shangguan, D., Mu, X., Wang, S., Jiang, L., and Qi, L. (2018) Simultaneous monitoring of mitochondrial temperature and ATP fluctuation using fluorescent probes in living cells. *Anal. Chem.*, **90**, 12553–12558.

Qiao, J., Hwang, Y.-H., Kim, D.-P., and Qi, L. (2020) Simultaneous monitoring of temperature and Ca^{2+} concentration variation by fluorescent polymer during intracellular heat production. *Anal. Chem.*, **92**, 8579–8583.

Qiao, J., Wu, D., Song, Y., Ji, W., Yue, Q., Mao, L., and Qi, L. (2021) Simultaneous monitoring of intracellular temperature and norepinephrine variation by fluorescent probes during norepinephrine reuptake. *Anal. Chem.*, **93**, 14743–14747.

Qiu, X., Zhou, Q., Zhu, X., Wu, Z., Feng, W., and Li, F. (2020) Ratiometric upconversion nanothermometry with dual emission at the same wavelength decoded via a time-resolved technique. *Nat. Commun.*, **11**, 4.

Rodríguez-Sevilla, P., Marin, R., Ximendes, E., del Rosal, B., Benayas, A., and Jaque, D. (2022) Luminescence thermometry for brain activity monitoring: a perspective. *Front. Chem.*, **10**, 941861.

Romshin, A. M., Zeeb, V., Glushkov, E., Radenovic, A., Sinogeikin, A. G., and Vlasov, I. I. (2023) Nanoscale thermal control of a single living cell enabled by diamond heater-thermometer. *Sci. Rep.*, **13**, 8546.

Sekiguchi, T., Sotoma, S., and Harada, Y. (2018) Fluorescent nanodiamonds as a robust temperature sensor inside a single cell. *Biophys. Physicobiol.*, **15**, 229–234.

Shen, F., Yang, W., Cui, J., Hou, Y., and Bai, G. (2021) Small-molecule fluorogenic probe for the detection of mitochondrial temperature in vivo. *Anal. Chem.*, **93**, 13417–13420.

Shen, Y., Lifante, J., Zabala-Gutierrez, I., de la Fuente-Fernández, M., Granado, M., Fernández, N., Rubio-Retama, J., Jaque, D., Marin, R., Ximendes, E., and Benayas, A. (2022) Reliable and remote monitoring of absolute temperature during liver inflammation via luminescence-lifetime-based nanothermometry. *Adv. Mater.*, **34**, 2107764.

Silva, P. L., Savchuk, O. A., Gallo, J., García-Hevia, L., Bañobre-López, M., and Nieder, J. B. (2020) Mapping intracellular thermal response of cancer cells to magnetic hyperthermia treatment. *Nanoscale*, **12**, 21647–21656.

Singhal, A. and Shaham, S. (2017) Infrared laser-induced gene expression for tracking development and function of single C. elegans embryonic neurons. *Nat. Commun.*, **8**, 14100.

Su, X., Wen, Y., Yuan, W., Xu, M., Liu, Q., Huang, C., and Li, F. (2020) Lifetime-based nanothermometry in vivo with ultra-long-lived luminescence. *Chem. Commun.*, **56**, 10694–10697.

Swietach, P. (2019) What is pH regulation, and why do cancer cells need it? *Cancer Metastasis Rev.*, **38**, 5–15.

Talbot, E. L., Kotar, J., Parolini, L., Michele, L. D., and Cicuta, P. (2017) Thermophoretic migration of vesicles depends on mean temperature and head group chemistry. *Nat. Commun.*, **8**, 15351.

Tian, F., Zhong, X., Zhao, J., Gu, Y., Fan, Y., Shi, F., Zhang, Y., Tan, Y., Chen, W., Yi, C., and Yang, M. (2021) Hybrid theranostic microbubbles for ultrasound/photoacoustic imaging guided starvation/low-temperature photothermal/hypoxia-activated synergistic cancer therapy. *J. Mater. Chem. B*, **9**, 9358–9369.

Tsai, P.-C., Epperla, C. P., Huang, J.-S., Chen, O. Y., Wu, C.-C., and Chang, H.-C. (2017) Measuring nanoscale thermostability of cell membranes with single gold-diamond nanohybrids. *Angew. Chem. Int. Ed.*, **56**, 3025–3030.

Tyrpak, D. R., Li, Y., Lei, S., Avila, H., and MacKay, J. A. (2021) Single-cell quantification of the transition temperature of intracellular elastin-like polypeptides. *ACS Biomater. Sci. Eng.*, **7**, 428–440.

Wang, Z., Ma, X., Zong, S., Wang, Y., Chen, H., and Cui, Y. (2015) Preparation of a magnetofluorescent nano-thermometer and its targeted temperature sensing applications in living cells. *Talanta*, **131**, 259–265.

Wei, Y., Liu, S., Pan, C., Yang, Z., Liu, Y., Yong, J., and Quan, L. (2020) Molecular antenna-sensitized upconversion nanoparticle for temperature monitored precision photothermal therapy. *Int. J. Nanomed.*, **15**, 1409–1420.

Wu, W., Shen, J., Banerjee, P., and Zhou, S. (2011) A multifunctional nanoplatform based on responsive fluorescent plasmonic ZnO-Au@PEG hybrid nanogels. *Adv. Funct. Mater.*, **21**, 2830–2839.

Wu, Y., Alam, M. N. A., Balasubramanian, P., Ermakova, A., Fischer, S., Barth, H., Wagner, M., Raabe, M., Jelezko, F., and Weil, T. (2021) Nanodiamond theranostic for light-controlled intracellular heating and nanoscale temperature sensing. *Nano Lett.*, **21**, 3780–3788.

Ximendes, E. C., Rocha, U., Jacinto, C., Kumar, K. U., Bravo, D., López, F. J., Rodríguez, E. M., García-Solé, J., and Jaque, D. (2016) Self-monitored photothermal nanoparticles based on core-shell engineering. *Nanoscale*, **8**, 3057–3066.

Xu, M., Zou, X., Su, Q., Yuan, W., Cao, C., Wang, Q., Zhu, X., Feng, W., and Li, F. (2018) Ratiometric nanothermometer in vivo based on triplet sensitized upconversion. *Nat. Commun.*, **9**, 2698.

Yang, C.-Y., Liao, C.-S., Tzeng, Y.-Y.,, and Chu, S.-W. (2011) Visualization of heat propagation in biological tissues with two-photon fluorescence microscopy. *Proc. SPIE*, **7903**, 790338.

Zhang, Z., Wang, J., and Chen, C. (2013) Near-infrared light-mediated nanoplatforms for cancer thermo-chemotherapy and optical imaging. *Adv. Mater.*, **25**, 3869–3880.

Zhang, W., El-Reash, Y. G. A., Ding, L., Lin, Z., Lian, Y., Song, B., Yuan, J., and Wang, X.-d. (2018) A lysosome-targeting nanosensor for simultaneous fluorometric imaging of intracellular pH values and temperature. *Microchim. Acta*, **185**, 533.

Zhang, H., Wu, Y., Gan, Z., Yang, Y., Liu, Y., Tang, P., and Wu, D. (2019) Accurate intracellular and *in vivo* temperature sensing based on CuInS$_2$/ZnS QD micelles. *J. Mater. Chem. B*, **7**, 2835–2844.

Zhou, M., Li, L., Xie, W., He, Z., and Li, J. (2021) Synthesis of a thermal-responsive dual-modal supramolecular probe for magnetic resonance imaging and fluorescence imaging. *Macromol. Rapid Commun.*, **42**, 2100248.

Zhou, H., Yao, W., Zhou, X., Dong, S., Wang, R., Guo, Z., Li, W., Qin, C., Xiao, L., Jia, S., Wu, Z., and Li, S. (2023) Accurate visualization of metabolic aberrations in cancer cells by temperature mapping with quantum coherence modulation microscopy. *ACS Nano*, **17**, 8433–8441.

Zhu, X., Li, J., Qiu, X., Liu, Y., Feng, W., and Li, F. (2018) Upconversion nanocomposite for programming combination cancer therapy by precise control of microscopic temperature. *Nat. Commun.*, **9**, 2176.

Appendix 1 Review and Feature Articles on Fluorescent Molecular Thermometers and Intracellular Thermometry in General

Developing highly sensitive fluorescent molecular thermometers and establishing their use in intracellular thermometry have rapidly activated these research areas for the past two decades. Fortunately, many research groups have contributed to the progress of intracellular thermometry, as summarized in the main chapters of this book. Accordingly, valuable review and account items on fluorescent molecular thermometers and intracellular thermometry have accumulated. As the author of this textbook, I strongly recommend a study of these review and account items for a comprehensive understanding, especially of history and researchers' personal viewpoints in this exciting research area. The following list is of general review items and feature articles, which are recommended reading but not mentioned in the main chapters in this textbook, in chronological order.

Book

Carlos, L. D. and Palacio, F. (eds.) (2016) *Thermometry at the Nanoscale: Techniques and Selected Applications*, Royal Society of Chemistry, Cambridge.

Dramićanin, M. (2018) *Luminescence Thermometry. Methods, Materials, and Applications*, Woodhead Publishing, Duxford.

Book Chapter

Chandrasekharan, N. and Kelly, L. A. (2004) Progress towards fluorescent molecular thermometers, in *Reviews in Fluorescence 2004* (eds. C. D. Geddes and J. R. Lakowicz), Kluwer Academic/Plenum Publishers, New York, pp. 21–40.

Arai, S. and Suzuki, M. (2018) Nanosized optical thermometers, in *Smart Nanoparticles for Biomedicine* (ed. G. Ciofani), Elsevier, Amsterdam, pp. 199–217.

Piñol, R., Brites, C. D. S., Silva, N. J., Carlos, L. D., and Millán, A. (2019) Nanoscale thermometry for hyperthermia applications, in *Nanomaterials for Magnetic and Optical Hyperthermia Applications* (eds. R. M. Fratila and J. M. de la Fuente), Elsevier, Amsterdam, pp. 139–172.

Review Articles

Uchiyama, S., de Silva, A. P., and Iwai, K. (2006) Luminescent molecular thermometers. *J. Chem. Educ.*, **83**, 720–727.

Brites, C. D. S., Lima, P. P., Silva, N. J. O., Millán, A., Amaral, V. S., Palacio, F., and Carlos, L. D. (2012) Thermometry at the nanoscale. *Nanoscale*, **4**, 4799–4829.

Jaque, D. and Vetrone, F. (2012) Luminescence nanothermometry. *Nanoscale*, **4**, 4301–4326.

McLaurin, E. J., Bradshaw, L. R., and Gamelin, D. R. (2013) Dual-emitting nanoscale temperature sensors. *Chem. Mater.*, **25**, 1283–1292.

Wang, X.-d., Wolfbeis, O. S., and Meier, R. J. (2013) Luminescent probes and sensors for temperature. *Chem. Soc. Rev.*, **42**, 7834–7869.

Jaque, D., del Rosal, B., Rodríguez, E. M., Maestro, L. M., Haro-González, P., and Solé, J. G. (2014) Fluorescent nanothermometers for intracellular thermal sensing. *Nanomedicine*, **9**, 1047–1062.

Sakaguchi, R., Kiyonaka, S., and Mori, Y. (2015) Fluorescent sensors reveal subcellular thermal changes. *Curr. Opin. Biotechnol.*, **31**, 57–64.

Binslem, S. A., Ahmad, M. R., and Awang, Z. (2015) Intracellular thermal sensor for single cell analysis – short review. *J. Teknol.*, **73**, 71–80.

Zhou, H., Sharma, M., Berezin, O., Zuckerman, D., and Berezin, M. Y. (2016) Nanothermometry: from microscopy to thermal treatments. *ChemPhysChem*, **17**, 27–36.

Bai, T. and Gu, N. (2016) Micro/nanoscale thermometry for cellular thermal sensing. *Small*, **12**, 4590–4610.

Uchiyama, S. and Gota, C. (2017) Luminescent molecular thermometers for the ratiometric sensing of intracellular temperature. *Rev. Anal. Chem.*, **36**, 20160021.

Nakano, M. and Nagai, T. (2017) Thermometers for monitoring cellular temperature. *J. Photochem. Photobiol. C*, **30**, 2–9.

Okabe, K., Sakaguchi, R., Shi, B., and Kiyonaka, S. (2018) Intracellular thermometry with fluorescent sensors for thermal biology. *Eur. J. Physiol.*, **470**, 717–731.

Qin, T., Liu, B., Zhu, K., Luo, Z., Huang, Y., Pan, C., and Wang, L. (2018) Organic fluorescent thermometers: highlights from 2013 to 2017. *Trends Anal. Chem.*, **102**, 259–271.

Mazza, M. M. A. and Raymo, F. M. (2019) Structural designs for ratiometric temperature sensing with organic fluorophores. *J. Mater. Chem. C*, **7**, 5333–5342.

Ogle, M. M., McWilliams, A. D. S., Jiang, B., and Martí, A. A. (2020) Latest trends in temperature sensing by molecular probes. *ChemPhotoChem*, **4**, 255–270.

Zhou, J., del Rosal, B., Jaque, D., Uchiyama, S., and Jin, D. (2020) Advances and challenges for fluorescence nanothermometry. *Nat. Methods*, **17**, 967–980.

Wang, F., Han, Y., and Gu, N. (2021) Cell temperature measurement for biometabolism monitoring. *ACS Sens.*, **6**, 290–302.

Chung, C. W. and Schierle, G. S. K. (2021) Intracellular thermometry at the micro-/nanoscale and its potential application to study protein aggregation related to neurodegenerative diseases. *ChemBioChem*, **22**, 1546–1558.

Bednarkiewicz, A., Drabik, J., Trejgis, K., Jaque, D., Ximendes, E., and Marciniak, L. (2021) Luminescence based temperature bio-imaging: status, challenges, and perspectives. *Appl. Phys. Rev.*, **8**, 011317.

Feng, G., Zhang, H., Zhu, X., Zhang, J., and Fang, J. (2022) Fluorescence thermometers: intermediation of fundamental temperature and light. *Biomater. Sci.*, **10**, 1855–1882.

Yang, N., Xu, J., Wang, F., Yang, F., Han, D., and Xu, S. (2022) Thermal probing techniques for a single live cell. *Sensors*, **22**, 5093.

Accounts, perspectives, and others

Inada, N. and Uchiyama, S. (2013) Methods and benefits of imaging the temperature distribution inside living cells. *Imaging Med.*, **5**, 303–305.

Katsnelson, A. (2017) Tiny temperature sensors. *ACS Cent. Sci.*, **3**, 364–366.

Yuexuan, Y. and Daocheng, W. (2020) Research shortcomings of fluorescent nanothermometers in biological and medical fields. *Nanomedicine*, **15**, 735–738.

Chung, C. W. and Schierle, G. S. K. (2021) Intracellular thermometry to study protein aggregation related to neurodegenerative diseases. *Trends Biochem. Sci.*, **46**, 251–252.

Appendix 2 Comprehensive Collection of Fluorescent Polymeric Thermometers Based on the Combination of a Thermo-responsive Polymer and an Environment-sensitive Fluorophore

After I published a series of papers on sensitive fluorescent polymeric thermometers based on the combination of a thermo-responsive polymer and an environment-sensitive fluorophore (see Section 2.11.3), an unexpectedly large number of research groups exploited this design concept in the development of novel fluorescent polymeric thermometers. Accordingly, various kinds of fluorescent monomers bearing an environment-sensitive fluorophore were newly synthesized. Furthermore, once the outstanding utility of a fluorescent polymeric thermometer was confirmed in intracellular thermometry in my laboratory in 2009, a similar functional evaluation of each fluorescent polymeric thermometer was performed in a laboratory worldwide. An accumulation of functional data concerning fluorescent polymeric thermometers has accelerated scientific progress on intracellular thermometry and polymer-based fluorescent sensors. Thus, fluorescent polymeric thermometers in the literature are summarized in Table A2.1. Readers should notice that this section mentions only linear-type (two-dimensional) fluorescent polymeric thermometers. Three-dimensional fluorescent nanogel thermometers having crosslinking units will be focussed on in Appendix 3.

Intracellular Thermometry with Fluorescent Molecular Thermometers, First Edition. Seiichi Uchiyama.
© 2024 Wiley-VCH GmbH. Published 2024 by Wiley-VCH GmbH.

Table A2.1 Fluorescent polymeric thermometers functioning in aqueous solution.

Entry	Thermo-responsive unit	Fluorescent unit/ mechanism[a]	Additional unit	λ_{ex} (nm)	λ_{em} (nm)	Parameter to be measured[b]	Functional temperature range (°C)	Parameter variation	Remarks	Ref.
1	NIPAM	C18-Py/A		336	379–399, 480	I_E/I_M	25–40	1.1–0.45		[1]
2	NIPAM	DBD-AE/B		444	533–566	FI	29–37	1–13.3		[2]
3	NIPAM	NBD-AE/B		469	522–530	FI	30–36	1–3.4		[2]
4	NIPMAM	DBD-AE/B		444	534–556	FI	45–54	1–7.1		[2]
5	NNPAM	DBD-AE/B		444	533–563	FI	18–24	1–16		[2]
6	NNPAM	MBC-AE/BC	DMAPAM	345	500	FI	5–56	1–0.28	In acidic buffer (pH 4)	[3]
7	NNPAM	DBD-AA/B	SPA	456	558	FI	30–65	1–3.2		[4]
8	NIPAM	RD/B		530	571	FI	10–33	0.19–1	FI suddenly decreases above 33 °C	[5,6]
9	NIPAM	DBD-AE/B		450	590	FI	37–45	1–3.9	Coating magnetite nanoparticles	[7]
10	NIPAM	DBD-AA/B		275	560	average τ_f	30–35	4.22–14.1 (ns)		[8]
11	NIPAM	HC/B		510	572	FI	25–40	1–35	At pH 6.4	[9]
12	NIPAM	CEA/A	Styrene	295	353	FI	25–50	1–0.014	Block copolymer	[10]

13	NIPAM	Pyrene (as an end group)/D	C60	340	375	FI	25–35	1–0.34		[11]
14	NIPAM	TPE/E		322	468	FI	25–34	1–1.9		[12]
15	NIPAM	BODIPY/F		490	575	FI	19–33	1–7.8		[13]
16	NIPAM	Fluorescein (labeling using FITC)/D		492	514	FI	20–50	1–0.7	Coating Au nanoparticles	[14]
17	NIPAM	Pyran-1/B		445	569–599	FI	20–50	1–5.5		[15]
18	NIPAM	DAAm/B		340	505–540	FI	30–40	1–3.3	Coating silica microbeads, position-dependent function	[16]
19	DEGMA	PyMMA/A		342	395, 467	I_E/I_M	5–30	0.81–0.40		[17]
20	VCa	Pyran-1/B		445	460	FI	25–50	1–2.5		[18]
21	PEO-PPO-PEO	tpyPtCl (as an endo group)/E		403	698	FI	15–31	1–19	Triblock copolymer	[19]
22	NIPAM	RhB-1/D		559	585	FI	20–40	1–0.27	Attaching to Ag core	[20]

(*Continued*)

Table A2.1 (*Continued*)

Entry	Thermo-responsive unit	Fluorescent unit/mechanism[a]	Additional unit	λ_{ex} (nm)	λ_{em} (nm)	Parameter to be measured[b]	Functional temperature range (°C)	Parameter variation	Remarks	Ref.
23	NNPAM	DBD-AA/B	SPA	405	560	Average τ_f	28–40	4.6–7.6 (ns)	Applied to intracellular temperature mapping in COS7 cells	[21]
24	NIPAM	AMC/G	Styrene	370	450	FI	33–36	1–0.36	Block copolymer, tested in MDCK cells	[22]
25	DEGA	4-DMN-1/B		435	535–542	FI	35–60	1–2.3	In PBS[c]	[23]
26	NNPAM	DBD-AA/B	APTMA	456	565	Average τ_f	10–35	6.0–8.6	In yeast cells	[24]
27	NNPAM, NIPAM	DBD-AA/B	APTMA	456	565	Average τ_f	25–40	5.2–8.7	In MOLT-4 cells	[24]
28	NIPAM	DEAC-1/H		400	475	FI	20–50	1–6.4	In PBS[c]	[25]
29	NIPAM	DBD-ED-1, square 660[d]/I		470	711–725	FI	31–37	1–3.8		[26]
30	NIPAM	Schiff base-1 (as an end group)/J	PEG	449	514	FI	35–75	1–5.6	Block copolymer, applied in *Bacillus thermophilus*	[27]
31	BzMa	VBK/I	OEGMA	330	Not indicated	FI	90–100	1–5	In ionic liquid $[C_2mim][NTf_2]$[e] (not in water), involving FRET from BzMa to VBK	[28]

32	NIPAM	NBD-AA/B, RhBAM (as a reference)		488	530, 571	FI ratio	32–39	1–1.5	Prepared by RAFT polymerization, tested in HeLa cells	[29]
33	NIPAM	TPP-A/K (as an end group)		363	517	FI	25–40	1–3.5		[30]
34	NIPAM	TPP-NI/K (as an endo group)		475	555	FI	25–40	1–3		[30]
35	NNPAM	DBThD-AA/B	APTMA	450	573	Average τ_f	25–40	4.8–7.9 (ns)	Tested in HeLa cells	[31]
36	NNPAM	DBThD-AA/B, BODIPY-AA (as a reference)	APTMA	458	515, 580	FI ratio	25–45	0.32–0.70	Tested in MOLT-4 and HEK293T cells	[32]
37	NIPAM	CMA, NBD-AE, RhB-1/I		365	422, 585	FI ratio	26–44	1.0–8.2	FRET system is constructed by three block copolymers with a different fluorophore, tested in HepG2 cells	[33]
38	NIPAM	NBD-AA/B, transferrin protein-stabilized Au nanocluster (as a reference)		488	545, 659	FI ratio	31–41	12–24	Tested in HeLa cells, chemical stimuli with ionomycin	[34]

(Continued)

Table A2.1 (*Continued*)

Entry	Thermo-responsive unit	Fluorescent unit/ mechanism[a]	Additional unit	λ_{ex} (nm)	λ_{em} (nm)	Parameter to be measured[b]	Functional temperature range (°C)	Parameter variation	Remarks	Ref.
39	VCa	NI-1/B		416	527–536	FI	10–60	1–8	Copolymers with a similar fluorescent monomer are also prepared	[35]
40	NNPAM	Ir-1/BJ, Ir-2/BJ	APTMA	405	482	τ_f	15–35	224–499 (ns)	Tested in HeLa cells and zebrafish larva	[36]
				405	470, 590	FI ratio	12–40	0.14–1.87		
41	NIPAM	MBC-AA/C		345	520	FI	30–45	1–0.55		[37]
42	PEO chains in V-EO7DCS	V-EO7DCS/A		353	540, 600	FI ratio	25–80	0.73–1.5		[38]
43	NIPAM	ANTH		403	530	FI	25–33	1–1.96		[39]
44	NIPAM	BODIPY-AA2/B		513	605	FI	36–48	1–7.4	Prepared by RAFT polymerization, tested in BHK cells	[40]
45	NIPAM	Curcumin-1/H (as an end group)		450	550	FI	39–57	1–5	Prepared by RAFT polymerization, introduced in HeLa cells	[41]

46	NIPAM	TFMCMA		340	436	FI ratio affected by FRET	30–50	0.11–0.31	Mixture of two copolymers, tested in HeLa cells	[42]
	NIPMAM	BOBPYBX		580	628					
47	PEO-PPO-PEO	AzOx, Naph (as end groups)/I		335	400, 550	FI ratio	20–65	0.39–0.70 (as FRET efficiency)	Attached to poly(propargyl acrylate) nanoparticle	[43]
48	NIPAM	Ir-2/BI, Eu-1	APTMA	405	470, 615	FI ratio affected by FRET	20–45	0.13–1.18	In PBS$^{c)}$, tested in HeLa cells and zebrafish	[44]
				405	480	τ_f	20–40	270–510 (ns)		
49	NIPAM	Acryloyl-α-CD-SP/B	Acryloyl-α-CD	365	625	FI	34–39	1–1.8	Inclusion of SP moiety into α-CD moiety works as a crosslinker	[45]
50	NIPAM	DACC	CTPP	365	454, 477	FI ratio	30.8–38.8	0.69–1.38	Tested in HeLa cells, applied for simultaneous monitoring of T and ATP	[46]
51	NIPAM	EM/E		430	607	FI	25–45	1–3.8		[47]
52	NIPAM	Pt-1/E	APTMA	365	500	FI	28–39	1–2.5	In PBS$^{c)}$ (pH 7.4), also tested in HepG2 cells	[48]

(Continued)

Table A2.1 (*Continued*)

Entry	Thermo-responsive unit	Fluorescent unit/mechanism[a]	Additional unit	λ_{ex} (nm)	λ_{em} (nm)	Parameter to be measured[b]	Functional temperature range (°C)	Parameter variation	Remarks	Ref.
53	NIPAM	Cy5 (as an end group), fluorescein (as labeled with FITC)	PEG	not indicated	520, 660	FI ratio	25–55	1.36–1.83	Coating Au nanoparticle, tested in MB49, HUVEC, and RAW264.7 cells	[49]
54	PyBEMA, DMAM	PyBEMA/A		344	470	FI ratio	40–45	1–1.1	Functional temperature range depends on unit ratio (PyBEMA:DMAM)	[50]
55	VCa	VCa (as a polymer)		340	415	FI	25–40	1–1.7	Tested in MCF-7 cells	[51]
56	NIPAM	NAI-DMAC, tBuODA/I		not indicated	460, 660	FI ratio	40–70	3.7–14.8		[52]
57	NIPAM	VPBA-C		488	738	FI	30–42	1–20	Simultaneous monitoring of Ca^{2+} with Fluo-4 AM, tested in HeLa cells	[53]

58	NIPAM	P33-B		370	399	FI affected by FRET	25–50		Three block copolymers are grafted on MoS$_2$ nanosheet, followed by incorporation in poly(ethylene glycol) hydrogel	[54]
	NIPAM-NIPMAM	P39-G	HEA		513					
	NIPMAM	RhB-1	HEA		581					
59	NIPAM	AANBD/B		488	542	FI	32–39	1–2.3	Prepared by RAFT polymerization, simultaneous monitoring of norepinephrine with PHE[f], tested in PC12 cells	[55]

a) A: excimer formation; B: sensitivity to polarity; C: sensitivity to hydrogen bonding; D: variation in distance to a quencher; E: aggregation-induced fluorescence; F: sensitivity to viscosity; G: decrease in light absorption; H: dissociation-induced fluorescence; I: variation in FRET (Förster resonance energy transfer) efficiency; J: restriction of rotation; K: aggregation-induced quenching.

b) I_E: emission intensity of excimer; I_M: fluorescence intensity of monomer; FI: fluorescence intensity; τ_f: fluorescence lifetime; FI ratio: fluorescence intensity ratio at two different wavelengths.

c) Phosphate-buffered saline.

d) Commercially available, but chemical structure is not specified.

e) 1-Ethyl-3-methylimidazolium bis(trifluoromethanesulfonyl)imide.

f) (E)-1-(4-Boronobenzyl)-2-(2-(1,3-dioxo-1H,3H-benzo[de]isochromen-6-yl)vinyl)pridin-1-ium bromide.

1 Ringsdorf, H., Venzmer, J., and Winnik, F. M. (1991) *Macromolecules*, **24**, 1678–1686.
2 Uchiyama, S., Matsumura, Y., de Silva, A. P., and Iwai, K. (2003) *Anal. Chem.*, **75**, 5926–5935.
3 Uchiyama, S., Takehira, K., Yoshihara, T., Tobita, S., and Ohwada, T. (2006) *Org. Lett.*, **8**, 5869–5872.
4 Gota, C., Uchiyama, S., and Ohwada, T. (2007) *Analyst*, **132**, 121–126.
5 Shiraishi, Y., Miyamoto, R., Zhang, X., and Hirai, T. (2007) *Org. Lett.*, **9**, 3921–3924.
6 Shiraichi, Y., Miyamoto, R., and Hirai, T. (2008) *J. Photochem. Photobiol. A: Chem.*, **200**, 432–437.
7 Herrera, A. P., Rodríguez, M., Torres-Lugo, M., and Rinaldi, C. (2008) *J. Mater. Chem.*, **18**, 855–858.
8 Gota, C., Uchiyama, S., Yoshihara, T., Tobita, S., and Ohwada, T. (2008) *J. Phys. Chem. B*, **112**, 2829–2836.
9 Shiraishi, Y., Miyamoto, R., and Hirai, T. (2008) *Langmuir*, **24**, 4273–4279.
10 Yan, Q., Yuan, J., Yuan, W., Zhou, M., Yin, Y., and Pan, C. (2008) *Chem. Commun.*, 6188–6190.
11 Hong, S. W., Kim, D. Y., Lee, J. U., and Jo, W. H. (2009) *Macromolecules*, **42**, 2756–2761.

12 Tang, L., Jin, J. K., Qin, A., Yuan, W. Z., Mao, Y., Mei, J., Sun, J. Z., and Tang, B. Z. (2009) *Chem. Commun.*, 4974–4976.

13 Wang, D., Miyamoto, R., Shiraishi, Y., and Hirai, T. (2009) *Langmuir*, **25**, 13176–13182.

14 Wongkongkatep, J., Ladadat, R., Lappermpunsap, W., Wongkongkatep, P., Phinyocheep, P., Ojida, A., and Hamachi, I. (2010) *Chem. Lett.*, **39**, 184–185.

15 Kim, S.-H., Hwang, I.-J., Gwon, S.-Y., and Son, Y.-A. (2010) *Dyes Pigm.*, **87**, 84–88.

16 Hattori, Y., Nagase, K., Kobayashi, J., Kikuchi, A., Akiyama, Y., Kanazawa, H., and Okano, T. (2010) *Chem. Phys. Lett.*, **491**, 193–198.

17 Pietsch, C., Vollrath, A., Hoogenboom, R., and Schubert, U. S. (2010) *Sensors*, **10**, 7979–7990.

18 Lee, E.-M., Gwon, S.-Y., Ji, B.-C., Bae, J.-S., and Kim, S.-H. (2011) *Fibers Polym.*, **12**, 288–290.

19 Hu, Y., Chan, K. H.-Y., Chung, C. Y.-S., and Yam, V. W.-W. (2011) *Dalton Trans.*, **40**, 12228–12234.

20 Liu, J., Li, A., Tang, J., Wang, R., Kong, N., and Davis, T. P. (2012) *Chem. Commun.*, **48**, 4680–4682.

21 Okabe, K., Inada, N., Gota, C., Harada, Y., Funatsu, T., and Uchiyama, S. (2012) *Nat. Commun.*, **3**, 705.

22 Qiao, J., Qi, L., Shen, Y., Zhao, L., Qi, C., Shangguan, D., Mao, L., and Chen, Y. (2012) *J. Mater. Chem.*, **22**, 11543–11549.

23 Inal, S., Kölsch, J. D., Chiappisi, L., Janietz, D., Gradzielski, M., Laschewsky, A., and Neher, D. (2013) *J. Mater. Chem. C*, **1**, 6603–6612.

24 Tsuji, T., Yoshida, S., Yoshida, A., and Uchiyama, S. (2013) *Anal. Chem.*, **85**, 9815–9823.

25 Inal, S., Kölsch, J. D., Sellrie, F., Schenk, J. A., Wischerhoff, E., Laschewsky, A., and Neher, D. (2013) *J. Mater. Chem. B*, **1**, 6373–6381.

26 Cheng, B., Wei, M.-Y., Liu, Y., Pitta, H., Xie, Z., Hong, Y., Nguyen, K. T., and Yuan, B. (2014) *IEEE J. Sel. Top. Quantum Electron.*, **20**, 6801214.

27 Zheng, Y., Li, G., Deng, H., Su, Y., Liu, J., and Zhu, X. (2014) *Polym. Chem.*, **5**, 2521–2529.

28 Zhang, C. and Maric, M. (2014) *Polym. Chem.*, **5**, 4926–4938.

29 Qiao, J., Chen, C., Qi, L., Liu, M., Dong, P., Jiang, Q., Yang, X., Mu, X., and Mao, L. (2014) *J. Mater. Chem. B*, **2**, 7544–7550.

30 Gu, P.-Y., Zhang, Y.-H., Chen, D.-Y., Lu, C.-J., Zhou, F., Xu, Q.-F., and Lu, J.-M. (2015) *RSC Adv.*, **5**, 8167–8174.

31 Hayashi, T., Fukuda, N., Uchiyama, S., and Inada, N. (2015) *PLoS ONE*, **10**, e0117677.

32 Uchiyama, S., Tsuji, T., Ikado, K., Yoshida, A., Kawamoto, K., Hayashi, T., and Inada, N. (2015) *Analyst*, **140**, 4498–4506.

33 Hu, X., Li, Y., Liu, T., Zhang, G., and Liu, S. (2015) *ACS Appl. Mater. Interfaces*, **7**, 15551–15560.

34 Qiao, J., Hwang, Y.-H., Chen, C.-F., Qi, L., Dong, P., Mu, X.-Y., and Kim, D.-P. (2015) *Anal. Chem.*, **87**, 10535–10541.

35 Enzenberg, A., Laschewsky, A., Boeffel, C., and Wischerhoff, E. (2016) *Polymers*, **8**, 109.

36 Chen, Z., Zhang, K. Y., Tong, X., Liu, Y., Hu, C., Liu, S., Yu, Q., Zhao, Q., and Huang, W. (2016) *Adv. Funct. Mater.*, **26**, 4386–4396.

37 Uchiyama, S., Remón, P., Pischel, U., Kawamoto, K., and Gota, C. (2016) *Photochem. Photobiol. Sci.*, **15**, 1239–1246.

38 Cui, J., Kwon, J. E., Kim, H.-J., Whang, D. R., and Park, S. Y. (2017) *ACS Appl. Mater. Interfaces*, **9**, 2883–2890.

39 Sasaki, S. and Konishi, G.-i. (2017) *RSC Adv.*, **7**, 17403–17416.

40 Gong, D., Cao, T., Han, S.-C., Zhu, X., Iqbal, A., Liu, W., Qin, W., and Guo, H. (2017) *Sens. Actuators B*, **252**, 577–583.

41 Yang, Q., Liu, H., Cheng, J., Hu, C., Zhang, S., Li, X., Zhao, H., Bai, L., Wang, S., and Wu, Y. (2017) *Int. J. Polym. Mater. Polym. Biomat.*, **66**, 907–914.

42 Ding, Z., Wang, C., Feng, G., and Zhang, X. (2018) *Polymers*, **10**, 283.

43 Klep, O., Bandera, Y., and Foulger, S. H. (2018) *Nanoscale*, **10**, 9401–9409.

44 Zhang, H., Jiang, J., Gao, P., Yang, T., Zhang, K. Y., Chen, Z., Liu, S., Huang, W., and Zhao, Q. (2018) *ACS Appl. Mater. Interfaces*, **10**, 17542–17550.

45 Zou, X., Xiao, X., Zhang, S., Zhong, J., Hou, Y., and Liao, L. (2018) *J. Biomater. Sci., Polym. Ed.*, **29**, 1579–1594.

46 Qiao, J., Chen, C., Shangguan, D., Mu, X., Wang, S., Jiang, L., and Qi, L. (2018) *Anal. Chem.*, **90**, 12553–12558.

47 Yang, J., Gu, K., Shi, C., Li, M., Zhao, P., and Zhu, W.-H. (2019) *Mater. Chem. Front.*, **3**, 1503–1509.

48 Lin, S., Pan, H., Li, L., Liao, R., Yu, S., Zhao, Q., Sun, H., and Huang, W. (2019) *J. Mater. Chem. C*, **7**, 7893–7899.

49 Wang, D., Zhou, M., Huang, H., Ruan, L., Lu, H., Zhang, J., Chen, J., Gao, J., Chai, Z., and Hu, Y. (2019) *ACS Appl. Bio Mater.*, **2**, 3178–3182.

50 Dong, Q., Sun, C., Chen, F., Yang, Z., Li, R., Wang, C., and Luo, C. (2019) *Polymers*, **11**, 1569.

51 Saha, B., Ruidas, B., Mete, S., Mukhopadhyay, C. D., Bauri, K., and De, P. (2020) *Chem. Sci.*, **11**, 141–147.

52 Christopherson, C. J., Mayder, D. M., Poisson, J., Paisley, N. R., Tonge, C. M., and Hudson, Z. M. (2020) *ACS Appl. Mater. Interfaces*, **12**, 20000–20011.

53 Qiao, J., Hwang, Y.-H., Kim, D.-P., and Qi, L. (2020) *Anal. Chem.*, **92**, 8579–8583.

54 Park, C. H., Kim, T., Lee, G. H., Ku, K. H., Kim, S.-H., and Kim, B. J. (2020) *ACS Appl. Mater. Interfaces*, **12**, 35415–35423.

55 Qiao, J., Wu, D., Song, Y., Ji, W., Yue, Q., Mao, L., and Qi, L. (2021) *Anal. Chem.*, **93**, 14743–14747.

Thermo-responsive unit

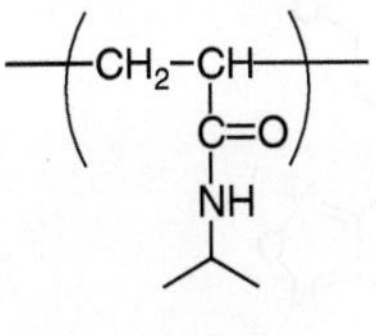

NIPAM

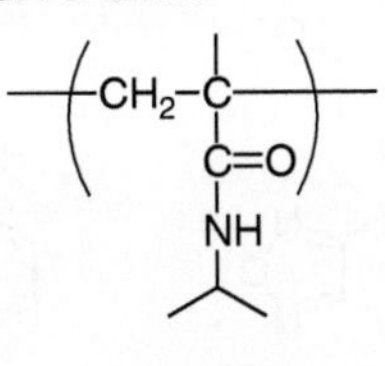

NIPMAM

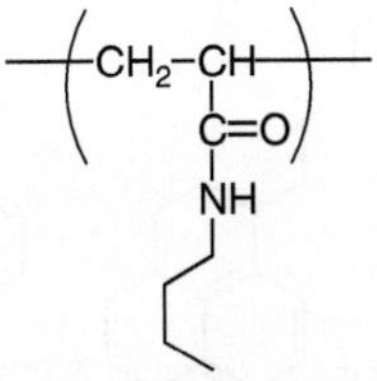

NNPAM

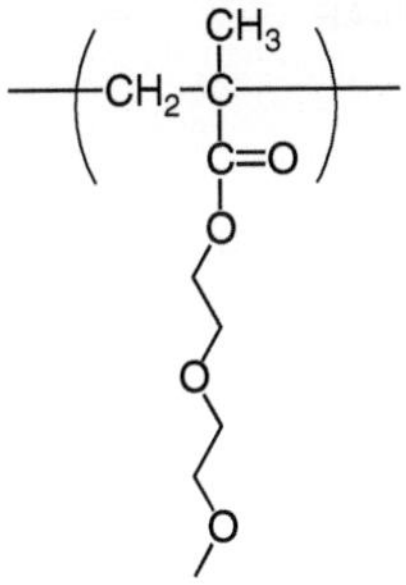

DEGMA

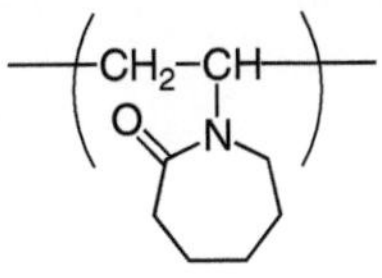

VCa

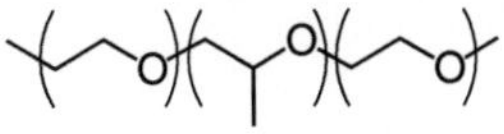

PEO-PPO-PEO

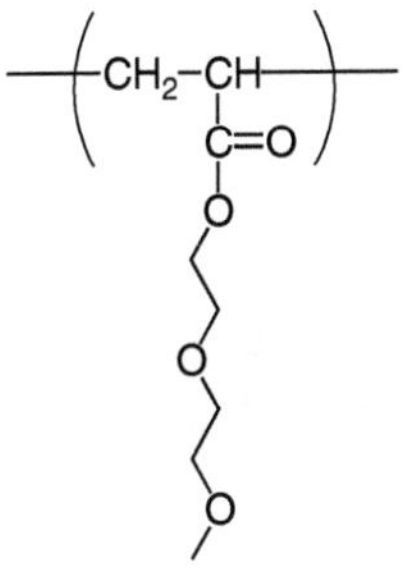

DEGA

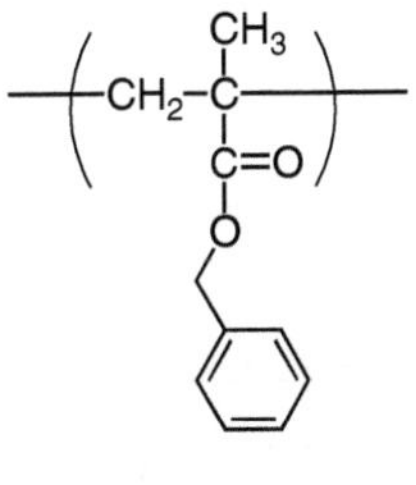

BzMA

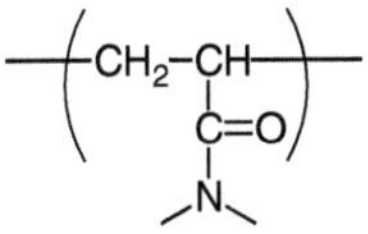

DMAM

Fluorescent unit 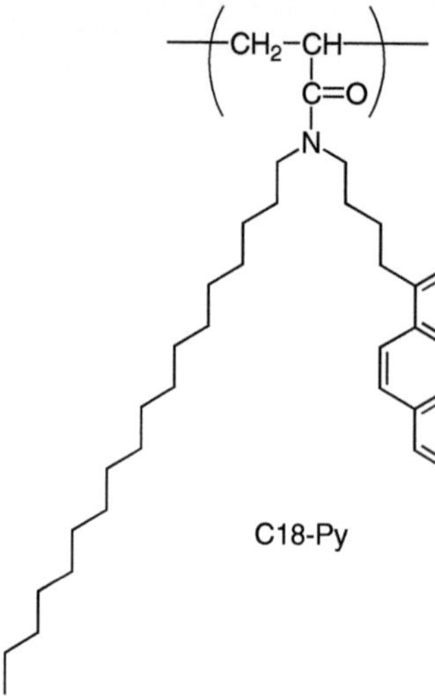
C18-Py 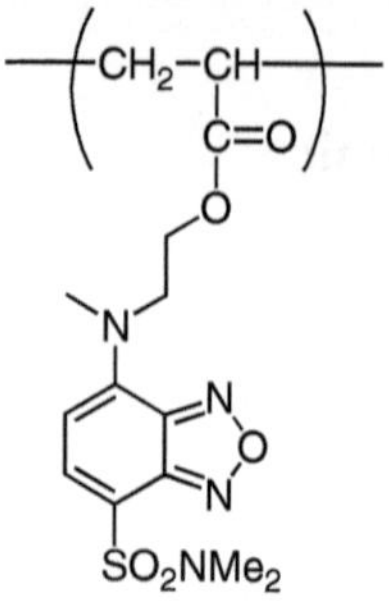
DBD-AE 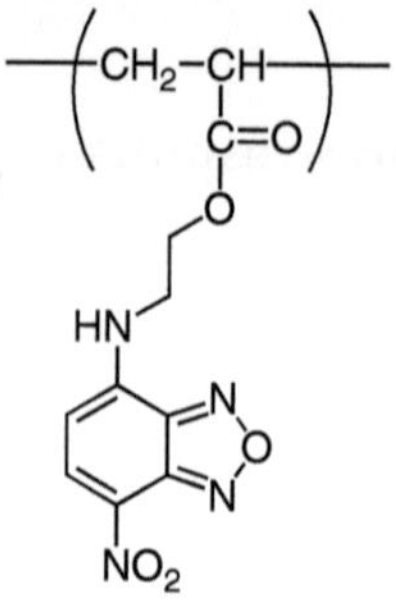
NBD-AE 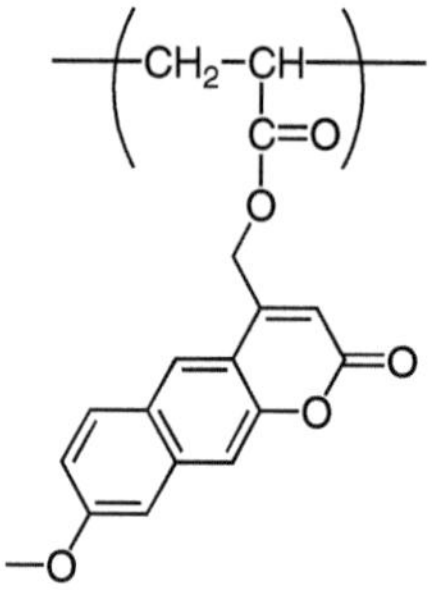
MBC-AE 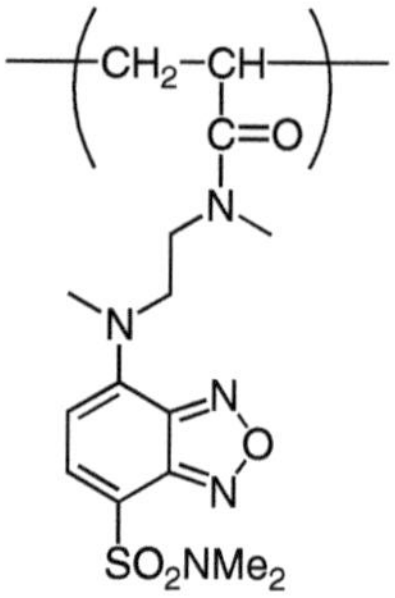
DBD-AA 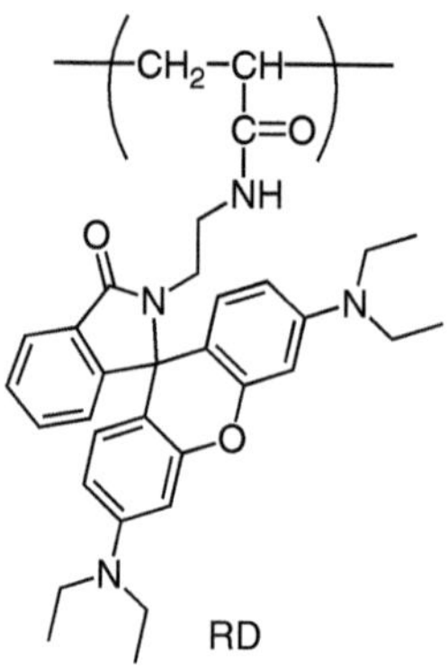
RD 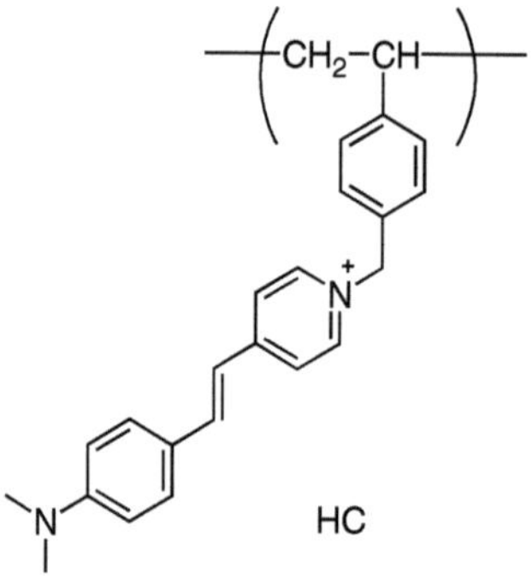
HC 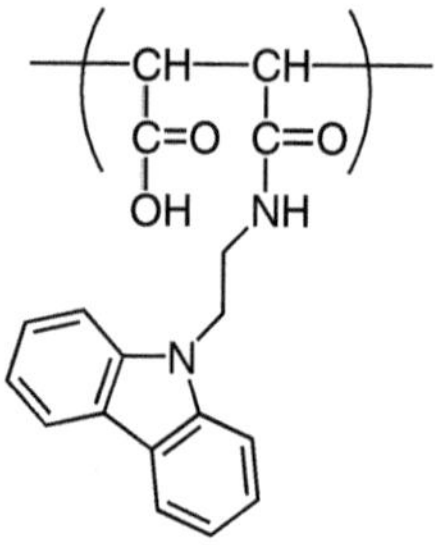
CEA 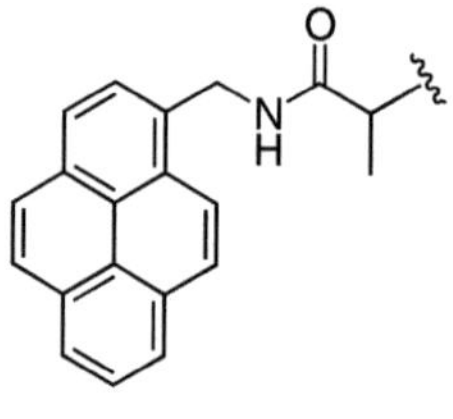
Pyrene (as end group) 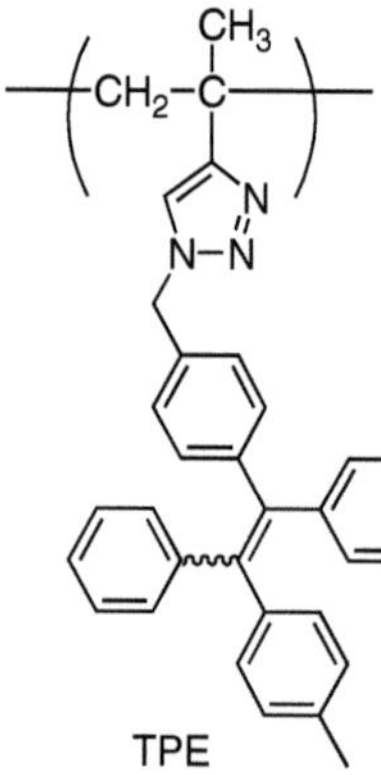
TPE 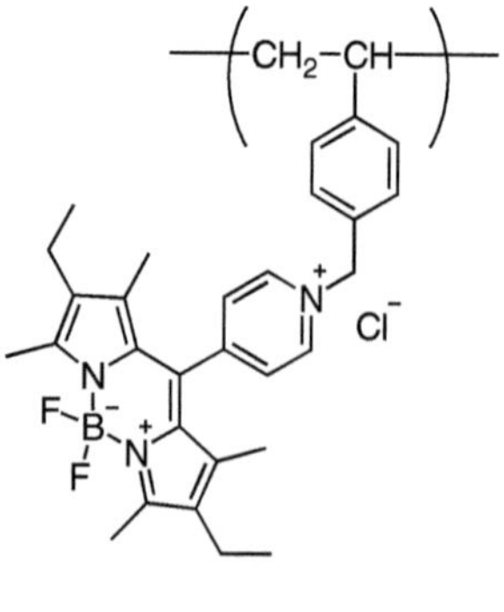
BODIPY 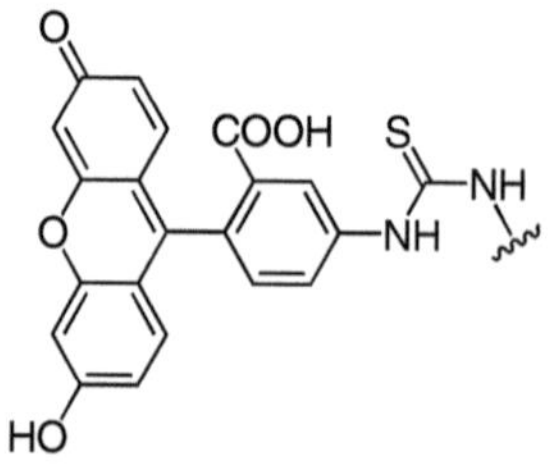
Fluorescein moiety (by FITC)

Fluorescent unit (*continued*)

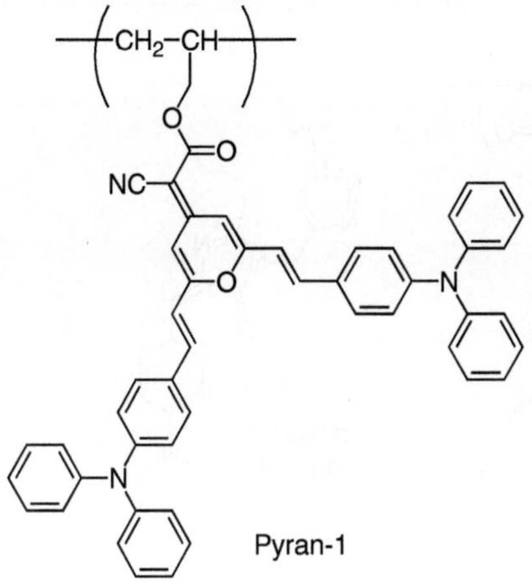

Pyran-1

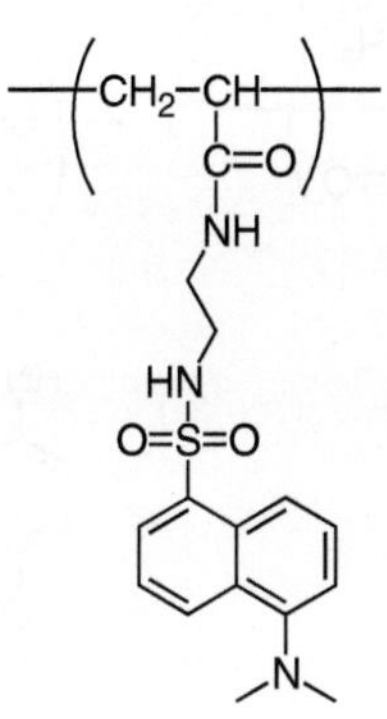

DAAm

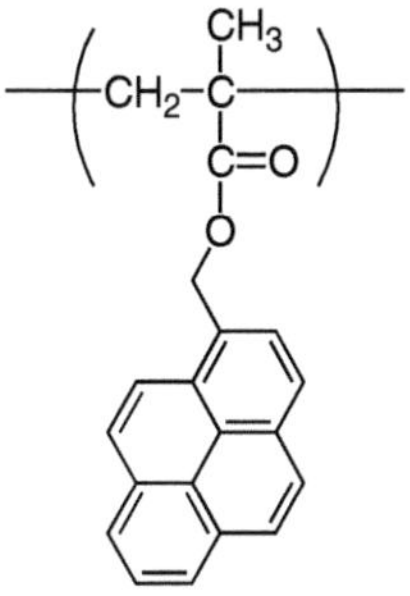

PyMMA

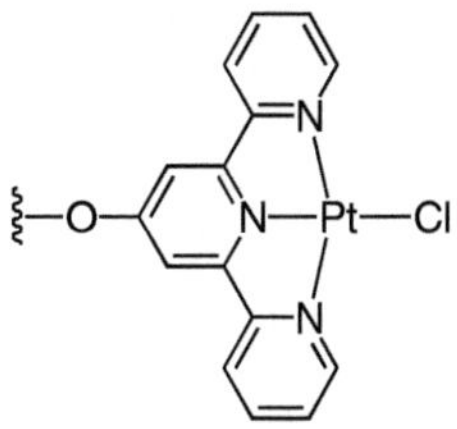

tpyPtCl (as end group)

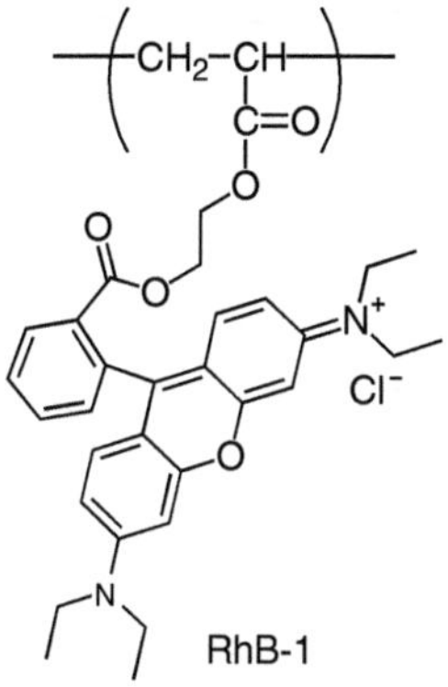

RhB-1

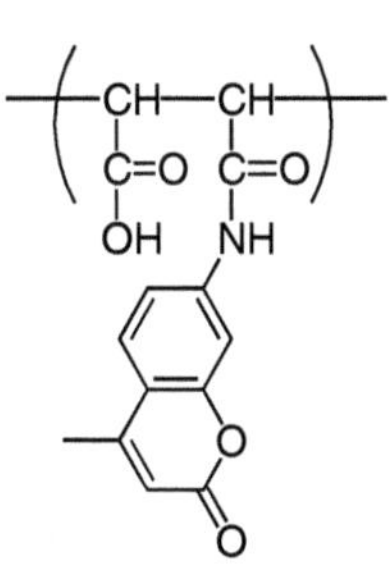

AMC

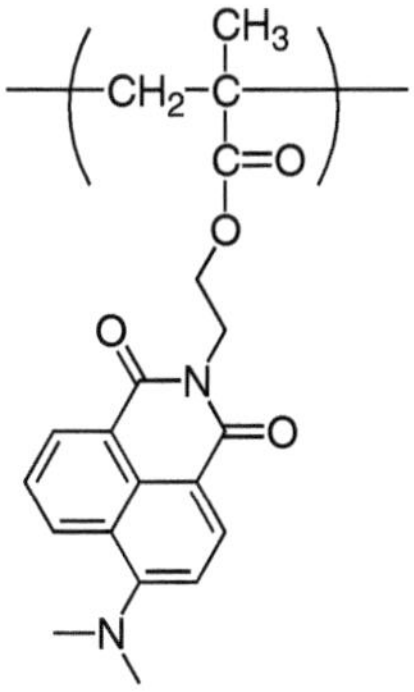

4-DMN-1

Fluorescent unit (*continued*)

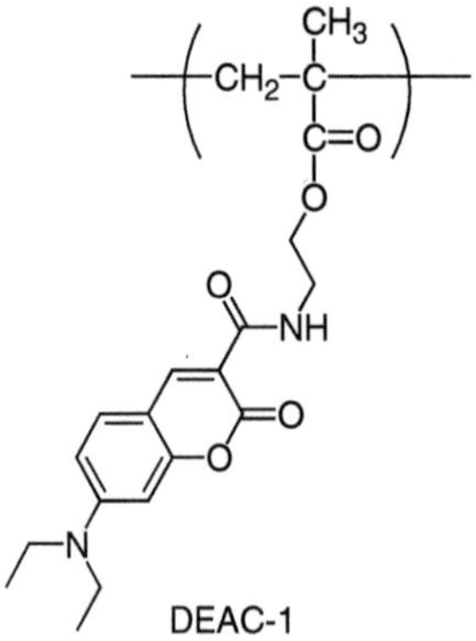

DEAC-1

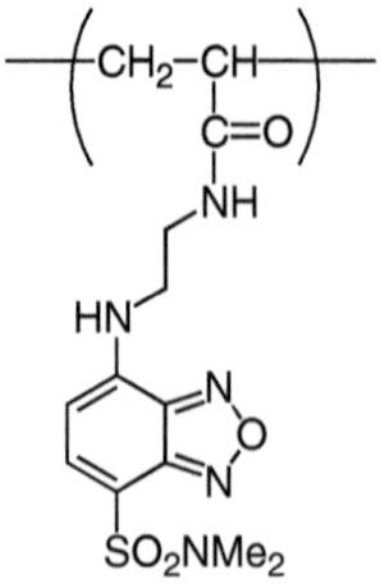

DBD-ED-1

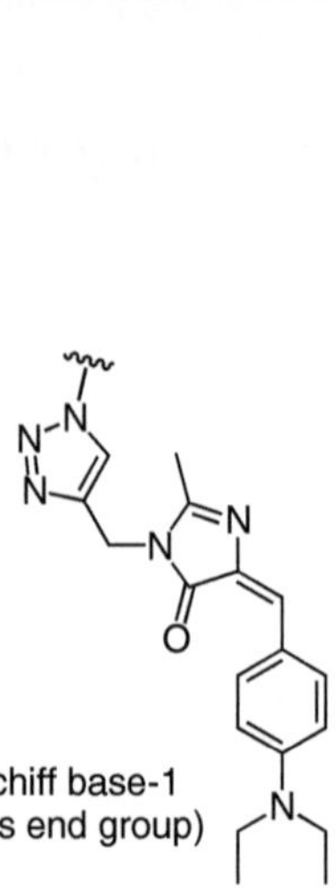

Schiff base-1
(as end group)

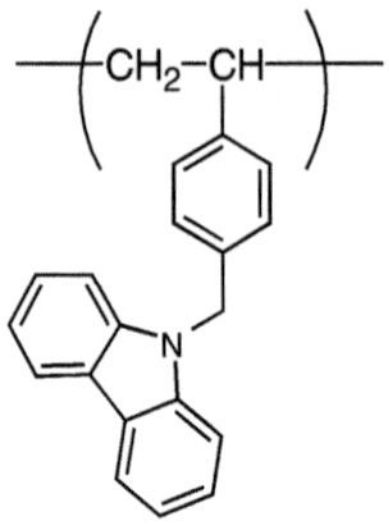

VBK

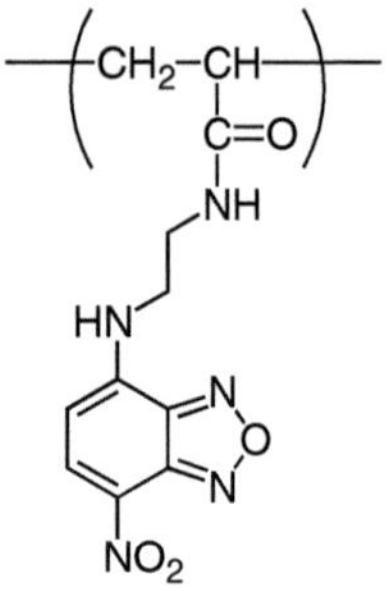

NBD-AA

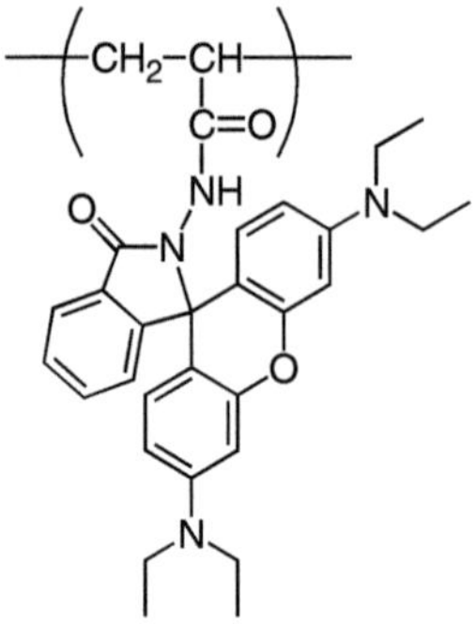

RhBAM

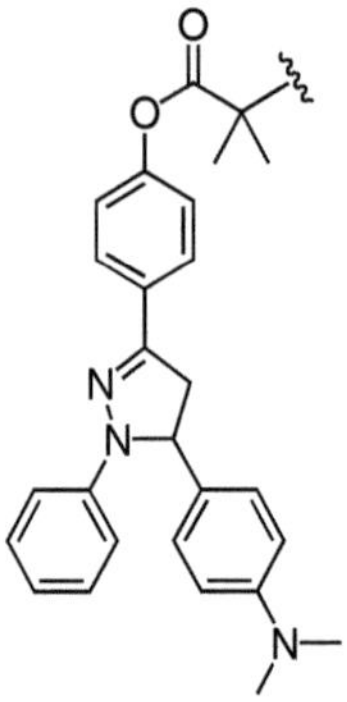

TPP-A

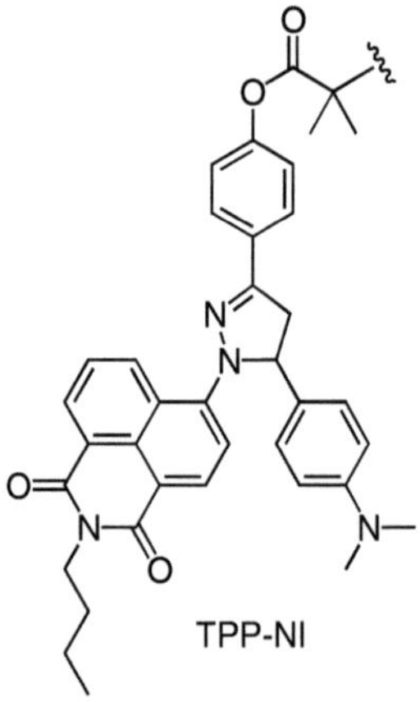

TPP-NI

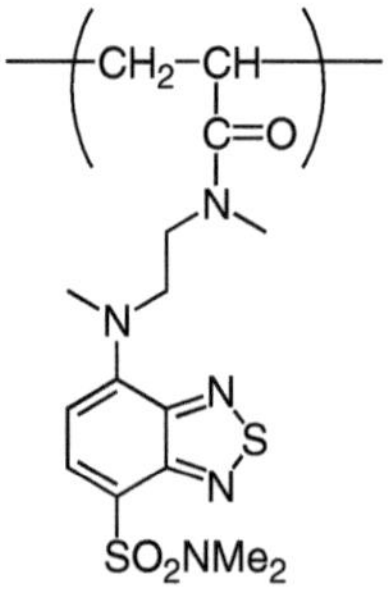

DBThD-AA

Fluorescent unit (*continued*)

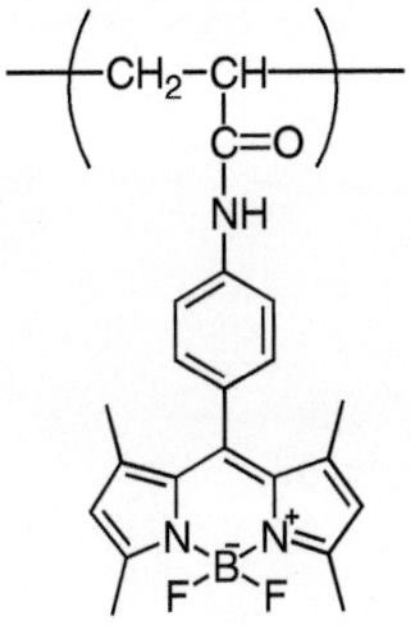

BODIPY-AA

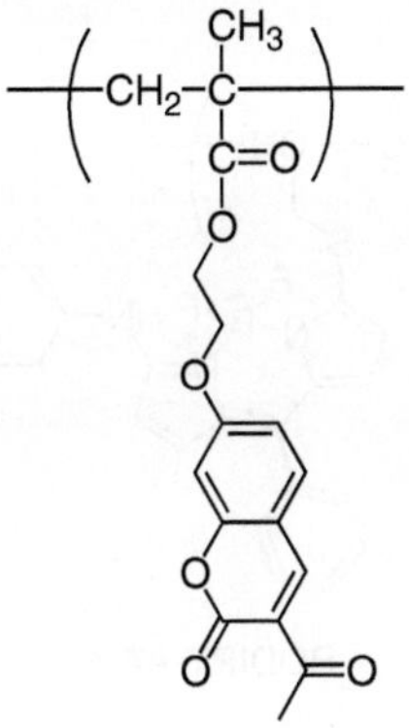

CMA

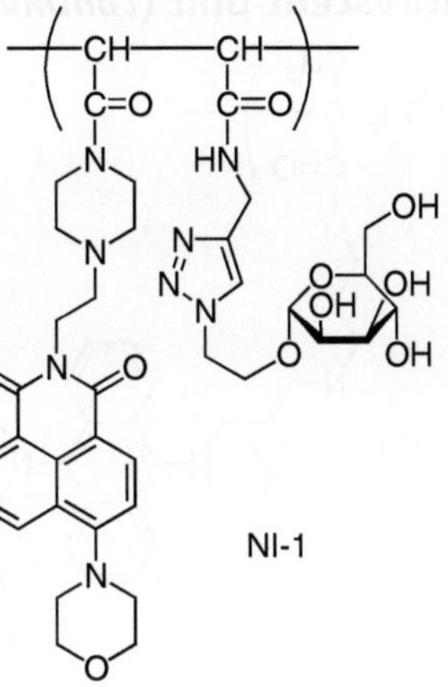

NI-1

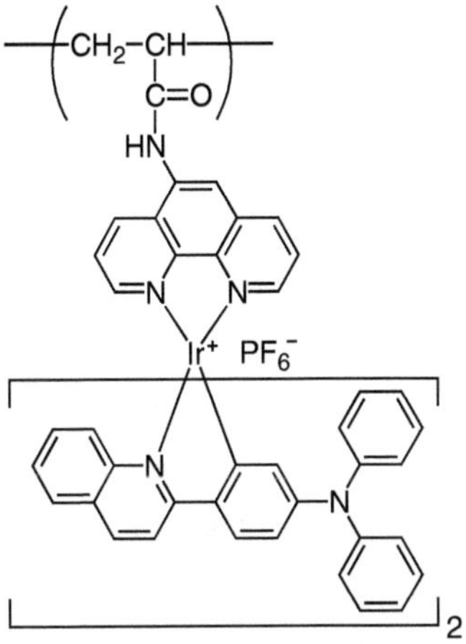

Ir-1

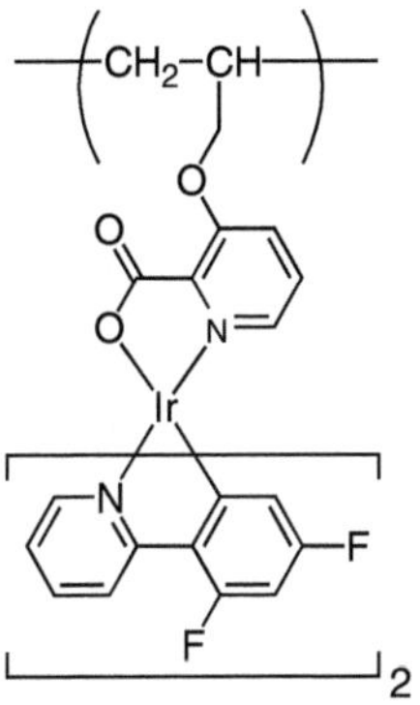

Ir-2

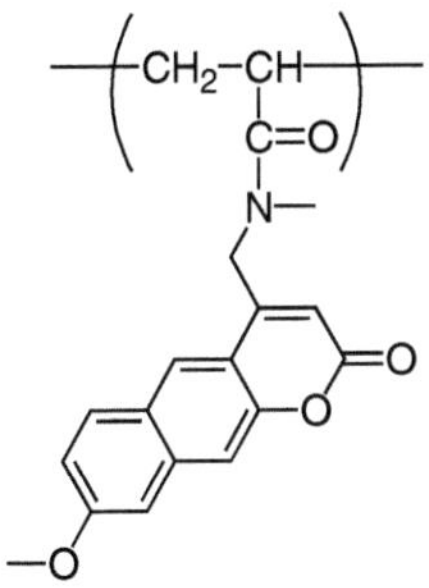

MBC-AA

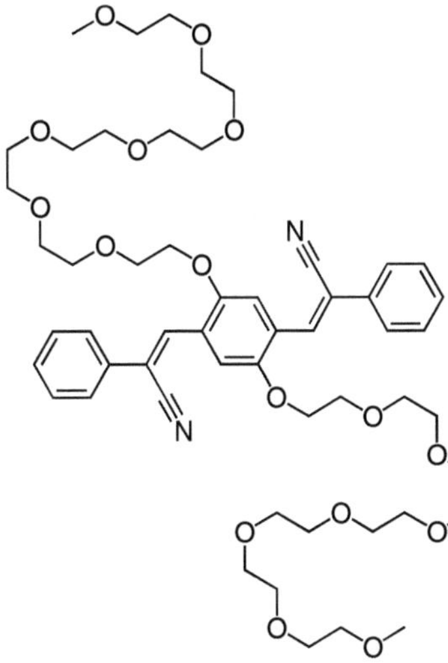

V-EO7DCS

Fluorescent unit (*continued*)

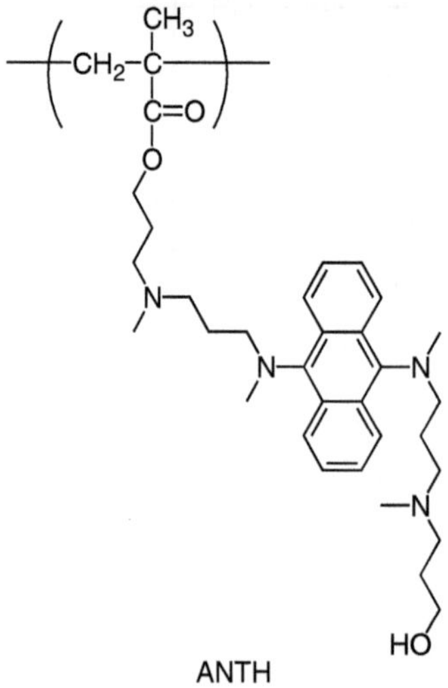

ANTH

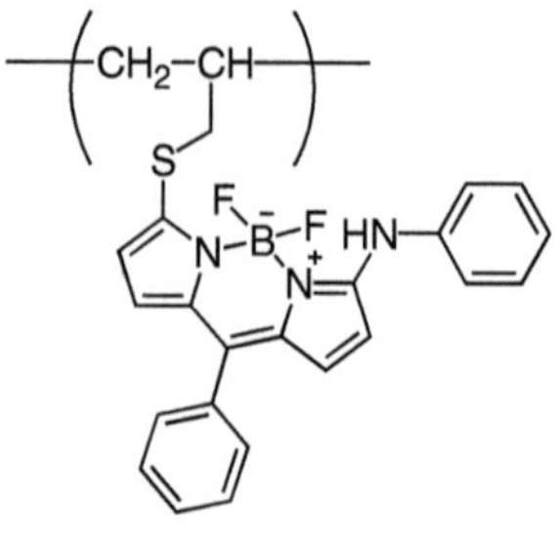

BODIPY-AA2

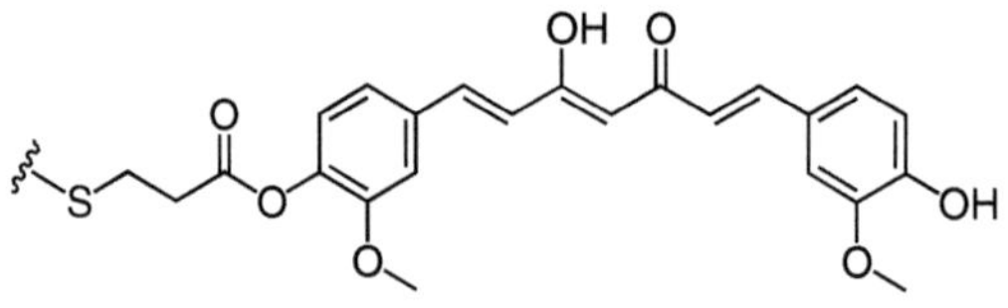

Curcumin-1
(as end group)

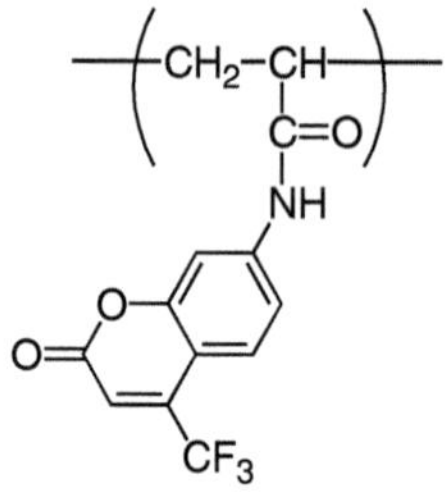

TFMCMA

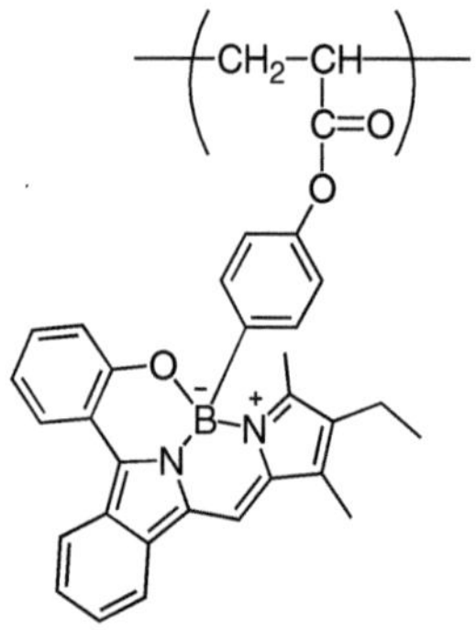

BOBPYBX

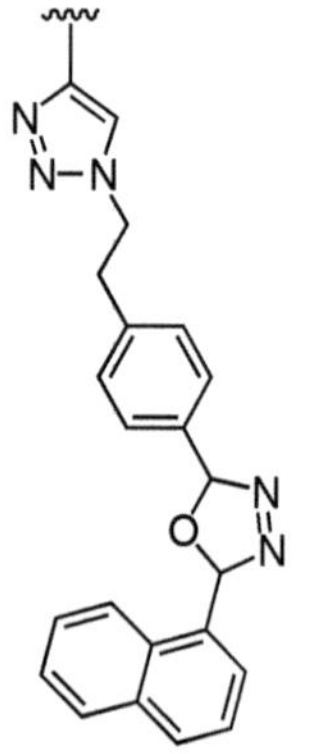

AzOx
(as end group)

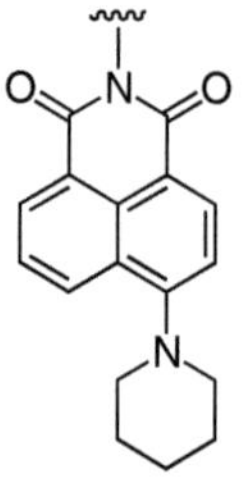

Naph
(as end group)

Fluorescent unit (*continued*)

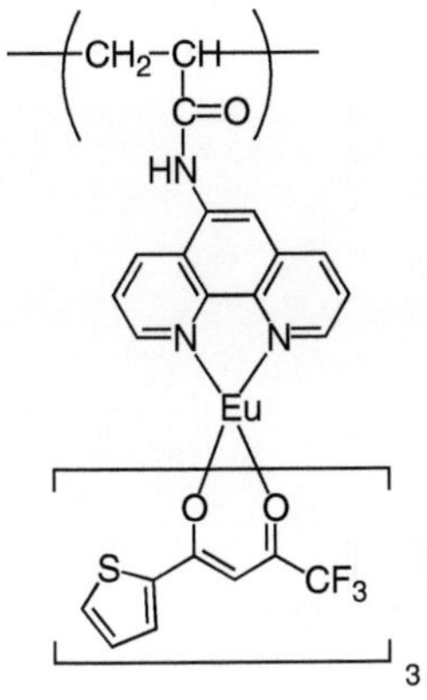

Eu-1

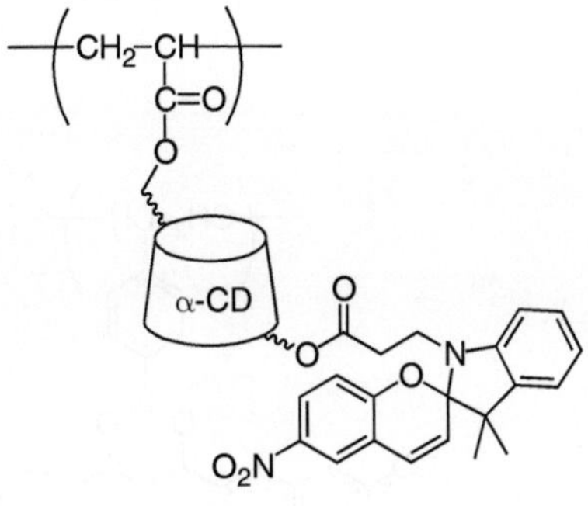

Acryloyl-α-CD-SP

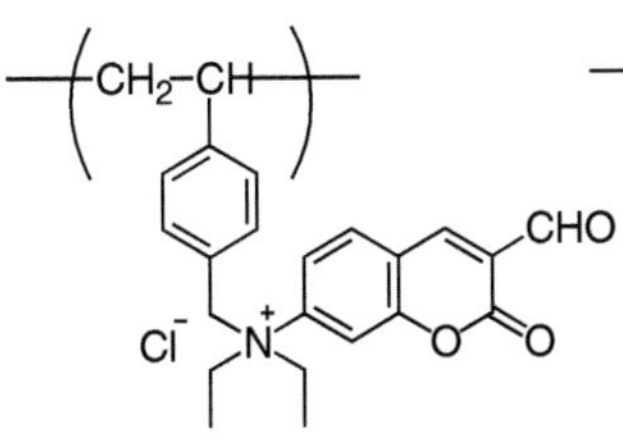

DACC

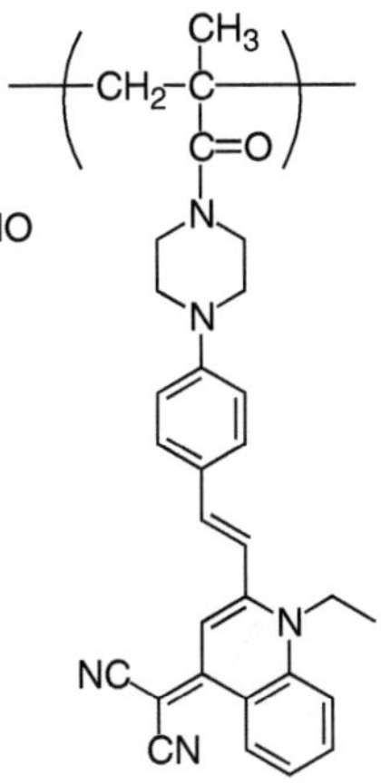

EM

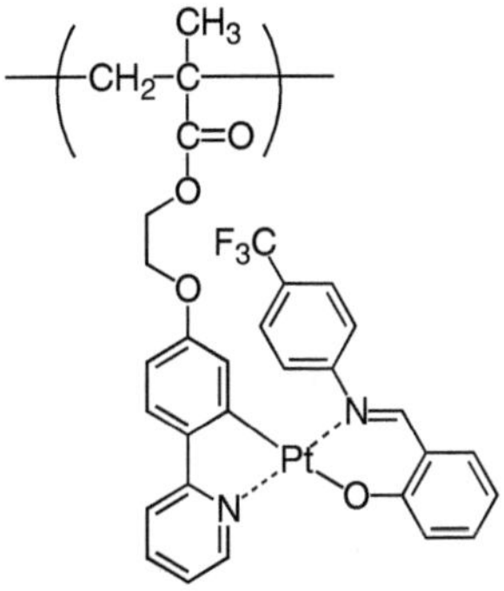

Pt-1

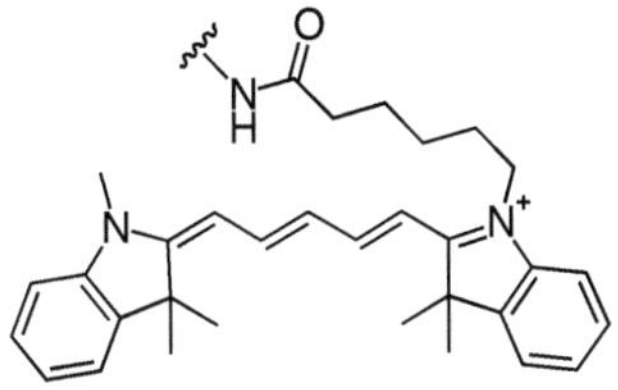

Cy5
(as end group)

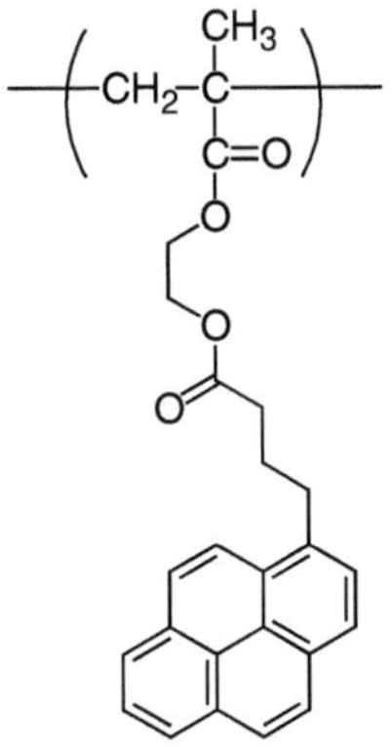

PyBEMA

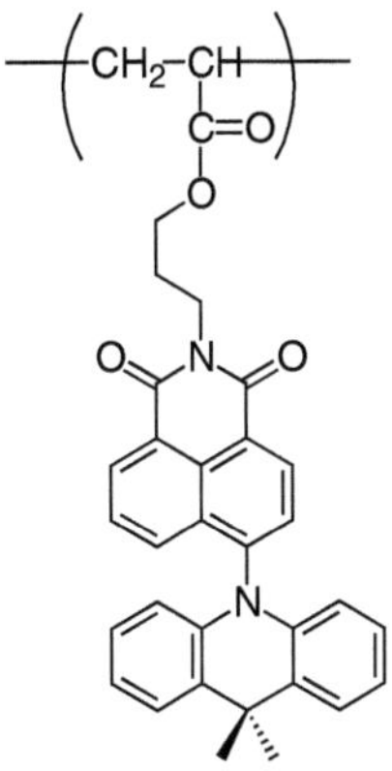

NAI-DMAC

Fluorescent unit (*continued*)

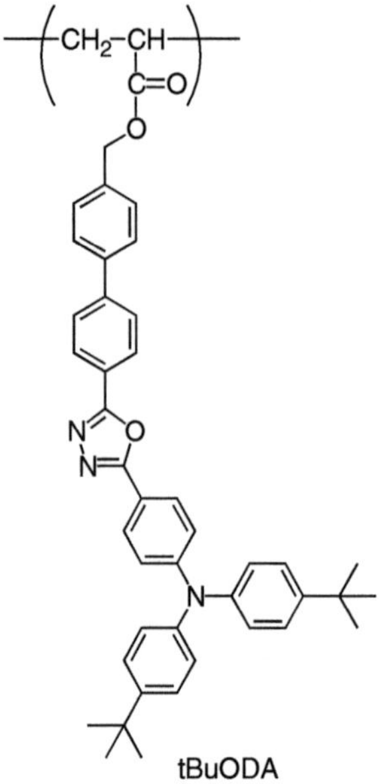

tBuODA

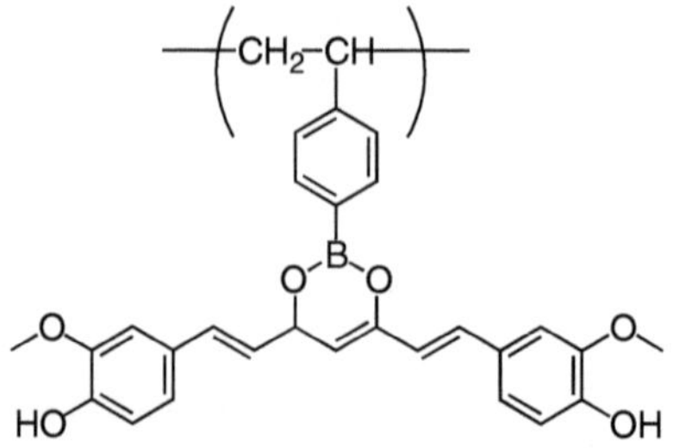

VPBA-C

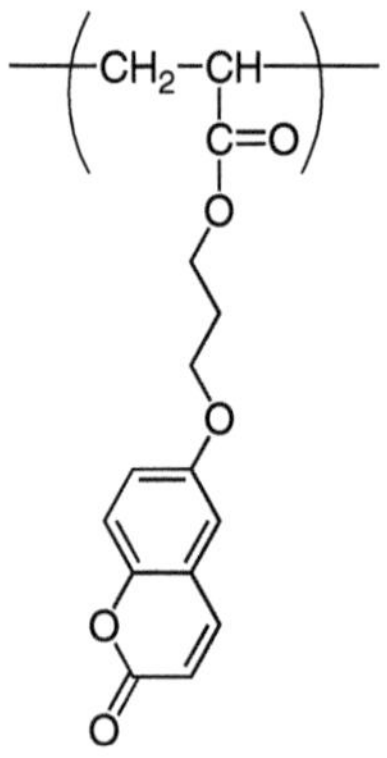

P33-B

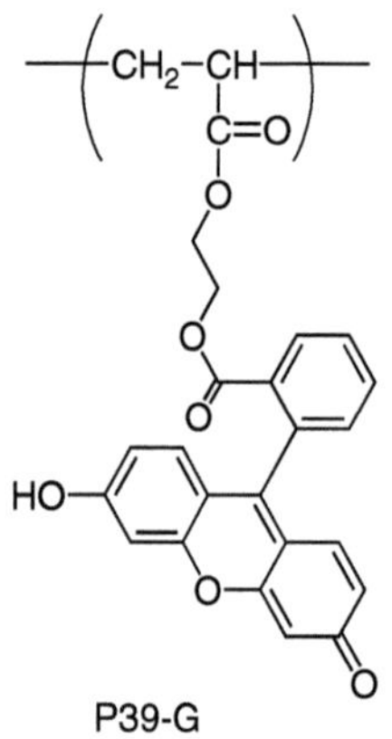

P39-G

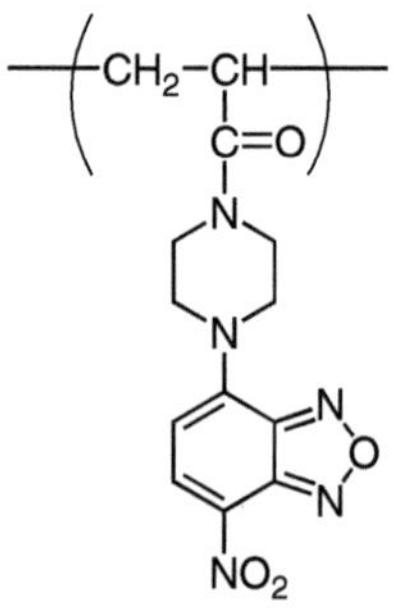

AANBD

Additional unit

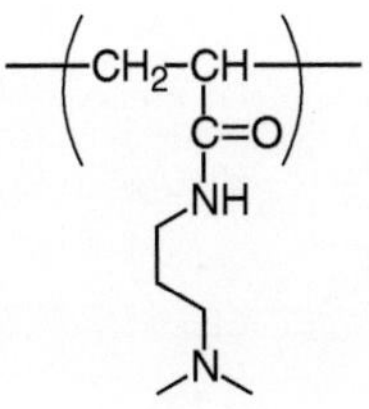

DMAPAM

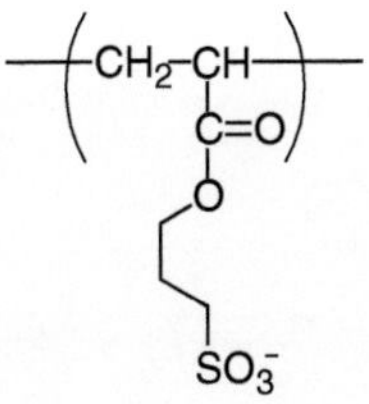

SPA

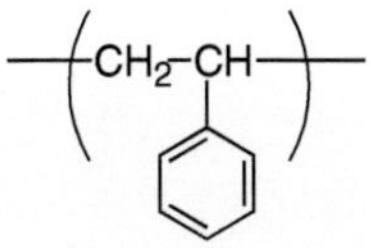

Styrene

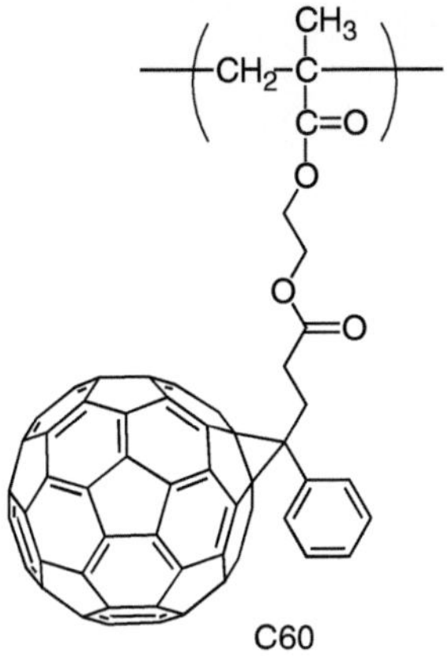

C60

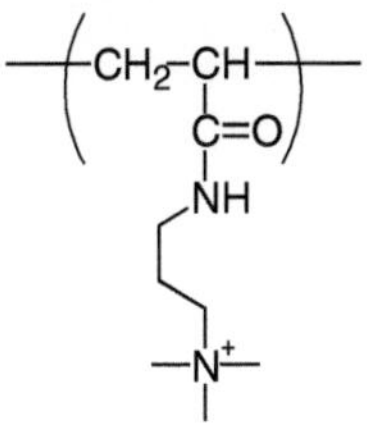

APTMA

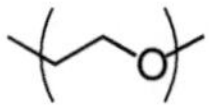

PEG

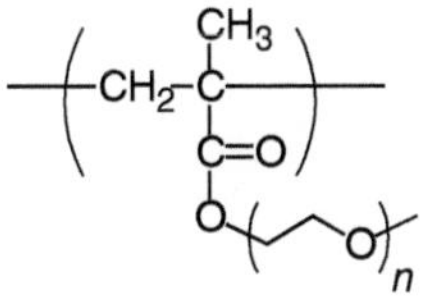

OEGMA

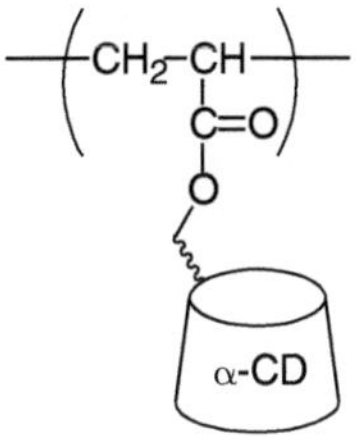

Acryloyl-α-CD

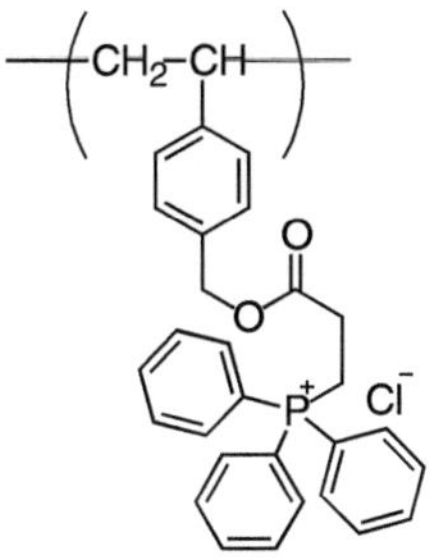

CTPP

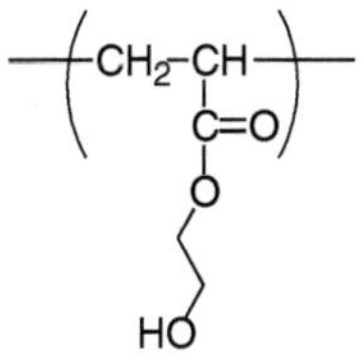

HEA

Appendix 3 Comprehensive Collection of Fluorescent Nanogel Thermometers Based on the Combination of a Thermo-responsive Polymer and an Environment-sensitive Fluorophore

As described in Section 2.11.3.5, gelation using crosslinking units is one of the modifications of fluorescent polymeric thermometers based on combining a thermo-responsive polymer and an environment-sensitive fluorophore. The resultant fluorescent nanogel thermometers with a diameter of submicrometer–micrometer acquire structural robustness. This characteristic is occasionally advantageous to escape from undesired chemical reactions (e.g., decomposition) when utilized for fluorescence thermometry. On the other hand, the spatial resolution of fluorescent nanogel thermometers is generally inferior to that of fluorescent polymeric thermometers because of the relatively larger size. All the fluorescent nanogel thermometers reported are summarized in Table A3.1.

Intracellular Thermometry with Fluorescent Molecular Thermometers, First Edition. Seiichi Uchiyama.
© 2024 Wiley-VCH GmbH. Published 2024 by Wiley-VCH GmbH.

Table A3.1 Fluorescent nanogel thermometers functioning in aqueous solution.

Entry	Thermo-responsive unit	Fluorescent unit/ mechanism[a]	Crosslinking unit	λ_{ex}(nm)	λ_{em}(nm)	Parameter to be measured[b]	Functional temperature range (°C)	Parameter variation	Remarks	Ref.
1	NIPAM	CdTe/AB	MBAM	not indicated	570–583	λ_{em}	25–41.1	570–583	Macrogel	[1]
2	NIPAM	DBD-AE/B	MBAM	444	533–557	FI	30–35	1–6.4		[2]
3	NIPMAM	DBD-AE/B	MBAM	444	533–557	FI	40–47	1–5.4		[2]
4	NNPAM	DBD-AE/B	MBAM	444	533–557	FI	18–22	1–10.7		[2]
5	NIPAM	VDP/B	MBAM	320	430–455	λ_{em}	25–34	455–430		[3]
6	NIPAM	DBD-AA/B	MBAM	456	560	FI	25–40	1–6.5	In 150 mM KCl solution, also tested in COS7 cells	[4]
7	NIPAM	3HF-AM/C	MBAM	355	436, 538	FI ratio	24–48	0.12–1	Two tautomers fluoresce at different wavelengths	[5]
8	DEGMA, OEGMA	ZnO quantum dots	PEGDMA	340	530	FI	15.5–50	1–4.5	Triple function (i.e., thermometer, heater, and DDS[c]), tested in B16F10 cells	[6]

9	NIPAM	DBThD-AA/B	MBAM	449	564	FI	23–35	1–5.9	In COS7 cell extract	[7]
10	NIPAM	DBD-ED, square 660[d]/AB	MBAM	470	718±7	FI	30–40	1–5.3	Involving FRET[e] from DBD-ED to square 660	[8]
11	NIPAM	RhB (labeled by RhB ITC)/D	Silica	543	not indicated	FI	25–41	1–0.27	Multicomponent nanoparticle consisting of Fe_3O_4, SiO_2, NIPAM, and Au in HeLa cells	[9]
12	NIPAM	DPTB/B Nile red (non-covalently)	MBAM	405	470, 625	FI ratio	32–56	1–13.1	Tested in NIH/3T3 cells	[10]
13	NIPAM	Fluorescein (labeled using FITC)/DE		490	518	FI	20–40	1–0.16	Ag nanoprism coated by NIPAM shell and fluorescein	[11]
14	NIPAM	Ru(bpy) monomer, Cy5 monomer/A	MBAM	454	610, 670	FI ratio	28–40	2.4–4.0	In phosphate buffer (pH 7.4)	[12]

(Continued)

Table A3.1 (*Continued*)

Entry	Thermo-responsive unit	Fluorescent unit/ mechanism[a]	Crosslinking unit	λ_{ex}(nm)	λ_{em}(nm)	Parameter to be measured[b]	Functional temperature range (°C)	Parameter variation	Remarks	Ref.
15	NIPAM	DBThD-AA/B	MBAM	440	570	FI	20–40	1–2.1	In MOLT-4 cells, cationic gel, also tested in HeLa cells	[13]
16	NIPAM	DBD-AA/B	MBAM	456	570	FI	25–45	1–2.3	In 150 mM KCl solution, also tested in HeLa cells	[14]

a) A: energy transfer; B: sensitivity to polarity; C: water-sensitive tautomerization; D: variation in distance between a fluorophore and a quencher; E: aggregation-induced quenching.
b) λ_{em}: maximum emission wavelength; FI: fluorescence intensity; FI ratio: fluorescence intensity ratio at two different wavelengths.
c) Drug delivery system.
d) Commercially available, but chemical structure is not specified.
e) Förster resonance energy transfer.

1 Li, J., Hong, X., Liu, Y., Li, D., Wang, Y., Li, J., Bai, Y., and Li, T. (2005) *Adv. Mater.*, **17**, 163–166.
2 Iwai, K., Matsumura, Y., Uchiyama, S., and de Silva, A. P. (2005) *J. Mater. Chem.*, **15**, 2796–2800.
3 Matsumura, Y. and Iwai, K. (2005) *Polymer*, **46**, 10027–10034.
4 Gota, C., Okabe, K., Funatsu, T., Harada, Y., and Uchiyama, S. (2009) *J. Am. Chem. Soc.*, **131**, 2766–2767.
5 Chen, C.-Y. and Chen, C.-T. (2011) *Chem. Commun.*, **47**, 994–996.
6 Wu, W., Shen, J., Banerjee, P., and Zhou, S. (2011) *Adv. Funct. Mater.*, **21**, 2830–2839.
7 Uchiyama, S., Kimura, K., Gota, C., Okabe, K., Kawamoto, K., Inada, N., Yoshihara, T., and Tobita, S. (2012) *Chem. Eur. J.*, **18**, 9552–9563.
8 Cheng, B., Wei, M.-Y., Liu, Y., Pitta, H., Xie, Z., Hong, Y., Nguyen, K. T., and Yuan B. (2014) *IEEE J. Sel. Top. Quantum Electron.*, **20**, 6801214.
9 Wang, Z., Ma, X., Zong, S., Wang, Y., Chen, H., and Cui, Y. (2015) *Talanta*, **131**, 259–265.
10 Liu, J., Guo, X., Hu, R., Xu, J., Wang, S., Li, S., Li, Y., and Yang, G. (2015) *Anal. Chem.*, **87**, 3694–3698.
11 Sugawa, K., Ichikawa, R., Takeshima, N., Tanoue, Y., and Otsuki, J. (2015) *Photochem. Photobiol. Sci.*, **14**, 870–874.
12 Pinaud, F., Millereux, R., Vialar-Trarieux, P., Catargi, B., Pinet, S., Gosse, I., Sojic, N., and Ravaine, V. (2015) *J. Phys. Chem. B*, **119**, 12954–12961.
13 Uchiyama, S., Tsuji, T., Kawamoto, K., Okano, K., Fukatsu, E., Noro, T., Ikado, K., Yamada, S., Shibata, Y., Hayashi, T., Inada, N., Kato, M., Koizumi, H., and Tokuyama, H. (2018) *Angew. Chem. Int. Ed.*, **57**, 5413–5417.
14 Hayashi, T., Kawamoto, K., Inada, N., and Uchiyama, S. (2019) *Polymers*, **11**, 1305.

Thermo-responsive unit 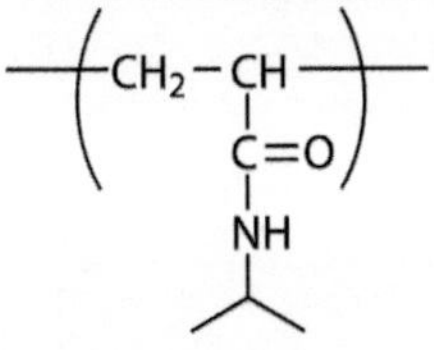
NIPAM 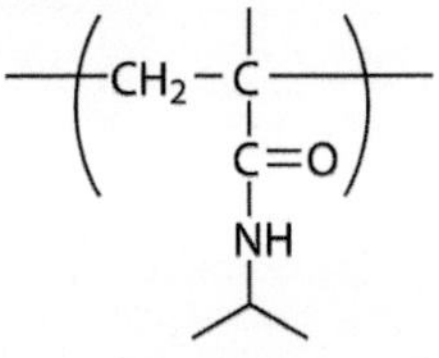
NIPMAM 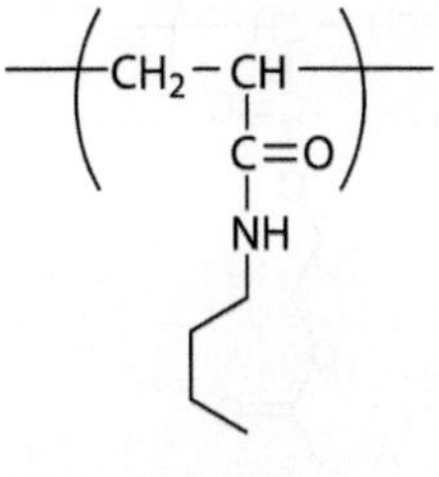
NNPAM 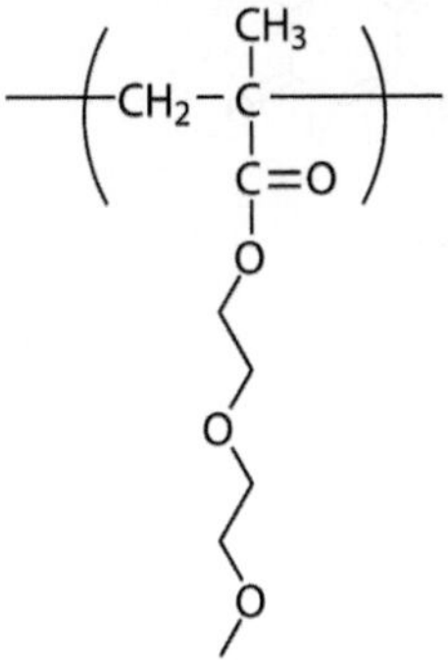
DEGMA 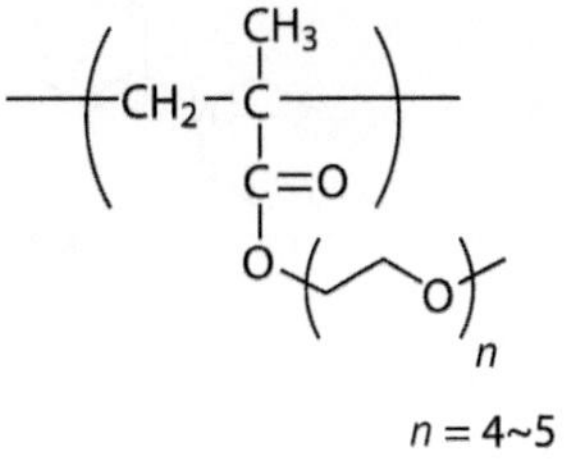
OEGMA
n = 4~5
Fluorescent unit 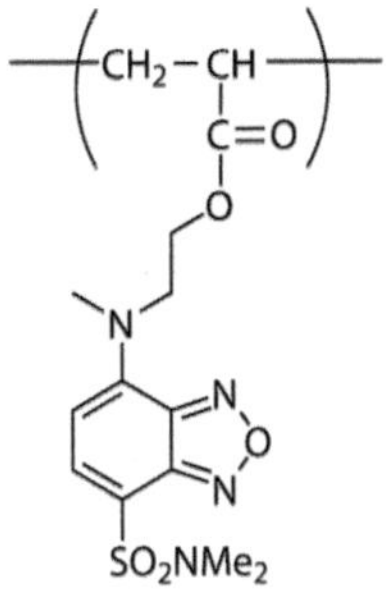
DBD-AE 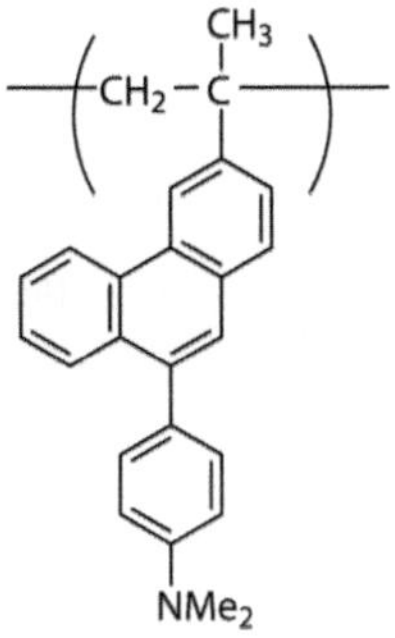
VDP 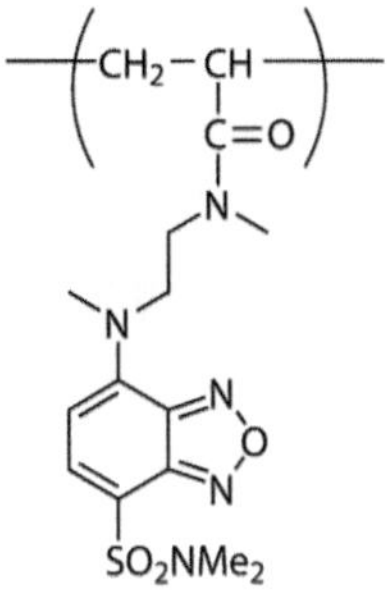
DBD-AA 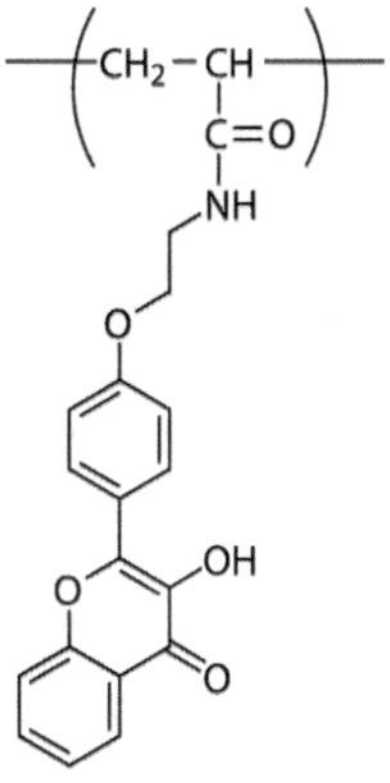
3HF-AM 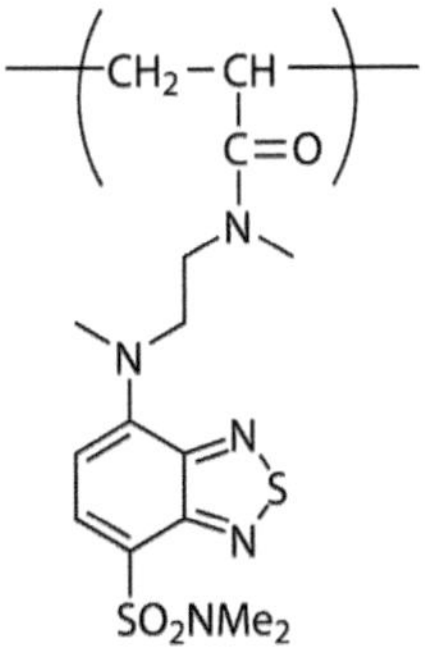
DBThD-AA 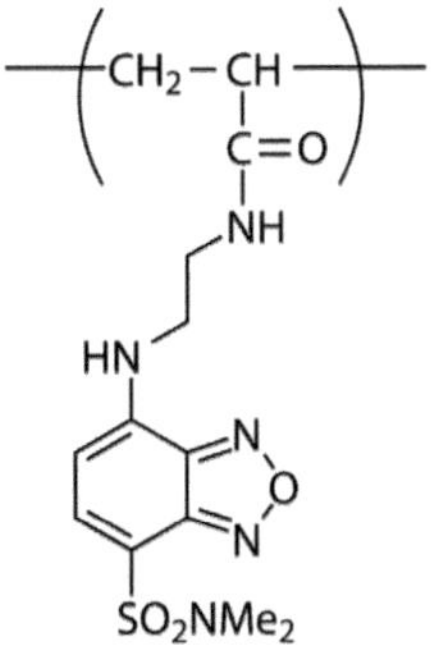
DBD-ED 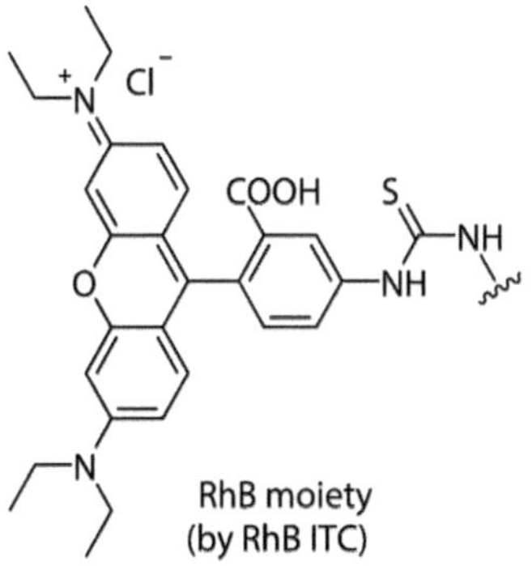
RhB moiety (by RhB ITC)

Fluorescent unit (*continued*)

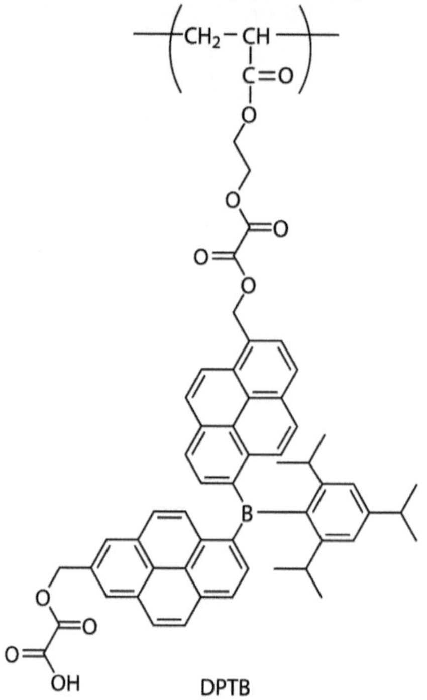

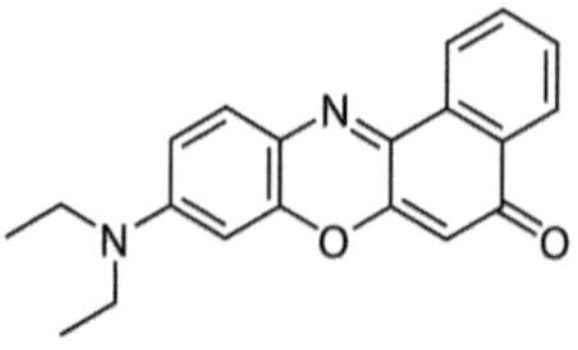

Nile red

DPTB

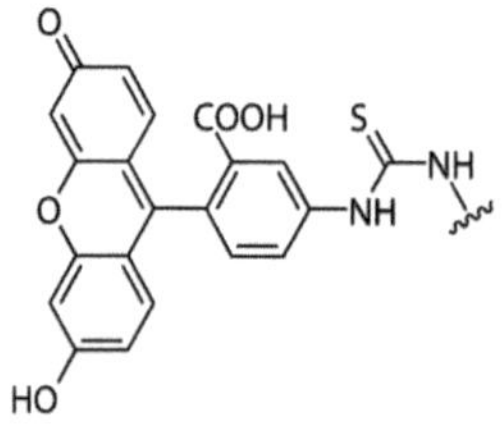

Fluorescein moiety
(by FITC)

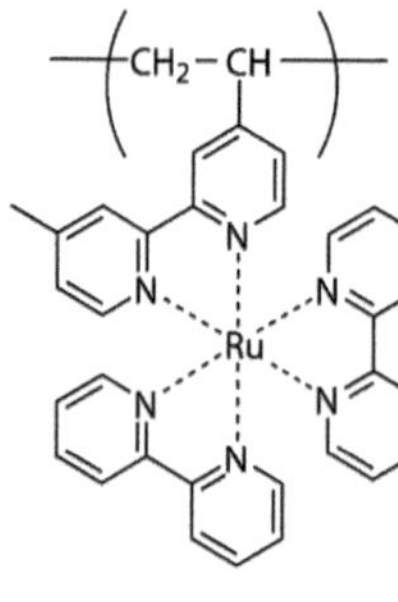

Ru(bpy) monomer

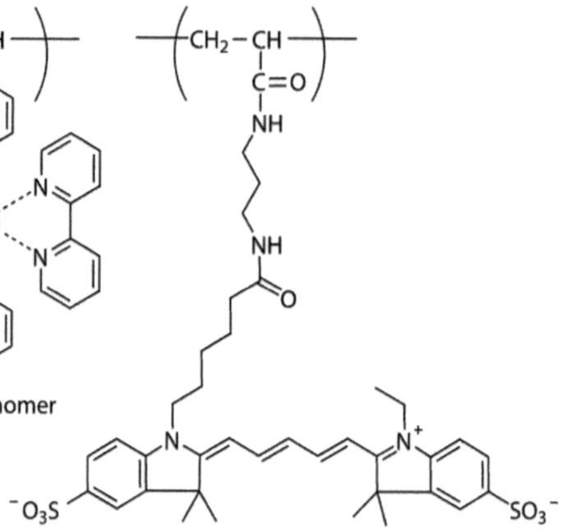

Cy5 monomer

Crosslinking unit

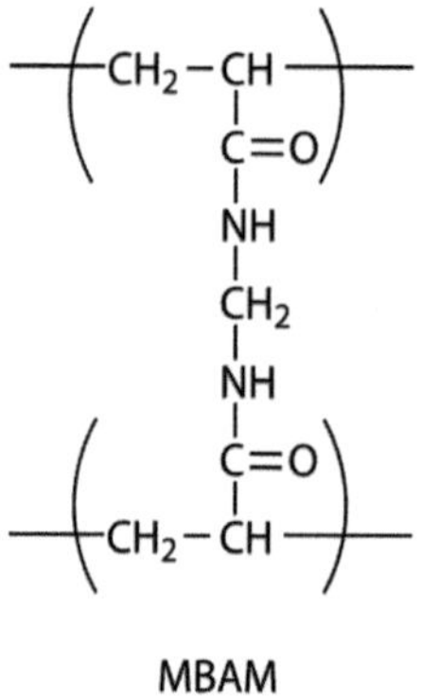

MBAM

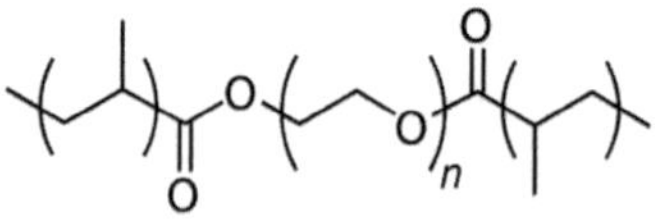

PEGDMA

$n = $ ca 4

Appendix 4 Comprehensive Collection of Non-covalent Fluorescent Temperature Sensing Systems Based on the Combination of a Thermo-responsive Polymer and an Environment-sensitive Fluorescent Compound

Non-covalent systems, i.e., the mixtures of a thermo-responsive polymer and an environment-sensitive fluorescent compound, are one options for fluorometric temperature sensing. Non-covalent systems can avoid the complicated synthesis of a fluorescent monomer. However, the reversibility and independence of fluorescence response would be sacrificed by the separation of the components and undesired interactions of a polymer or a fluorescent compound with the surrounding molecules. Reports on non-covalent systems composed of a thermo-responsive polymer and an environment-sensitive fluorescent compound are summarized in Table A4.1.

Table A4.1 Fluorescent temperature sensing systems based on the combination of a thermo-responsive polymer and an environment-sensitive fluorescent compound.

Entry	Thermo-responsive polymer	Fluorescent compound/mechanism[a]	λ_{ex}(nm)	λ_{em}(nm)	Parameter to be measured[b]	Functional temperature range (°C)	Parameter variation	Remarks	Ref.
1	PNIPAM	S0522/A	543	565–572	λ_{em}	30–32.5	565–572		[1]
2	PNIPAM	Perylene nanocrystals/BC	Not indicated	440	FI	30–65	1–18		[2]
3	PNIPAM	Quinacridone nanocrystals/BC	Not indicated	550	FI	30–65	1–14		[2]
4	PNIPAM	Zinc phthalocyanine nanocrystals/BC	Not indicated	680	FI	30–65	0–1		[2]
5	POSS-*b*-PNIPAM block copolymer[c]	1,1,2,2-Tetraphenylethylene/D	330	448	FI	25–50	Not monotonous change		[3]
6	PNIPAM	Nile red/B	532	632	FI	31–41	1–21	Sodium alginate is also mixed and the whole mixture is coated by chitosan	[4]
7	PNIPAM gel	Pyrene/E	350	373, 458	FI ratio[d]	11–40	0.1–0.9	Macrogel	[5]
8	PEO-*b*-poly(AAm-*co*-AN-*co*-DMA)	EuW$_{10}$/B	254	580–660	Total fluorescence area	5–55	1–0.27	UCST[e]-type polymer	[6]
9	PNIPAM gel	Folic acid-modified silicon nanocrystals/B	404	476	FI	17–88	1–0.23	In water, also tested in HeLa and MCF-7 cells	[7]

10	Cysteine-modified, chitosan-grafted PNIPAM gel	Palmatine/D	Not indicated	530	FI	24–40	1–29	In a pH 7.4 solution, also tested in HeLa cells	[8]
11	Dextran-*g*-poly(NIPAM)	PDI[f]/F	520	563	FI	24–47	1–1.46	Au nanoparticles are also mixed	[9]
12	Poly(NIPAM-*co*-TPSS)	TPEBT[g]/D	420	640	FI	32–40	1–0.55	Mainly localized in lysosomes when introduced in HeLa cells	[10]

a) A: aggregation in a hydrophobic environment; B: sensitivity to polarity; C: prevention of self-quenching; D: aggregation-induced emission; E: excimer formation; F: variation in FRET (Förster resonance energy transfer) efficiency.

b) λ_{em}: maximum emission wavelength; FI: fluorescence intensity; FI ratio: fluorescence intensity ratio at two different wavelengths.

c) POSS: polyhedral oligomeric silsesquioxane.

d) FI at 373 nm divided by (FI at 373 nm + FI at 458 nm). e) Upper critical solution temperature.

f) N,N-Carboxymethyl-1,6,7,12-tetrachloroperylenediimide.

g) 3-Ethyl-2-[4-(1,2,2-triphenylvinyl)styryl]-benzothiazol-3-ium iodide.

1 Kolaric, B., Sliwa, M., Vallée, R. A. L., and Van der Auweraer, M. (2009) *Colloids Surf. A*, **338**, 61–67.

2 Baba, K., Kasai, H., Nishida, K., and Nakanishi, H. (2011) *Jpn. J. Appl. Phys.*, **50**, 010202.

3 Li, M., Song, X., Zhang, T., Zeng, L., and Xing, J. (2016) *RSC Adv.*, **6**, 86012–86018.

4 Barbieri, M., Cellini, F., Cacciotti, I., Peterson, S. D., and Porfiri, M. (2017) *J. Mater. Sci.*, **52**, 12506–12512.

5 Hamasaki, A., Sato, N., Kubo, K., Katsuki, A., and Ozeki, S. (2019) *Chem. Lett.*, **48**, 902–905.

6 Zhang, J.-L., Tan, J.-Y., Wan, X.-H., and Zhang, J. (2019) *Chinese J. Polym. Sci.*, **37**, 1113–1118.

7 Li, Y., Zhang, L., Shi, Y., Huang, J., Yang, Y., and Ming, D. (2020) *Polymers*, **12**, 2565.

8 Xu, L., Liang, X., You, L., Yang, Y., Fen, G., Gao, Y., and Cui, X. (2021) *Int. J. Biol. Macromol.*, **189**, 316–323.

9 Yeshchenko, O. A., Kutsevol, N. V., Tomchuk, A. V., Khort, P. S., Kuziv, Y. I., Hudhomme, P., and Krupka, O. M. (2022) *Opt. Mater.*, **131**, 112753.

10 Yin, N., Lin, B., Huo, F., Shu, Y., and Wang, J. (2022) *Anal. Chem.*, **94**, 12111–12119.

Thermo-responsive polymer

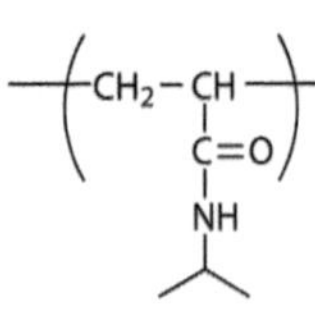

PNIPAM

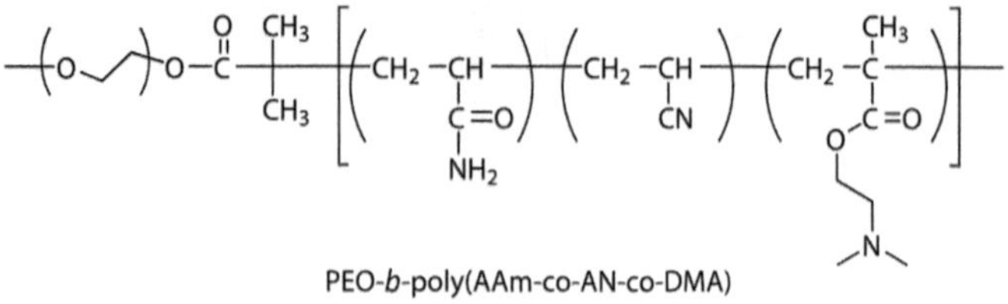

PEO-*b*-poly(AAm-co-AN-co-DMA)

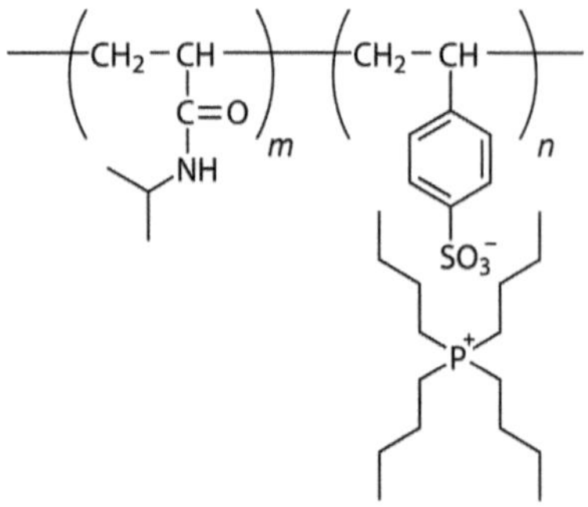

Poly(NIPAM-*co*-TPSS)

Fluorescent compound

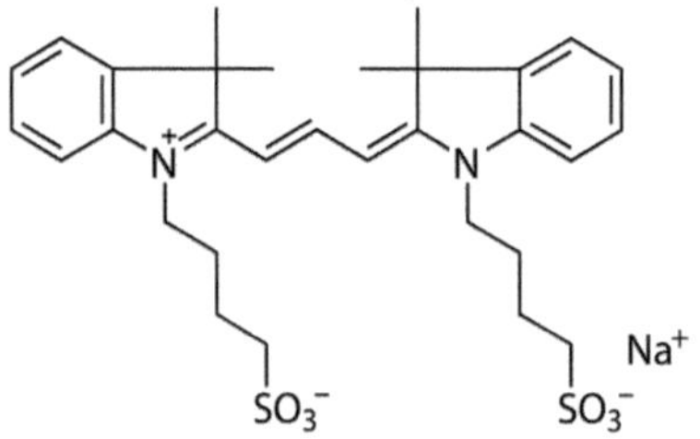

S0522

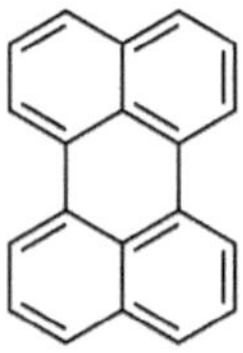

Perylene

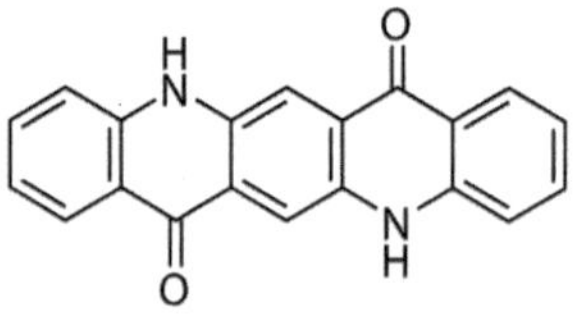

Quinacridone

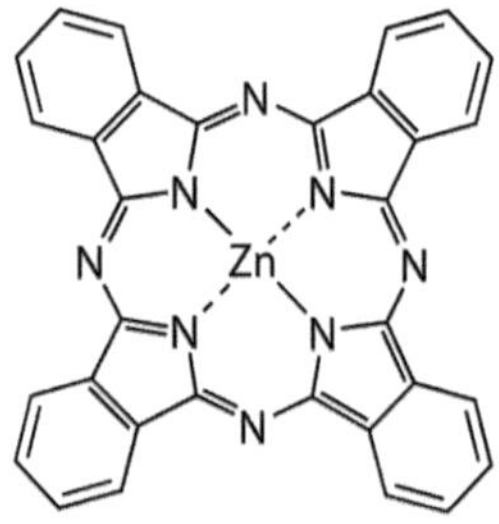

Zinc phthalocyanine

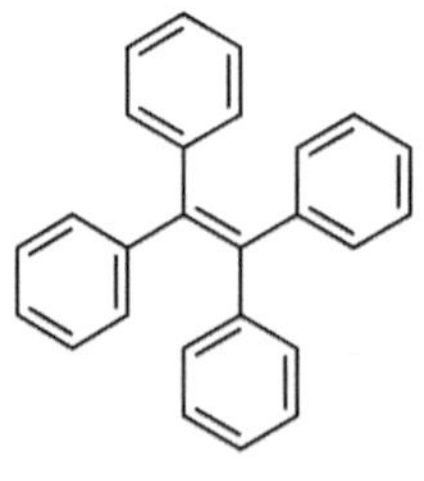

1,1,2,2-Tetrapheylethylene

Fluorescent compound (*continued*)

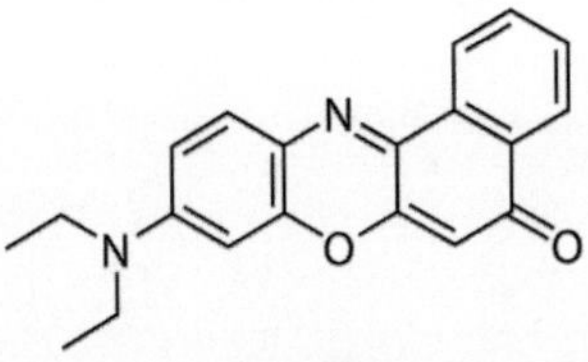

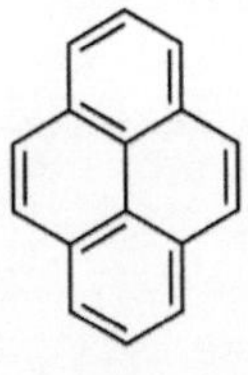

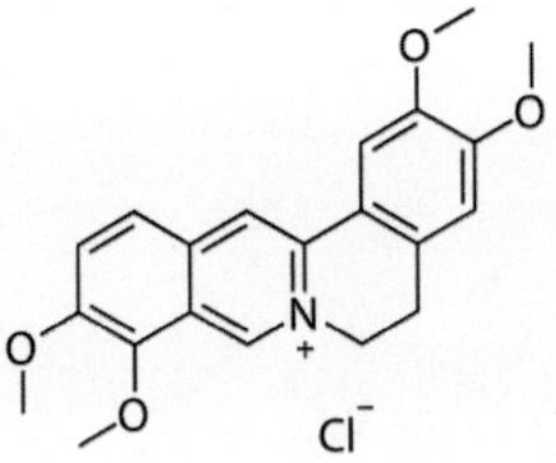

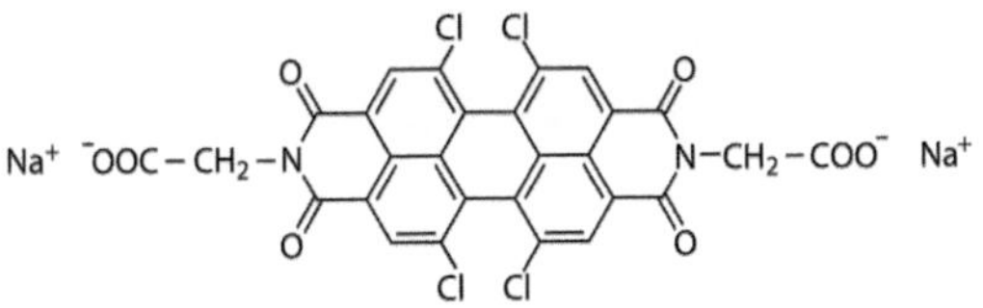

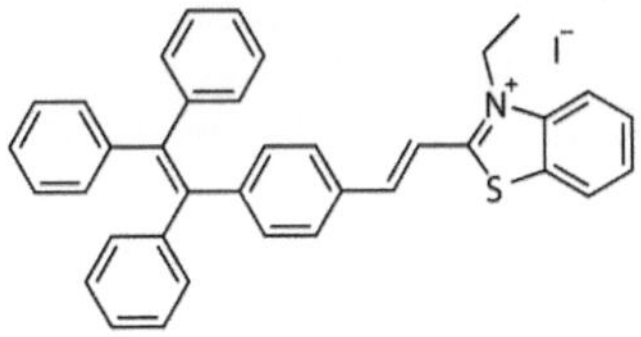

Appendix 5 Comprehensive Collection of Fluorescent Polymeric Logic Gates Based on the Combination of a Thermo-responsive Polymer and an Environment-sensitive Fluorophore

A straightforward addition of extra functions is one of the advantages of fluorescent polymeric thermometers based on combining a thermoresponsive polymer and an environment-sensitive fluorophore. For instance, the introduction of crosslinking units creates fluorescent nanogel thermometers that are more chemically stable. In addition, cationic units supply fluorescent polymeric thermometers that can spontaneously enter live cells. Similarly, introducing responsive units to a chemical or physical factor other than temperature into fluorescent polymeric thermometers constructs fluorescent polymeric logic gates which respond to multiple parameters, i.e., temperature and the selected element. An excellent book [1] and review items [2,3] are available for details on molecular logic gates. An additional chemical factor (e.g., pH or metal ion) or a physical one (e.g., light irradiation) works as one of the inputs for fluorescent polymeric logic gates. A three-input logic operation is also possible. In some cases, a fluorescent unit also serves as a responsive unit towards the selected second factor by bearing a selective responsive moiety in the structure. The functions of fluorescent polymeric logic gates were occasionally assessed in living cells. Table A5.1 summarizes the structures and functional types (e.g., inputs and output) of fluorescent polymeric logic gates based on a thermo-responsive polymer. The classification (called a truth table) of two-input logic gates is shown in Table A5.2 for reference.

1 de Silva, A. P. (2013) *Molecular Logic-based Computation*, Royal Society of Chemistry, Cambridge.
2 de Silva, A. P. and Uchiyama, S. (2007) Molecular logic and computing. *Nat. Nanotechnol.*, **2**, 399–410.
3 Yao, C.-Y., Lin, H.-Y., Crory, H. S. N., and de Silva, A. P. (2020) Supra-molecular agents running tasks intelligently (SMARTI): recent developments in molecular logic-based computation. *Mol. Syst. Des. Eng.*, **5**, 1325–1353.

Intracellular Thermometry with Fluorescent Molecular Thermometers, First Edition. Seiichi Uchiyama.
© 2024 Wiley-VCH GmbH. Published 2024 by Wiley-VCH GmbH.

Table A5.1 Fluorescent polymeric logic gates functioning in aqueous solution.

Entry	Thermo-responsive unit	Fluorescent unit/ mechanism[a]	Second input	Receptor unit toward second input	λ_{ex} (nm)	λ_{em} (nm)	Parameter to be measured[b]	Logic type[c]	Remarks	Ref.
1	NTBAM	DBD-AE/A	pH	DMAPAM	444	530	FI	AND		[1]
2	NIPAM	Fluorene monomer/A	pH	Fluorene monomer	340	388	FI	AND		[2]
3	NIPMAM	MDCPDP/A	pH	MDCPDP	460	620	FI	INHIBIT	In water-ethanol (5:1, v/v), Cu^{2+} is a third input	[3]
4	NIPAM	PhenUMA/A	Cu^{2+}	PhenUMA	280	452	FI	INHIBIT	Nanogel with crosslinking MBAM[d] units	[4]
5	NIPAM	NBD-AE, SPMA/AB	hν	SPMA	480	530, 618	FI ratio	AND	Attached to silica nanoparticles	[5]
6	NIPAM	FL/A	pH	FL	490	515	FI	INHIBIT		[6]
7	NIPAM	RhBHA	pH	RhBHA	500	584	FI	(NS)	Block copolymer with PEO[e] units, Hg^{2+} is a third input	[7]
8	NIPAM	NBD-AE, RhBAM, SPMA/AB	pH	RhBAM	470	518, 580	FI ratio	(NS)	Block copolymer with PS[f] units, hν is a third input	[8]
9	NIPAM	NPTUA/A	Hg^{2+}	NPTUA	390	482&528 or 457&511	FI ratio	(NS)	Irreversible response to Hg^{2+}, nanogel with crosslinking MBAM[d] units	[9]

10	NIPAM	NBD-AE, RhBEA/AB	hν	DMNA	470	527, 588	FI ratio	INHIBIT	Irreversible response to hν, nanogel with crosslinking MBAM[d] units	[10]
11	NIPAM	NBD-AE, RhBEA/AB	glucose	APBA	470	532, 587	FI ratio	INHIBT	pH is a possible third input, nanogel with crosslinking MBAM[d] units	[11]
12	NIPAM	Pyran monomer/A	Ni^{2+}	Pyran monomer	400	560	FI	INHIBIT		[12]
13	NIPAM	NBD-AE, RhBHA/AB	Hg^{2+}	RhBHA	470	518, 585	FI ratio	(NS)	Block copolymer including PS[f] units, Cu^{2+} and pH are other inputs	[13]
14	DEGMA	NBD, RhB (as an end cap)/B	pH	RhB	470	525, 593	FI ratio	INHIBIT		[14]
15	NIPAM	PtPorphyrin	O$_2$	PtPorphyrin	not indicated	660	FI	INHIBIT	Also tested in culture media for the growth of *Escherichia coli*	[15]
16	NIPAM	Dansyl monomer/A	Cu^{2+}	NAIDA	330	508	FI	AND		[16]
17	NIPAM	Fluorescein (labeled by FITC), RhBAM/B	pH	Fluorescein, RhBAM	495	522, 582	FI ratio	(NS)	Block copolymer containing OEGMA[g] units	[17]

(Continued)

Table A5.1 (*Continued*)

Entry	Thermo-responsive unit	Fluorescent unit/mechanism[a]	Second input	Receptor unit toward second input	λ_{ex} (nm)	λ_{em} (nm)	Parameter to be measured[b]	Logic type[c]	Remarks	Ref.
18	NIPAM	Pyran monomer/A	$h\nu$	SPO	460	560	FI	(NS)	Cu^{2+} and H^+ work as additional inputs	[18]
19	NIPAM	Pyran monomer 2/A	pH	DMAPAM	440	not indicated	FI	AND		[19]
20	NIPAM	AHBTA (attached by ionic interaction)/A	pH	AHBTA	390	450	FI	AND	BVP[h] units in copolymer interact with the carboxylic group of AHBTA	[20]
21	NIPMAM	CPMA/A	pH	CPMA	380	480	FI	AND	Prepared by RAFT[i] polymerization	[21]
22	NIPAM	Naphthalimide monomer/A	pH	Naphthalimide monomer	410	525	FI	(NS)	Nanogel with crosslinking MBAM[d] units, introduced into HeLa cells	[22]
23	NIPAM	AlexaFluor555, AlexaFluor647/B	Cu^{2+}	Bipy	525	565, 670	FI ratio	INHIBIT	Two copolymers labeled by a different fluorophore are mixed	[23]
24	NIPAM	NBD-AE, RhBEA/AB	$h\nu$	DMNA	470	520, 585	FI ratio	INHIBIT	Mixture of two block copolymers containing PEO[e] units	[24]
25	DEGMA	Polythiophene/C	pH	DMAEMA	417	541	FI	AND	Graft copolymer	[25]
26	NIPAM	Fluoranthene/A	pH	Fluoranthene	390	480	FI	(NS)		[26]
27	NIPAM	PBI, MANI/B	$h\nu$	NP	not indicated	523	FI	INHIBIT	Block copolymer including styrene units	[27]

28	NIPAM	NDPOE	pH	NDPOE	483	520	FI	(NS)	Nanogel with crosslinking MBAM[d] units	[28]
29	NIPAM	FL/A	pH	FL	440	515	FI	INHIBIT	Uptake by RAW264.7 cells was monitored	[29]
30	NIPAM	CO/A	pH	CO	350	460	FI	NOR	Uptake by RAW264.7 cells was monitored	[29]
31	NIPAM	M6AzCOONa/D	pH	M6AzCOONa	360	430	FI	(NS)	hν is a third input, block copolymer	[30]
32	NIPAM	TPE/E	pH	DPA	350	470	FI	(NS)	Block copolymer, function was tested in MCF-7 cells	[31]
33	NIPAM	BMPN, R6GEM/B	pH	BMPN, R6GEM	400	520, 555	FI ratio	(NS)	Fe^{3+} is a third input	[32]

a) Response mechanism towards a temperature variation. A: sensitivity to polarity; B: variation in FRET efficiency; C: sensitivity to water molecules; D: restriction of rotation; E: aggregation-induced fluorescence.
b) FI: fluorescence intensity; FI ratio: fluorescence intensity ratio at two different wavelengths.
c) See Table A5.2. NS: not specified.
d) N,N′-Methylenebisacrylamide.
e) Poly(ethylene oxide).
f) Polystyrene.
g) Oligo(ethylene glycol) methyl ether methacrylate.
h) 1-Benzyl-4-vinylpyridine bromide.
i) Reversible addition-fragmentation chain transfer.

1 Uchiyama, S., Kawai, N., de Silva, A. P., and Iwai, K. (2004) *J. Am. Chem. Soc.*, **126**, 3032–3033.
2 Yang, C.-C., Tian, Y., Jen, A. K.-Y., and Chen, W.-C. (2006) *J. Polym. Sci. A, Polym. Chem.*, **44**, 5495–5504.

(Continued)

Table A5.1   (*Continued*)

3   Guo, Z., Zhu, W., Xiong, Y., and Tian, H. (2009) *Macromolecules*, **42**, 1448–1453.
4   Liu, T., Hu, J., Yin, J., Zhang, Y., Li, C., and Liu, S. (2009) *Chem. Mater.*, **21**, 3439–3446.
5   Wu, T., Zou, G., Hu, J., and Liu, S. (2009) *Chem. Mater.*, **21**, 3788–3798.
6   Kobayashi, H., Nishikawa, M., Sakamoto, C., Nishio, T., Kanazawa, H., and Okano, T. (2009) *Anal. Sci.*, **25**, 1043–1047.
7   Hu, J., Li, C., and Liu, S. (2010) *Langmuir*, **26**, 724–729.
8   Li, C., Zhang, Y., Hu, J., Cheng, J., and Liu, S. (2010) *Angew. Chem. Int. Ed.*, **49**, 5120–5124.
9   Li, C. and Liu, S. (2010) *J. Mater. Chem.*, **20**, 10716–10723.
10   Yin, J., Hu, H., Wu, Y., and Liu S. (2011) *Polym. Chem.*, **2**, 363–371.
11   Wang, D., Liu, T., Yin, J., and Liu, S. (2011) *Macromolecules*, **44**, 2282–2290.
12   Lee, E.-M., Gwon, S.-Y., Ji, B.-C., and Kim, S.-H. (2011) *J. Lumin.*, **131**, 2004–2009.
13   Hu, J., Dai, L., and Liu, S. (2011) *Macromolecules*, **44**, 4699–4710.
14   Wan, X. and Liu, S. (2011) *J. Mater. Chem.*, **21**, 10321-10329.
15   Zhou, X., Su, F., Tian, Y., Johnson, R. H., and Meldrum, D. R. (2011) *Sens. Actuat. B*, **159**, 135–141.
16   Du, J., Yao, S., Seitz, W. R., Bencivenga, N. E., Massing, J. O., Planalp, R. P., Jackson, R. K., Kennedy, D. P., and Burdette, S. C. (2011) *Analyst*, **136**, 5006–5011.
17   Hu, J., Zhang, X., Wang, D., Hu, X., Liu, T., Zhang, G., and Liu, S. (2011) *J. Mater. Chem.*, **21**, 19030–19038.
18   Lee, E.-M., Gwon, S.-Y., Ji, B.-C., Wang, S., and Kim, S.-H. (2012) *J. Lumin.*, **132**, 665–670.
19   Lee, E.-M., Gwon, S.-Y., Hwang, I.-J., Son, Y.-A., and Kim, S.-H. (2012) *Spectrochim. Acta A*, **92**, 33–36.
20   Cui, K., Zhu, D., Cui, W., Lu, X., and Lu, Q. (2012) *J. Phys. Chem. C*, **116**, 6077–6082.
21   Liu, G., Zhou, W., Zhang, J., and Zhao, P. (2012) *J. Polym. Sci. A, Polym. Chem.*, **50**, 2219–2226.
22   Yin, L., He, C., Huang, C., Zhu, W., Wang, X., Xu, Y., and Qian, X. (2012) *Chem. Commun.*, **48**, 4486–4488.
23   Yao, S., Jones, A. M., Du, J., Jackson, R. K., Massing, J. O., Kennedy, D. P., Bencivenga, N. E., Planalp, R. P., Burdette, S. C., and Seitz, W. R. (2012) *Analyst*, **137**, 4734–4741.
24   Wu, Y., Hu, H., Hu, J., Liu, T., Zhang, G., and Liu, S. (2013) *Langmuir*, **29**, 3711–3720.
25   Das, S., Chatterjee, D. P., Samanta, S., and Nandi, A. K. (2013) *RSC Adv.*, **3**, 17540–17550.
26   Guo, Y., Yu, X., Xue, W., Huang, S., Dong, J., Wei, L., Maroncelli, M., and Li, H. (2014) *Chem. Eng. J.*, **240**, 319–330.
27   Li, D., Munyentwali, A., Wang, G., Zhang, M., and Xing, S. (2015) *Dyes Pigm.*, **117**, 92–99.
28   Eftekhari-Sis, B. and Ghahramani, F. (2015) *Des. Monomers Polym.*, **18**, 460–469.
29   Yamada, A., Hiruta, Y., Wang, J., Ayano, E., and Kanazawa, H. (2015) *Biomacromolecules*, **16**, 2356–2362.
30   Ren, H., Chen, D., Shi, Y., Yu, H., and Fu, Z. (2016) *Polymer*, **97**, 533–542.
31   Zhao, Y., Wu, Y., Chen, S., Deng, H., and Zhu, X. (2018) *Macromolecules*, **51**, 5234–5244.
32   Kong, F., Lin, M., and Qiu, T. (2018) *Polymer*, **151**, 117–124.

Thermo-responsive unit

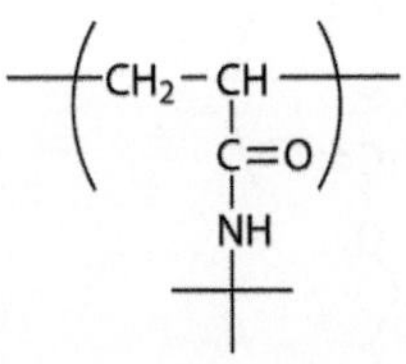

NTBAM 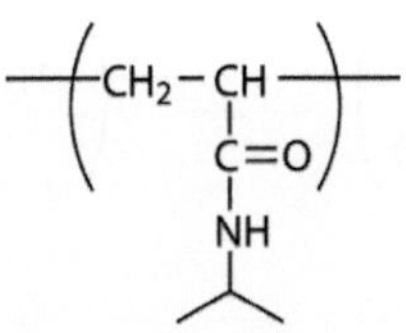   NIPAM 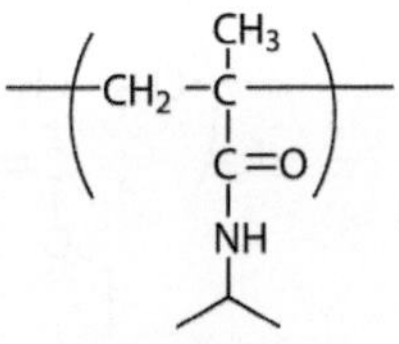   NIPMAM 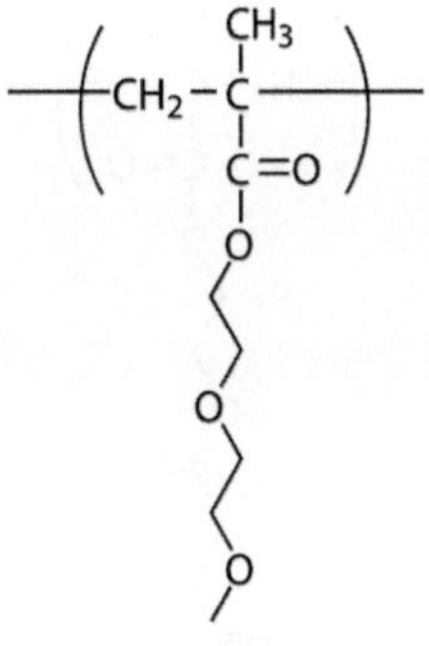   DEGMA

Fluorescent unit

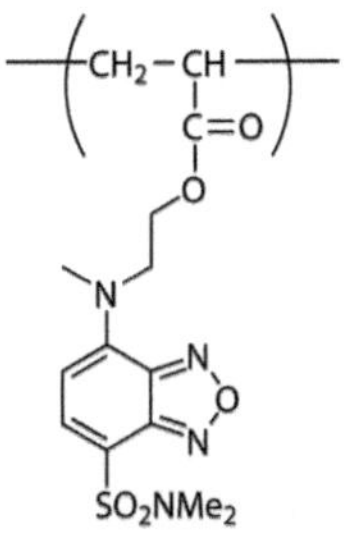

DBD-AE

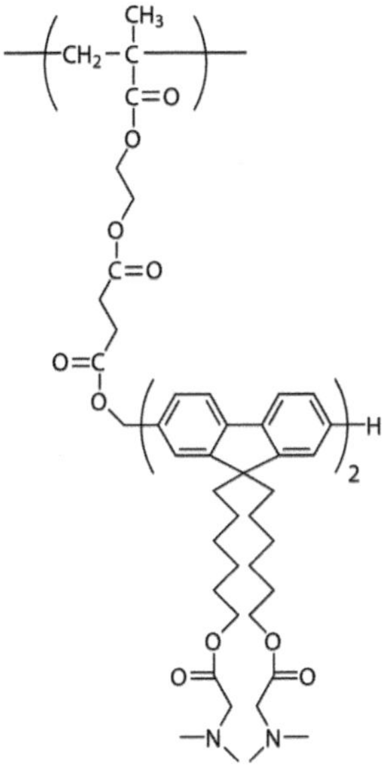

fluorene monomer

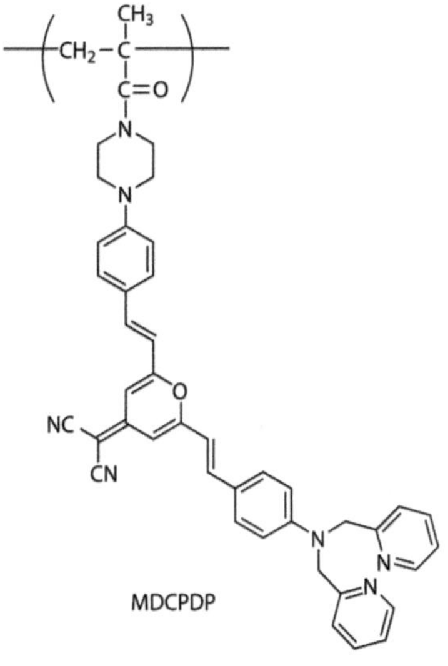

MDCPDP

Fluorescent unit (*continued*)

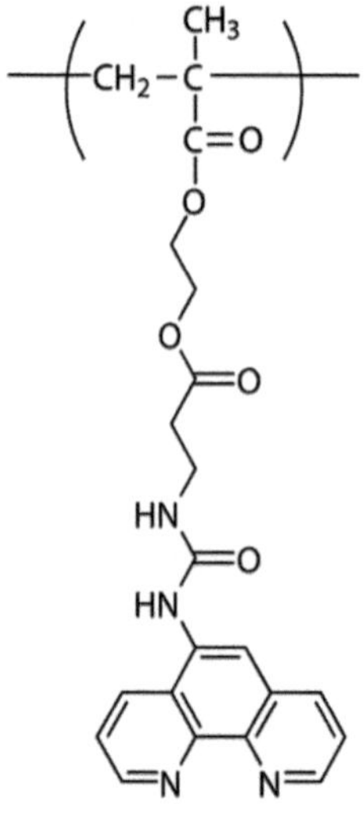

PhenUMA

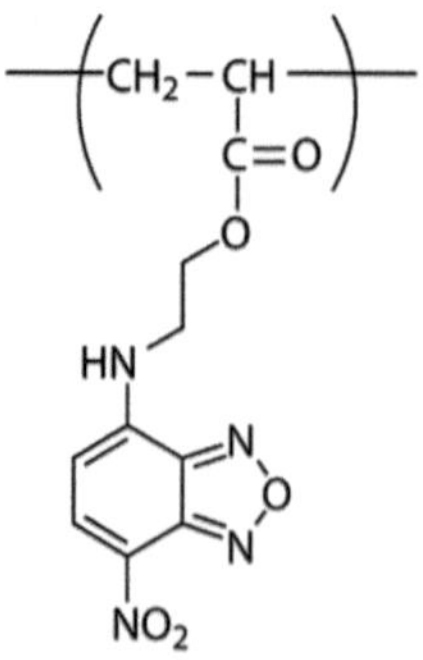

NBD-AE

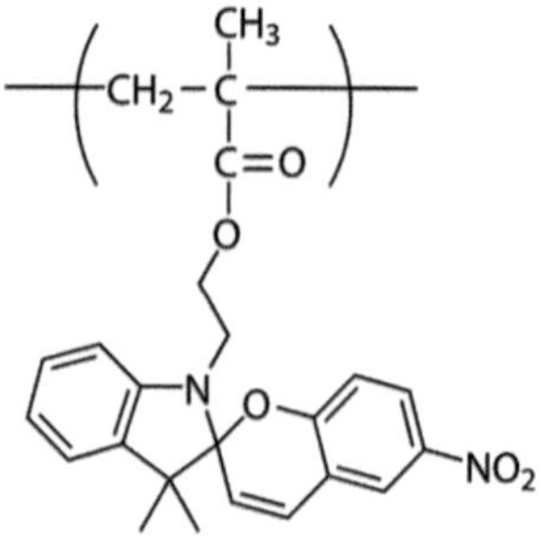

SPMA

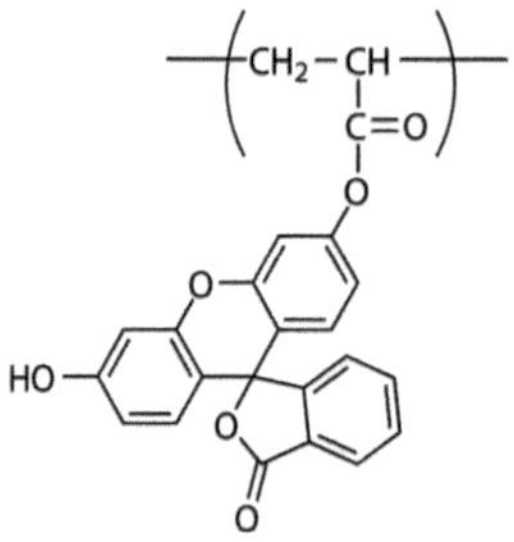

FL

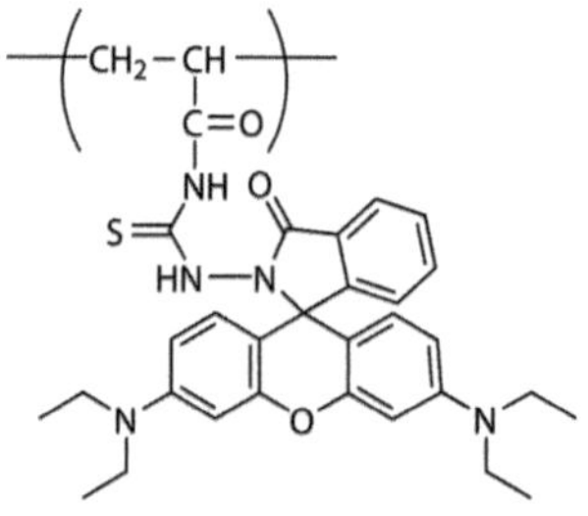

RhBHA

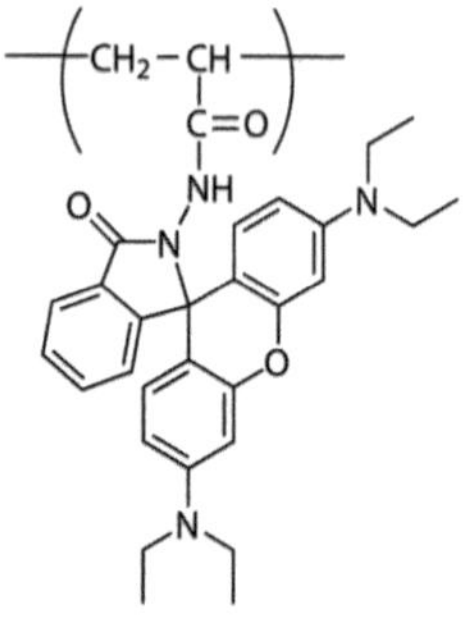

RhBAM

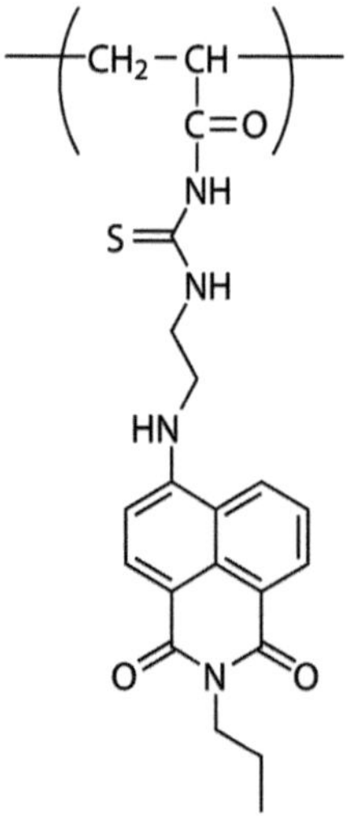

NPTUA

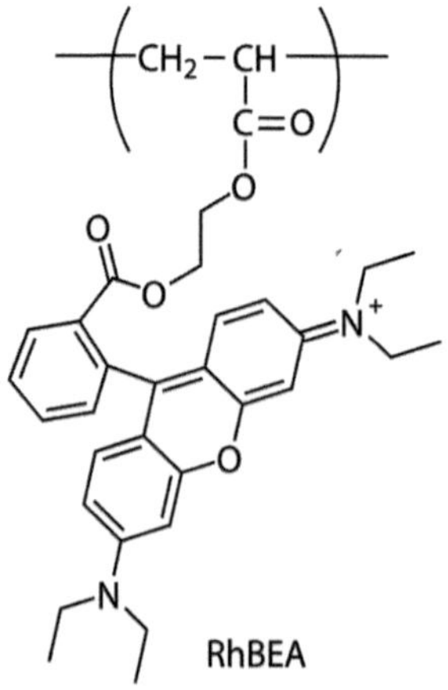

RhBEA

Fluorescent unit (*continued*)

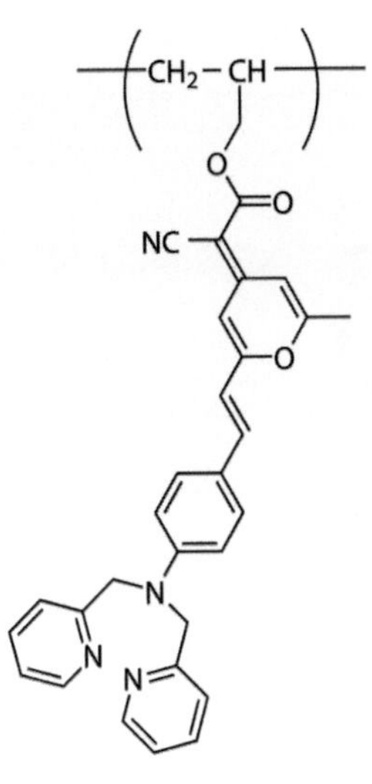

Pyran monomer

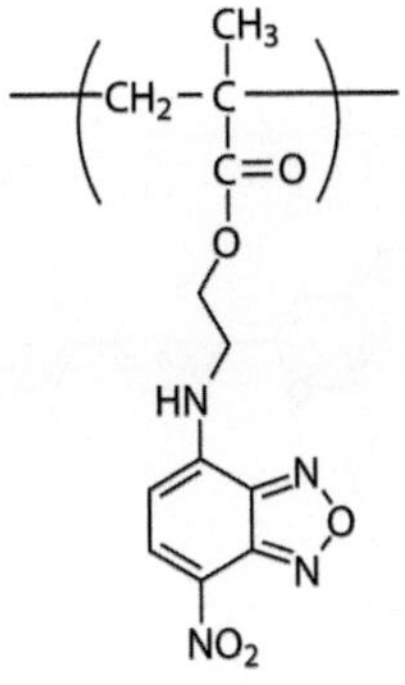

NBD

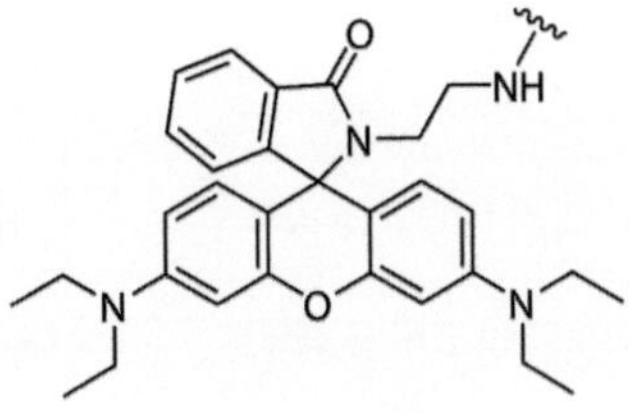

RhB
(as end group)

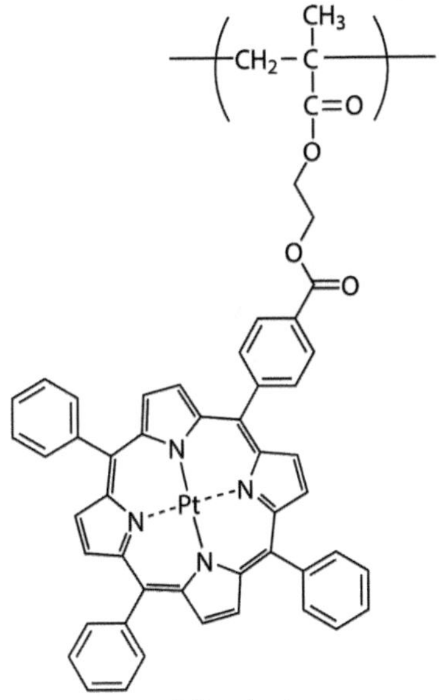

PtPorphyrin

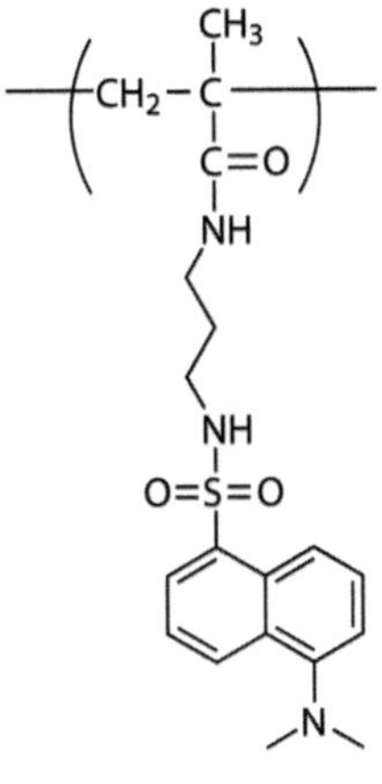

Dansyl monomer

Fluorescent unit (*continued*)

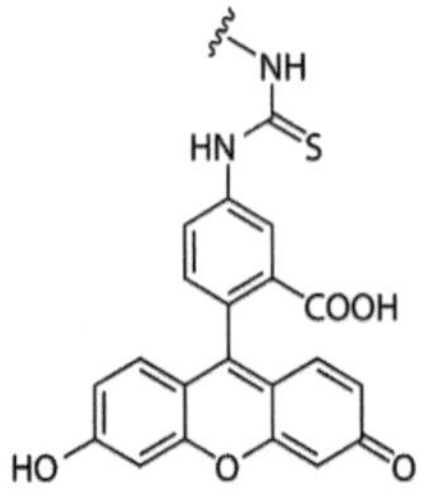

FITC

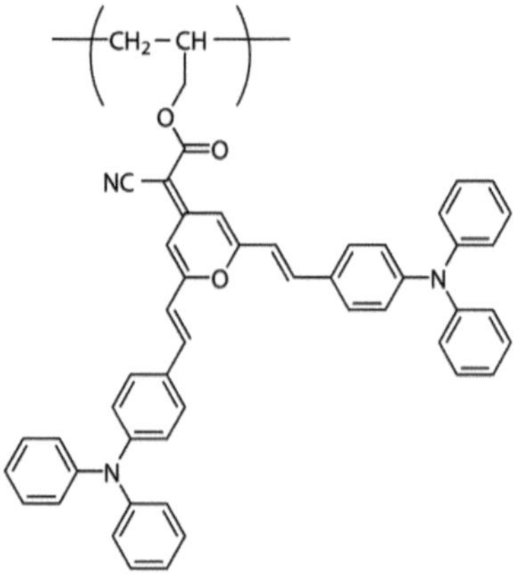

Pyran monomer 2

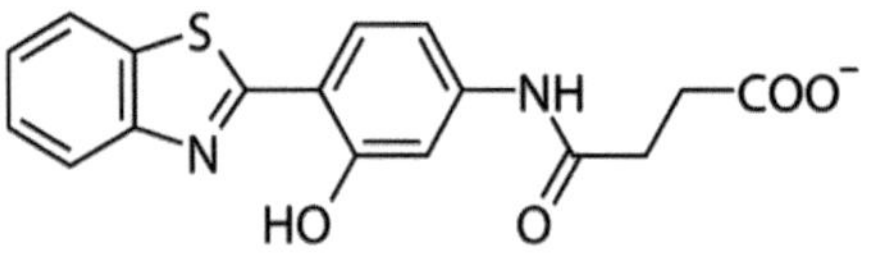

AHBTA

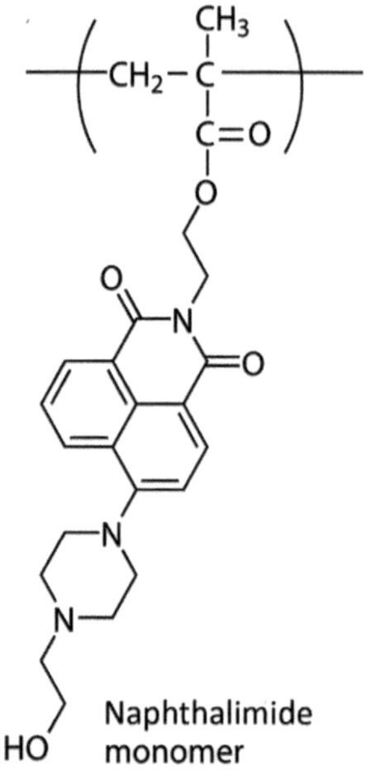

CPMA

Naphthalimide monomer

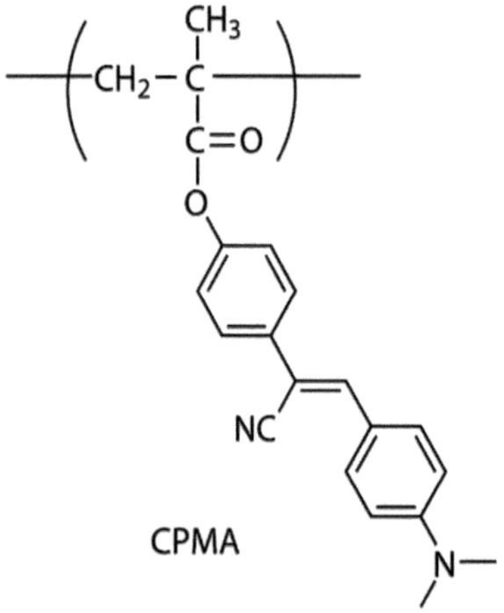

Fluorescent unit (*continued*)

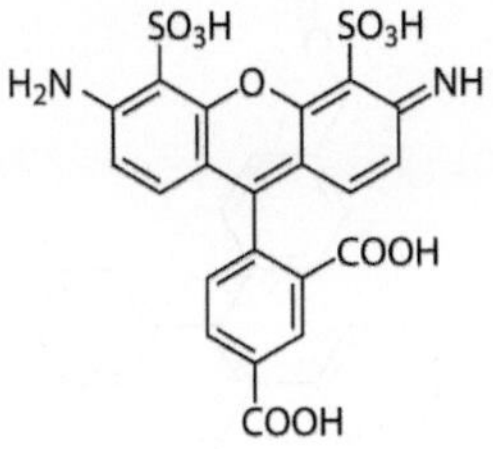

AlexaFluor555

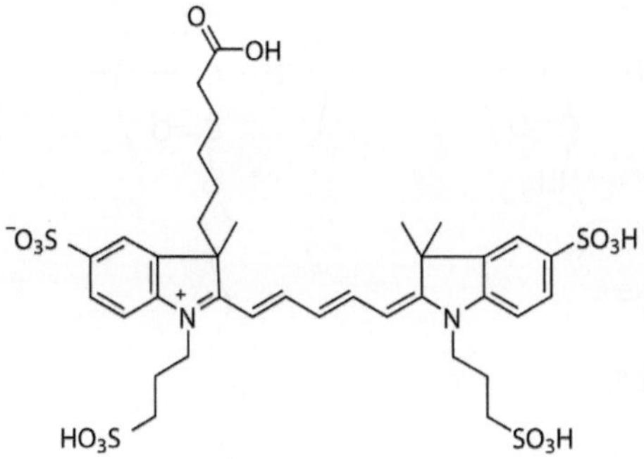

AlexaFluor647

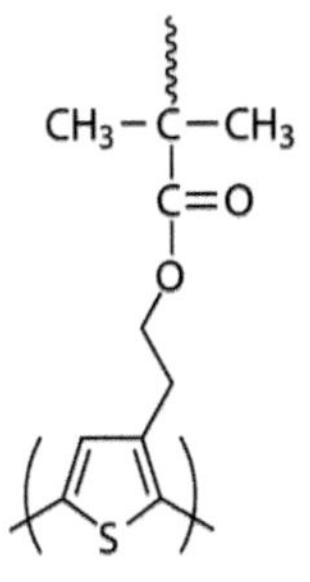

Polythiophene

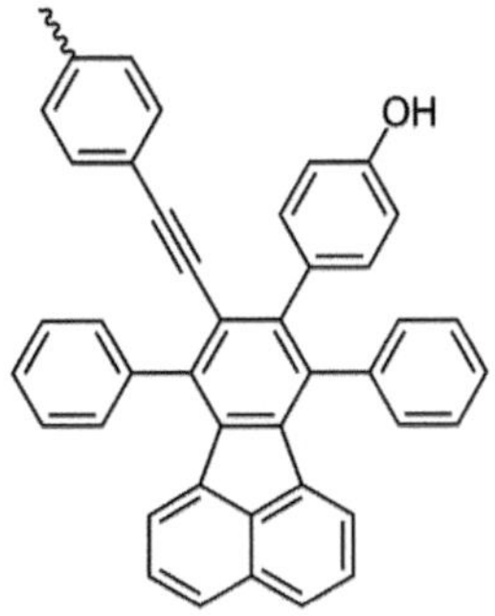

Fluoranthene

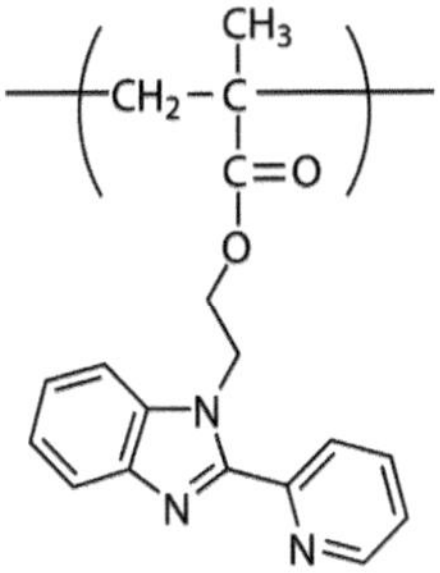

PBI

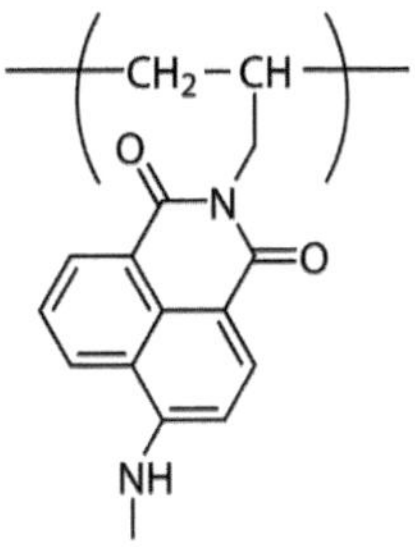

MANI

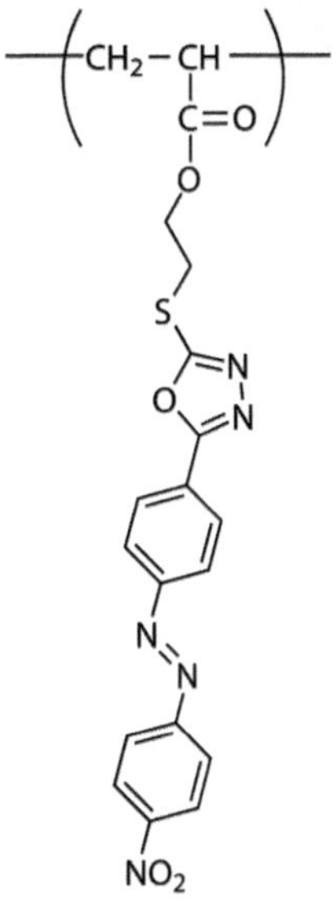

NDPOE

Fluorescent unit (*continued*)

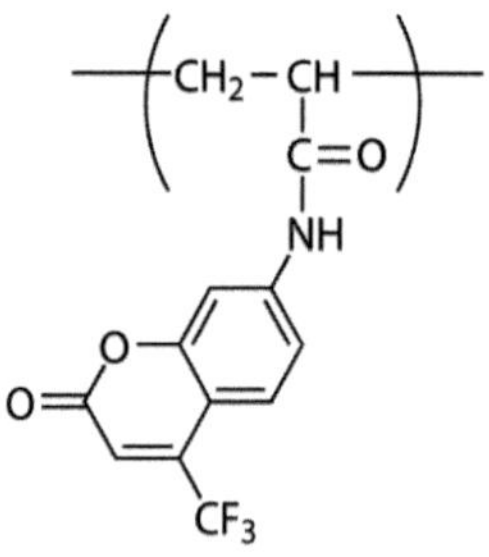

CO

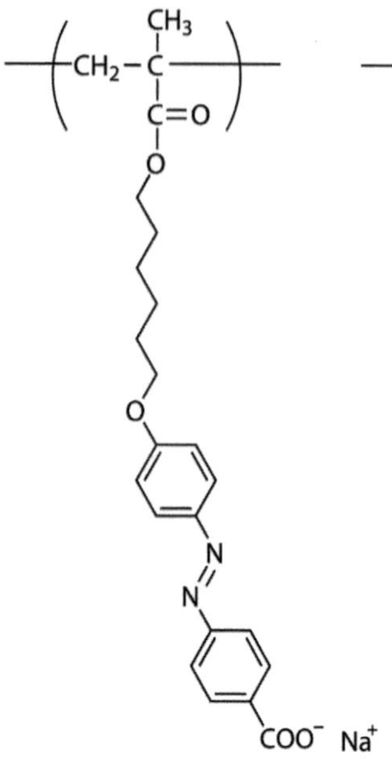

M6AzCOONa

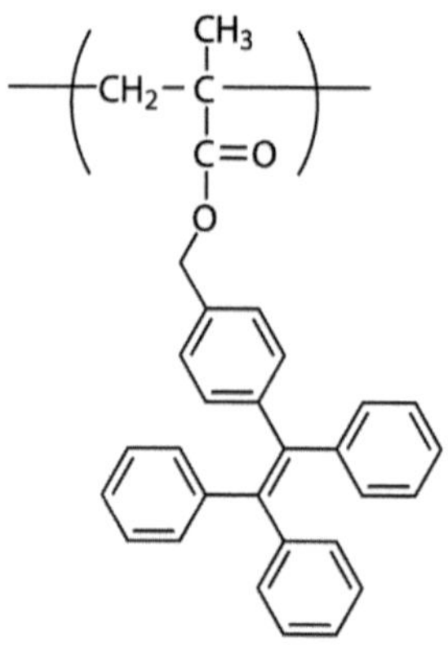

TPE

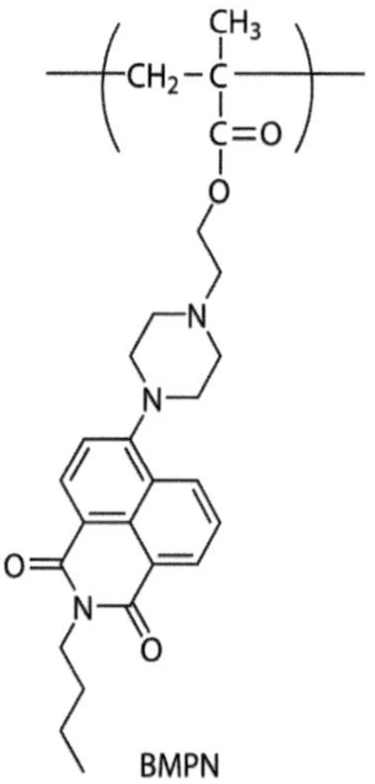

BMPN

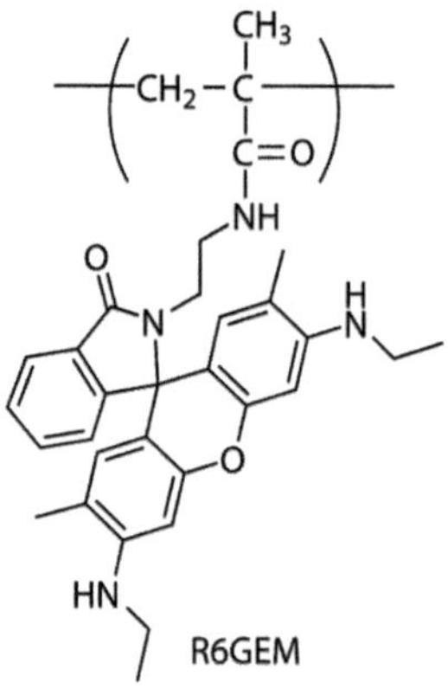

R6GEM

Receptor unit

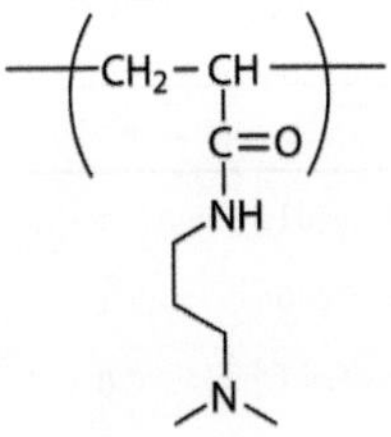

DMAPAM

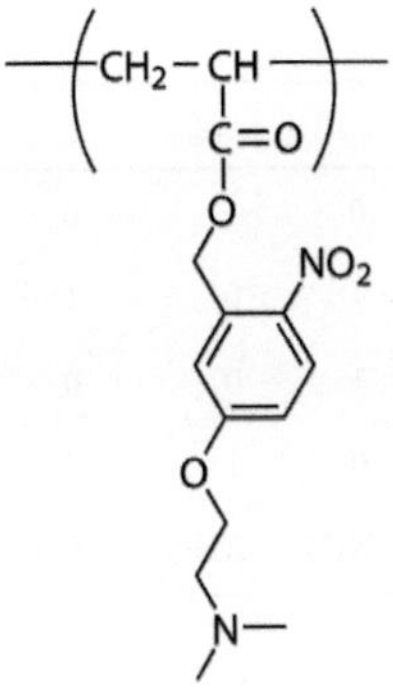

DMNA

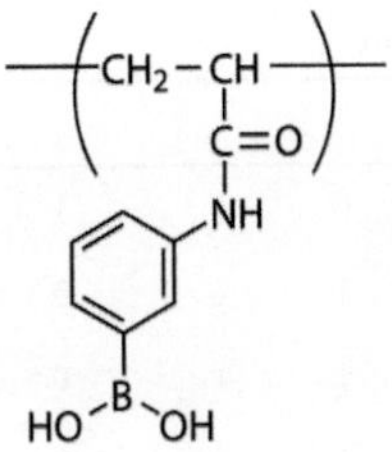

APBA

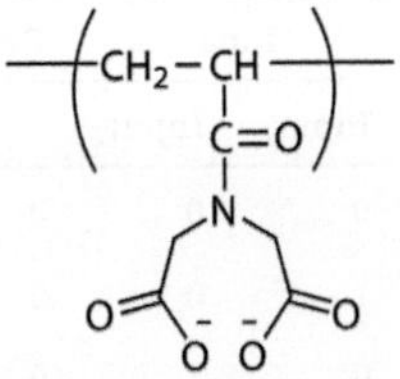

NAIDA

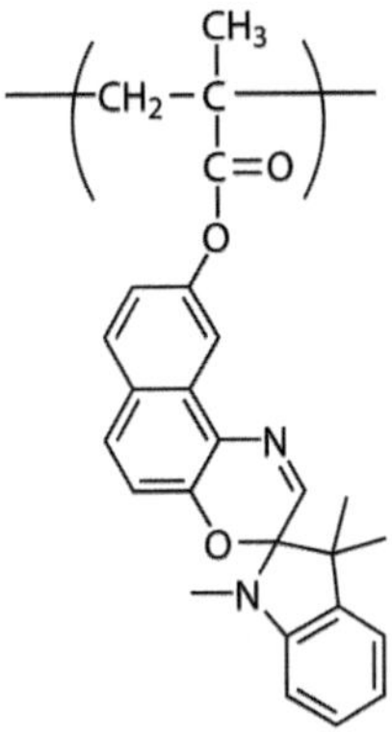

SPO

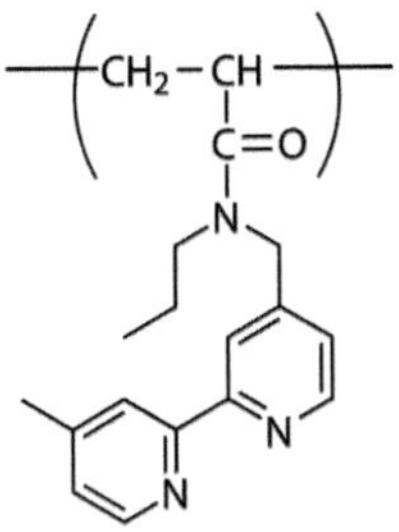

Bipy

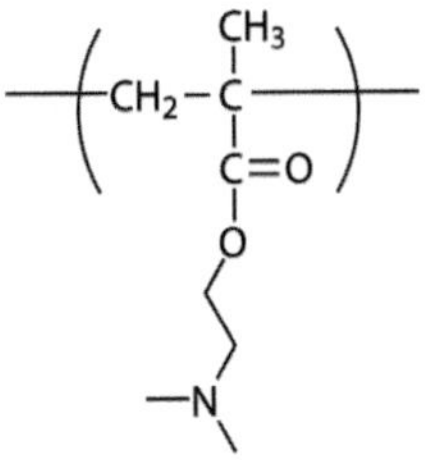

DMAEMA

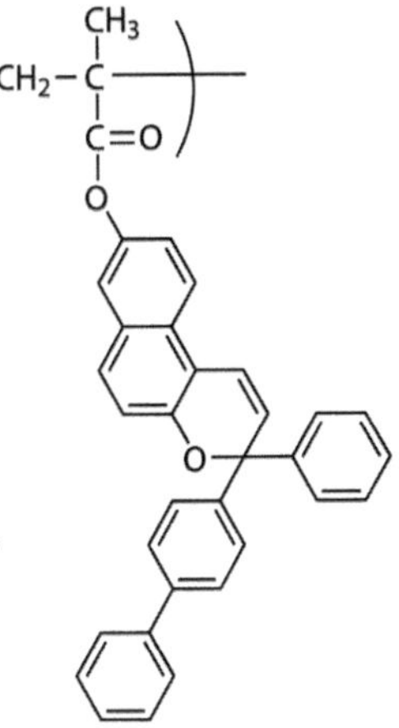

NP

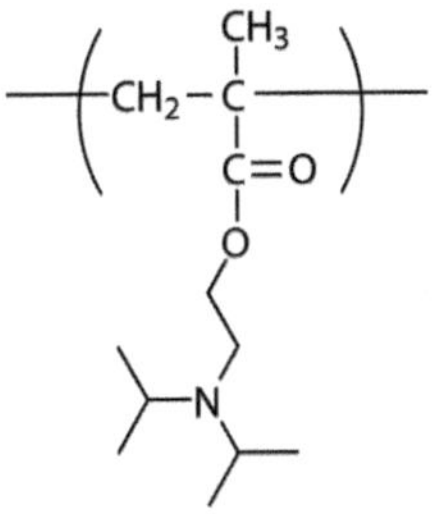

DPA

Table A5.2 A truth table for double input–single output logic gates.[a]

Input		Output									
Input$_1$	Input$_2$										
0	0	0	0	1	1	0	1	0	0	1	1
1	0	0	1	1	0	1	0	1	0	0	1
0	1	0	1	1	0	1	0	0	1	1	0
1	1	1	1	0	0	0	1	0	0	1	1
Name		AND	OR	NAND	NOR	XOR	XNOR	INHIBIT		IMPLICATION	

a) '0' and '1' represent 'low' and 'high', respectively.

Appendix 6 Solid Inorganic Nanostructures Showing Temperature-dependent Emission Properties

As with small organic molecules and polymers, inorganic metal species such as trivalent lanthanide and transition metal ions also show temperature-dependent fluorescence (or emission) properties (i.e., intensity, wavelength, and lifetime). Inorganic materials doped with these ions also emit temperature-dependent fluorescence. In particular, the latter materials have significant advantages in mechanical, chemical, and thermal stabilities over organic molecule-based fluorescent thermometers. In addition, the functional temperature ranges of inorganic molecule-based thermometers are generally much wider than those of organic molecule-based thermometers. Until now, more than 300 papers have been published on the luminescent characteristics of inorganic metal phosphors and the effects of host material structures on phosphor's luminescence in the solid state. Several examples (e.g., Eu^{3+} complex and Er^{3+} doped nanoparticle) have been utilized in intracellular thermometry with encapsulation or surface modification to improve the dispersibility in aqueous media, highlighted in this book's central chapters. Instead, there are many inorganic candidates for highly useful fluorescent thermometers applied to intracellular thermometry once the biocompatibility issues (i.e., dispersibility in the intracellular environments and cytotoxicity) are solved. Such inorganic nanostructures showing temperature-dependent emission characteristics in the solid state are out of the subject area of this textbook. So, selected well-organized review articles on lanthanide and transition metal ions with temperature-dependent photophysical properties are listed here in chronological order.

Review articles

Allison, S. W. and Gillies, G. T. (1997) Remote thermometry with thermographic phosphors: instrumentation and applications. *Rev. Sci. Instrum.*, **68**, 2615–2650.

Rocha, J., Brites, C. D. S., and Carlos, L. D. (2016) Lanthanide organic framework luminescent thermometers. *Chem. Eur. J.*, **22**, 14782–14795.

Brites, C. D. S., Balabhadra, S., and Carlos, L. D. (2019) Lanthanide-based thermometers: at the cutting-edge of luminescence thermometry. *Adv. Optical Mater.*, **7**, 1801239.

Ansari, A. A., Parchur, A. K., Nazeeruddin, M. K., and Tavakoli, M. M. (2021) Luminescent lanthanide nanocomposites in thermometry: chemistry of dopant ions and host matrices. *Coord. Chem. Rev.*, **444**, 214040.

Marciniak, L., Kniec, K., Elżbieciak-Piecka, K., Trejgis, K., Stefanska, J., and Dramićanin, M. (2022) Luminescence thermometry with transition metal ions. A review. *Coord. Chem. Rev.*, **469**, 214671.

Appendix 7 Tips for Intracellular Thermometry with Fluorescent Polymeric Thermometers Developed in the Author's Laboratory

Among the various fluorescent polymeric thermometers that are based on the combination of a thermoresponsive polymer with an environment-sensitive benzofurazan fluorophore and were developed in the author's laboratory, a fluorescent polymeric thermometer (Diffusive Thermoprobe) (Okabe et al., 2012), a fluorescent nanogel thermometer (Particulate Thermoprobe) (Uchiyama et al., 2012), a cationic fluorescent polymeric thermometer (Cellular Thermoprobe for Fluorescence Lifetime) (Hayashi et al., 2015), and a ratiometric fluorescent polymeric thermometer (Cellular Thermoprobe for Fluorescence Ratio) (Uchiyama et al., 2015) are commercially available from a Japanese reagent company, Funakoshi Co., Ltd. (https://www.funakoshi.co.jp/exports) and their worldwide dealers as catalog numbers FDV-0002~FDV-0005. Researchers who intend to utilize these fluorescent polymeric thermometers can refer to experimental protocols described in the original literature (Okabe et al., 2012; Uchiyama et al., 2012; Hayashi et al., 2015; Uchiyama et al., 2015) or the attached documents provided by Funakoshi (https://www. funakoshi.co.jp/exports_contents/46141 (Diffusive Thermoprobe, FDV-0002), https://www.funakoshi.co.jp/exports_contents/46138 (Particulate Thermoprobe, FDV-0003), https://www.funakoshi.co.jp/exports_contents/80404 (Cellular Thermoprobe for Fluorescence Lifetime, FDV-0004), https://www.funakoshi.co.jp/ exports_contents/80405 (Cellular Thermoprobe for Fluorescence Ratio, FDV-0005)). In addition, detailed experimental protocols are available for intracellular thermometry of HeLa (human epithelial carcinoma) cells with the cationic fluorescent polymeric thermometer (Inada et al., 2019) and that of brown adipocytes with the ratiometric fluorescent polymeric thermometer (Tsuji et al., 2023).

References

A front web page of a reagent company, Funakoshi Co., Ltd. https://www.funakoshi. co.jp/exports (accessed 24 Dec 2023).

A product description of Diffusive Thermoprobe (FDV-0002) (2014). https://www. funakoshi.co.jp/exports_contents/46141 (accessed 24 Dec 2023).

A product description of Particulate Thermoprobe (FDV-0003) (2014). https://www.funakoshi. co.jp/exports_contents/46138 (accessed 24 Dec 2023).

A product description of Cellular Thermoprobe for Fluorescence Lifetime (FDV-0004) (2015). https://www.funakoshi.co.jp/exports_ contents/80404 (accessed 24 Dec 2023).

A product description of Cellular Thermoprobe for Fluorescence Ratio (FDV-0005) (2017). https://www.funakoshi.co.jp/exports_ contents/80405 (accessed 24 Dec 2023).

Hayashi, T., Fukuda, N., Uchiyama, S., and Inada, N. (2015) A cell-permeable fluorescent polymeric thermometer for intracellular temperature mapping in mammalian cell lines. *PLoS ONE*, **10**, e0117677.

Inada, N., Fukuda, N., Hayashi, T., and Uchiyama, S. (2019) Temperature imaging using a cationic linear fluorescent polymeric thermometer and fluorescence lifetime imaging microscopy. *Nat. Protoc.*, **14**, 1293–1321.

Okabe, K., Inada, N., Gota, C., Harada, Y., Funatsu, T., and Uchiyama, S. (2012) Intracellular temperature mapping with a fluorescent polymeric thermometer and fluorescence lifetime imaging microscopy. *Nat. Commun.*, **3**, 705.

Tsuji, T., Kajimoto, K., and Inada, N. (2023) Measurement of intracellular temperature in brown adipocytes using a cationic fluorescent polymeric thermometer, in *Thermogenic Fat: Methods and Protocols* (ed. I. J. Lodhi), *Methods Mol. Biol.*, **2662**, Springer Nature, New York, pp. 87–102.

Uchiyama, S., Kimura, K., Gota, C., Okabe, K., Kawamoto, K., Inada, N., Yoshihara, T., and Tobita, S. (2012) Environment-sensitive fluorophores with benzothiadiazole and benzoselenadiazole structures as candidate components of a fluorescent polymeric thermometer. *Chem. Eur. J.*, **18**, 9552–9563.

Uchiyama, S., Tsuji, T., Ikado, K., Yoshida, A., Kawamoto, K., Hayashi, T., and Inada, N. (2015) A cationic fluorescent polymeric thermometer for the ratiometric sensing of intracellular temperature. *Analyst*, **140**, 4498–4506.

Index

Note: *Italic* page numbers refer to *figure* and **Bold** page numbers reference to **tables**.

Intracellular Thermometry with Fluorescent Molecular Thermometers, First Edition. Seiichi Uchiyama.
© 2024 Wiley-VCH GmbH. Published 2024 by Wiley-VCH GmbH.